Human Medicinal Agents from Plants

ACS SYMPOSIUM SERIES **534**

Human Medicinal Agents from Plants

A. Douglas Kinghorn, EDITOR
University of Illinois at Chicago

Manuel F. Balandrin, EDITOR
NPS Pharmaceuticals, Inc.

Developed from a symposium sponsored
by the Division of Agricultural and Food Chemistry
at the 203rd National Meeting
of the American Chemical Society,
San Francisco, California,
April 5–10, 1992

American Chemical Society, Washington, DC 1993

Library of Congress Cataloging-in-Publication Data

Human medicinal agents from plants / [edited by] A. Douglas Kinghorn, Manuel F. Balandrin.

p. cm.—(ACS symposium series, ISSN 0097–6156; 534)

"Developed from a symposium sponsored by the Division of Agricultural and Food Chemistry of the American Chemical Society at the 203rd Meeting of the American Chemical Society, San Francisco, California, April 5–10, 1992."

Includes bibliographical references and index.

ISBN 0–8412–2705–5

1. Materia medica, Vegetable—Congresses. 2. Medicinal plants—Congresses. 3. Pharmacognosy—Congresses.

I. Kinghorn, A. Douglas. II. Balandrin, Manuel F., 1952– . III. American Chemical Society. Division of Agricultural and Food Chemistry. IV. American Chemical Society. Meeting (203rd: 1992: San Francisco, Calif.) V. Series.

RS164.H88 1993
615′.32—dc20 93–22180
CIP

The paper used in this publication meets the minimum requirements of American National Standard for Information Sciences—Permanence of Paper for Printed Library Materials, ANSI Z39.48–1984. ♾

Second printing 1995

1993 Advisory Board

ACS Symposium Series

M. Joan Comstock, *Series Editor*

Foreword

THE ACS SYMPOSIUM SERIES was first published in 1974 to provide a mechanism for publishing symposia quickly in book form. The purpose of this series is to publish comprehensive books developed from symposia, which are usually "snapshots in time" of the current research being done on a topic, plus some review material on the topic. For this reason, it is necessary that the papers be published as quickly as possible.

Before a symposium-based book is put under contract, the proposed table of contents is reviewed for appropriateness to the topic and for comprehensiveness of the collection. Some papers are excluded at this point, and others are added to round out the scope of the volume. In addition, a draft of each paper is peer-reviewed prior to final acceptance or rejection. This anonymous review process is supervised by the organizer(s) of the symposium, who become the editor(s) of the book. The authors then revise their papers according to the recommendations of both the reviewers and the editors, prepare camera-ready copy, and submit the final papers to the editors, who check that all necessary revisions have been made.

As a rule, only original research papers and original review papers are included in the volumes. Verbatim reproductions of previously published papers are not accepted.

M. Joan Comstock
Series Editor

Contents

Anticancer and Cancer Chemopreventive Agents from Plants

ANTI-INFECTIVE AND ANTIMICROBIAL CHEMOTHERAPEUTIC AGENTS FROM PLANTS

PROMISING PLANT-DERIVED NATURAL PRODUCTS WITH MULTIPLE BIOLOGICAL ACTIVITIES

Preface

NATURAL SUBSTANCES ONCE SERVED AS THE SOURCES of all drugs and medicinal agents, and higher plants provided most of these therapeutic entities. The serious and systematic study of the chemistry and biological properties of natural products, which began in earnest at the start of the 19th century, has been a major factor in the development of modern synthetic organic chemistry. As the many interesting biologically active compounds described in this book exemplify, plants continue to retain their historical significance as important sources of new drugs, "lead" compounds for structural modification and optimization, and specific biochemical and pharmacological probes. For all of these reasons, biologically active constituents of plants have served as sources of inspiration for generations of medicinal and organic chemists. In the foreseeable future, plants will continue to provide humankind with valuable agents of potential use in the investigation, prevention, and treatment of diseases such as cancer, acquired immunodeficiency syndrome (AIDS) and other viral infections, malaria, and schistosomiasis; disorders of the cardiovascular and central nervous systems; and many others.

Powerful new chemical and biological technologies permit receptor isolation and characterization so that drug design principles can be applied rapidly to "fast-track" natural product leads as never before. The structural determination of novel plant constituents can be performed with minimal delay by using a combination of sophisticated spectroscopic and X-ray crystallographic techniques. High-throughput automated bioassays are widely available, so that a detailed biological profile can be obtained easily on just a few milligrams of a natural product. Thus, there is every indication that the direct utility and promise of plants for the improvement of human health will continue well into the 21st century.

The publication of this volume is timely because during the closing stages of its preparation, the plant-derived drug Taxol (paclitaxel) was approved by the U.S. Food and Drug Administration (FDA) for the treatment of refractory ovarian cancer [Holden, C. *Science (Washington, D.C.)* **1993**, *259*, 181]. Although semisynthetic natural product derivatives have recently been approved (i.e., the podophyllotoxin derivatives etoposide and teniposide), paclitaxel is the first naturally occurring plant-derived drug product to gain FDA approval in more than a quarter of a century.

Paclitaxel has attracted widespread interest in the United States not only for its therapeutic potential but also because of the need to preserve the native stands of the Pacific yew, the source plant of paclitaxel, and at the same time ensure an adequate supply of the drug.

Several chapters in this book touch on the crucial natural-product supply issue; information is also provided on a number of compounds that hold the potential for being the next approved plant-derived drugs after paclitaxel. Contributions to this book address each of the various aspects involved in plant-derived drug discovery and development, including botany and taxonomy, phytochemistry, biological evaluation, and regulation. Attention is paid to the value of the tropical rain forests in affording important biologically active compounds, as well as to the urgent need to avoid further species extinction and consequent loss of biodiversity, which threaten our future ability to discover new drugs from these parts of the world. Current research endeavors in laboratories in academic, governmental, and large and small industrial settings are spotlighted. Much of the book is devoted to accounts of promising research results and current research strategies in the discovery of plant drugs for the treatment of cancer, AIDS, malaria, and other diseases. Some recent initiatives of the U.S. government in supporting plant drug discovery are included. This book shows that excellent progress in finding new therapeutic entities from plants is being made not only in North America but also in countries in western Europe, and in Japan, India, and China.

The symposium on which this book is based was generously supported by the Division of Agricultural and Food Chemistry of the American Chemical Society and was cosponsored by the American Society of Pharmacognosy and the Society for Economic Botany. We are very grateful to the following companies, which made generous donations in support of the symposium: Bristol–Myers Squibb Pharmaceutical Research Institute; Glaxo Group Research Ltd.; Murdock Healthcare, Inc.; NPS Pharmaceuticals, Inc.; Phytopharm Ltd.; Schering–Plough Research Institute; Shaman Pharmaceuticals, Inc.; and Wyeth-Ayerst Laboratories. We thank the scientists who served as referees for the chapters in this book. Finally, we are very saddened to record the untimely passing of three of the symposium participants and chapter authors, Alwyn H. Gentry, Daniel L. Klayman, and Matthew Suffness.

A. DOUGLAS KINGHORN
Program for Collaborative Research
in the Pharmaceutical Sciences
College of Pharmacy
University of Illinois at Chicago
Chicago, IL 60612

MANUEL F. BALANDRIN
Department of Medicinal Chemistry
NPS Pharmaceuticals, Inc.
University of Utah Research Park
Salt Lake City, UT 84108

March 23, 1993

Current Role and Importance of Plant-Derived Natural Products in Drug Discovery and Development

Chapter 1

Plant-Derived Natural Products in Drug Discovery and Development

An Overview

Manuel F. Balandrin[1], A. Douglas Kinghorn[2], and Norman R. Farnsworth[2]

[1]Department of Medicinal Chemistry, NPS Pharmaceuticals, Inc., University of Utah Research Park, Salt Lake City, UT 84108
[2]Program for Collaborative Research in the Pharmaceutical Sciences, College of Pharmacy, University of Illinois at Chicago, Chicago, IL 60612

The natural world once served as the source of all medicinal agents, with higher plants constituting by far the principal sources of these. Today, higher plants continue to retain their historical significance as important sources of novel compounds useful directly as medicinal agents, as model compounds for synthetic or semisynthetic structure modifications and optimization, as biochemical and/or pharmacological probes, and as sources of inspiration for generations of synthetic organic medicinal chemists. Plant-derived compounds which have recently undergone development include the anticancer agents, Taxol (paclitaxel) and camptothecin, the Chinese antimalarial drug, artemisinin, and the East Indian Ayurvedic drug, forskolin. These and many other examples serve to illustrate the continuing value of plant-derived secondary metabolites as viable compounds for modern drug product development.

Historical Perspective

Higher plants have served humankind as sources of medicinal agents since its earliest beginnings. In fact, natural products once served as the source of all drugs. Today, natural products (and their derivatives and analogs) still represent over 50% of all drugs in clinical use, with higher plant-derived natural products representing *ca.* 25% of the total. On numerous occasions, the folklore records of many different cultures have provided leads to plants with useful medicinal properties (*1-11*). In the past two centuries, the chemical investigation and purification of extracts of plants purported to have medicinal properties, and those used as toxins and hunting poisons in their native habitats, have yielded numerous purified compounds which have

0097–6156/93/0534–0002$06.00/0

proven to be indispensable in the practice of modern medicine (*3,9,12*). For example, the curare alkaloids were obtained from South American vines that had long been used by natives to make arrow poisons, and African *Strophanthus* species and Calabar beans yielded medicinally useful cardiac glycosides and physostigmine, respectively, which were originally used as arrow and ordeal poisons in their native habitats. The East Indian snakeroot, *Rauvolfia serpentina* (L.) Benth. ex Kurz, has been used for centuries as a native East Indian medicinal plant, and its main active principle, reserpine, is now used in western medicine as an antihypertensive and tranquilizer. Similarly, other bioactive and poisonous plants with extensive folklore histories have yielded the cardiac (*Digitalis*) glycosides, Δ^9-tetrahydrocannabinol, the opiates (codeine, morphine), the *Cinchona* alkaloids (quinine, quinidine), the solanaceous tropane alkaloids (atropine, *l*-hyoscyamine, scopolamine), pilocarpine, ephedrine, cocaine, theophylline, vincristine, vinblastine, taxol, and other well-known and useful drugs (*3,9-11*).

The Role of Plant-Derived Natural Products in Modern Medicine

The commercial value of drug products still derived directly from higher plants is considerable and should not be underestimated. For example, in 1980 American consumers paid about $8 billion for prescription drugs derived solely from higher plant sources (see Table I). From 1959 to 1980, drugs derived from higher plants represented a constant 25% of all new and refilled prescriptions dispensed from community pharmacies in the United States (this does not take into account non-prescription drug products or drugs used exclusively in hospital settings). Plant-derived drugs thus represent stable markets upon which both physicians and patients rely. In addition, worldwide markets in plant-derived drugs are difficult to estimate, but undoubtedly amount to many additional billions of dollars (*13-21*). Some important plant-derived drugs and intermediates that are still obtained commercially by extraction from their whole-plant sources are listed in Table I.

Plants continue to be important sources of new drugs, as evidenced by the recent approvals in the United States of several new plant-derived drugs, and semisynthetic and synthetic drugs based on plant secondary compounds. For example, Taxol (paclitaxel), an anticancer taxane diterpenoid derived from the relatively scarce Pacific or western yew tree, *Taxus brevifolia* Nutt., has recently (December, 1992) been approved in the United States for the treatment of refractory ovarian cancer (see chapter by Kingston, this volume). Etoposide is a relatively new semisynthetic antineoplastic agent based on podophyllotoxin, a constituent of the mayapple (also known as American mandrake), *Podophyllum peltatum* L., which is useful in the chemotherapeutic treatment of refractory testicular carcinomas, small cell lung carcinomas, nonlymphocytic leukemias, and non-Hodgkin's lymphomas (*22-27*). Atracurium besylate is a relatively new

Table I. Some Examples of Economically Important Plant-Derived Drugs and Intermediates that are Still Obtained Commercially from Whole-Plant Sources (*4,9,13-16,40*)

Compound or Class	Botanical Sources	Therapeutic Category/Use
A. *Steroids*		
1. Hormones (derived from diosgenin, hecogenin, and stigmasterol)	*Dioscorea* spp. (Mexican yams); soybean-derived stigmasterol	Oral contraceptives and other steroid drugs and hormones
2. *Digitalis* glycosides (digoxin, digitoxin)	*Digitalis purpurea* L., *D. lanata* Ehrhart (foxgloves)	Cardiotonic glycosides (cardenolides)
B. *Alkaloids*		
1. Belladonna-type solanaceous tropane alkaloids (atropine, *l*-hyoscyamine, scopolamine)	*Atropa belladonna* L. (belladonna), *Datura metel* L., *D. stramonium* L. (jimson weed), *Hyoscyamus niger* L. (henbane), *Mandragora officinarum* L. (European mandrake), and other solanaceous species	Anticholinergics (parasympatholytics)
2. Opium alkaloids (codeine, morphine)	*Papaver somniferum* L. (opium poppy)	Analgesics, antitussive
3. Reserpine	*Rauvolfia serpentina* (L.) Bentham ex Kurz (East Indian snakeroot)	Antihypertensive, psychotropic
4. *Catharanthus* (*Vinca*) alkaloids (vinblastine, vincristine)	*Catharanthus roseus* (L.) G. Don (Madagascan rosy periwinkle)	Anticancer

Table I. Continued

Compound or Class	Botanical Sources	Therapeutic Category/Use
5. Physostigmine	*Physostigma venenosum* Balfour (Calabar bean)	Cholinergic (parasympathomimetic)
6. Pilocarpine	*Pilocarpus jaborandi* Holmes (jaborandi) and related species	Cholinergic (parasympathomimetic)
7. *Cinchona* alkaloids (quinine, quinidine)	*Cinchona* spp. (*Cinchona* bark)	Antimalarial, cardiac antiarrhythmic
8. Colchicine	*Colchicum autumnale* L. (autumn crocus)	Antigout
9. Cocaine	*Erythroxylum coca* Lamarck (coca leaves)	Local anesthetic
10. *d*-Tubocurarine	*Strychnos toxifera* Bentham, *Chondodendron tomentosum* Ruiz et Pavon (curare)	Skeletal muscle relaxant
11. Taxol	*Taxus brevifolia* Nutt. (western or Pacific yew)	Anticancer

synthetic skeletal muscle relaxant which is structurally and pharmacologically related to the curare alkaloids (*26,27*). In addition, synthetic Δ^9-tetrahydrocannabinol (originally derived from the marijuana plant, *Cannabis sativa* L.) and some of its synthetic analogs (e.g., nabilone) have recently been approved in the U.S. for the treatment of the nausea associated with cancer chemotherapy (*6,26,28,29*). Cannabinoids are also being developed for use in glaucoma and in neurological disorders (e.g., epilepsy and dystonia), and as antihypertensives (cardiovascular agents), antiasthmatics (bronchodilators), and potent analgesics (*29*).

Plant-derived drugs which are currently undergoing development and testing include the Chinese drug artemisinin (qinghaosu) and several of its derivatives, which are newly discovered rapidly acting antimalarial agents derived from *Artemisia annua* L. (*26,30-32*; also see chapter by Klayman, this volume), and forskolin, a naturally occurring labdane diterpene with antihypertensive, positive inotropic, and adenyl cyclase-activating properties (*33,34*; also see chapter by de Souza, this volume). Forskolin is derived from the root of *Coleus forskohlii* Briq., a plant used in East Indian folk medicine and cited in ancient Hindu and Ayurvedic texts (*34*).

In addition to the compounds mentioned above, other bioactive plant secondary metabolites are being investigated for their potential utility. For example, the medicinally active organosulfur compounds of garlic and onions are currently being investigated and evaluated as potentially useful cardiovascular agents (*35,36*; also see chapter by Lawson, this volume), and ellagic acid, β-carotene, and vitamin E (tocopherols) are being tested and evaluated for their possible utility as prototype anti-mutagenic and cancer-preventing agents (*26*; also see chapter by Pezzuto, this volume).

In the future, advances in our understanding of immunology and related areas should permit the development of new selective and sensitive bioassays to guide the isolation of bioactive natural products (*37,38*). These continuing developments should provide the means for identifying new plant-derived antiviral, antitumor, immunostimulating, and adaptogenic agents (adaptogens reportedly increase stress tolerance) (*38*). Potential adaptogenic crude drugs worthy of detailed investigation include American ginseng (*Panax quinquefolius* L.), Asiatic ginseng (*Panax ginseng* C.A. Mey), and eleuthero or Siberian ginseng (*Eleutherococcus senticosus* Maximowicz) (*38,39*).

The Role of Plant-Derived Natural Products as Lead Compounds for the Design, Synthesis, and Development of Novel Drug Compounds

In addition to the biologically active plant-derived secondary metabolites (mostly toxins) which have found direct medicinal application as drug entities, many other bioactive plant compounds have proven useful as "leads" or model compounds (templates) for drug syntheses or semisyntheses (*1,2,9,13-20,40-44*). Natural compounds of pharmaceutical importance that were once obtained from higher plant sources, but which are now produced commercially largely by synthesis, include caffeine, theophylline, theobromine, ephedrine, pseudoephedrine, emetine, papaverine, L-dopa, salicylic acid, and Δ^9-tetrahydrocannabinol. In addition, β-carotene, a plant primary metabolite which may be useful in the prevention and/or treatment of certain cancers (*45*; also see chapter by Pezzuto, this volume), is currently produced synthetically on a commercial scale. However, despite these numerous examples of economically synthesizable natural products, it is frequently forgotten that plant secondary compounds can, and often do, serve additionally as chemical models or templates for the design and total synthesis of new drug entities. For example, the belladonna alkaloids (e.g., atropine), physostigmine, quinine, cocaine, gramine, the opiates (codeine and morphine), papaverine, and salicylic acid have served as models for the design and synthesis of anticholinergics, anticholinesterases, antimalarial drugs, benzocaine, procaine, lidocaine (Xylocaine) and other local anesthetics, the analgesics pentazocine (Talwin), propoxyphene (Darvon), methadone, and meperidine (Demerol), verapamil (a near-anagram of the words *Papaver* and papaverine), and aspirin (acetylsalicylic acid), respectively (*3,42,43,46,47*).

Similarly, the study of synthetic analogs of khellin, a furanochromone derived from the fruits of *Ammi visnaga* (L.) Lam. and formerly marketed in the United States as a bronchodilator and coronary medication, led to the preparation and development of disodium cromoglycate, also known as cromolyn sodium. Cromolyn is now a major drug used as a bronchodilator and for its antiallergenic properties. Related synthetic studies based on the benzofuran moiety eventually led to the development of amiodarone, which was originally introduced in Europe as a coronary vasodilator for angina, but which was subsequently found to have a more useful application in the treatment of a specific type of arrhythmia, the Wolff-Parkinson-White syndrome, and for arrhythmias resistant to other drugs. It was introduced as a drug in the U.K. in 1980 and most recently in the United States (*48*).

In still another example, the guanidine-type alkaloid, galegine, was found to be the active principle of goat's rue (*Galega officinalis* L.), and was used clinically for the treatment of diabetes. It had been known for some time previous to this discovery that guanidine itself had antidiabetic properties, but was too toxic for human use. After hundreds of synthetic compounds were prepared, metformin, a close relative to galegine, was developed and marketed as a useful antidiabetic drug (*48*). These examples and many others serve to illustrate the continuing value and importance of plant-derived secondary metabolites as model compounds for modern drug product development.

The Depletion of Genetic Resources and the Urgent Need for Conservation Efforts

In spite of impressive recent advances in available extraction technology, separation science (chromatographic techniques), and analytical and spectroscopic instrumentation, we still know surprisingly little about the secondary metabolism of most of the world's higher plant species. This is especially true in the case of tropical rain-forest floras. Although the tropics contain most of the world's plant species, it has been estimated that more than half of these are unknown to science (having never been described) and that most have never been surveyed for chemical constituents (*49-56*; also see chapters by Gentry and Soejarto, this volume). For example, it has been estimated that nothing is known about the chemistry of the vast majority of the plant species comprising the immense flora of Brazil (*53*). The same is probably true of the floras of most of the other countries in tropical Latin America. This paucity of knowledge is alarming in view of the current rate of extinction and decimation of tropical floras and ecosystems, especially forests, before their plants have been adequately catalogued and studied. If the current trends of destruction of tropical forest habitats and general global simplification of the biota continue at their present rates, biochemists, ethnobotanists, molecular biologists, organic chemists, pharmacognosists, pharmacologists, taxonomists, and other scientists interested and involved in medicinal plant research may have only a few decades remaining in which to survey and sample the diverse chemical constituents of a large part of the plant kingdom for potentially useful novel bioactive compounds (*49-52, 54-57*). Under such circumstances, it is virtually certain that many significant opportunities for successful drug product development will be lost.

Prospects for the Future of Plant-Derived Natural Products in Drug Discovery and Development

In spite of numerous past successes in the development of plant-derived drug products, it has been estimated that only 5 to 15% of the *ca.* 250,000 existing species of higher plants have been systematically surveyed for the presence of biologically active compounds (*14,16,58-61*). Moreover, it is often the case that even plants that are considered to have been "investigated" have been screened for only a single type (or, at best, a few types) of biological activity (*14*). The best example of an extensive, but narrow, screening program is the National Cancer Institute's search for antitumor agents from higher plants. Over a 25-year period (1960-1986), an enormous number of plant extracts representing approximately 35,000 plant species were tested solely for cytotoxic and/or antitumor activity using only a few different bioassays (*14,37,61-63*; also see chapter by Cordell and coworkers, this volume). The hard-won successes of this massive screening effort are now apparent in the successful development of anticancer agents such as Taxol and derivatives (see chapter by Kingston, this volume) and camptothecin and analogs (see chapter by Wall and Wani, this volume). However, during the course of this screening effort, naturally occurring compounds potentially useful as new drugs for other ailments or conditions (e.g., analgesic, antiarthritic, antipsychotic, and psychotropic agents) were overlooked. Thus, since at least 85% of the world's species of higher plants have not been adequately surveyed for potentially useful biological activity, it appears that the plant kingdom has received relatively little attention as a resource of potentially useful bioactive compounds. Because many plant secondary metabolites are genus- or species-specific, the chances are therefore good to excellent that many other plant constituents with potentially useful biological properties remain undiscovered, uninvestigated, and undeveloped. Furthermore, there is the hope that in the future, the process of plant drug discovery and development by way of mass screening will be greatly facilitated and made more efficient by using new automated multiple biological screening methods which are now becoming available and which require only minimal amounts of test samples for evaluation.

Summary and Conclusions

Plant-derived natural products have long been and will continue to be extremely important as sources of medicinal agents and models for the design, synthesis, and semisynthesis of novel substances for treating humankind's diseases. Many of the medicinally important plant-derived pharmaceuticals have been instrumental and essential in ushering in the era of modern medicine and therapeutics, and some of these substances, such as morphine, have attained the official status of strategic materials. However, despite these many important past contributions from the plant kingdom, a great many plant species have never been described and remain unknown to science, and relatively few have been surveyed systematically to any extent for biologically active chemical constituents. Thus, it is reasonable to expect that new plant sources of valuable and pharmaceutically interesting materials remain to be discovered and developed. The continuing interest of at least some sectors of the pharmaceutical industry in plant-derived drugs is demonstrated by, for example, the recent investigation by a major pharmaceutical company of a number of traditional Chinese herbal medicines used to treat cancer, cardiovascular diseases, and central nervous system disorders (*64*), and by a collaborative program instituted by another pharmaceutical giant to systematically investigate a representative sample of the tropical flora of Costa Rica. Regrettably, if the current trends of destruction of tropical forests and general biotic simplification continue at their present rates, scientists interested and involved in medicinal plant research may have only a few decades remaining in which to investigate much of the rich diversity of the plant kingdom for useful new bioactive compounds, and many opportunities for successful drug product development will almost certainly be lost. It is therefore imperative that endangered, fragile, and over-exploited genetic resources be preserved to the greatest extent possible for future generations which may have at their disposal the tools (both technical and intellectual) necessary to successfully exploit and manage these species more intelligently.

Literature Cited

1. Bohonos, N.; Piersma, H.D. *BioScience* **1966**, *16*, 706-714, 729.
2. *Plants in the Development of Modern Medicine*; Swain, T., Ed.; Harvard University Press: Cambridge, MA, 1972.
3. Goldstein, A.; Aronow, L.; Kalman, S.M. *Principles of Drug Action: The Basis of Pharmacology*, 2nd ed.; John Wiley and Sons: New York, NY, 1974; pp 741-761.
4. Lewis, W.H.; Elvin-Lewis, M.P.F. *Medical Botany*; John Wiley and Sons (Wiley-Interscience): New York, NY, 1977.
5. Balandrin, M.F.; Klocke, J.A.; Wurtele, E.S.; Bollinger, W.H. *Science* **1985**, *228*, 1154-1160.
6. Duke, J.A.; Balandrin, M.F.; Klocke, J.A. *Science* **1985**, *229*, 1036-1038.
7. *Folk Medicine: The Art and the Science*; Steiner, R.P., Ed.; American Chemical Society: Washington, DC, 1986.
8. Balandrin, M.F.; Klocke, J.A. In *Medicinal and Aromatic Plants I*; Bajaj, Y.P.S., Ed.; Biotechnology in Agriculture and Forestry 4; Springer-Verlag: Berlin, FRG, 1988, Vol. 4; pp 3-36.
9. Tyler, V.E.; Brady, L.R.; Robbers, J.E. *Pharmacognosy*, 9th ed.; Lea and Febiger: Philadelphia, PA, 1988.
10. Kinghorn, A.D. In *Phytochemical Resources for Medicine and Agriculture*; Nigg, H.N.; Seigler, D., Eds.; Plenum Press: New York, NY, 1992; pp 75-94.
11. Kinghorn, A.D. In *Discovery of Natural Products with Therapeutic Potential*; Gullo, V.P., Ed.; Butterworths: Boston, MA, 1993.
12. Geissman, T.A.; Crout, D.H.G. *Organic Chemistry of Secondary Plant Metabolism*; Freeman, Cooper: San Francisco, CA, 1969; pp 8-11.
13. Farnsworth, N.R. In *Phytochemistry*; Miller, L.P., Ed.; Van Nostrand Reinhold: New York, NY, 1973, Vol. 3; pp 351-380.
14. Farnsworth, N.R.; Morris, R.W. *Am. J. Pharm.* **1976**, *148*, 46-52.
15. Farnsworth, N.R. In *Crop Resources*; Seigler, D.S., Ed.; Academic Press: New York, NY, 1977; pp 61-73.
16. Farnsworth, N.R.; Bingel, A.S. In *New Natural Products and Plant Drugs with Pharmacological, Biological or Therapeutical Activity*; Wagner, H.; Wolff, P., Eds.; Springer-Verlag: Berlin, FRG and New York, NY, 1977; pp 1-22.
17. Phillipson, J.D. *Pharm. J.* **1979**, *222*, 310-312.
18. Farnsworth, N.R. *Econ. Bot.* **1984**, *38*, 4-13.
19. Farnsworth, N.R. In *Natural Products and Drug Development*;

Krogsgaard-Larsen, P.; Christensen, S.B.; Kofod, H., Eds.; Alfred Benzon Symposium 20; Munksgaard: Copenhagen, 1984; pp 17-30.
20. Farnsworth, N.R.; Soejarto, D.D. *Econ. Bot.* **1985**, *39*, 231-240.
21. Principe, P.P. In *Economic and Medicinal Plant Research*; Wagner, H.; Hikino, H.; Farnsworth, N.R., Eds.; Academic Press: London, UK, 1989; pp 1-17.
22. Horwitz, S.B.; Loike, J.D. *Lloydia* **1977**, *40*, 82-89.
23. Cabanillas, F. *Drugs Fut.* **1979**, *4*, 257-261.
24. Issell, B.F.; Crooke, S.T. *Cancer Treat. Rev.* **1979**, *6*, 107-124.
25. Radice, P.A.; Bunn, P.A., Jr.; Ihde, D.C. *Cancer Treat. Rep.* **1979**, *63*, 1231-1240.
26. *The Merck Index*, 11th ed.; Budavari, S.; O'Neil, M.J.; Smith, A.; Heckelman, P.E., Eds.; Merck: Rahway, NJ, 1989.
27. Hussar, D.A. *Am. Pharm.* **1984**, *NS24*(3), 23-40.
28. Anon. *Am. Pharm.* **1985**, *NS25*(12), 10.
29. *Cannabinoids as Therapeutic Agents*; Mechoulam, R., Ed.; CRC Press: Boca Raton, FL, 1986.
30. Klayman, D.L.; Lin, A.J.; Acton, N.; Scovill, J.P.; Hoch, J.M.; Milhous, W. K. *J. Nat. Prod.* **1984**, *47*, 715-717.
31. Klayman, D.L. *Science* **1985**, *228*, 1049-1055.
32. Nair, M.S.R.; Acton, N.; Klayman, D.L.; Kendrick, K.; Basile, D.V.; Mante, S. *J. Nat. Prod.* **1986**, *49*, 504-507.
33. de Souza, N.J.; Dohadwalla, A.N.; Reden, J. *Med. Res. Rev.* **1983**, *3*, 201-219.
34. Anon. *Am. Pharm.* **1984**, *NS24*(12): 30.
35. Block, E.; Ahmad, S.; Jain, M.K.; Crecely, R.W.; Apitz-Castro, R.; Cruz, M. R. *J. Am. Chem. Soc.* **1984**, *106*, 8295-8296.
36. Block, E. *Scient. Am.* **1985**, *252*(3), 114-119.
37. Suffness, M.; Douros, J. *J. Nat. Prod.* **1982**, *45*, 1-14.
38. Tyler, V.E. *Econ. Bot.* **1986**, *40*, 279-288.
39. Carlson, A.W. *Econ. Bot.* **1986**, *40*, 233-249.
40. Farnsworth, N.R. *J. Pharm. Sci.* **1966**, *55*, 225-276.
41. Sandberg, F.; Bruhn, J.G. *Bot. Notiser* **1972**, *125*, 370-378.
42. *Design of Biopharmaceutical Properties through Prodrugs and Analogs*; Roche, E.B., Ed.; American Pharmaceutical Association, Academy of Pharmaceutical Sciences: Washington, DC, 1977.
43. *Anticancer Agents Based on Natural Product Models*; Cassady, J.M.; Douros, J.D., Eds.; Academic Press: New York, NY, 1980.
44. Trease, G.E.; Evans, W.C. *Pharmacognosy*, 12th ed.; Bailliere Tindall: London, UK, 1983.
45. Ames, B.N. *Science* **1983**, *221*, 1256-1264.
46. Gund, P.; Andose, J.D.; Rhodes, J.B.; Smith, G.M. *Science* **1980**, *208*, 1425-1431.
47. *Principles of Medicinal Chemistry*, 3rd ed.; Foye, W.O., Ed.; Lea and Febiger: Philadelphia, PA, 1989.
48. Sneader, W. *Drug Discovery: The Evolution of Modern Medicines*; John Wiley and Sons: Chichester, UK, 1985; pp 165-174, 192-226.

49. Heywood, V.H. *Pure Appl. Chem.* **1973**, *34*, 355-375.
50. *Tropical Botany*; Larsen, K.; Holm-Nielsen, L.B., Eds.; Academic Press: New York, NY, 1979.
51. Myers, N. *The Sinking Ark*; Pergamon Press: New York, NY, 1979.
52. Myers, N. *Conversion of Tropical Moist Forests*; National Academy of Sciences: Washington, DC, 1980.
53. Gottlieb, O.R.; Mors, W.B. *J. Agric. Food Chem.* **1980**, *28*, 196-215.
54. Myers, N. *Impact Sci. Soc. (UNESCO)* **1984**, *34*, 327-333.
55. Myers, N. *The Primary Source: Tropical Forests and Our Future*; Norton: New York, NY, 1984.
56. Congressional Office of Technology Assessment (OTA). *Technologies to Sustain Tropical Forest Resources*; U.S. Congress, OTA-F-214; U.S. Government Printing Office: Washington, DC, 1984.
57. Bates, D.M. *Econ. Bot.* **1985**, *39*, 241-265.
58. von Reis Altschul, S. *Drugs and Foods from Little-Known Plants: Notes in Harvard University Herbaria*; Harvard University Press: Cambridge, MA, 1973.
59. von Reis Altschul, S. *Scient. Am.* **1977**, *236*(5), 96-104.
60. von Reis, S.; Lipp, F.J., Jr. *New Plant Sources for Drugs and Foods from the New York Botanical Garden Herbarium*; Harvard University Press: Cambridge, MA, 1982.
61. Spjut, R.W. *Econ. Bot.* **1985**, *39*, 266-288.
62. Frei, E., III. *Science* **1982**, *217*, 600-606.
63. Curtin, M.E. *Bio/Technology* **1983**, *1*, 649-657.
64. Shaik, F. *Chem. Week* **1986**, *139*(8), 16.

RECEIVED March 23, 1993

Chapter 2

Tropical Forest Biodiversity and the Potential for New Medicinal Plants

Alwyn H. Gentry†

Missouri Botanical Garden, P.O. Box 299, St. Louis, MO 63166

Tropical forests are likely to contain a disproportionate share of the earth's plants with interesting pharmacologically active constituents. Although most tropical plant species have not yet been investigated chemically, several are already sources of important medicines and there are hints that others may become so. Many other tropical plants have local medicinal uses documented by ethnobotanical data. Interactions between tropical plants and their natural predators are often complex and tend to involve unusual biodynamically active chemicals with pharmacological potential. Lianas, which often biosynthesize compounds with noteworthy biodynamic properties, occur almost exclusively in tropical forests as do several of the plant families richest in alkaloids and other medicinally effective secondary compounds.

It is becoming increasingly well known that the earth's species are disproportionately concentrated in the 7-8% of the planet's land surface that is covered by tropical forests (e.g., *1*). It has also become almost axiomatic that we have only the vaguest idea of how many kinds of plants and animals with which we share our world, due largely to our incomplete knowledge of the species of tropical forests. A few groups, like birds, where only about 55 new species have been discovered in the past 25 years (T. Parker, personal communication), are relatively well known, even in the tropics, but in many other taxa only a tiny fraction of their anticipated species have been discovered. Insects are the most glaring example of this abysmal taxonomic inadequacy; current estimates of numbers of species of insects range over more than an order of magnitude from 5 million (*2*) or 6 million (*3*), to 10 (*2*), 30 (*4*), or even 50 or 80 (*5*) million, as compared to the mere 750,000-827,000 described insect species (*2,6*), with this discrepancy being due to wildly differing estimates of the virtually unknown tropical forest entomofauna. Estimates of the number of undescribed higher

†Deceased

0097–6156/93/0534–0013$06.00/0

plants tend to converge around a figure of 10,000, with nearly all of these being tropical (*7-9*), which may be contrasted with the *ca.* 250,000 currently known species. However, a recent attempt to calculate fungal diversity came up with a minimal estimate of 1.5 million, potentially even as high as 13.5 million, as compared to the 69,000 known species (*10*). Again, the tropics represent a virtual black box taxonomically speaking.

Medicinal Potential of Tropical Plants

Given our inability to even describe, much less catalogue, the biodiversity of tropical forests, it is obvious that we know next to nothing about the potential uses of the plethora of tropical forest species. While it does not necessarily follow that the high numbers of chemically uninvestigated tropical plant species should translate into numerous new medicines, there are tantalizing hints that this might, in fact, be the case. For example, about 75, or 38%, of the *ca.* 200 species of Bignoniaceae of northwestern South America are recorded to have specific uses, excluding horticulture, mostly for medicinal purposes (*8*). Included in this total are at least six useful species which have not yet been described (*8,11*). If these numbers are extrapolated, perhaps half the species of the other plant families might also be expected to have specific medicinal uses, or about 40,000 or more of the estimated 90,000 neotropical plant species. This compares with only about 600 medicinal agents of plant origin which have ever been officially approved for use in the U. S. (see chapter by Tyler, this volume).

Estimates of how higher plant species are distributed on the planet suggest that half to two-thirds probably occur in the tropics (*9,12*). We might well anticipate that most pharmacologically active compounds will come from tropical forests merely because most of the world's plant species grow there. By the same argument, the Neotropics, with many more plant species than either tropical Australasia or tropical Africa (*9,13*) might also be expected to have many more medicinal plants.

Although relatively few species of tropical plants have been investigated for possible medicinal effectiveness, those few that have entered the world's pharmacopeia have already had a major impact on developed-world health care. Farnsworth and Soejarto (*14*) have estimated that a quarter of all U. S. prescription medicines include natural plant chemicals as active ingredients. While about half of that 25% figure is due to biologically active chemicals from temperate zone plants (*15*), that still implies that tropical plants are already major suppliers of human medicine. According to Duke (*15*), the value of these tropical plant products to developed-world medicine may well be on the order of $6.25 billion a year. If the value of tropical plants as the major source of medicinal agents for the majority of the world's population that lives in undeveloped countries is included, then they might well approach $900 billion/year in value according to Duke's (*15*) estimates. Most of these biologically active ingredients are obtained from natural sources rather than by synthesis, as Soejarto and Farnsworth (*16*) have emphasized. Indeed, many of them do not yet appear to have been synthesized [e.g., tubocurarine (*17*), azadirachtin (*18*)]. As additional tropical plants are examined chemically and new biologically active chemicals are added to our pharmacopeial armamentarium, these monetary values are likely to increase significantly.

The recent realization of the commercial promise of tropical rain forest plants as possible sources of new medicines has led in the last few years to a rapidly growing interest in harnessing this potential. At least one company, Shaman Pharmaceuticals, has been set up specifically to transform tropical plant chemicals into modern medicines (*19*). Established pharmaceutical companies are also becoming increasingly intrigued by such possibilities, as evidenced by the recent signing of an agreement between Merck, Sharp & Dohme and Costa Rica's national biodiversity inventory, INBios, to chemically survey that country's plants having been the most publicized example (*20*). The sales of the anticancer drugs, vincristine and vinblastine, obtained from Madagascar's rosy periwinkle, currently amount to $100 million a year, with 80% of this representing profits for Eli Lilly, the proprietor of the extraction process (*21*). Many other tropical plants are significant medicines, including pilocarpine from Brazilian species of *Pilocarpus jaborandi* (Rutaceae), a cholinergic that is the drug-of-choice for treating glaucoma, and the antihypertensive and psychotropic reserpine from *Rauwolfia serpentina* (Apocynaceae). The muscle relaxant curare (mostly from *Chondodendron tomentosum*, Menispermaceae) until recently was widely used in open-heart surgery although it is now largely replaced by synthetic drugs such as atracurium (*22*). Even more important is quinine from Andean species of *Cinchona* (Rubiaceae), the first antimalarial drug, which is now making a comeback as chloroquine-resistant strains of malaria become more prevalent.

It is noteworthy that only about 4% of plant-based drugs are produced by synthesis (*16*), with natural sources generally proving cheaper and more efficient. Pesticides differ markedly from plant-based medicines in that about 98% are synthetics, with *ca.* 1% of the pesticide market accounted for by biological control and natural pesticides from a few tropical plants making up the other 1%. Poisons are appropriate to consider here, since they tend to be the same or similar biologically active compounds as medicines, except for being given in larger doses. Moreover, poisoning of disease vectors is a direct contribution to medicine where tropical plants are already making significant contributions, e.g. Endod, from a tropical *Phytolacca* species, for the control of the snail vectors of schistosomiasis (*15*) (see chapter by Maillard and colleagues, this volume). Among tropical plants being used as natural pesticides are *Ryania*, *Derris* (rotenone), and (mostly subtropical) *Chrysanthemum* (pyrethrin). The latter alone accounted for 356 tons of U. S. imports in 1990, worth nearly $44/kg.

In addition to the already established drugs based on tropical plants, there are many more plants that are widely used medicinally in tropical countries, but whose use in first-world medicine has not yet become established. A good example is "sangre de grado", the red latex of *Croton lechleri* (Euphorbiaceae), which is widely used in Amazonia to promote the healing of wounds and as a vaginal wash (*23*; R. Ortiz, personal communication). Twenty years ago, a principle that promoted blood clotting was isolated from "sangre de grado" in the U. S., but, despite its promise, was never developed into a medicine due to the excessive cost of getting FDA approval (S. McDaniel, personal communication). Another very widespread folk medicine in Amazonia is *Ficus insipida* Willd., locally called "oje" in Peru, which contains the enzyme ficin. "Oje" is used as a vermifuge and was commercially exported for this purpose earlier in this century (*24*); the dried latex is currently exported as a digestive aid and meat tenderizer (*25*). At Iquitos, Peru, "sangre de grado" and "oje" are employed even by middle-class city dwellers who otherwise use only store-purchased pharmaceuticals (R. Ortiz, personal communication).

Peru balsam, *Myroxylon balsamiferum* (Leguminosae), is another folk medicine that has achieved a small export market in Amazonia and elsewhere. It has antiseptic properties and, diluted with castor oil, is used for bedsores and chronic ulcers, and when in suppositories it is used for the treatment of hemorrhoids (*15*). Export of curare (*Chondodendron tomentosum*), a muscle relaxant used in open-heart surgery, is a significant local business around Tarapoto, Peru, although the value of the drug has decreased dramatically in recent years due to the availability of alternative drug therapy, as indicated previously.

Many other tropical plants whose medicinal properties may or may not have been established have also achieved export status locally. As a field botanist working in some of the most out-of-the-way corners of the world, I have often been surprised to discover local export industries for some specific pharmacological product. For example, in the Tsaratana region of north-central Madagascar, where no one under 12 had ever seen a European face before my arrival, the only vehicle in a stretch of more than 500 km was a giant quarry truck used to transport *Pygium* spp. (Rosaceae) bark to the provincial capital for shipment to Paris for the preparation of a medicine for prostate cancer. In Peru's Chanchamayo Valley, export of *Uncaria* spp. (Rubiaceae) to Germany for a prostate cancer medicine reached seven tons a year two years ago (C. Diaz, personal communication), although recent tests by NCI have shown no antineoplastic activity (W. Lewis, personal communication). Around several Amazonian population centers, the bark of *Tabebuia* spp. (Bignoniaceae) has been so overexploited as sources of remedies for cancer and diabetes, that the trees are locally extinct or very rare. *Tabebuia* bark products are widely available in health food stores in the U. S. and, in addition to their folk-medicinal use against cancer, are currently being promoted for use against candidiasis-type fungal infections (R. Troth-Ovrebo, personal communication).

In addition to such bona fide medicines, there are many tropical plants which contain other kinds of biodynamically active compounds with medicine-like properties. An Amazonian example is "guarana" (*Paullinia cupana*), the fruits of which have a much higher content of caffeine than does coffee and which is widely used as a stimulating beverage in Amazonian Brazil. While the cocaine derived from *Erythroxylon coca* is mostly used illicitly, the unprocessed leaves of the coca plant contain other, less potent alkaloids that are also used to flavor Coca-Cola, are still widely (and non-addictively) masticated as a stimulant in western South America, and are locally considered the medicine-of-choice for the treatment of dysentery and stomachache (*26*). Moreover, the estimated $100 billion per year street value of cocaine in the U. S. (*15*) is surely a testimonial to the economic value of biologically active tropical rain forest drugs, albeit one of ill-repute.

While I have emphasized the Neotropics, there are also many medicinal plants in tropical Asia and Africa (e.g., *27-30*). An example is *Dysoxylum binectariferum* (Meliaceae), that contains rohitukine, a prospective anticancer drug that functions as a tyrosine kinase inhibitor, attacking only cells undergoing active growth (see chapter by de Souza, this volume).

The examples cited above suggest that there can be little doubt that there are many more potentially useful tropical drug plants. The question is how to find them. The standard procedure of searching for new drugs has involved primarily the random screening of plant extracts, as undertaken by the U. S. National Cancer Institute (NCI) program (see chapter by Cragg and colleagues, this volume). In the case of many pharmaceutical companies, this kind of search has typically been conducted by screening compounds related to known pharmacologically active compounds. Such processes are time-consuming and expensive. It has been estimated that 12 years and $231 million are needed to bring a new drug to market (*31*).

Ethnobotanical Clues

The use of ethnobotanical data can provide a valuable short cut by indicating plants with specific folk-medicinal uses which might be likely sources of biologically active chemicals. There are numerous chemically uninvestigated tropical plants with reported ethnomedicinal uses to choose from, with each new ethnobotanical paper including many new local medicinal uses of plants (e.g., *32-39*). Folk medicines have the added advantage that, in essence, native use of these drugs constitutes a kind of pre-existing clinical testing.

Although eminently logical, this approach seems to have been little employed until very recently. For example, when I published an article outlining what I considered to be impressive ethnobotanical evidence that the roots of *Martinella* spp. (Bignoniaceae) cure conjunctivitis and other eye ailments (*40*), I anticipated being contacted by chemists or drug company representatives interested in following up on the pharmaceutical possibilities. However, only in the last year did such a contact actually occur (S. King, personal communication).

Some folk-medicinal uses of plants, like that of the garlic-smelling *Mansoa* species (Bignoniaceae) for a wide variety of unrelated ailments, may or may not have largely psychological value. Others, such as the use in Amazonia of the penis-shaped branches of *Maquira coriacea* (Moraceae; "capinuri") to restore sexual vigor, no doubt reflect the "doctrine of signatures". One reason "chuchuhuasi" bark (*Maytenus krukovii*) is so popular as a general tonic in Amazonia may be that it comes dissolved in high-proof alcohol. On the other hand, much of the folklore knowledge clearly reflects real medicinal properties. Indeed, 74% of our plant-derived major medicines were discovered by following up empirical folkloric leads (*16*).

Unfortunately, this knowledge is rapidly being lost. A dramatic example of how folk-medicine use and knowledge are disappearing comes from the Jivaro Indians of Amazonian Peru. The Jivaro have traditionally chewed the leaves of plants of the genera *Neea* (Nyctaginaceae) and *Calatola* (Icacinaceae) to prevent tooth decay (*41*). A side effect of this practice is that the teeth are stained black. Today, the Jivaro villages present a curious dental mix of old tradition-retaining Indians with well-preserved black teeth and largely toothless young Indians who have rejected the traditional tooth care and the resultant unfashionably black teeth (*41*; W. Lewis, personal communication). In one generation, a well-established medicinal-plant tradition has been abandoned. Medicinal lore held only by native "curanderos" is in even more peril of being lost, as aging traditional healers often can no longer find interested younger tribesmen to whom to pass on their knowledge (*42*).

Worse, the plants themselves are being lost, both from overexploitation and from rampant deforestation. Phillips (*43*) reports on the disappearance of *Ficus insipida* near major Amazonian population centers due to over-harvesting. *Myroxylon* species have been virtually exterminated in western Ecuador (personal observation). Even the forests in which the medicinal plants once grew are disappearing. A good example comes from the Santo Domingo area of western Ecuador, where the Colorado Indians have thus far managed to avoid the loss of their knowledge of medicinal plants. Indeed, many of them have successfully melded this knowledge with the advance of civilization, setting up herbal healing centers which are visited by the ill from all over Ecuador. Unfortunately, the forests of western Ecuador have virtually disappeared during the past twenty years with only 4% of the original forested area remaining today (*44*). As a result, the only source for the medicinals needed by the Colorado Indian "curanderos" is the 1 km^2 Rio Palenque Science Center, the last patch of uncut forest in their part of the world, where they now come to beg the owner's permission to extract medicinal plants (C. Dodson, personal communication).

An additional consideration with respect to the use of folk-medicinal knowledge to search for new medicines is that there is also much that local aboriginal tribes do *not* know about their local plants. An example is *Melloa* sp. (Bignoniaceae), a liana that ranges from Mexico to Argentina, but whose very specific and effective use to paralyze crabs seems to be a closely guarded secret of a few "campesino" families around Cartagena, Colombia (*8,11*). *Petrea* (Verbenaceae) is a widespread liana genus which is an important medicinal plant in the Guianas where its ash is used to relieve aches and pains (*45*), but elsewhere is not used medicinally (except for a single report from the opposite side of South America of a very different use as a contraceptive) (R. Rueda, personal communication). Thus, loss of even a single tribe's or individual shaman's knowledge may result in an irreplaceable loss of folk-medicinal lore.

Ecological Hints

In addition to looking at native uses of tropical forest plants, we might profitably take advantage of ecological clues as to possible sources of biologically active chemicals. One place to concentrate searches for useful chemicals might be in plant taxa that have been involved in "coevolutionary arms races" with specific groups of insect predators. Some of the most dramatic examples come from interactions of various families or genera of butterflies with their larval food plants. Typically involving alkaloid-rich plant families that are unpalatable to most generalist predators, each plant species has a specific predator species whose caterpillar is able to detoxify its chemical poisons. The adult butterflies, distasteful from the chemicals sequestered by their larvae, are frequently models in complex mimicry systems. The strong selective pressure on the host plants has evidently been responsible for the production of complex arrays of novel chemical compounds in different host taxa. Examples include passionvines (Passifloraceae) with the subfamily Heliconiinae (*46*), pipevines (Aristolochiaceae) with the swallowtail genera *Battus* and *Parides* in the neotropics and with the birdwing butterflies in southeast Asia (*47*), Apocynaceae and Asclepiadaceae with the Danainae (*48*), and Solanaceae with the Ithomiidae (*49*). It is unlikely to be due to chance that most of these plant families are already well known for their rich array of biodynamically active compounds. I anticipate that chemical investigation of additional taxa of these families and of other plants involved in similar biochemical interactions with their predators might well uncover many more medicinally useful chemicals. It should also be noted that such entwined ecological interactions are far more widespread and complex in tropical forests, where these taxa have many species as compared to only one or two species of *Battus*, Danainae, or Heliconiinae (and none of Ithomiidae) in temperate North America.

Another likely source of new biologically active compounds might be lianas in general. Climbing plants, by their very nature, are mostly diffusely branched, fast-growing and light-demanding. Thus, their leaves are likely to be more scattered throughout the canopy, shorter-lived, and generally less apparent to herbivores (*sensu 50*) than those of many other mature forest plants. Therefore, ecological theory (*50-52*) might predict that vines should be expected to concentrate more of their resources in specific highly active "qualitative" (i.e., toxic) defensive compounds rather than energetically expensive broad-spectrum "quantitative" (i.e., digestibility-reducing) defenses like tannins and lignins (*53*). It is precisely these low-molecular weight toxic compounds that tend to be biodynamic and medicinally effective (*40,43*). Moreover, small-molecule qualitative defenses are likely to be both effective at lower concentrations and more readily synthesizable in the laboratory than are the polysaccharide and polyphenolic quantitative defensive compounds. In addition, because of their structural constraints, vines tend to have a larger surface to volume ratio (which could favor toxic compounds, which tend to be concentrated near the plant's surface) and may be weight-limited by the ability of host trees to support them (which should favor low-weight toxic protective compounds) (*43,53,54*).

Circumstantial evidence suggests that a disproportionately large array of the already-known medicinally useful compounds from tropical forests come from lianas (cf., *43*). In addition to the numerous medicinal uses of lianas noted above and cited by Phillips (*43*), it is suggestive to consider the case of hallucinogens. By far the most important Amazonian hallucinogen is "ayahuasca" (also known as "caapi" and "yaje"), prepared from various lianas of the family Malpighiaceae. The most important of these is *Banisteriopsis caapi* (Spruce ex Griseb.) Morton, which contains β-carboline alkaloids (*55*); the various hallucinogenic preparations contain dimethyltryptamines, which are presumably synergistic with the *B. caapi* alkaloids (*56,57*). Among less well-known, but more widely used as hallucinogens locally are the "kosibo" snuffs of the Paumari Indians of Brazil (*58*) from scandent *Tanaecium nocturnum* (Barb. Rodr.) Bur. & K. Schum. and "chamairo" used alone or as a coca additive by several upper Amazonian tribes (*59*) from scandent *Mussatia hyacinthina* (Standl.) Sandw. of the same family.

Another hint of the correlation of biodynamic properties with scandent habit comes from a consideration of monocots, generally under-represented in tropical pharmacopeias. The only two monocot families with significant ethnomedicinal uses in Amazonia are the scandent families Dioscoreaceae and Smilacaceae. The former is widely appreciated for its steroidal sapogenins, which are used as starting points in the synthesis of cortisone, estrogen, and oral contraceptives; the latter, which contains similar chemical compounds, is the source of "sarsaparilla", used as a condiment as well as having numerous medical applications, especially in the treatment of venereal diseases (*57*).

There are indications that alkaloids, at least, are disproportionately concentrated in plants endemic to the tropics (*60,61*), although our limited attempt to correlate alkaloid abundance with the liana habit proved equivocal (*53*). Lianas are an essentially tropical phenomenon (*62*), and to whatever degree scandent habit promotes the evolution of biodynamically active molecules, one might expect tropical forests to be exceptionally rich in potential medicines. It is perhaps not coincidental that one of the NCI's most interesting new anti-AIDS compounds, michellamine B, from *Ancistrocladus* sp. (Ancistrocladaceae), which is unusual in inhibiting the growth of HIV-2 as well as resistant strains of HIV-1 (see chapter by Cardellina and coworkers, this volume), comes from a tropical forest liana. Nor is it likely to be coincidental that all of the plant taxa involved in the putative "coevolutionary arms races" noted above are largely or entirely scandent (*63*).

Taxonomic Correlations

Taxonomy can also provide useful clues. Certain families are known to be rich in the kinds of specific secondary compounds that tend to be medicinally effective, as has been emphasized by Lewis & Elvin-Lewis (*22*). For example, most hallucinogenic compounds come from the closely related tubiflorous families Rubiaceae, Loganiaceae, Apocynaceae, Solanaceae, and Convolvulaceae. Biodynamically active alkaloids are concentrated in the same families of the tubiflorous alliance (e.g., Rubiaceae, Solanaceae, Apocynaceae), in the Leguminosae, and in the magnoliid family, Ranunculaceae, and its close relatives Berberidaceae, Menispermaceae, and Papaveraceae. To judge from the tabulations of Lewis & Elvin-Lewis (*22*), a disproportionate number of the known drug plants from these taxa are from temperate zone representatives as compared to the percentage of total temperate zone species, presumably reflecting the bias of the temperate zone-oriented medical literature. Thus, medicinal screening of the numerous tropical members of these taxonomic alliances might yield more positive results than would random screening.

There are also families rich in biodynamic compounds as indicated by the ethnobotanic literature, but which have been little studied chemically, perhaps in part because they are almost exclusively tropical in their distribution. The Bignoniaceae is a good example. The preeminent neotropical liana family (*62*), the Bignoniaceae might be anticipated on that basis alone (cf. above) to be rich in biodynamically active compounds. Moreover, as judged from the ethnobotanical record, Bignoniaceae does indeed have a wealth of biologically active compounds (*11*). For example, Lewis and coworkers (*35*) focussed on the ethnomedicinal importance of this family among the Jivaro, in contrast to the sparse mention of the family in their standard text (*22*). Not surprisingly, 10 of the 12 Bignoniaceae genera used medicinally by the Jivaro were lianas (*35*). Overall, I have records of 27 genera of lianas and 19 genera of trees of this family that have been used medicinally (*11* and unpublished data). Various Bignoniaceae have been recorded as cures for almost the whole gamut of human diseases including asthma, cancer, diabetes, diarrhea, epilepsy, headaches, influenza, leishmaniasis, malaria, hepatitis, rabies, syphilis, and tuberculosis, as well as ailments of the eyes, ears, teeth, nose and throat, skin, stomach and intestines, liver, and joints. Some species are used to treat broken bones, as contraceptives and to cause abortion, as skin cosmetics, to remove pimples, to combat hemorrhaging, hiccoughs, flatulence, and snakebite, and as tranquilizers to combat aggressiveness and other behavioral disorders (*11*). While some of the reported uses may be spurious, some clearly reflect chemical properties. Especially interesting are repeated uses of the same or closely related taxa for the same ailment by different indigenous groups in different parts of Latin America, for example *Martinella* spp. for conjunctivitis (*40*), *Jacaranda* spp. for venereal diseases, and *Callichlamys* spp. for leishmaniasis.

The discrepancy between the numerous ethnomedicinal uses of a tropical family like the Bignoniaceae and the low incidence of drugs derived from members of the family in the pharmacopeias of the developed world is striking. Such patterns point in the same direction as the rest of the available evidence. It is almost certain that many more effective and useful drugs remain to be discovered in rain forest plants.

Conclusion

As international concern about the rapid demise of tropical rain forests builds, many conservationists have focussed on the idea of extractive reserves as a useful strategy (e.g., *64*). Extractive reserves represent a conservational compromise, with incentives for long-term rain forest preservation derived from the sustainable harvesting of forest products. It has been shown that the long-term economic value of the products of intact rain forests can be considerable, in fact much higher than that of so-called development schemes based on replacement of the natural forest by cattle-ranching or plantation forestry (*65*). Medicinal plants can contribute a significant part of that value (*66*). Much of the value of the rain forest lies in its very biodiversity, the multitude of species of which it is composed. One of the key tenets of extractive reserves is that the best utilization is of resources of relatively low-volume, high-value products like fruits and latex whose harvest is much less destructive than lumbering. In this context, potentially high-value medicinal plants, especially if they are disproportionately represented by relatively small plants like lianas (or herbs), could be a cornerstone in making the concept of extractive reserves a viable one, especially if some of the value added by processing can be captured locally. Implementation of such schemes could provide excellent new opportunities for mutually advantageous collaborations between drug companies and conservationists. Is it an impossible dream to hope that through medicinal plants the biodiversity of tropical forests might be able to save the world from cancer or AIDS and at the same time contribute to its own salvation?

Acknowledgments

This work was supported by a Pew Fellowship and, indirectly, by a grant from the National Science Foundation (BSR-8607113). I thank W. Lewis, M. Plotkin, and O. Phillips for reviewing the manuscript.

Literature Cited

1. Wilson, E. O. *Biodiversity*; National Academy Press: Washington, D.C., 1988.
2. Gaston, K. J. *Conserv. Biol.* **1991**, *5*, 283-296.
3. Thomas, C. D. *Nature* **1990**, *347*, 237.
4. Erwin, T. *Coleopterists Bull.* **1982**, *36*, 74-75.
5. Stork, N. E. *Biol. J. Linn. Soc. London* **1988**, *35*, 321-337.
6. Ehrlich, P. R.; Wilson, E. O. *Science* **1991**, *253*, 758-762.
7. Gentry, A. H. *Oikos* **1992**, *63*, 19-28.
8. Gentry, A. H. In *Sustainable Harvest and Marketing of Rain Forest Products*; Plotkin, M.; Famolare, L., Eds.; Island Press: Washington, D.C., in press.
9. Raven, P. H. In *Conservation of Threatened Plants*; Simmons, B. *et al.*, Eds.; Plenum Press: New York, 1976; pp 155-179.
10. Hawksworth, D. L. *Mycol. Res.* **1991**, *95*, 641-655.
11. Gentry, A. H. *Ann. Missouri Bot. Gard.* **1992**, *79*, 52-64.
12. Arroyo, M. T. K.; Raven, P. H.; Sarukhan, J. In *An Agenda of Science for Environment and Development into the 21st Century*; Doode, J. C. *et al.*, Eds.; Cambridge University Press: Cambridge, U. K., in press.
13. Gentry, A. H. *Ann. Missouri Bot. Gard.* **1982**, *69*, 557-593.
14. Farnsworth, N. R.; Soejarto, D. D. *Econ. Bot.* **1985**, *39*, 231-240.
15. Duke, J. A. In *Proceedings of Second National Herb Growing and Marketing Conference*, Simon, J. E.; Grant, L., Eds.; Purdue Research Foundation: West Lafayette, IN, 1987; pp 1-5.
16. Soejarto, D. D.; Farnsworth, N. R. *Perspect. Biol. Med.*, **1989**, *32*, 244-256.
17. Mabberley, D. J. *The Plant Book*; Cambridge University Press: Cambridge, U. K., 1987.
18. Stone, R. *Science*, **1992**, *255*, 1070-1071.
19. Riggle, D. *Business*, **1992**, *1*, 26-29.
20. Gamez, R. In *Conservancia y Manejo de Recursos Naturales en Latin America*; Dirzo, R.; Pinero, D.; Arroyo, M., Eds.; Univ. Nac. Autonoma Mexico: Mexico City, in press.
21. Farnsworth, N. R. In *Biodiversity*; Wilson, E. O., Ed.; National Academy Press, Washington, D.C., 1988; pp 83-97.
22. Lewis, W.; Elvin-Lewis, M. *Medical Botany: Plants Affecting Man's Health*; Wiley and Sons: New York, 1977.
23. Vasquez, R. *Plantas Utiles de la Amazonia Peruana;* Proyecto Flora del Peru: Iquitos, Peru, 1989.
24. Padoch, C. *Adv. Econ. Bot.* **1987**, *5*, 74-89.
25. Hansson, A.; Veliz, G.; Nuquira, C.; Amren, M.; Arroyo, M.; Arevalo, G. *J. Ethnopharmacol.*, **1986**, *17*, 105-138.
26. Plowman, T. *Adv. Econ. Bot.* **1984**, *1*, 62-111.
27. Ayensu, E. *Medicinal Plants of West Africa*; Reference Publications: Algonac, MI, 1978.

28. Oliver-Bever, B. *Medicinal Plants in Tropical West Africa*; Cambridge University Press: Cambridge, U. K., 1986.
29. Pushpangadan, P.; Atal, C. K. *J. Ethnopharmacol.* **1984**, *11*, 59-77.
30. Jacobs, M. *Flora Malesiana Bull.* **1982**, *35*, 3768-3782.
31. Vagelos, R. *Science* **1991**, *252*, 1080-1084.
32. Ayala, F. *Adv. Econ. Bot.* **1984**, *1*, 1-8.
33. Boom, B. M. *Adv. Econ. Bot.* **1987**, *4*, 1-68.
34. Garcia-Barriga, H. *Flora Medicinal de Colombia*, Imprenta Nacional: Bogota, Colombia, 1975.
35. Lewis, W.; Elvin-Lewis, M.; Gnerre, M. C. In *Medicinal and Poisonous Plants of the Tropics*, Leeuwenberg, A., Ed.; Pudoc: Wageningen, Netherlands, 1987.
36. Prance, G.; Balee, W., Boom, B. *Conserv. Biol.* **1987**, *1*, 296-310.
37. Reynel, C.; Alban, J.; Leon, J.; Diaz, J. *Etnobotanica Campa-Ashaninca*; Fac. Cien. Forest. Univ. Nac. Agraria, La Molina: Lima, Peru, 1990.
38. Schultes, R. E.; Raffauf, R. *The Healing Forest*; Dioscorides Press: Portland, OR, 1990.
39. Vickers, W. T.; Plowman, T. *Fieldiana Bot., n.s.* **1984**, *15*, 1-63.
40. Gentry, A. H.; Cook, K. *J. Ethnopharmacol.* **1984**, *11*, 337-343.
41. Lewis, W.; Elvin-Lewis, M. *Adv. Econ. Bot.* **1984**, *1*, 53-61.
42. Plotkin, M. In *Sustainable Harvest and Marketing of Rain Forest Products*; Plotkin, M.; Famolare, L., Eds.; Island Press: Washington, D.C., in press.
43. Phillips, O. *Econ. Bot.* **1990**, *44*, 534-536.
44. Dodson, C. H.; Gentry, A. H. *Ann. Missouri Bot. Gard.* **1991**, *78*, 273-295.
45. DeFillipps, R. In *Sustainable Harvest and Marketing of Rain Forest Products*; Plotkin, M.; Famolare, L., Eds.; Island Press: Washington, D.C., in press.
46. Gilbert, L. In *Coevolution of Plants and Animals*; Gilbert, L.; Raven, P., Eds.; University of Texas Press: Austin, TX, 1975; pp 210-240.
47. DeVries, P. J. *The Butterflies of Costa Rica*; Princeton University Press: Princeton, NJ, 1987.
48. Boppre, M. *Entomol. Exp. Appl. (Amsterdam)* **1978**, *24*, 64-77.
49. Brown, K. S., Jr. *Ann. Missouri Bot. Gard.* **1987**, *74*, 359-397.
50. Feeny, P. P. *Rec. Adv. Phytochemistry* **1976**, *10*, 1-40.
51. Coley, P. D.; Bryant, J. P.; Chapin, F. S., III. *Science* **1985**, *230*, 895-899.
52. Feeny, P. P. In *Coevolution of Animals and Plants*; Gilbert, L.; Raven, P., Eds.; University of Texas Press: Austin, TX, 1975; pp 3-19.
53. Hegarty, M. P.; Hegarty, E. E.; Gentry, A. H. In *The Biology of Vines*; Putz, F.; Mooney, H., Eds.; Cambridge University Press: Cambridge, U. K., 1991; pp 287-310.
54. Putz, F. E. *Biotropica* **1983**, *15*, 185-189.
55 Schultes, R. E. *Bot. Mus. Leafl. Harvard Univ.* **1978**, *26*, 311-332.
56. Ayala, F.; Lewis, W. H. *Econ. Bot.* **1978**, *32*, 154-156.
57. Phillips, O. In *The Biology of Vines;* Putz, F.; Mooney, H., Eds.; Cambridge University Press: Cambridge, U. K., 1991; pp 419-467.
58. Prance, G.; Campbell, D.; Nelson, B. *Econ. Bot.* **1977**, *31*, 129-139.
59. Plowman, T. *Bot. Mus. Leafl. Harvard Univ.* **1980**, *28*, 253-261.

60. Levin, D. A. *Am. Natur.* **1976**, *110*, 261-284.
61. Levin, D. A.; York, B. M.; *Biochem. Syst. Ecol.* **1978**, *6*, 61-76.
62. Gentry, A. H. In *The Biology of Vines;* Putz, F.; Mooney, H., Eds.; Cambridge University Press: Cambridge, U. K., 1991; pp 3-46.
63. Gentry, A. H. In *Historia Natural de Panama*; D'Arcy, W.; Correa, M., Eds.; Missouri Botanical Garden: St. Louis, MO, 1985; pp 29-42.
64. Fearnside, P. *BioScience* **1989**, *39*, 387-393.
65. Peters, C. M.; Gentry, A. H.; Mendelssohn, R. *Nature* **1989**, *339*, 655-656.
66. Balick, M.; Mendelssohn, R. *Conserv. Biol.* **1992**, *6*, 128-130.

RECEIVED March 31, 1993

Chapter 3

Phytomedicines in Western Europe

Potential Impact on Herbal Medicine in the United States

Varro E. Tyler

Department of Medicinal Chemistry and Pharmacognosy, School of Pharmacy and Pharmacal Sciences, Purdue University, West Lafayette, IN 47907–1333

Laws and regulations governing the sale of phytomedicines (herbal remedies) in several European countries, particularly Germany, are compared to those in the United States. This provides an understanding as to why the development of medicines prepared from long-known plant drugs flourishes there with substantial support from the pharmaceutical industry and the medical community, while such research languishes here. The composition and therapeutic value of several popular European phytomedicinals, including those obtained from ginkgo, echinacea, chamomile, feverfew, valerian, milk thistle, St. John's wort, saw palmetto (sabal), hawthorn, and melissa (lemon balm) are discussed. Unless the regulatory attitude towards phytomedicines changes in the United States, it is likely that even the most exciting European developments will have little impact in this country and that many useful drugs widely used in Europe will continue to be unavailable here.

Phytomedicines are defined as crude vegetable drugs (herbs) and the galenical preparations (extracts, fluidextracts, tinctures, etc.) made from them. In general, the plants from which they are obtained fall into a group of some 13,000 species that have been used for at least a century as traditional medicines by people in various cultures around the world (*1*). Several hundred of these are not obscure plants with ambiguous uses whose existence is known only to ethnobotanists. Instead, they are classic drugs, once commonly used, some of which have already re-entered the materia medica as a result of modern investigations. It is the group that is certainly the most likely to yield additional significant drugs when subjected to further scrutiny. Phytomedicines are not unconventional drugs, even in the United States. Over the years, more than 600 botanical items have received official recognition in the various editions of *The United States Pharmacopoeia* (2).

0097–6156/93/0534–0025$06.00/0

Comparison of Laws and Regulations

To understand the effects which recent developments in the area of phytomedicines in Western Europe may have on herbal medicine in the United States, it is necessary to have some knowledge of the laws and regulations that govern the sale of such products there and here and, specifically, how these laws and regulations have affected plant-derived drug development on both continents. As can well be imagined, this is a very complex situation which differs appreciably in each of the countries involved.

Reuter has summarized the European situation most succinctly (*3*) by noting that in Germany, approval of herbal drugs is based on the results of clinical and pharmacological studies; in other European countries, such as France and the United Kingdom, the traditional use of herbal drugs is generally considered to be sufficient proof of product efficacy and safety. However, even in Germany where substantive evidence is required, its quality and quantity is much less than is required for new chemical entities in the United States. The process has been characterized as one requiring and providing "reasonable certainty" rather than "absolute proof" of safety and efficacy (*4*).

Much of the research on phytomedicinals in Germany is conducted in-house by pharmaceutical companies, but much is also carried out under grants and contracts by institutes of the universities. There is considerable concern among researchers and other persons in the latter that, as the date for European union approaches, this sort of support will be lost. Bureaucrats tend to reduce things to the lowest common denominator, and Germans fear that the lax regulations currently extant in Britain and France which allow the sale of phytomedicinals without substantial testing will prevail throughout all of the countries involved. This would be analogous to the reduction in the quality of beer that was recently imposed on Germany. If phytomedicinals can be sold without the need for evaluation, research funds devoted to this purpose are sure to dry up, and progress in the field will be severely limited.

In this country, no deviation from the most rigid criteria for determining drug safety and efficacy is permitted, even for phytomedicines long in use. It was clearly the intent of the United States Congress to "grandfather" from such requirements all drugs marketed in the U.S. prior to 1962. Because the vast majority of herbs now in use here were at one time official in *The United States Pharmacopeia* or *The National Formulary* and therefore fall into that category, it may seem unusual that they are not presently viewed in this way by the Food and Drug Administration (FDA). However, that agency, by a clever interpretation of administrative law, simply ruled that pre-1962 drugs could continue to be sold but would be considered misbranded if any therapeutic claims, not proven by post-1962 standards, were made on the label (5).

Effects of Regulatory Philosophies

Phytomedicines in the United States have been placed in a kind of never-never land between drugs and foods. Although most are basically drugs by almost any definition applied to them, they cannot legally be sold as such unless millions of dollars are spent to prove them safe and effective. Since exclusivity of sales rights cannot be assured through the patent mechanism for drugs already used for centuries or even millennia, no one was willing to hazard such an investment.

Phytomedicines, therefore, continue to be marketed under the guise of foods, nutritional supplements, or the like, but without therapeutic claims. No legal standards of quality exist or are now enforced, and in the case of certain expensive herbal products, the consumer has less than a 50% chance of receiving a product that conforms to the labeling in terms of quality or quantity (*6*).

The most serious effect of the FDA regulations on pre-1962 drugs has been to discourage research in this country, particularly in the pharmacologic and therapeutic areas, but also in the area of chemistry, on the classic phytomedicinals. Among the most recent drugs of plant origin marketed in this country are the catharanthus alkaloids vinblastine and vincristine, which were discovered in 1958 and 1961, respectively (*7*). There are high hopes that the antitumor drug, taxol, will soon be marketed, but first a reliable synthetic or cell-culture production method may be necessary. Still, a 30-year hiatus between new natural plant-derived drug introductions is much too long.

Consequently, research in the field has shifted from the United States to other countries, and in Western Europe, Germany has played a leading role for two reasons. In the first place, phytomedicinals are extremely popular in that country. In 1989, the market for phytomedicinals in Western Europe was estimated to be 2.2 billion U.S. dollars, and Germany alone accounted for 70% of that figure (*8*). Incidentally, these statistics, compiled by the European Scientific Cooperative on Phytotherapy (ESCOP) are viewed by many as quite conservative. One of the principal difficulties in compiling such data is the fact that, in some countries, phytomedicinals are prescription-only drugs. In others, they are over-the-counter drugs sold only in pharmacies; in still others, they are sold in so-called drug stores (not pharmacies) or health food stores.

The second reason why Germany leads the world in phytomedicinal research is that, following the establishment of a product's safety, only reasonable certainty of its efficacy is required to market it there. The *Bundesgesundheitsamt* (Federal Health Agency) established in 1978 a Commission E that was assigned the task of evaluating some 1,400 herbal drugs corresponding to 600-700 different plant species (*9*). This evaluation is based on data from the published literature, information supplied by practicing physicians, reports from patients, pharmacological investigations, and limited clinical studies. Such a procedure has resulted in a series of monographs that assess the safety and utility of each drug. These provide, in summary form, what is probably the best information available today on the proper use of many phytomedicinals. Needless to say, this has encouraged even the smaller pharmaceutical manufacturers—and there are many in Germany—to sponsor appropriate research in the hope that the data obtained will facilitate a favorable evaluation by Commission E.

Selected European Phytomedicines

What, then, are some of the phytomedicines that have been researched in Western Europe in recent years but which are, as yet, available in the United States only as crude herbs for making teas or as finished products sold under the euphemisms of foods or nutritional supplements? For purposes of this chapter, ten phytomedicines have been chosen arbitrarily, using as criteria not only their utility or potential utility as drugs but, in addition, the belief that some of them, not well-known to American scientists, are eminently worthy of further investigation. Two omissions, which some may find curious, are ginseng and garlic. In the case of ginseng, it is believed that until substantial clinical studies in human beings are conducted, further investigations of its chemistry and pharmacology are of academic interest only. As for garlic, the findings of Americans, such as Block and Lawson (see this volume), have done much to promote interest in it (*10*). So, in this unusual case, Americans, not Western Europeans, have led the way in recent phytomedicinal research.

Ginkgo. Without question, a concentrated extract of the leaves of *Ginkgo biloba* L., a popular ornamental tree that is truly a living fossil, is the most important phytomedicinal agent to be marketed in Europe during the last decade. Its popularity may be judged by the 5.4 million prescriptions written for it in Germany in 1988, more than for any other pharmaceutical. In addition, it is available in Western Europe as an over-the-counter drug (*11*). Ginkgo biloba extract (GBE) is prepared by extracting the dried, green leaves with an acetone-water mixture under partial vacuum. After removal of the solvent, the extract is adjusted to a potency of 24% w/w flavonoids, mostly a complex mixture of mono- and diglycosides formed by glucose and rhamnose with kaempferol, quercetin and isorhamnetin as genins. The aglycone quercetin is itself also present. GBE also contains 6% w/w of terpenes, principally a unique group of diterpenes known as ginkgolides A, B, and C, as well as the sesquiterpene, bilobalide (*12*).

What makes this complex phytomedicinal so interesting is its apparent effectiveness in treating ailments associated with decreased cerebral blood flow, particularly in geriatric patients. Such conditions, including memory loss, headache, tinnitus, and depression, respond to the vasodilation and improved blood flow induced by GBE in both the arteries and capillaries. There is also evidence that the extract is an effective free-radical scavenger. Side effects are usually minimal but may include restlessness, diarrhea, nausea, and vomiting (*13*).

GBE is marketed in this country as a food supplement without labeled claims of therapeutic value. As is customary with phytomedicinals here, ancillary claims made in catalogs and books about such products are mostly pure hyperbole. Assertions that GBE offers significant protection against strokes or the development of Alzheimer's disease or that it will reverse the aging process and promote increased longevity are, of course, unproven. GBE products of several manufacturers are available in the United States. The original one, and the one on which practically all of the scientific and clinical studies have been conducted, is manufactured by the German firm of Willmar Schwabe. In Western Europe it is marketed as Tebonin® or Rökan®; in this country it is sold as Ginkgold®.

Echinacea. Introduced into medicine in 1871 by a patent medicine vendor in Nebraska who learned of its value as a "blood purifier" from the Indians, echinacea--or cone flower--has had a checkered therapeutic history. Originally, the rhizome and root of *Echinacea pallida* (Nutt.) Nutt. (pale-flowered echinacea) and *E. angustifolia* DC. (narrow-leaved cone flower) were extensively employed here, usually in the form of a tincture (hydroalcoholic solution), as anti-infective agents. However, with the advent of sulfa drugs and other chemotherapeutic agents, the use of echinacea underwent a rapid decline, and the drug was dropped from official status in 1950 (*14*).

In the meantime, Dr. Gerhard Madaus had introduced the plant into Germany. His seed, purchased in Chicago as *E. angustifolia*, turned out to be *E. purpurea* (L.) Moench, the common garden variety of echinacea. Consequently, most of the research in Western Europe has been conducted with this species, much of it with the freshly expressed juice of the aboveground portion of the flowering plant (*15*).

As early as 1939, it was observed that various diseases resulting from infection with microorganisms responded favorably to echinacea therapy, despite the fact that this phytomedicinal did not exert a direct action on the disease-producing pathogen. Although it was postulated in 1941 that constituents in the plant seemed to stimulate the body's own healing powers, it was not until 1971 that serious research on echinacea's nonspecific immunostimulant properties got underway. As a result of intensive research, particularly during the last decade, it now appears that echinacea stimulates the activity of various types of phagocytes, inhibits hyaluronidase, increases the number of fibroblasts, raises properdin levels, and may cause increased production of interferon (*16*). All of these activities would result in enhanced resistance to disease.

The search for the identity of the active principles of echinacea has led to the publication of more than 200 papers on the subject since 1940. Most of the chemical research has been conducted in Germany, and the extensive literature on the subject has been summarized (*17*). It is now generally believed that, with respect to increased phagocytosis, high-molecular-weight polysaccharides are definitely active, but their activity is enhanced by components of the alkamide fraction (principally isobutylamides) and by chicoric acid (*18*).

About 150 conventional (i.e., non-homeopathic) phytomedicinals containing echinacea are currently marketed in Germany. Those preparations intended for external use (ointments, lotions, creams) are employed for the treatment of wounds, burns, and various inflammatory conditions. Oral preparations (tinctures, extracts) are used to increase immunity and appear to be most useful in preventing or moderating the symptoms of colds or flu. Parenteral preparations are also available. Use of the drug in its various forms has been approved by Commission E (*19*). Side effects are not known on local or oral administration.

Chamomile. Chamomile, or more specifically German chamomile, consists of the dried flower heads of *Matricaria recutita* L. (formerly designated *M. chamomilla* L. pro parte); it is without question the most enigmatic plant drug. Designated as the medicinal plant of the year in 1987 in Europe (*20*) where scores of phytomedicinals containing it are used to treat everything from gastrointestinal spasms to skin irritations, the remedy is known in this country primarily as a pleasant-tasting tea. This herb is perhaps the best example of the wide chasm separating medicinal practice in Western Europe and the United States.

Europeans utilize chamomile internally and externally for its anti-inflammatory, antispasmodic, antibacterial, and mild sedative properties. Commission E has declared it effective for all of these actions except as a sedative (*21*). A wide variety of chamomile extracts, tinctures, ointments, and teas is available commercially in Europe. It is one of the first herbs for which a European monograph was proposed by the European Scientific Cooperative on Phytotherapy (*22*). The drug is an ancient one, and the literature dealing with it is extensive. A 1986 review of the literature on chamomile and its near relative, Roman chamomile [*Chamaemelum nobile* (L.) All., formerly designated *Anthemis nobilis* L.], listed 220 references (*23*).

Much of the biological activity of chamomile is associated with the blue-colored volatile oil yielded by the plant on distillation in amounts of about 0.5% w/w. The principal anti-inflammatory and antispasmodic constituents include the terpenoids (-)-α-bisabolol, bisabolol oxides A and B, and matricin. The latter colorless compound is converted into the blue artifact, chamazulene, when the oil is distilled (*24*). This azulene derivative also possesses some anti-inflammatory activity. A large number of flavonoids, including apigenin, luteolin, and quercetin, contribute to the anti-inflammatory and antispasmodic properties of chamomile. Several coumarins, such as umbelliferone and its methyl ether, herniarin, also possess antispasmodic properties (*25*). The total effects of the flower head are thus due to a very complex mixture of polar and nonpolar constituents.

Contraindications to the use of chamomile are not known. Side effects may include allergic reactions, and some authors place great emphasis on this potential hazard. However, only five such cases were reported in the literature between 1887 and 1982 (*26*). Recognizing that an estimated 1 million cups of chamomile tea are consumed daily worldwide, this is indeed a very low incidence of allergenicity (*27*).

Feverfew. Utilized as a folk remedy for the treatment of headache and related conditions for some 2000 years, feverfew, the leaves of *Tanacetum parthenium* (L.) Schultz Bip., has recently gained considerable popularity as an effective treatment for migraine. This has been the result of clinical studies carried out in Britain in the 1980's. Relatively small doses of freeze-dried, powdered leaves (*ca.* 60 mg) produced significant decreases in the frequency and severity of migraine headache and its side effects of nausea and vomiting (*28*).

The active principles of feverfew are a number of sesquiterpene lactones, of which parthenolide is the predominant one, constituting about 84% w/w of the total sesquiterpene fraction (*29*). Regulatory agencies in various countries have proposed standards ranging from not less than 0.1 to not less than 0.2% w/w of parthenolides in whole, dried feverfew leaf. Parthenolide apparently functions as a serotonin antagonist, inhibiting the release of that vasoconstrictor from blood platelets.

In the absence of standardized, dried leaf preparations, many persons consume the fresh leaves, chewing them before swallowing. This may result in the development of aphthous ulcers and, more commonly, inflammation of the oral mucosa and tongue, often with swelling of the lips. These conditions subside when administration is terminated. No chronic toxicity tests have been carried out with feverfew, but no serious side effects have been observed, even in long-term users (*30*).

Feverfew is available as a food supplement in this country in the form of compressed tablets or capsules containing 300 to 380 mg of the powdered herb.

Dosage recommended by the manufacturer is up to six tablets or capsules per day. This is many times the effective daily dose, but perhaps this large excess is an attempt to insure some activity with low-quality plant material. A recent study in Canada showed that no commercial feverfew products in North America contained even 50% of the minimum parthenolide concentration (0.2% w/w) characteristic of a quality phytomedicinal (*31*).

Valerian. Who could imagine that a plant with mild, but effective, tranquilizing and sedative properties could have been widely used in medicine for a millennium; further, that today in Germany alone more than 100 phytomedicinals containing it are marketed; and, most amazing of all, that the active principles responsible for its therapeutic effects remain unidentified? Such a plant actually exists. This unbelievable situation aptly summarizes the failure of modern-day scientists to coordinate the chemical and pharmacological testing of classical plant drugs. The drug is valerian, the root and rhizome of *Valeriana officinalis* L. Other species of the genus possess similar activity and are similarly used (*32*).

Over the years, valerian's depressant effects on the central nervous system (CNS) have been attributed to its volatile oil and especially to its contained sesquiterpenes, valerenic acid and valeranone. More recently, it was believed that a mixture of unstable iridoid compounds known as the valepotriates was responsible. However, in 1988, investigators published the results of an extensive study in rats which indicated that, while the crude drug was effective, these principles were not (*33*). Here the matter stands. All we can be certain of at this time is that the fresh rhizome and root, or that which has been recently dried at low temperature (< 40° C), exhibits the most effective CNS depressant activity (*34*).

The German Commission E has declared valerian to be an effective treatment for restlessness and sleep disturbances resulting from nervous conditions. In the United States, the drug is available as a coarse powder for making tea, as an encapsulated powder, and as a tincture. All these are sold as food supplements, not as phytomedicinals. Significant side effects have not been reported nor has acute toxicity been demonstrated, either in animals or humans (*35*).

Milk Thistle. About 20 years ago, scientists succeeded in isolating from the milk thistle, *Silybum marianum* (L.) Gaertn., a crude mixture of antihepatotoxic principles designated silymarin. Subsequently, silymarin was shown to consist of a large number of flavonolignans, including principally silybin, accompanied by isosilybin, dehydrosilybin, silydianin, silychristin, and others (*36*).

Studies on small animals have shown that silymarin protects the liver against a variety of toxins including those of the deadly *Amanita phalloides* (Vaill.) Secr. mushroom (*37*). Human trials have also been encouraging for several liver conditions, including hepatitis and cirrhosis of various origins (*38*). Results of both types of studies suggest that silymarin protects intact liver cells, or cells not yet irreversibly damaged, by acting on the cell membranes to prevent the entry of toxic substances. Protein synthesis is also stimulated, thereby accelerating the regeneration process and the production of liver cells. As a result of these findings, the German Commission E has endorsed the use of milk thistle fruits as a supportive treatment for inflammatory liver conditions and cirrhosis.

Silymarin is very poorly soluble in water, so aqueous preparations (teas) of the fruit are ineffective (*39*). Silymarin is also poorly absorbed (23-47%) from the gastrointestinal tract (*40*), so the drug is best administered parenterally. Oral use requires a concentrated product. Milk thistle is currently marketed in this country as a food supplement in the form of capsules containing 200 mg each of a concentrated extract representing 140 mg of silymarin. Toxic effects from the consumption of milk thistle have apparently not been reported.

St. John's Wort. Only a portion of the potential of even well-known drug plants has been examined. A case in point is St. John's wort, *Hypericum perforatum* L., the leaves and tops of which have long been used internally as a diuretic and treatment for menstrual disorders and externally, usually in an oil base, for wound healing (*41*).

More recently, the plant has gained a considerable reputation in Europe as an effective treatment for nervousness, sleep disturbances, and, particularly, depression. At first, such activity was attributed to hypericin, a naphthodianthrone pigment, but it is now believed that various flavonoids and xanthones play a significant role in its antidepressant action (*42*). These compounds function as monoamine oxidase (MAO) inhibitors. Tests on small animals and preliminary tests in humans have confirmed the activity (*43,44*), and Commission E has approved the use of St. John's wort for the treatment of psychotic disturbances, depression, anxiety, and nervous unrest.

However, the most important property of this ancient plant is perhaps still to be realized. In 1988, investigators showed that hypericin and pseudohypericin exhibited dramatic antiretroviral activity and low toxicity at effective doses (*45*). Phase I clinical trials of hypericin are now underway at New York University Medical Center (*46*), and it appears that St. John's wort may eventually yield a useful treatment for retroviral-induced diseases such as acquired immunodeficiency syndrome (AIDS).

Hypericin is known to induce photosensitivity in livestock. While this is not normally a problem in humans taking recommended doses of the herb, fair-skinned consumers are cautioned against excessive exposure to sunlight (*47*). St. John's wort is available in this country both in capsules and as a fluidextract. Naturally, these are sold only as food supplements, not as drugs.

Saw Palmetto (Sabal). On August 27, 1990, the Food and Drug Administration banned the sale in this country of all nonprescription drugs used for the treatment of benign prostatic hypertrophy (BPH). The reasons given were twofold: 1) no evidence had been presented to them that the products were effective, and 2) people who used them might delay proper medical treatment as their condition worsened (*48*). What the FDA overlooked was the considerable evidence in Western Europe that certain phytomedicinals are effective in treating BPH and that persons using them experience an appreciable increase in their comfort level due to their edema-protective and anti-inflammatory effects. It may be argued that their use no more prevents proper subsequent treatment of BPH than does aspirin if taken in the early stages of pneumonia. This assumes, of course, that an accurate diagnosis of the condition has been made by a qualified physician.

Several plants favorably affecting BPH have been approved by the German Commission E. Perhaps the most popular of these is saw palmetto, or sabal, the fruits of *Serenoa repens* (Bartr.) Small. An extract of the fruits exerts at least some of its activity, not through estrogenic effects *per se*, but because of its antiandrogenic properties. The beneficial effects include increased urinary flow, reduced residual urine, increased ease in commencing micturition, and decreased frequency of urination (*49*). The chemical constituents responsible for this activity have not been identified, but they occur in the fraction extracted by nonpolar solvents (*50*).

A number of phytomedicinals containing saw palmetto extract are currently marketed in Europe. In this country, only the crude drug is available in cut or powdered form for use as a tea because all OTC drug preparations containing it have been banned by the FDA. This is unfortunate because, whatever the active constituents are, they are lipophilic, and little or no benefit would result from the consumption of any aqueous preparation of saw palmetto.

Hawthorn. The leaves and flowers or fruits of the hawthorn, *Crataegus laevigata* (Poir.) DC. and related species, constitute one of the most widely utilized heart remedies in Germany today. Hawthorn apparently brings about its beneficial effects on the heart in two ways. First, it produces a dilation of the blood vessels, particularly the coronary vessels, thus improving circulation and strengthening the heart indirectly. In addition, it exerts a positive inotropic effect directly on the heart muscle (*51*).

The action of hawthorn develops slowly over a period of days or even weeks. Toxic effects are uncommon, even at large doses. The drug thus provides a useful treatment for certain milder forms of heart disease (*52*). Such usage has been approved by the German Commission E (*53*), and scores of phytomedicinals containing hawthorn are marketed in Western Europe.

Two groups of constituents, the simple flavonoids, such as hyperoside and rutin, as well as the oligomeric procyanidins, are apparently the principal active constituents in hawthorn (*54*). Many of the phytomedicinal products are standardized on the basis of their oligomeric procyanidin content. This assures the consumer of a quality product of known potency.

Hawthorn was selected as the 1990 medicinal plant of the year by the Union of German Druggists (*55*). (Druggists in Germany sell herbal teas and various proprietary products, but they are not pharmacists.) In spite of the fact that it is available in this country as a food supplement, both in tablet form and as a powdered herb, it is a product that should be used by the laity only with considerable caution, if at all. The self-treatment of cardiac problems based on the self-diagnosis of such problems should not be encouraged. It is unfortunate that an agent as potentially useful as standardized hawthorn extract is not available here as a drug to be utilized on the advice of a competent physician following appropriate professional diagnosis.

Melissa (Lemon Balm). The last drug to be considered in this series is one whose therapeutic utility is essentially unknown in this country, but which is both widely used and highly valued in Western Europe. It is melissa, or lemon balm, the leaves of *Melissa officinalis* L., a volatile-oil-containing herb that was selected as the medicinal plant of the year in Europe in 1988. The plant is familiar to bee keepers all over the world because the odor of its essential oil resembles that of the pheromone produced by bees which consequently find the plant very attractive (*56*).

Melissa volatile oil is used as a sedative, spasmolytic, and antibacterial agent. The sedative action is attributed largely to citronellal with other volatile terpenes such as citronellol, geraniol, caryophyllene, linalool, citral, and limonene, with the phenylpropanoid eugenol also contributing to the effect (*57*). Some of these same constituents are responsible for the plant's antispasmodic and antibacterial properties.

The German Commission E has approved melissa only as a calmative and a carminative, but a cream containing the plant extract is widely marketed as a local treatment for cold sores and related conditions caused by the herpes simplex virus. Here the activity is attributed to a complex mixture of tannins that is contained in the plant to the extent of about 5%. Tests indicate that treatment with the cream not only reduces the time required to heal the herpes lesions but also extends the interval between recurrent episodes (*58*).

Proposal for Reform

The extensive use in Western Europe of useful phytomedicines such as these just discussed, and many like them, seems to have had relatively little impact on the generally negative attitude of scientists and clinicians toward herbal medicine in the United States. In Germany, at least, phytomedicine seems to become ever more scientific, with proven product safety and efficacy a desired goal. Here, herbal medicine is not only stifled by regulatory neglect but is debased, even by some supposedly knowledgeable people, to the status of unconventional medicine, along with homeopathy, Ayurvedic treatment, anthroposophical therapy, crystal power, quigong, and numerous other examples of the power of the placebo effect.

Rational herbal therapy does not belong here. It is as conventional as the use of digitalis for cardiac decompensation or plantago seed for constipation; both, incidentally, are official drugs in the current edition of *The United States Pharmacopeia* (*2*). The problem is a combination of adverse educational, regulatory, and financial circumstances that conspires to keep useful phytomedicines out of the American health-care system.

If this continues unchanged, herbalism in this country will also remain unchanged—a medieval system in which drugs, not even called drugs, are separated by the Atlantic Ocean from their modern, scientific counterparts in Western Europe. Scientists are dedicated problem solvers, so this chapter should not be concluded without indicating how this unfortunate situation might be changed. All of us know that the best way to effect change is through politics, and once it seemed that public opinion would force our government to change its outmoded outlook on phytomedicinals in a relatively short period of time.

In 1986, this author wrote (*4*), "Their [consumers'] powerful political influence attained before the end of this century will force the government to act . . . ." This prediction now seems faulty, probably, in large measure, because certain overly enthusiastic herbal advocates lacking scientific knowledge have, by their outrageous hyperbole, caused the government to view all things associated with herbs as nonscientific and, therefore, not worthy of consideration. When one reads much of the herbal literature available in this country, it is not difficult to see how this impression might be gained.

Initially, it was believed that the best way to differentiate between rational herbalism and faulty herbalism was to label the latter paraherbalism. However, many paraherbalists are not anxious to adopt this designation with its pejorative implication. It seems, therefore, more appropriate now to reverse the tactics and refer to herbs utilized on a rational basis as phytomedicinals and the practice of using such herbs as phytomedicine. Adoption of this terminology would go far to remove the stigma now associated with herbalism.

While it is true that not many busy scientists will take time from their work to organize a political movement, it seems plausible that, once made aware of the value of phytomedicinals, they can explain those advantages to others, thereby markedly increasing consumer awareness of the useful remedies to which Americans are now denied access. Increased public awareness soon translates into public demand and, eventually, into political activism.

Research on classical natural products is not carried out extensively in this country in the pharmaceutical industry because of the difficulty in obtaining patent protection and the subsequent lack of profitability. But many academic scientists are little concerned about these factors. Some of the plant drugs just discussed present intriguing problems that could be resolved by academic scientists, if funding could be obtained from appropriate granting agencies, such as the National Institutes of Health. Unfortunately, aside from its anticancer and anti-AIDS programs, the NIH has been loathe to support phytomedicinal research, but perhaps the current taxol breakthrough will serve to alter perceptions and bring about a change in this attitude.

In summary, it is obvious that phytomedicinals can be a significant source of new drugs and drug products. That is already the case in Western Europe. Scientists themselves need to be made aware of this situation so that they, in turn, can create public awareness. This will not only change the antiquated regulatory philosophy that now exists but should provide increased funding for natural product research as well. Consumers and scientists in this country urgently need both.

Literature Cited

1. Dragendorff, G. *Die Heilpflanzen der verschiedenen Völker und Zeiten*; Ferdinand Enke: Stuttgart, Germany, 1898; p 2.
2. Boyle, W. *Official Herbs: Botanical Substances in the United States Pharmacopoeias*; Buckeye Naturopathic Press: East Palestine, OH, 1991; pp 14-53.
3. Reuter, H.D. *J. Ethnopharmacol.* **1991**, *32*, 187-193.
4. Tyler, V.E. *Econ. Bot.* **1986**, *40*, 279-288.
5. Harlow, D.R. *Food, Drug, Cosmet. Law J.* **1977**, *32*, 248-272.
6. Liberti, L.E.; Der Marderosian, A. *J. Pharm. Sci.* **1978**, *67*, 1487-1489.
7. Tyler, V.E. *Planta Med.* **1988**, *54*, 95-100.
8. Bielderman, B.J. *European Phytotelegram* **1991**, No. 3, 14-15.

9. Keller, K. *J. Ethnopharmacol.* **1991**, *32,* 225-229.
10. Reuter, H.D. *Z. Phytother.* **1991**, *12*, 83-91.
11. Tyler, V.E. *Nutr. Forum* **1991**, *8*, 23-24.
12. Drieu, K. In *Rökan*: *Ginkgo biloba*; Fünfgeld, E.W., Ed.; Springer-Verlag: Berlin, Germany, 1988; pp 32-36.
13. Hänsel, R. *Phytopharmaka,* 2nd ed.; Springer-Verlag: Berlin, Germany, 1991; p 69.
14. Claus, E.P. *Pharmacognosy,* 3rd ed.; Lea & Febiger: Philadelphia, PA, 1956; pp 77-78.
15. Foster, S. *Echinacea*: *The Purple Coneflowers*; American Botanical Council: Austin, TX, 1991; 7 pp.
16. Hobbs, C. *The Echinacea Handbook*; Eclectic Medical Publications: Portland, OR, 1989; pp 6-51 - 6-68.
17. Bauer, R.; Wagner, H. *Echinacea*; Wissenschaftliche Verlagsgesellschaft: Stuttgart, Germany, 1990; 182 pp.
18. Bauer, R.; Remiger, P.; Jurcic, K.; Wagner, H. *Z. Phytother.* **1989**, *10,* 43-48.
19. Kommission E. Bundesgesundheitsamt *Bundesanzeiger* **1989**, January 5.
20. Carle, R.; Isaac, O. *Z. Phytother.* **1987**, *8,* 67-77.
21. Kommission E. Bundesgesundheitsamt *Bundesanzeiger* **1984**, December 5 and **1990**, March 6.
22. Anon. *Proposals for European Monographs on the Medicinal Use of Frangulae Cortex, Matricariae Flos, Sennae Folium, Sennae Fructus, Valerianae Radix*; European Scientific Cooperative on Phytotherapy: Brussels, Belgium, 1990; pp 5-8.
23. Mann, C.; Staba, J. In *Herbs, Spices, and Medicinal Plants: Recent Advances in Botany, Horticulture, and Pharmacology*, Vol. 1; Craker, L.E.; Simon, J.E., Eds.; Oryx Press: Phoenix, AZ, 1986; pp 235-280.
24. Steinegger, E.; Hänsel, R. *Lehrbuch der Pharmakognosie und Phytopharmazie,* 4th ed.; Springer-Verlag: Berlin, Germany, 1988; pp 308-311.
25. Schilcher, H. *Die Kamille*; Wissenschaftliche Verlagsgesellschaft: Stuttgart, Germany, 1987; pp 61-78.
26. Hausen, B.M.; Busker, E.; Carle, R. *Planta Med.* **1984**, *50,* 229-234.
27. Foster, S. *Chamomile*: *Matricaria recutita & Chamaemelum nobile*; American Botanical Council: Austin, TX, 1990; p 6.
28. Foster, S. *Feverfew: Tanacetum parthenium*; American Botanical Council: Austin, TX, 1991; 8 pp.
29. Awang, D.V.C. *Can. Pharm. J.* **1989**, *122,* 266-269.
30. Anon. *Lawrence Review of Natural Products* **1990**, June, 1-2.
31. Awang, D.V.C.; Dawson, B.A.; Kindack, D.G.; Crompton, C.W.; Heptinstall, S. *J. Nat. Prod.* **1991**, *54,* 1516-1521.
32. Foster, S. *Valerian*: *Valeriana officinalis*; American Botanical Council: Austin, TX, 1990; 8 pp.
33. Krieglstein, J.; Grusla, D. *Deut. Apoth. Ztg.* **1988**, *128,* 2041-2046.
34. Béliveau, J. *Can. Pharm. J.* **1986**, *119*, 24-27.
35. Haas, H. *Arzneipflanzenkunde*; B.I. Wissenschaftsverlag: Mannheim, Germany, 1991; pp 56-57.

36. Wagner, H.; Seligmann, O. In *Advances in Chinese Medicinal Materials Research*; Chang, H.M.; Yeung, H.W.; Tso, W.-W.; Koo, A., Eds.; World Scientific Publishing: Singapore, 1985; pp 247-256.
37. Foster, S. *Milk Thistle, Silybum marianum*; American Botanical Council: Austin, TX, 1991; 7 pp.
38. Ferenci, P.; Dragosics, B.; Dittrich, H.; Frank, H.; Benda, L.; Lochs, H; Meryn, S.; Base, W.; Schneider, B. *J. Hepatol.* **1989**, *9,* 105-113.
39. Merfort, I.; Willuhn, G. *Deut. Apoth. Ztg.* **1985**, *125,* 695-696.
40. Haas, H. *Arzneipflanzenkunde*; B.I. Wissenschaftsverlag: Mannheim, Germany, 1991; pp 102-103.
41. Hobbs, C. *HerbalGram* **1989**, No. 18/19, 24-32.
42. Hölzl, J. *Deut. Apoth. Ztg.* **1990**, *130,* 367.
43. Okpanyi, S.N.; Weischer, M.L. *Arzneim.-Forsch.* **1987**, *37*(I), 10-13.
44. Müldner, H.; Zöller, M. *Arzneim.-Forsch.* **1984**, *34*(II), 918-920.
45. Meruelo, D.; Lavie, G.; Lavie, D. *Proc. Natl. Acad. Sci. USA* **1988**, *85,* 5230-5234.
46. Persinos, G.J., Ed. *Washington Insight* **1991**, *4*(4), 6.
47. Foster, S. *Health Foods Bus.* **1991**, *37*(5), 22-23.
48. Anon. *Am. Pharm.* **1990**, *NS30,* 321.
49. Harnischfeger, G.; Stolze, H. *Z. Phytother.* **1989**, *10,* 71-76.
50. Anon. *Lawrence Review of Natural Products* **1990**, Sept., 1-2.
51. Weiss, R.F. *Herbal Medicine*; AB Arcanum: Gothenburg, Sweden, 1988; pp 162-169.
52. Hamon, N.W. *Can. Pharm. J.* **1988**, *121,* 708-709, 724.
53. Kommission E. Bundesgesundheitsamt *Bundesanzeiger* **1984**, January 1 and **1988**, May 5.
54. Hänsel, R. *Phytopharmaka,* 2nd ed; Springer-Verlag: Berlin, Germany, 1991; pp 33-36.
55. Anon. *Deut. Apoth. Ztg.* **1989**, *129*, 2353-2354.
56. Koch-Heitzmann, I.; Schultze, W. *Z. Phytother.* **1988**, *9*, 77-85.
57. Wagner, H.; Sprinkmeyer, L. *Deut. Apoth. Ztg.* **1973**, *113*, 1159-1166.
58. Wölbling, R.H.; Milbradt, R. *Therapiewoche* **1984**, *34*, 1193-1200.

RECEIVED January 27, 1993

Chapter 4

Biological and Chemical Diversity and the Search for New Pharmaceuticals and Other Bioactive Natural Products

James D. McChesney

Research Institute of Pharmaceutical Sciences, School of Pharmacy, University of Mississippi, University, MS 38677

Higher plants are remarkable in their ability to produce a vast array of diverse metabolites varying in chemical complexity and biological activity. Natural products have historically served as templates for the development of many important classes of drugs. Further, efforts are now being focused increasingly on the potential application of higher plant-derived natural products as agrochemicals. Evaluation of the factors limiting the promise plant-derived natural products have as new agrochemicals and pharmaceuticals, stimulated us to establish the Center for the Development of Natural Products at The University of Mississippi. The Center will coordinate a multidisciplinary effort to discover, develop and commercialize plant-derived natural products as drugs and agrochemicals. The partnership of the Research Institute of Pharmaceutical Sciences of The University of Mississippi, The United States Department of Agriculture, and select pharmaceutical and agrochemical companies should ensure our success in this endeavor.

Human survival has always depended on plants. Early man relied entirely on them for food, medicine, and much of his clothing and shelter, and his empirically acquired botanical skills should not be underestimated. All of the world's major crops were brought in from the wild and domesticated in prehistoric times.

Aside from their value as sources of food, drugs or industrial raw materials, plants are important to man in many other ways. One can hardly imagine modern society without soaps and toiletries, perfumes, condiments and spices, and similar materials, all of plant origin, which enhance our standard of living. Furthermore, the roles of forests and other types of natural vegetation in controlling floods and erosion, climate regulation, and in providing recreational facilities are of immeasurable value.

An adequate food supply is, and always has been, man's most outstanding need. That food shortages are often due to political, socio-economic, unequal resource allocation and distribution or even cultural issues is recognized. However, in many cases, shortages are caused by attempts to employ inappropriate Western technology or to grow inappropriate crops. As

0097–6156/93/0534–0038$06.00/0

reported in a recent National Academy of Sciences study,(1) man has used some 3,000 plant species for food throughout history. At least 150 of them have been, to some extent, commercially cultivated. However, over the centuries the tendency has been to concentrate on fewer and fewer of these. Today, nearly all of the people in the world are fed by about 20 crops: cereals such as wheat, rice, maize, millet, and sorghum; root crops such as potato, sweet potato, and cassava; legumes such as peas, beans, peanuts (groundnuts), and soybeans; and sugar cane, sugar beet, coconuts, and bananas. These plants are the main bulwark between mankind and starvation. It is a very small bastion.

Yet, as the prospect of food shortages becomes more acute, people must depend increasingly on plants rather than animals for the protein in their diet. Reliance on a small number of plants carries great risk, for monocultures are extremely vulnerable to catastrophic failure brought about by disease, variations in climate, or insect infestations. As is well recognized, research to increase the yield of the established high-protein food plants must be continued. At the present time, our most effective methodology for increasing crop productivity is through the utilization of chemical substances which reduce the effects of environmental and biological stresses upon crop plants. These chemical substances, pesticides and growth regulators, must have selective activity and be non-persistent in the environment.

Paralleling man's need for food is his need for agents to treat his ailments. The practice of medicine by physicians today in the United States and Western Europe is very different from the practice of their predecessors. This is mostly because modern doctors have available a large array of medicines with specific curative effects. However, we still lack specific curative agents for a number of important diseases. Some 800 million to 1 billion people, nearly one-fifth of the world's population, suffer from tropical diseases: malaria, schistosomiasis, trypanosomiasis, leprosy, leishmaniasis, etc. Even in the United States, heart disease, cancer, viral diseases (for example, AIDS), antibiotic-resistant infections, and many others still lack adequate treatment.

The areas of research and development need outlined above, reinforce the need for the study of chemical substances found in plants (natural products) and the exploitation of those substances to meet the needs of mankind. If these research and development prerequisites are to be met adequately, it will only be as a result of a broadly based, multidisciplinary effort coordinated in philosophy and practical approach. Such an endeavor must enlist the expertise of many scientific areas in a systematic and coordinated effort. This broadly based, multidisciplinary effort could function appropriately if coordinated by the establishment of a "Center for the Development of Natural Products".

The Research Institute of Pharmaceutical Sciences at the University of Mississippi is forming a partnership with the Federal government and the pharmaceutical and agrochemical industries by establishing the Center for the Development of Natural Products. The Center will take advantage of the strengths of each partner to discover, develop and commercialize products derived from natural sources, especially higher plants.

The Center will use an integrated and multidisciplinary approach to discover, develop and commercialize new pharmaceutical and agrochemical products. This process will begin with the exploitation of knowledge about traditional uses (a source of discovery of numerous medications on the market today), carry through to the development and commercialization of products derived from higher plants, and finally, come full-circle by developing the plant sources of these products into alternative high-value crops for farmers. One unique feature of the Center is an organizational structure that emphasizes the importance of commercialization; a division of the Center will be applications-oriented and will be devoted solely to commercialization to ensure that marketable products and processes developed in the Center become commercialized.

The potential for developing new sources of valuable plant chemicals is largely unexplored and the benefits from doing so unexploited. Plants are known sources of medicines, insecticides, herbicides, medicines, and other useful substances; developing new industries and crops based upon plant extracts and extraction residues provides opportunities for agricultural and industrial expansion that will benefit farmers, consumers and industry, both in the developing as well as the developed nations.

New-crop and plant-product development will:

- provide consumers with new products, including new drugs to treat diseases,
- provide less environmentally hazardous pesticides,
- diversify and increase the efficiency of agricultural production,
- improve land resource use,
- offer increased economic stability to farmers,
- create new and improve existing agriculturally related industries,
- increase employment opportunities,
- provide industries with alternative and sustainable sources of raw materials.

Chemicals derived from higher plants have played a central role in the history of humankind. The Age of Discovery was fostered by explorations to find more economic trade routes to the East to bring back plant-derived spices and other products. Indeed, the discovery of the New World, whose 500th Anniversary we have just celebrated, was a direct consequence of that effort. The prototype agent for the majority of our classes of pharmaceuticals was a natural product of plant origin. Numerous examples could be cited. However, with the discovery and development of fermentation-based natural products beginning in the early 1940's and the increasing sophistication of synthetic organic chemistry, interest in plant-derived natural products as prototypes for pharmaceuticals and agrochemicals waned greatly during the decades of the 1960's through 1980's. Today a renewed interest in the potential of substances found in higher plants to provide prototypes for new pharmaceuticals, agrochemicals and consumer products is being evidenced. Until recently, efforts to realize the potential of plant-derived natural products have been very modest and largely restricted to discovery programs centered in academic settings. A number of convergent factors are bringing about a renaissance in higher plant-related natural products research and development. Some of those factors are: advances in bioassay technology; advances in separation and structure elucidation technology; advances in our understanding of biochemical and physiological pathways; the biotechnology revolution; historical success of the approach; loss of practitioners of traditional medicine; loss of biological diversity; loss of chemical diversity; world wide competition.

Efforts to develop new, clinically effective pharmaceutical agents traditionally have relied primarily on one of five approaches, most of which utilize existing agents in some manner: (1) derivatization of existing agents, (2) synthesis of additional analogs of existing agents, (3) use of combination therapy of existing agents with other drugs, (4) improvement of delivery of existing agents to the target site, (5) discovery of new prototype pharmaceutical agents. While approaches 1-4 are important and need to be continued because they seek to optimize the use of existing agents and information in the most effective manner, there is an urgent need for the development of totally new, prototype agents which do not share the same toxicities, cross resistance or mechanism of action as existing agents. Natural products have, in the past, provided a rich source of such prototype bioactive compounds and it is essential that the search for new drugs follow this approach. The major advantage of this strategy is the liklihood of identifying new **prototype** drugs with quite different chemical structures and mechanisms of action and hence, lower likelihood of similar toxicities and cross resistance. Clearly, higher plants represent a bountiful source of new prototypic bioactive agents which must be examined.

The fundamental element of a drug discovery program is the bioassay(s) utilized to detect substances with the desired biological activity. The bioassay protocol selected for the discovery of new prototype drugs must meet a variety of requirements. In addition to the expected criteria of ease of operation and low to moderate cost, the assay must show specificity and sensitivity. An important requirement of the assay is its ability to serve as a guide during the bioassay-directed phase of purification of agents showing activity. This is especially true in the discovery of substances from natural sources, since these materials are likely to be present in very low concentrations in very complex mixtures. Only a combination of procedures meet these demanding criteria to serve as primary screens for biological activity. Other important program elements must be coupled to the appropriate bioassay. Selection and procurement of novel sources of natural products as well as capability to accomplish bioassay-directed purification and structure elucidation must be established in conjunction with the primary bioassay procedures. Initially detected activity must be confirmed in suitable secondary and tertiary assays which will help define the potential of the substance for clinical utility. Finally, a "portfolio" of information about the substance must be accumulated upon which to make a judgement about its potential for successful development to a clinically useful agent. For example, something must be known about its general toxicity, pharmacokinetics, mechanism of action, analog development, etc. Based on these considerations, The University of Mississippi Center for the Development of Natural Products has formulated an innovative program of collaborative multidisciplinary research to discover and develop new **prototype** pharmaceutical and agrochemical agents from higher plants. An overview of the general strategy is shown in Figure 1.

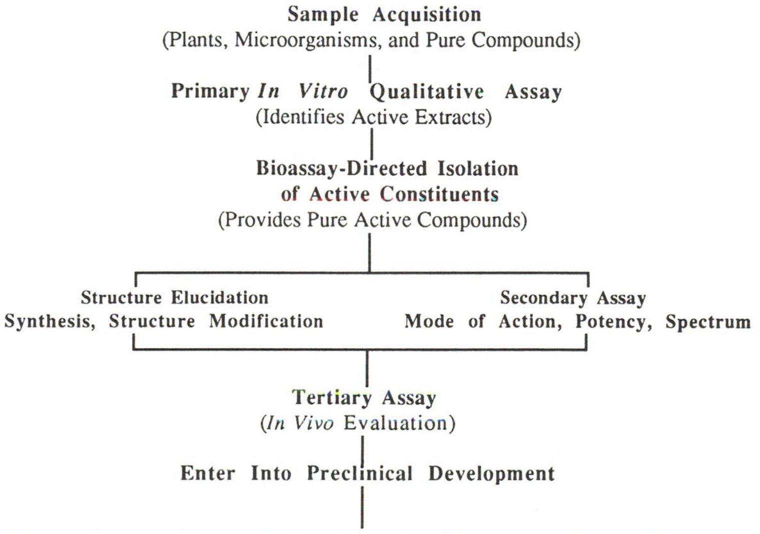

Figure 1. General Strategy for Prototype Drug Discovery

The common thread of all research endeavors of the Center is the identification and commercial development of specific plant-derived substances having a desired, selective biological activity. The research efforts of the Center emphasize two broadly defined areas: improved human health and increased agricultural productivity. The differentiation between these two broadly defined areas results primarily from the selection of the specific bioassays utilized to identify prospective materials. Bioassays which delineate activity useful for the treatment of disease are employed to find prototype pharmaceuticals. Bioassays which measure activities which would reduce biological or environmental stress on food plants or animals are utilized to identify prospective materials with utility to increase agricultural productivity. As outlined in Figure 1, a general Discovery-Development-Commercialization Model directs the efforts of the Center.

STEP 1. Preliminary Economic Assessment.
Preliminary economic assessment consists primarily of examinations of the magnitude of various market segments (therapeutic needs), the market potential of classes of products, the potential for commercialization of classes of products, etc. Additional analyses accompany each stage of scientific and product development. In this way we are assured of optimal investment of Center resources for the discovery and development of new pharmaceuticals and agrochemicals.

STEP 2. Identification of Plants Likely to Possess Desired Biological Activity.
Plant genera and species likely to possess a desired biological activity are identified through one or more of several sources:

a. Databases (e.g., NAPRALERT, USDA EBL Database, DIALOG, etc.),
b. Research conducted by the Collaborative Programs for the Study of Traditional Practices,
c. Systematic Evaluation of Promising Genera and Families.

STEP 3. Collection of Plant Material.
Plant material will be collected by a Center field botanist or by arrangement with other laboratories and groups throughout the world.

STEP 4. Verification of Plant Identificaion and Preparation of Voucher Specimens.
Taxonomic authentification is extremely important because plants are often misidentified by folk users and in the literature. Plant taxonomists working at the Center and at collaborating botanical institutions perform this function.

STEP 5. Preparation of Extracts of the Plant Material.
Standard procedures for the preparation of extracts are followed.

STEP 6. Biological Evaluation of the Extracts.
Biological evaluation of the extracts is conducted using assays performed in the Center or through collaborative arrangements with other laboratories. Standardized bioassays utilizing molecularly based in vitro assays as well as intact organisms are used.

Steps 2 through 6 are repeated, as necessary, for those plants showing potentially useful biological activity. This confirmation of biological activity is extremely important. It saves the effort and expense of fractionation and purification of those compounds whose activity cannot be verified (replicated). Many drug discovery programs have failed because they did not pay careful attention to reconfirmation and verification of biological activity in plant material.

STEP 7. Bulk Collection, Extraction, Fractionation and Purification. Plant species showing reconfirmed activity are collected in bulk and large-scale extractions are performed. Extracts will be fractionated and the active principles purified and characterized chemically. This fractionation and purification is guided by careful biological evaluation of all fractions; fractions continuing to show biological activity are carried on to isolation and purification of the active chemical constituent(s).

STEP 8. Identification of the Active Principles.
The chemical structure of each purified, biologically active principle is determined utilizing, as appropriate, various spectroscopic, chemical or x-ray crystallographic means.

STEP 9. Definition the Toxicity and Pharmacology of the Active Principles (Preliminary Determination).
The pure, chemically characterized active principles are systematically evaluated for toxicity and biological activity using specific in vivo and in vitro bioassays (Secondary and Tertiary Assays). These evaluations provide an indication of the spectrum of biological activity of each new compound, a measure of its potency, and often yield information on the mode of action of the material. The in vivo tertiary assays give indication of the efficacy of the agent and its promise for further development.

STEP 10. Determination Structure-Activity Relationships.
In most cases the active chemical constituent of the plant material will serve as a "lead" for the development of a more desirable molecule having greater efficacy or fewer or less severe side effects. The lead compound is modified chemically, and the efficacy and toxicity of these analogs are determined. The Center's Molecular Modeling Laboratory will play an especially important support role in this endeavor.

STEP 11. Evaluation of Status for Commercial Development.
By this step, a portfolio of information on the prototype will be developed. This information is evaluated (usually in cooperation with one of our corporate sponsors) and a decision is reached as to whether or not to proceed to full commercial development.

STEP 12. Conduct Clinical Trials or Field Trials.
Clinical trials (pharmaceuticals) or field trials (agricultural products) will be conducted in accordance with regulatory guidelines promulgated by the appropriate govermental agency. Ordinarily, by the time a product is ready for clinical or extensive field trials, the Center will have identified a strong corporate partner who will participate fully in the continued development of the product.

STEP 13: Commercialization.
Commercialization is actually much more than a final step in the development of a natural product. It represents the maximum return of benefit to our constituents - a new product to treat disease or increase agricultural productivity is available for their use.

STEP 14. Conduct Agronomic Studies.
Agronomic studies are conducted to determine optimal growth conditions and to increase the yield of the desired chemical constituent of the plant. The ultimate goal of this activity is to develop alternative cash crops. An important consideration or nagging concern which is responsible for the significant reluctance on the part of the private sector to initiate a higher plant based natural products research and development program is the issue of natural product chemical supply.

With the advances in bioassay, separation and structure elucidation technologies mentioned above, we are now able to discover and identify substances with potential utility from plant preparations quickly and with very modest amounts of material. The sensitivity of the bioassays routinely in use today and the desired potency for new prototypes of less than 1 μg/ml, make it easily possible to detect substances present at concentrations of 0.001% or less of the dry weight of the plant.

Ordinarily 50 mg or so of pure chemical substance is sufficient to determine its complete structure, including stereochemistry. In order to obtain 50 mg of a chemical which is present at 0.001% of the dry weight of the biomass, one needs at least 5 kg In practice, however, one might actually need as much as 10 kg of biomass since no isolation procedure is 100% efficient.

Herbaceous plants are roughly 75-80% water when fresh, so to obtain 5 to 10 kg of dry plant material one must collect 20 to 50 kg of fresh plant material. For woody plants, the situation may be somewhat more complex since we usually process a specific part of the plant - bark, leaves, fruits, roots, or wood. For the leaves and fruits the situation is like that for herbaceous plants. In the case of wood, bark or roots, the dry weight yield from fresh weight is often much higher, sometimes as high as 75%. So for woody plant parts one may need 20 to 50 kg of leaves of fruits; 10 to 25 kg of roots or bark and 8-15 kg of wood in order to get sufficient substance to accomplish isolation and chemical characterization of the active principle.

At this stage, one reaches a decision point - is the structure novel? Does the substance represent a potential new prototype? In order to assess the full potential of the substance so that a valid decision can be made as to whether to proceed further with development, quantities are needed for an array of evaluations: secondary and confirmatory bioassays; preliminary toxicology and initial in vivo evaluation. Ordinarily at least 400-500 mg of material are needed for this next phase of evaluation. This represents roughly a 10-fold increase over the original quantities required for detection of the lead compound and its chemical characterization. To isolate and purify the necessary quantities of the bioactive natural product may require that 100 - 500 kg of biomass be processed. It is already at this point that concerns about supplies of the natural product begin to surface. If one must invest in bulk collection of plant material and its subsequent processing to obtain quantities for these next evaluations, what must one do for the larger quantities, perhaps as much as two kilograms, required to take the drug candidate into preclinical evaluation and subsequent clinical trials, and ultimately to provide for actual marketing and clinical use of the new drug, maybe tens or hundreds of kilograms?

Two more or less extreme examples will serve to illustrate the range of quantities of active drug substance that would be required to provide for clinical use of a plant-derived drug. Let us assume that the condition to be treated is an acute condition, that the agent is relatively potent, i.e., 2 grams are required for therapy, and that only a modest patient population is expected - 10,000 patients per year. To provide for this market roughly 20 kg of drug substance would be required. If again we assume a worst case of 0.001% w/w of active substance isolated from our biomass, then 2,000,000 kg of biomass will be necessary to provide the quantity of drug needed to meet market demand. This may seem like a great deal, but 2,000,000 kg of biomass is roughly the same as 75,000 bushels of wheat, a quantity produced by even a modest-sized family farm in America. Now, if we assume the plant-derived drug would be used to treat a chronic condition in a much larger population, would that make a great difference? For a chronic condition, a moderately potent agent might require administration of 50 mg per day per patient which is equivalent to 20 g per year per patient. If 100,000 patients could benefit from this drug, then 2,000 kg of active drug substance would be needed to meet market demand. This is 100 times that of the example above, or in other words, 200,000,000 kg of biomass would need to be processed to provide the required material. Looked at another way, that is equivalent to 7,500,000 bushels of wheat, a quantity easily produced in an average wheat-producing county of a wheat-producing state such as Kansas.

Clearly these are quantities of biomass which can be readily produced and processed, if thought and planning for a system for drug plant production is initiated.

Are there any actual examples which might illustrate the point? First, let me comment on two plant-derived drugs which are problems for society, marijuana and cocaine(2). The best estimates on marijuana production worldwide in 1990 placed the figure at 30,000 tons (27,000,000 kg). The U.S. Drug Enforcement Administration estimated that U.S. production was at least 5,000 tons, more than 4,500,000 kg. At an average price of $1,200 to $2,000 per pound, there is apparently incentive enough for growers to risk even imprisonment, to provide for the market demand. Cocaine, a purified drug from the leaves of the coca plant, was produced in quantities of at least 1,000 tons worldwide in 1990. With prices of $6,000 to $18,000 per pound wholesale, this was enough incentive to cause illicit producers to grow and process enough drug to meet market demand. It is to be emphasized that these levels of natural product drug production are in place, even though governments are making a concerted effort to stop or eradicate the growing of these plants.

Any system for production of a plant-derived natural product must meet certain criteria. It must be economic, sustainable, reliable and non-environmentally impacting. These over-arching criteria can only be met if a careful and systematic evaluation of each step of the production system is made. One must discover and develop a superior source of the natural product which contains a consistently high yield of agent, an effective production and harvesting system, appropriate technology for processing and storage of biomass, and an economic and efficient extraction and purification system which minimizes waste-product generation.

Wild populations of medicinal plants may not be a reliable source of a drug entity and their harvesting may be counter-productive to the development of a reliable, cost-effective, long-term production system of a clinically utilized drug. Several unpredictable and uncontrollable phenomena preclude establishment of a stable cost of production of the drug entity harvested from wild populations: forest fires; annual climatic variations; natural variation and presence of chemotypes of wild populations; increasing pressures to protect and regulate the harvesting of wild plants; high cost of collection of scattered plants; and transportation to processing facilities and lack of assured accessibility of populations which occur on public and privately held lands. Reliance on the harvesting of wild plants for production may lead to uncontrollable interruption of drug supply.

Even more critical is the fact that wild harvesting risks the destruction of germplasm essential for the future cultivation of the plant for drug production. This includes genes for disease and pest resistance, hardiness, and tolerance to full sunlight, drought and flooding as well as genes for high growth rates and high levels of drug production. The preservation of these wild genes can be critical to the development of long-term, cost-effective supplies, whether produced by cultivated plants, tissue culture, or genetically modified microorganisms. Because of the critical role wild germplasm will serve in future production strategies, the preservation of wild populations should be considered an essential component of the development strategy for drug production.

Lack of a stable and reliable supply of plant-derived drug at a predictable cost will even more significantly impede clinical utilization of the agent. Procurement of drug by harvest of limited wild populations may not be an appropriate strategy to provide the agent once clinical utility is established. Development of a sustainable, economic and reliable source is imperative.

The appropriate alternative may be production from a cultivated plant source. Two strategies may be taken to accomplish this: (1) to bring into cultivation the currently recognized source of drug; (2) to evaluate and select currently cultivated varieties for drug or drug precursor content. Strategy One is fraught with all of the problems and uncertainties associated with the introduction of a new plant into cultivation, a process which has been successful in bringing into cultivation only about 3,000 species of the estimated 300,000 to 500,000 plant species known to occur in the world. We will not detail all of those uncertainties here. However, as our understanding of agronomics and plant biology has increased, we can have ever-increasing confidence in our ability to bring into cultivation the plant source of a drug. Even with success, a period of several years may be required for this strategy to provide a reliable source for drug production. The necessity of recovering the investments made during the development period will impact the economics of this strategy to provide a source for the drug.

In contrast, Strategy Two, selection of currently cultivated varieties for drug production, presents many advantages. A proven cultural system is in place. An additional advantage is the known genetic origin and uniformity of cultivated plants. Finally, cultivation will provide high plant densities in defined locations which will significantly reduce collection and transportation costs.

The greatest advantage of this strategy is, however, its flexibility and responsiveness to demand. This system can be placed into production for drugs faster and more reliably than any other production system. A diversity of cultivars may be currently available, already in cultivation. Let us assume that a cultivar can be found which will provide drug at an isolated yield of 0.03% w/w of the weight of the biomass. To produce the required 20 kg of drug to meet marketing needs, it will be necessary to collect and process approximately 70,000 kilograms of biomass. If one were to collect as little as 10 g of dry biomass per plant, we would need to harvest biomass from 7,000,000 plants. If those plants are planted on 3 foot by 12 inch spacings (4,500 plants per acre), then less than 1,600 acres would provide all the biomass required to meet the clinical need. Even if 100 times this quantity is ultimately needed for clinical use, the biomass can be quickly attainable. This scale of production is easily accomplished in a relatively short period of time.

Careful and thoughtful analysis of the issue of supply of sufficient quantities of a plant-derived drug to meet development and clinical needs, brings one to the realization that this is not an issue which should cause us to preclude the carrying forward of such plant-derived agents through the development process to clinical use. Indeed, it is clear that adequate supplies of most plant-derived drugs can be assured with only modest additional research and investment.

In conclusion, the Center for the Development of Natural Products at the University of Mississippi has established a broad-based, multidisciplinary effort to discover, develop and commercialize natural products, especially those derived from higher plants, as pharmaceuticals and agrochemicals. We have given systematic evaluation to those constraints which have impeded traditional plant-derived natural products development, and have designed strategies to overcome those historical limitations. With the establishment of the Center as a partnership of the Research Institute of Pharmaceutical Sciences of The University of Mississippi, The United States Department Agriculture and select pharmaceutical and agrochemical companies, we are confident that the promise of plant-derived natural products as pharmaceuticals and agrochemicals will be realized.

Literature Cited

1. *Underexploited Tropical Plants with Promising Economic Value*; Report of an Ad Hoc Panel of the Advisory Committee on Technology Innovation, Board on Science and Technology for International Development, Commission on International Development; National Academy of Sciences: Washington, DC, 1975.
2. *The Supply of Illicit Drugs to the United States*; The National Narcotics Intelligence Consumers Committee Report; Drug Enforcement Administration, Office of Intelligence: Washington, DC, 1990.

RECEIVED March 5, 1993

Chapter 5

The Renaissance of Plant Research in the Pharmaceutical Industry

Melanie J. O'Neill and Jane A. Lewis

Biochemical Targets and Natural Products Discovery Departments, Glaxo Group Research Ltd., Greenford Road, Greenford, Middlesex UB6 0HE, United Kingdom

The current vogue in the pharmaceutical industry for highly automated, sensitive target screening as a major strategy in drug discovery, has prompted renewed interest in natural product sources of biologically active compounds. A small, but very impressive range of medicines has been derived from screening microbial fermentation broths and these continue as favourite input samples. However, the desire to optimise sample diversity has turned attention to the potential of other organism types, so that after many years of their virtual exclusion from modern drug discovery programmes, higher plants have a place once again. Establishing a plant research initiative creates practical challenges which are not encountered with fermented organisms. Nevertheless, surmounting the obstacles can build a programme which has significant impact, not only in high-throughput screening, but also in alternative approaches to drug discovery. This communication will address the project management considerations of conducting research on higher plants within the pharmaceutical industry.

Other chapters in this book have emphasised the contribution of higher plants to medicine, both traditional and orthodox, since time immemorial. Indeed, over half the world's 25 best selling pharmaceuticals for 1991 owe their origin to one of a range of natural source materials (Table I) (*1,2*). Nature's ability to synthesise highly diverse chemicals, some of which possess exquisitely selective biological activities, is well known to today's scientists involved in drug discovery. Over the last decade, high-throughput screening against mechanistically pure physiological targets has come to the forefront in the search for new chemical entities. Both

0097–6156/93/0534–0048$06.00/0

TABLE I. The World's 25 Best Selling Pharmaceuticals

Position 1991	Product	Therapeutic Class	Sales $m
1	Ranitidine	H_2 antagonist	3,032
2	[a]Enalapril	ACE inhibitor	1,745
3	[a]Captopril	ACE inhibitor	1,580
4	[a]Diclofenac	NSAID	1,185
5	Atenolol	β-antagonist	1,180
6	Nifedipine	Ca^{2+} antagonist	1,120
7	Cimetidine	H_2 antagonist	1,097
8	[a]Mevinolin	HMGCoA-R inhibitor	1,090
9	[a]Naproxen	NSAID	954
10	[a]Cefaclor	β-lactam antibiotic	935
11	Diltiazem	Ca^{2+} antagonist	912
12	Fluoxetine	5HT reuptake inhibitor	910
13	Ciprofloxacin	Quinolone	904
14	Amlodipine	Ca^{2+} antagonist	896
15	[a]Amoxycillin/ clavulanic acid	β-lactam antibiotic	892
16	Acyclovir	Anti-herpetic	887
17	[a]Ceftriaxone	β-lactam antibiotic	870
18	Omeprazole	H^+ pump inhibitor	775
19	Terfenadine	Anti-histamine	768
20	[a]Salbutamol	β_2-agonist	757
21	[a]Cyclosporin	Immunosuppressive	695
22	[a]Piroxicam	NSAID	680
23	Famotidine	H_2 antagonist	595
24	Alprazolam	Benzodiazepine	595
25	[a]Oestrogens	HRT	569

[a] Natural product derived

synthetic compounds and natural product extracts are trawled in what is essentially a commercial race to find likely candidate molecules with a predetermined mode of action. The chances of winning this race are likely to be enhanced by looking at large numbers of samples and optimising their diversity.

Until quite recently within the pharmaceutical industry, the natural product extracts used for screening were almost exclusively microbial, with filamentous bacteria making up the bulk. A series of notable pharmaceuticals, including ß-lactam antibiotics, endectocides, immunosuppressives and cholesterol-lowering agents, have emerged from screening microbial samples. In contrast it has been difficult to think of such medicines emerging by screening higher plants. This is not because plants do not have any new, interesting compounds to offer, but rather that few organisations have included them as major components of the natural product samples screened.

Screening Natural Products

If we take an 'assay' to mean a technique which allows a response against a biological target to be measured and estimated, then a 'screen' implies a large number of assays, often, though not always, tuned so as to indicate simply the presence or absence of a response. The target may be a particular cell type, an enzyme believed to be a key regulator of a specific biosynthetic pathway, a receptor-ligand interaction or a molecule involved in gene transcription. Not all screens are suitable for natural product input samples. In those that are, the target is packaged in such a way as to be robust (i.e. to be relatively insensitive to ubiquitous non-selective toxins or pigments present in the extracts which can create background noise) and the assay is devised so as to allow a high degree of automation and the use of small sample sizes. A clear strategy needs to be in place for following up any primary screen hits and ultimately for selecting extracts as candidates for fractionation, isolation and identification of active components.

Plants as Screen Input Material

In choosing natural product extracts for screening, the major difference between higher plant and microbial samples boils down to relative accessibility. This relates to acquisition of primary samples for screening, obtaining additional material for secondary evaluation and sustaining the resource for exploratory development.

A large number of microbial organisms can be obtained from a single soil sample or from the surface of a single leaf or piece of bark. A variety of culture conditions can be used to encourage the organism to elaborate a diverse range of chemicals, any one of which may possess the biological properties required to moderate the target screen. As scientific interest in a metabolite mounts and once optimal culture conditions have been established in-house, the secondary metabolites are (theoretically) on-tap. Scale-up fermentations can then be undertaken when additional quantities of secondary metabolites are required for further biological evaluation.

A different situation holds true for higher plant samples. For instance, a much larger initial sampling site is needed to acquire a range of different organisms and if small-scale sampling is undertaken at the outset, large-scale recollections may be required to support isolation and structure elucidation and further evaluation of active components. Awaiting recollections from far flung corners of the Earth can delay progress in research, and environmental and genetic variability may result in difficulties in recovering initially promising activities. In most cases, plant cell culture is not a viable alternative in terms of secondary metabolite production. These reasons alone are sufficient to deter many organisations from including higher plants in their screening programmes.

The situation with plants becomes much less complicated if we are realistic about what we expect to get from natural products screening in general. Referring again to Table I, only 2 of the 25 top-selling pharmaceuticals, cyclosporin and mevinolin, are themselves a natural product, whereas 12 are natural product derived. High-throughput screening is a means of finding lead structures which provide starting points for medicinal chemistry. Rather than looking for new pharmaceuticals, the aim is to find novel pharmacophores which can be synthesised and refined in the laboratory. So, on the whole, current research programmes require relatively small-scale samples of plants.

Sources and Numbers of Plant Samples. The feasibility of a plant screening programme initially depends on effective procurement strategies. The preferred route is through collaborations with national or international organisations which possess the authority to collect and supply and the expertise to classify the plants taxonomically. In such collaborations, a legal agreement should be drawn up to protect both supplier and recipient. The recipient of plant samples may wish to ensure further supply of samples if required and the plant supplier needs to be assured of benefit in the event of a commercial product being developed from any sample provided. A full policy document which gives guidelines at the outset to potential suppliers of samples will help promote confidence in a two-way exchange of information and expertise. Such a document may provide that collaborators represent an established institution and possess both taxonomical expertise and appropriate authority to collect plant material. In addition to reimbursement for efforts in collection, classification and sample preparation, an agreement will be drawn up to define a further financial benefit to the supplier in the event of development of a commercial product as a consequence of screening the plant samples supplied. The magnitude of such a payment will reflect the net sales of product and the overall contribution of the supplier. Implicit in the agreement is that the supplier returns a significant portion of any such financial return back to the source country for its material benefit, perhaps in the area of training or education.

The next issue relates to the method of selection of plant samples. Random acquisition of available species is one approach. The chances of success in identifying a compound possessing interesting biological activity are likely to increase by testing large numbers of samples across a wide range of target screens. Maximum diversity in the botanical and geographical sources of the samples collected is desirable. An alternative selection strategy relies on the use of ethnobotanical information to identify particular species reported to induce significant physiological effects. This approach is likely to involve relatively small numbers of closely related plant species. Detailed biological evaluation is more appropriate than high-throughput screening for these samples. A third rationale for selection relates to samples which are reported to contain chemicals of specific interest to on-going in-house projects. It goes without saying that it is much easier to access 'random' samples than selected species. However, one of the rewards of establishing a global network of suppliers is an enhanced likelihood of accessing species of particular interest and also of ensuring that recollections of plant material can be carried out if required.

Sample Presentation. The practicalities of screening large numbers of plant samples from a wide range of geographical locations makes the use of dried plant material the preferred option. It has to be tolerated that some constituents may be lost or degraded in the drying process, although expert collectors will take care to minimise this problem. Plants are usually crudely chopped to ease drying and reduce bulk before transportation. Careful labelling and packaging, plus full drying of samples is essential if material is to arrive at its destination in a usable condition. Import licences and phytosanitary certificates may also be prerequisites for any supply agreement.

In the majority of high-throughput screening programmes, any plant part or combination of parts is acceptable, providing that a sufficient quantity of material is available. It is well known that many plants concentrate certain secondary metabolites in specific organs and, therefore, in terms of anticipated chemical diversity, it is quite legitimate to screen several samples from the same phenotype. A record of harvest date will allow any effects of seasonal variation to be catered for should a recollection become necessary.

Sample Preparation. If certain types of chemicals are required, selective extraction procedures can be used, but a more pragmatic approach to sample preparation must be adopted for high-throughout screening. The requirement is to optimise the chances of detecting any chemical type of biologically active molecule present in the plant extract with the exception of non-selective interfering agents. Rather a tall order! A key project management decision is whether to use the given resources to screen a variety of solvent extracts from each same plant sample or whether to screen one extract from an increased number of plant samples. The decision will need to take into account factors such as potential sample supply, capacity, range and selectivity of the screens and potential for automation.

Should an ethnomedical selection procedure be used, then it is likely that there will be recommended methods for preparation of extracts as used by the patient. When relatively small numbers of plants are being studied for specific therapeutic areas, then it is worthwhile expending effort in preparing a series of extracts according to individual protocols. When specific chemical entities are required, phytochemical procedures reported in the literature can be followed.

Selection Criteria and Progression of 'Hits'

The term 'screen' implies a cursory look at many samples rather than an in-depth investigation of a few. To be effective, natural product screens must be able to pick out the extracts containing selectively active potential lead compounds. In an ideal world, such extracts would be the only types highlighted, but inevitably some screens are also sensitive to certain non-selective toxins and all screens show up false positives from time to time. The key is to establish a strategy for progression (Figure 1) which will allow efforts to focus on the most interesting extracts as soon as possible.

False Positives. In the case of higher plant extracts, the majority of false positives can be attributed to polyphenols, detergents such as saponins, certain pigments or fatty acids. Whilst the phenols will affect highly purified enzyme-based targets, the detergents primarily disrupt membranes in cellular targets or dislodge substrates adsorbed onto assay wells and the pigments interfere with read-out in colorimetric or quenched assays. Fatty acids show activity through a variety of mechanisms. Removal of these 'undesirable' compounds from plant extracts is preferable before primary screening, but it is frequently easier to run virgin extracts through the primary screen and introduce measures to discriminate at a secondary stage in the process.

In most cases, a pragmatic approach is adopted in estimating whether the 'undesirable' compound titre can account for the potency of the extract. For example, activity due to polyphenols can be inferred by assaying before and after treatment with polyvinylpyrrolidone (*3*) and estimated by colorimetric analysis after ferric chloride or potassium iodate treatment. TLC and GC techniques are available to identify common fatty acids (*4*). Biological controls can exclude both detergent-like activities and pigments which interfere with read-out. These procedures can, of course, result in discarding co-occurring 'desirable' compounds.

Non-Selective Toxins

Plant extracts also may contain less ubiquitous, but equally non-selective active compounds. These can be dealt with by comparing activities of the extracts across a broad range of primary screen targets or by introducing specific secondary screens.

Known Selectives

The natural product screening process aims to find novel compounds, but known compounds with a previously unknown selective mode of action can also be very valuable leads. An advantage of higher plant samples over microbials is that often there is considerable phytochemical and pharmacognostical information in the literature relating to plant species. Databases such as NAPRALERT (*5*) can give early clues on the nature of active moieties and can help in planning the initial

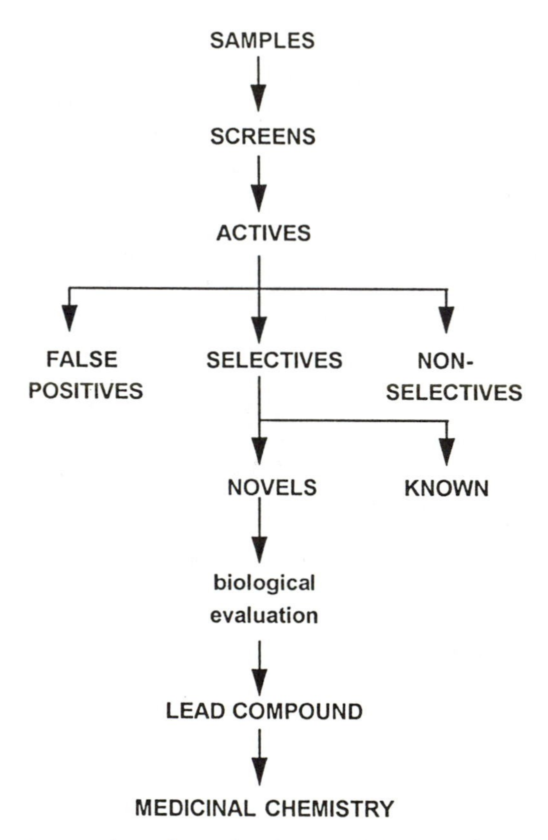

Figure 1. Progression of samples through high-throughput screens

stages in fractionation so as to purify the active components as expeditiously as possible.

Clearly there is a need to prevent re-invention of the wheel. The chances of doing this can be minimised by establishing local databases of chromatographic-spectroscopic features of plant extracts or purified metabolites which are known to show activity in the screen.

Chemical Leads from Plants

What sort of compound are we looking for from plants? A good example is khellin, a low-molecular-weight furanochromone which is the principal active constituent of *Ammi visnaga*. This plant has long been used in Egyptian traditional medicine as a remedy for bronchial asthma and cardiac irregularities. Khellin provided the lead for chemical programmes which gave rise to two important medicines: the anti-asthmatic drug cromoglycate (*6*) and the anti-arrythmic agent amiodarone (*7-9*).

Conclusions

The resurgence of interest in higher plants in the pharmaceutical industry has resulted largely from the advent of high-throughput screening as a key route to finding new drugs with new modes of action. Crucial to the success of such programmes is access to a diverse range of input materials and an effective, perhaps quite ruthless, strategy for following up primary actives. These factors, in turn, depend upon establishing effective collaborations between plant suppliers and those engaged seriously in drug discovery.

Literature Cited

1. Phillips and Drew *Global Pharmaceutical Review*, **1992,** July issue
2. Sneader,W. *Drug Discovery: The Evolution of Modern Medicines*; John Wiley & Sons: Chichester, U.K., 1985
3. Loomis,W.D.; Battaile, J. *Phytochem.* **1966**, *5*, 423-438
4. Harborne, J.B. *Phytochemical Methods*; Chapman and Hall; London, 1984
5. Loub, W.D.; Farnsworth, N.R.; Soejarto, D.D.; Quinn, M.L. *J. Chem. Info. Comp. Sci.* **1985**, *25*; 99-103
6. Gilman, A.G.; Rall, T.W.; Nies, A.S.; Taylor, P. *The Pharmacological Basis of Therapeutics*; Pergamon Press, Oxford, U.K., **1990**, p630
7. Chalier, R.; Deltour, G.; Tondour, R.; Binon, F. *Arch. Int. Pharmacodyn.* **1962**, *139*, 255-264
8. Vastesaeger, M.; Gillot, P.; Rasson, G. *Acta Cardiol.* (Brux), **1967**, *22*, 483-500.
9. Singh, B.N.; Venkatesh, N.; Nadermanee, K.; Josephson, M.A.; Kannan, R. *Prog. Cardiovas. Dis.* **1989**, *31*, 249-280

RECEIVED May 21, 1993

Chapter 6

Secondary Metabolism in Plant Tissue Cultures Transformed with *Agrobacterium tumefaciens* and *Agrobacterium rhizogenes*

G. H. Neil Towers and Shona Ellis

Botany Department, University of British Columbia, Vancouver, British Columbia V6T 1Z4, Canada

Leaves, roots, flowers and fruits consist of complex cellular tissues containing compartmentalized phytochemicals. Cell, callus and organ cultures of plants are capable of producing most types of secondary compounds, including phenylpropanoids, polyketides, terpenoids, alkaloids, and carbohydrates, and offer excellent materials for biosynthetic studies because they consist of populations of actively growing cells. The biosynthesis of phytochemicals in plants and in tissue cultures is, unfortunately, often dependent on the presence of differentiated tissues or the development of specialized cells (e.g., many sesquiterpene lactones of the Asteraceae are produced in trichomes). Callus cultures, exhibiting a greater degree of cellular differentiation, can be obtained by genetic transformation using *Agrobacterium tumefaciens*. These transgenic tissues may produce secondary compounds which are otherwise difficult to obtain in normal callus culture. For example, transgenic callus of some species in the Asteraceae readily synthesize thiarubrines, whereas normal callus usually does not. On the other hand, transgenic cell suspension cultures do not produce these compounds either. Many desirable pharmaceuticals are produced in plant roots and the use of genetically stable, fast-growing hairy root cultures obtained with *Agrobacterium rhizogenes*, now offers very real opportunities for commercial exploitation by the pharmaceutical industry.

0097–6156/93/0534–0056$06.75/0

Plant cells can be cultured successfully as undifferentiated or partially differentiated cell suspensions, callus or completely differentiated organs, depending on nutrient supplies, hormonal levels, light regimes, temperature and access to oxygen and carbon dioxide. In many instances, the regeneration of complete plants from these cultures can also be achieved. The production of secondary metabolites, on a large scale, in culture systems, is an interdisciplinary task requiring an understanding of plant physiology, phytochemistry and chemical engineering. The desired phytochemicals may be confined within the cells or excreted into the medium or may be distributed between both of these compartments. In the last case, all or only a particular compound of a series of related compounds may be excreted. If most of the desired product is excreted, cell or callus immobilization may be usefully employed.

Cultures may also be used for "biotransformation" studies. A recent example is the conversion, by suspension-cultured cells of *Curcuma zedoaria*, of germacrone (**1**) and germacrone epoxide into a series of eudesmane (**2**) and guaiane (**3**) sesquiterpenes, typical of that species (*1*). A second example, on a somewhat larger scale, is the conversion, over an eight-day period, of steviol to related glycosides and esters using a cell suspension of *Eucalyptus perriniana*, in a 10 liter fermenter (*2*).

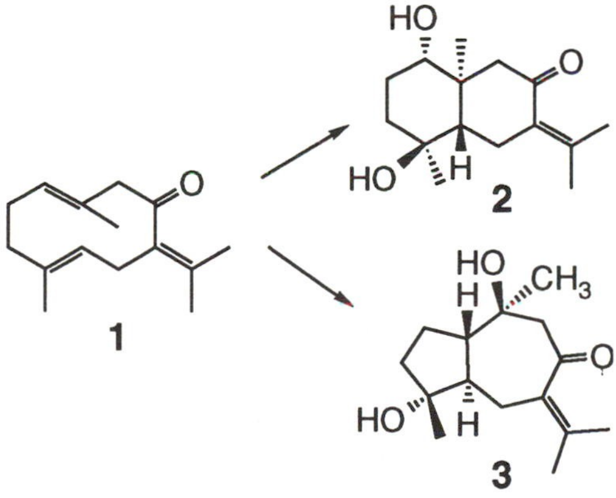

Although large-scale production of secondary metabolites by cell-culture techniques continues to attract research interest, there are currently only very few genuine examples of actual commercial production of phytochemicals on a large scale. These are the production of shikonin and related naphthoquinones, used in the treatment of burns in China and Japan, by Mitsui Petrochemicals, from cultures of *Lithospermum erythrorhizon* (*3,4*), and the production of sanguinarine by elicited cell cultures of *Papaver somniferum* (*5,6*).

There are disadvantages in the use of undifferentiated cultures such as callus and suspension cultures. The genetic and karyological instability of such cultures are well known to all those who have worked in this field (*7,8*). Binarova and Dolezel (*9*) have analyzed the stability of the nuclear DNA content in alfalfa embryogenic cell-suspension cultures by flow cytometry and, among other conclusions, suggest that short sub-culture intervals help to maintain karyological stability and long sub-culture intervals, where the cells remain for a long time in a stationary phase, may lead to increases in ploidy level. After sub-culture and mitotic stimulation, a mixoploid population is often obtained. This is referred to as endoreduplication. These instabilities may be manifested in many ways. Growth rates may be reduced, the biosynthesis of the desired chemical may be inconsistent, or it may cease completely. Root cultures offer the advantage of being genetically more stable and usually produce the secondary compounds of the normal roots of the species, e.g., emetine and cephaeline from *Cephaelis ipecacuanha* (*10*) or ginsenosides from *Panax ginseng* (*11*). We have found that root cultures of *Ajuga reptans* produce large amounts of ecdysterone offering good material for biosynthetic work (Dreier, S.; Towers, G.H.N., unpublished data).

Unfortunately, many desired phytochemicals or natural products are produced in highly specialized cells or groups of cells, e.g., glands and canals, so that often undifferentiated cell cultures are not useful. Absolutely essential secondary compounds, e.g., lignin or carotenoids, may be produced in these types of uniform cell cultures. Large-scale fermenters for the production of kilogram quantities of plant cells have been available for the last decade, and growth conditions for the biosynthesis of a number of important pharmaceuticals have been or are being established, suggesting that this area of research is poised for commercialization. From the viewpoint of both the phytochemist and the chemical engineer, desirable requirements for any of these culture systems are rapid growth and the production of the required phytochemicals in high and reproducible yields. As biochemical pathways are often non-linear, however, they are inherently chaotic systems and this may pose problems over long periods of culture. Special techniques such as the use of root cultures instead of suspension cultures, induction of stress e.g., the elicitation or use of cell immobilization techniques, may be required. In addition to these approaches, plant material may be genetically transformed by species of soil bacteria in the genus *Agrobacterium,* leading to a greater degree of differentiation in callus cultures and the production of the desired secondary compounds. It is this last technique which we wish to highlight in this paper.

Agrobacterium: Nature's Genetic Engineer

Species of gram-negative soil bacteria in the family Rhizobiaceae are known for their pathogenesis in many plant species. *Agrobacterium tumefaciens*, the causative agent of crown-gall disease in plants, gives rise to tumours which form at the crown of the plant, a common site of infection. A lesser-studied species, *A. rhizogenes,* causes hairy-root disease, which is characterized by a massive proliferation of roots at the infection site.

It has been observed that *A. tumefaciens* does not have to be present for the growth and maintenance of the tumour. This implies that there is some kind of "factor" introduced by the bacterium which influences plant-cell growth, referred to as the tumour-inducing (Ti) principle. It was found that this principle was the result of a stable integration, into the plant cell nuclear genome, of bacterial DNA (*12*). The piece of DNA which is transferred is part of a large plasmid (159-200 kpb), later termed the Ti-plasmid. A similar plasmid was found for *A. rhizogenes* (*13*) and came to be known as the Ri-plasmid (root-inducing plasmid). Three major components of these plasmids are required for the incorporation of DNA into the genome of the plant cell. First, there is the DNA to be transferred called transfer DNA (T-DNA) (*14*). Secondly, there are the virulence (*vir*) genes, comprising six complementation groups found on the Ti- and Ri-plasmids, which are responsible for the transfer of the T-DNA. The third component is a cluster of chromosomal genes which are probably important in cell-membrane dynamics (*15*).

The process of transformation begins with the injury of a plant cell. Upon wounding, the plant cells excrete low-molecular-weight phenolic compounds which attract the bacterium and stimulate the expression of the *vir* genes. A number of these "signal" compounds have been identified. These are simple phenols, with a syringyl nucleus, such as acetosyringone and hydroxyacetosyringone, esters of sinapic acid and even certain chalcones (*16,17*). Similar compounds with a guaiacyl nucleus are also virulence inducers (*16,17*). The production of acetosyringone (**4**) and acetovanillone may be restricted to solanaceous species (*18,19*). This induction by specific chemicals may be used to advantage in obtaining transgenic plants or plant-tissue cultures of species which are otherwise resistant to infection. Mathews *et al.* (*20*), working with *Atropa belladonna*, found that there was a higher incidence of transformation when *A. tumefaciens* was cultured with acetosyringone. Soybean cells or whole plants, when treated with *A. tumefaciens* in the presence of acetosyringone (**4**) or syringaldehyde (**5**), were also successfully transformed (*21*). *Zea mays* has been transformed, however, without the use of *vir*-inducers, by co-cultivation of shoot apices with *A. tumefaciens* (*22*).

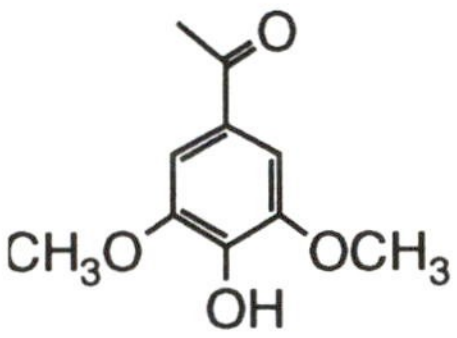

O CHO

CH_3O OCH_3 CH_3O OCH_3
OH OH

4 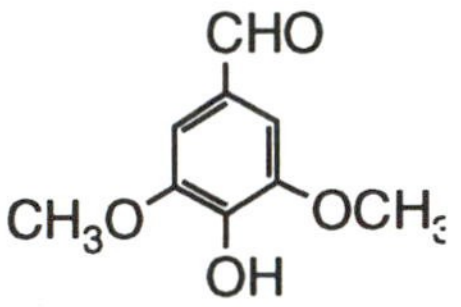 **5**

How the T-DNA is incorporated into the nuclear genome of the plant cell is not known. The T-DNA varies within and between *Agrobacterium* species. The nopaline type of *A. tumefaciens* T-DNA is one contiguous piece of DNA, whereas the octopine type consists of two pieces, termed the T_L-DNA (left) and the T_R-DNA (right). The agropine type found in *A. rhizogenes* also consists of two pieces of T-DNA, whereas the mannopine type has only one, which has homology with the T_L-DNA of the agropine-type. Nopaline (**6**), octopine (**7**), agropine (**8**), and mannopine (**9**) refer to a group of relatively rare amino acids known as opines which are encoded for by the corresponding T-DNAs. They are metabolized by the bacteria by virtue of their opine-catabolizing genes, which are found on the Ti- and Ri-plasmids.

Following integration, the plant cell is programmed for a different growth pattern. The T-DNA from *A. tumefaciens* encodes for the auxin, indole acetic acid (IAA), and the cytokinin, isopentenyl adenine. The interaction of these two known plant growth regulators causes the proliferation of cells resulting in tumour formation. The T_R-DNA of *A. rhizogenes* shows homology for the auxin encoding genes and not for the cytokinin gene. Because auxins are known to induce roots, whereas cytokinins inhibit their induction, it was originally believed that the over-production of IAA causes the induction and subsequent proliferation of roots formed at the infection site. When *A. tumefaciens* strains, in which the cytokinin-encoding gene was mutated, were used to transform plant tissue, it was observed that roots did form at the site of infection (*23*). Therefore, it was concluded that auxin production was central to the development of the roots in hairy-root disease.

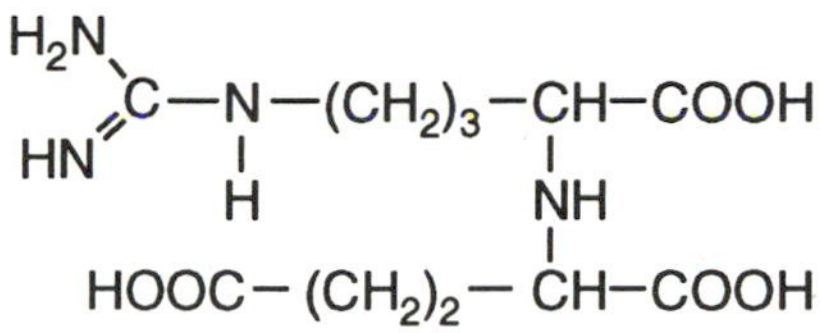

H_2N
$C{-}N{-}(CH_2)_3{-}CH{-}COOH$
HN | H | NH
$HOOC{-}(CH_2)_2{-}CH{-}COOH$

6

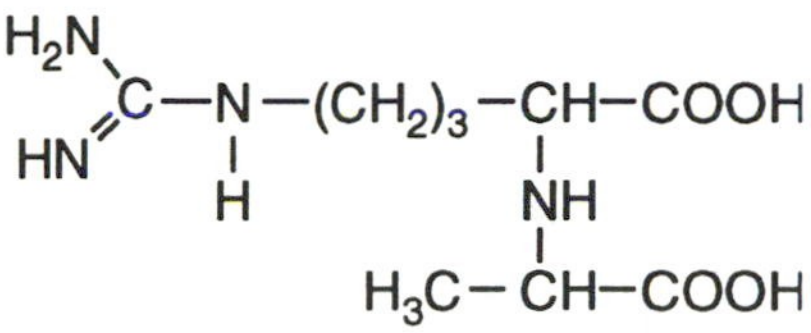

H_2N
$C{-}N{-}(CH_2)_3{-}CH{-}COOH$
HN | H | NH
$H_3C{-}CH{-}COOH$

7

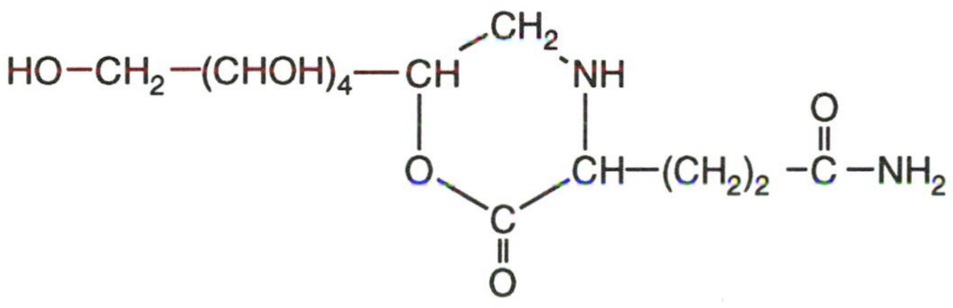

CH_2
$HO{-}CH_2{-}(CHOH)_4{-}CH$ NH
O $CH{-}(CH_2)_2{-}C(=O){-}NH_2$
$C{=}O$

8

$HO{-}CH_2{-}(CHOH)_4{-}CH_2$
NH
$HOOC{-}CH{-}(CH_2)_2{-}C(=O){-}NH_2$

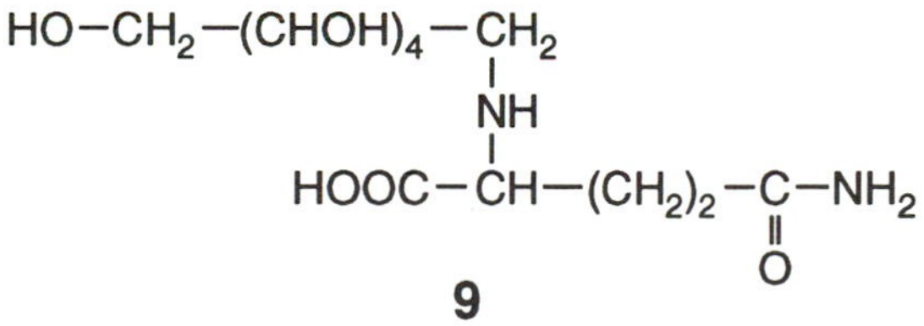

9

The phenotypic characteristics of hairy-roots include: reduced apical dominance (which leads to much lateral branching); numerous root hairs; a high growth rate; and plagiotropic growth. Yet, when plants were transformed with T_R-DNA alone (the DNA encoding for the production of auxin), root induction was seldom observed (*24*). In those cases where rhizogenesis did occur, the number of root primordia, as well as the rate of growth were much reduced. Interestingly, it was observed that the phenotype generally attributed to this infection was altered.

The first group to examine the involvement of the T_L-DNA was that of White *et al.* (*25*) who found that four genes on this segment were responsible for the phenotype of hairy-roots. These were called the *rol* (root loci) genes, *rol*A, *rol*B, *rol*C, and *rol*D. In 1986 Slightom *et al.* (*26*) sequenced the T_L-DNA and found that these genes corresponded to open reading frames which were numbered 10, 11, 12, and 15. They also confirmed the lack of homology between this DNA and that of the *A. tumefaciens.* The hairy-roots were therefore induced in a slightly different way than was the case with crown gall induction. The growth patterns of the tumour apparently are the result of the over-production of growth regulators. However, with *A. rhizogenes,* using unregulated genes such as cauliflower mosaic virus promoter, it is not the over-production of auxin which causes root formation to occur (*27*), but rather an increase in the sensitivity of the cells to the auxin by virtue of the *rol* genes (*28*).

There has been little study to determine the differences between adventitious root induction by transformation and by induction using growth regulators. Luza and Sutter (*29*), working with apple shoots, suggested that transgenic roots arise exogenously, that is, from dedifferentiated cortical tissue or from callus. Roots induced by treatment with naphthalene acetic acid (NAA) were found to be initiated endogenously, with connections between the adventitious root and the vascular tissue of the mother explant.

Crown-gall tumor tissues, as mentioned previously, grow independently of cytokinins and indole acetic acid. Nopaline strains of A. *tumefaciens* have a gene, *tzs*, that encodes a cytokinin prenyl transferase. It is constitutively expressed at low levels and, as a result, the bacteria excrete low levels of the cytokinin, zeatin, thus influencing cell proliferation in the plant. Interestingly, zeatin excretion by the bacterium can be increased 100-fold by acetosyringone (*30*).

In this process of "natural" genetic engineering by *Agrobacterium,* it is clear that not all cells are transformed and that both transformed and non-transformed cells will survive in continued culturing systems without phytohormones because of the supply of these by the transformed cells. For example, chimeric transgenic cowpea (*Vigna unguiculata*) cells can be produced from mature embryos by co-cultivation with an oncogenic hypervirulent strain of *A. tumefaciens* (*31*). A technique for single-cell cloning of crown-gall tumor tissue from protoplasts has been described (*32*). Protoplasts may be co-cultivated with the bacterium to produce transgenic cells which may be selected for their ability to grow in the absence of hormones (*33*). In potato transformation, a medium has been found which induces the rapid formation of morphologically normal shoots from potato discs without an intervening callus stage, thus lowering the frequency of chromosomal variations (*34*).

The potential for using *Agrobacterium* in biotechnology has to some extent been realized. *Agrobacterium tumefaciens* has been used extensively as a vector to introduce foreign DNA into other plant species. The objective of this paper is to discuss the potential of transgenic tissues for the production of useful phytochemicals.

Agrobacterium tumefaciens and Secondary Metabolism

De Cleene (*35*) was probably the first to recognize that transgenic tissue displayed differences in secondary metabolism when he showed elevated levels of ascorbic acid and glutathione in crown-gall tissues of tomatoes and beets. The transfer of secondary metabolites in tissue cultures was assumed to be a carry-over from the tissues from which the tumors originated (*36,37*).

Our early attempts to study the biochemistry and photobiology of polyynes and related sulfur derivatives, compounds typical of the Asteraceae, with callus and suspension cultures, were completely unsuccessful (*38*). We were unable to induce their biosynthesis by varying culture media, light regimes, temperature or hormonal levels. In one instance, very low concentrations of thiophenes, such as α-terthienyl (**10**) were obtained in long-term callus cultures of *Tagetes patula* grown in SH medium with 4 mg/l NAA. This material, which had partially differentiated into roots, produced essentially the same spectrum of thiophenes as are found in normal roots of this species (*39,40*). Others had also failed to obtain normal amounts of polyynes (*41,42*) or thiophenes (*43*) in short-term callus cultures. Eventually, we discovered that phenylheptatriyne (**11**) could be induced in tissue cultures of *Bidens pilosa* by the addition of fungal culture filtrates (*38*). Similarly, root cultures of *Bidens sulphureus* show a dramatic increase in tridec-1,3,11-ene-5,7,9-yne when exposed to mycelial extracts of the fungal pathogens, *Phytophthora* spp. and *Pythium* sp. (*44*). More unusual perhaps, was our discovery that callus prepared from crown galls of *B. pilosa* and other species in this genus yielded characteristic polyynes such as phenylheptatriyne in contrast to normal callus which did not (*38,39*).

We then turned our attention to a group of red, antifungal dithiane heterocycles which occur in a number of genera in the Asteraceae and which we called thiarubrines, such as thiarubrine A (**12**) and thiarubrine B (**13**). Because of their color, we were able to obtain, by selective transfer of a line transformed with *A. tumefaciens*, callus which accumulated these compounds in

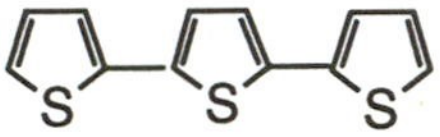

S S S **10**

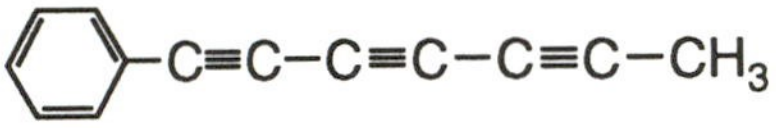

$-C{\equiv}C-C{\equiv}C-C{\equiv}C-CH_3$ **11**

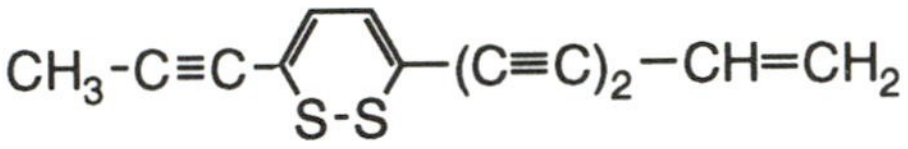

$CH_3-C{\equiv}C-$ S-S $-(C{\equiv}C)_2-CH{=}CH_2$ **12**

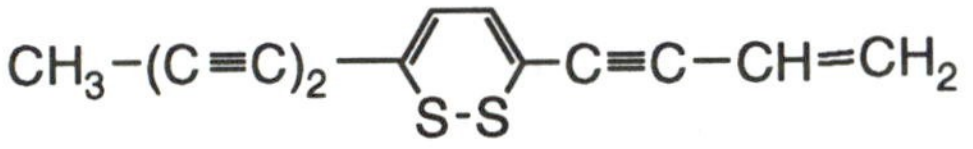

$CH_3-(C{\equiv}C)_2-$ S-S $-C{\equiv}C-CH{=}CH_2$ **13**

amounts equal to those found in natural roots (*45*). Histological studies of these transgenic callus cultures revealed that they contained well organized, lignified xylem, with pockets of thiarubrines in intercellular spaces. Addition of auxins to these cultures resulted in the disappearance of the thiarubrines. Similarly, cell suspensions, prepared from this callus, rapidly switched off production of these compounds. This observation of a correlation between differentiation and secondary metabolism was in agreement with the work of others (*46*). Because of the toxicity of the thiarubrines, as well as their very low solubility in water, it is not likely that they will be produced in undifferentiated cell cultures unless special methods are adopted, e.g., a biphasic system.

Some alkaloids are produced quite readily in non-transformed cultures. Callus cultures of *Trigonella foenum-graecum* produce 3-4 times more of the plant growth hormone, trigonelline (**14**) (*47*), than do the seeds of the plant, and 12-13 times more than occur in roots or shoots. Higher levels of alkaloids are also produced by suspension cultures of *Cinchona ledgeriana* transformed with *A. tumefaciens* (*48*), especially when grown in the dark rather than when grown in light. Untransformed cultures did not show this "dark" enhancement of alkaloid production. Also senecionine *N*-oxide and related *N*-oxides, which accumulate in the flower heads of *Senecio vulgaris*, are not formed in cell-suspension cultures, crown-gall tumor cultures or shoot-like teratomas obtained by transformation with *A. tumefaciens*. They are biosynthesized, however, in root cultures, lending support to the hypothesis that they are normally biosynthesized in roots and transported to the aerial parts of the plant (*49*).

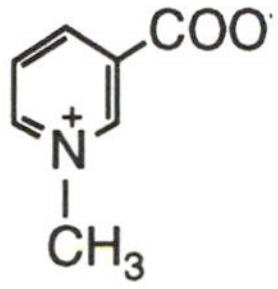

$$\text{N-methylpyridinium-3-carboxylate (}COO^-\text{, }\overset{+}{N}\text{-}CH_3\text{)}$$

14

Even though *A. tumefaciens* offers semi-differentiated tissue systems with biochemical potential, the yields of many phytochemicals are still not close to yields obtained in plants. The production of low concentrations of digitoxin in *A. tumefaciens*-transformed *Digitalis lanata* cell cultures has been reported (*33*). *Agrobacterium rhizogenes,* in contrast, lends itself well to the induction of secondary metabolite accumulation because the tissues incited are fully differentiated roots which often produce similar chemical profiles to those of natural roots.

Hairy-Root Cultures and Plant Secondary Metabolism

Studies of plant tissue cultures originated in the classical work of Philip White, who used root cultures (*50*), and the roots of many plant species have since been grown aseptically. Roots are often the site of synthesis and the source of biochemicals such as alkaloids, and it has been found that root cultures always produce the alkaloids that are found in roots of the plant. Hairy-roots (HR) can be produced on leaves, petioles or stems. They can be put into culture and usually grow much faster than normal roots. As with *A. tumefaciens*-induced transgenic cultures, they do not require an exogenous supply of hormones. They may, however, require special aeration in liquid medium. For example, we have found that without vigorous shaking of Ginseng hairy-root cultures, they rapidly revert to callus growth. However, Mugnier (*51*) has extended the range of hairy-root cultures so that in his laboratory almost 100 species of plants have been transformed with *A. rhizogenes*. His cultures have remained viable for 2-6 years on Murashige and Skoog medium. Unfortunately, he found that many woody species and monocotyledons could not be transformed in this manner.

Our limited success with thiophene and thiarubrine production in transgenic cultures, derived form *A. tumefaciens,* led to the studies of production in HR cultures of numerous species eg. in root cultures of *Bidens* spp. (*52*). The concentrations of these polyynes were found to be approximately 10% higher than in normal root cultures. Transgenic cultures exhibited a higher sensitivity to kinetin than normal root cultures, reverting to callus readily. Dedifferentiation of the hairy-roots was correlated with a decrease in thiophene production.

Thiarubrines have been isolated from hairy-root cultures of numerous species of the Asteraceae. *Chaenactis douglasii* produces thiarubrine A and B in hairy-root culture. The selection of high-yielding lines made this culture system suitable for examining the biosynthetic pathways of these two compounds (*53*). Thiarubrines are easily converted into the corresponding thiophenes upon irradiation. There are many species in the Asteraceae which biosynthesize thiophenes, whereas there are only a very few species that produce thiarubrines. Constabel and Towers (*54*) suggest that, even though thiophenes can be produced from thiarubrines by irradiation, the two classes of compound are not produced sequentially in the plant, but instead have different biosynthetic pathways. Hairy-root cultures offer a good system for this type of biosynthetic analysis.

Ambrosia chamissonis produces a wider array of thiarubrines and thiophenes than does *C. douglasii* Hairy-root cultures have chemical profiles which match those of the roots of the original species (Ellis, S. M.; Towers, G. H. N., unpublished data). Anatomical studies of the hairy-roots of *Ambrosia* and *Chaenactis* spp. reveal that the thiarubrines are sequestered in canals which run between the two layers of the double endodermis. In root cultures of *Ambrosia,* the more polar thiarubrines appear in the medium. This is apparently due to the formation of lateral roots which are initiated in the pericycle and break through the endodermis, rupturing the canals in the process.

Mukundan and Hjortso (*55*) studied the production of thiophenes in root cultures of *Tagetes erecta* and found that the overall production in normal root cultures was less than in hairy-root cultures. This was calculated on the basis of the yield per flask as opposed to the amount of thiophene per gram of root material. The chemical profile of the transformed roots matched that of the intact roots. This was also confirmed by Parodi *et al.* (*56*), who quantified both bithiophenes and benzofurans in hairy-root cultures of *Tagetes patula.*

It has become apparent that the chemistry of root cultures mimics that of the parent roots and that HR cultures can produce elevated amounts of secondary metabolites. The cultures are more stable than normal root cultures. Selection in sub-culturing is important for the production of high-yielding cultures. An aspect which has not been extensively examined is the chemistry of transgenic cultures derived from different strains of *A. rhizogenes*. The agropine type strains possess T_L- and T_R-DNA, whereas those of the mannopine type only possess T_L-DNA. This means that the latter does not encode for the production of auxin.

Croes *et al.* (*57*) examined the role of auxin in root culture of *T. patula* and showed that the addition of high concentrations of IAA increased the production of root primordia, and that the agropine-type strains (A4 and 15834) of *A. rhizogenes* produced even more branching. The mannopine type (TR-7) of transformed roots had much less in the way of secondary branching, probably because auxin was not encoded for. Croes *et al.* (*57*) found that thiophenes accumulated in severed root tips. It was concluded that, since this is the main site of auxin biosynthesis, auxin is important for the production of thiophenes. This is in variance with the findings of Norton and Towers (*52*), who found that exogenously supplied auxin inhibited the production of thiophenes. In another system, exogenous auxin was found to be inhibitory to thiarubrine production in hairy-root cultures of *Chaenactis douglasii* (*53*).

Flores *et al.* (*44*) established "green roots" by incubating HR cultures in light. The green roots had normal anatomy, but contained chloroplasts in the cortical cells. Green roots of *Tagetes* and *Bidens* spp., grown in light, exhibited an increase in growth rate as well as yields of thiophenes. This also contradicts the findings of Norton and Towers (*52*), who found light to be inhibitory to the production of thiophenes. It is therefore apparent that different environmental factors are required for the biosynthesis of thiophenes in different types of cultures. Because thiarubrines are sensitive to irradiation, cultures have to be maintained in the dark (*58*).

Two new polyynes, lobetyolin (**15**) and lobetyol (**16**), have been isolated from *Lobelia inflata* hairy-root cultures (*59*). The authors neglected to mention if these compounds were present in the plants from which the cultures were initiated.

Alkaloid biosynthesis and production has been examined extensively in transgenic-root cultures. Studies with the potato suggest that transformation with A. *rhizogenes* can be a suitable system for genetically stable plant regeneration from cells (*60*).

However, there are often alterations in morphology of *Nicotiana hesperis* plants, spontaneously regenerated from hairy-roots. The plants have wrinkled leaves, smaller flowers, reduced fertility and reduced apical dominance. This pattern has been observed in other species and has been attributed to the effects of Ri T-DNA (*61*).

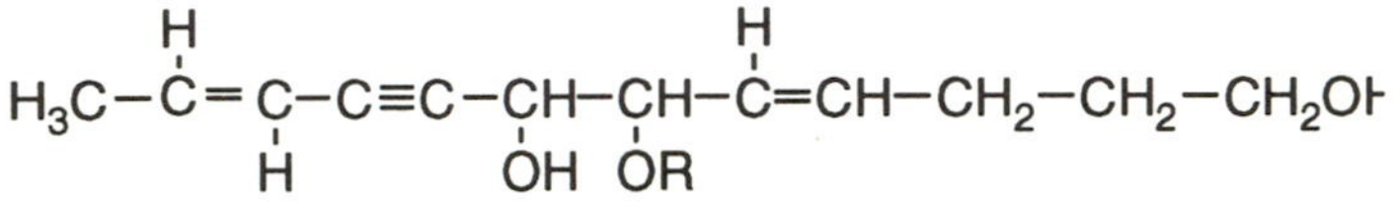

$$H_3C-CH=CH-C\equiv C-CH(OH)-CH(OR)-CH=CH-CH_2-CH_2-CH_2OH$$

15 R = glc

16 R = H

The amounts of atropine and scopolamine (**17**) in hairy-root cultures of *Atropa belladonna* were the same or even higher than those of normal root cultures or roots of plants grown in the field (*62*). The transgenic roots produced high levels of these alkaloids under non-selective conditions for over a year or more in contrast to callus cultures which produced the alkaloids, in high amounts only during the selection period, the amounts becoming progressively lower after several sub-cultures under non-selective conditions. The growth rate of the hairy-root cultures and production of hyoscyamine (**18**) of *Hyoscyamus muticus* remained stable after six years of culture (*63*). Hairy-roots of *Datura stramonium* grew in the absence of added phytohormones and the cell mass increased 55-fold during 28 days of incubation (*64*). Hyoscyamine production was comparable to that of roots from pot-grown plants and accounted for 0.3% of the dry matter. Both T_L-DNA and T_R-DNA were identified in hairy-root tissue. Hyoscyamine-6ß-hydroxylase, an enzyme in the biosynthetic pathway of these alkaloids, has been characterized in root cultures of *Hyoscyamus* spp. (*65*), illustrating the usefulness of cultures in enzymological studies.

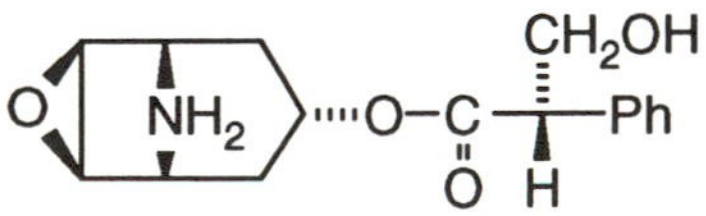

17 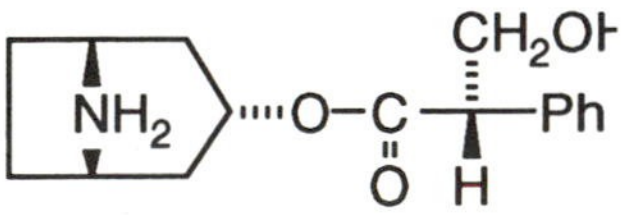 **18**

Hairy-roots have also been established from three different *Datura stramonium* seedlings (*66*) and, from one of them, four different strains were isolated. After 22 days of growth, a high but variable rate of biomass increase was observed, as well as marked differences in the contents of scopolamine in these cultures. Transformation also increased biomass and tropane alkaloid production in root cultures of *A. belladonna* and *Calystegia sepium* (*67*). The capacity of these roots for increased growth and alkaloid production did not diminish over 6 years of continuous culture.

Sauerwein and Shimomura (*68*) examined how various culture conditions affected the production of the five main tropane alkaloids in hairy-root cultures of *Hyoscyamus albus*. They could enhance the growth rate of the cultures by modifying sucrose levels, but this did not lead to enhancement of the tropane alkaloids. Increasing the amount of KNO_3 in Woody Plant Medium (*69*) optimized alkaloid biosynthesis. Shimomura *et al.* (*70*) examined the production of alkaloids in transgenic and adventitious root cultures of various solanaceous species. They found that the highest alkaloid yielding *H. albus* culture was adventitious, whereas *H. niger* transgenic cultures produced higher yields of the alkaloids. Hairy-root cultures demonstrated the advantage of more rapid growth. They emphasized that although the chemical composition of the cultures was dependent upon the species, the media composition also plays an important role.

Interesting studies on the biosynthesis of tobacco alkaloids have been carried out with hairy-root cultures of *Nicotiana* species. Nicotinic acid and nicotinamide, when fed to hairy-roots of *N. rustica*, were found to be toxic (the 50% phytostatic dose being 2.4 and 9 mM, respectively) while nicotine (**19**) was not toxic below 10 mM. Nicotinic acid (up to 5 mM) was phytostatic rather than phytotoxic (*71*). Roots exposed to increasing nicotinic acid or nicotinamide levels had altered alkaloid accumulation patterns relative to the controls. The principal effects were to increase the intracellular and extracellular levels of anatabine (**20**) and nicotine, with a marked preponderance of anatabine.

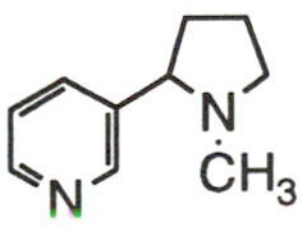

N CH$_3$ N N H N N H N

19 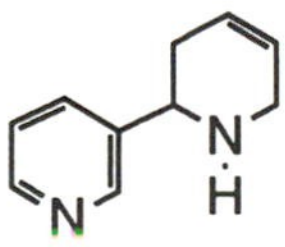 **20** 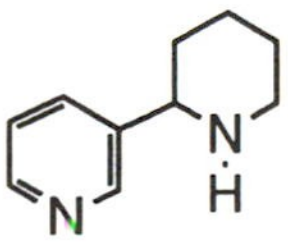 **21**

In another study, unlabeled cadaverine did not diminish the incorporation into anabasine (**21**) of ^{14}C from L-(U-^{14}C)-lysine supplied to hairy-root cultures of *N. hesperis* despite a stimulation of anabasine production (*61*). This finding, in agreement with previous observations, would seem to indicate that free cadaverine is not an intermediate in the biosynthesis of anabasine from lysine. However, experimental result was explained by invoking cellular compartmentalization.

Transformed roots of *Catharanthus roseus* were obtained from surface-sterilized leaves (*6*). The levels of serpentine (**22**), ajmalicine (**23**) and catharanthine (**24**) resembled those found in roots of intact plants although a significant degree of short-term variation in alkaloid productivity was noted. Vinblastine was also produced in very small amounts (0.05 μg/g dry wt.). Alkaloid production in this species is known to be sensitive to growth conditions, i.e., environmental factors such as light, elicitors or stresses. Forty-seven clones of hairy-root cultures of the same species showed wide variations in growth rate and morphology, but the spectra of alkaloids were qualitatively similar to that of the corresponding roots (*72*). Novel alkaloids have been obtained from hairy-root cultures of *C. trichophyllus* (*73*), although normal and some hairy-roots showed similar alkaloid compositions (*74*). A new piperidone alkaloid, hyalbidone (**25**), was identified in a strain of transgenic roots of *H. albus* (*75*).

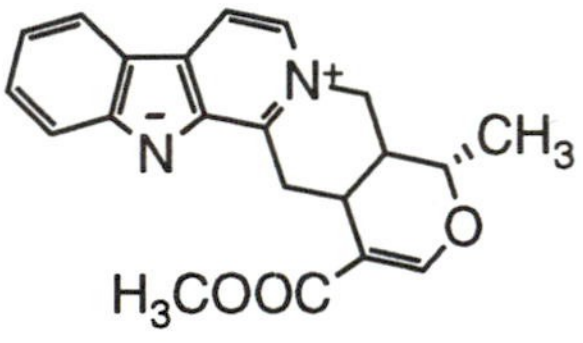

22 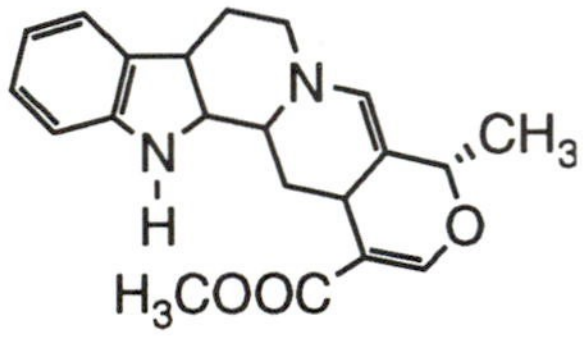 **23**

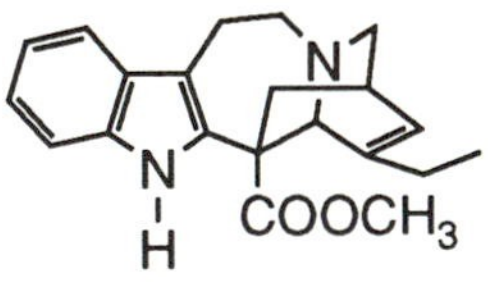

24 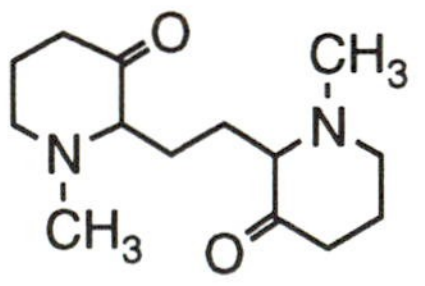 **25**

Root cultures of *Coreopsis tinctoria*, transformed with *A. rhizogenes*, produced the same or even higher levels of substituted phenylpropanoids (e.g., isobutyryloxyeugenol isobutyrate) than normal roots (*76*). Shikonin (**26**) is a secondary metabolite produced in large amounts from callus and suspension cultures of

Lithospermum erythrorhizon (*77*). To produce shikonin, a two-phase culture system is employed. Because of the genetic instability of non-differentiated tissue, Shimomura *et al.* (*78*) have examined the production of shikonin in hairy-roots cultures. They were able to extract this compound from the medium without changing the culture conditions. The production was enhanced by using an adsorbent (XAD-2 column) and 5.2 mg per day could be obtained from a 2-liter batch. Anthraquinone production in transgenic root cultures of *Rubia tinctorum* has been examined in relation to the influence of phytohormones and sucrose concentrations in the medium (*79*). Elevated sucrose levels in phytohormone-free medium resulted in the best growth and highest anthraquinone production. Medium which was supplemented with 5 μM IAA and reduced sucrose enhanced growth and chemical production, but to a lesser degree.

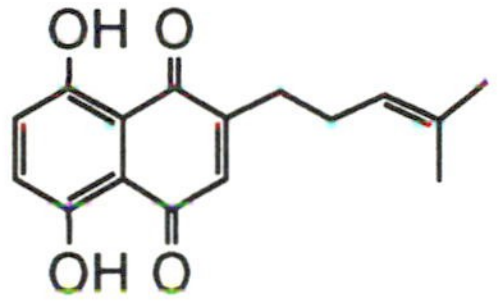

OH O

OH O

26

Hairy-roots of *Panax ginseng* synthesize the same ginsenosides as normal roots but in greater amounts. For instance, Furuya reported (*80*) 2.4 times the quantity og ginsenosides found in native roots and twice as much as in normal cultured roots on a dry weight basis. The hairy-root cultures of *P. ginseng* (*81*) is qualitatively and quantitatively similar in its ginsenoside composition to that of commercial Ginseng, but commercialization presents an unexpected kind of problem. Apparently some dealers in Ginseng may be reluctant to handle the laboratory product, considering that it is not "natural". Until this bias is overcome, it may not be an economically viable production system. It should be mentioned that laboratory Ginseng consists of thin roots, resembling closely the cheaper type of material which is prepared for much of the Ginseng trade in teas and tonics. The factor (or factors) concerned with the initiation of cambial activity in roots is not known at this time. Thick roots have never been produced in cultures from any plant sources. This surely is an invitation to the more mercenarily-inclined morphogeneticists; large roots of Ginseng, with desirable human-like shapes, fetch extremely high prices in Asian markets.

The same levels of saponins were found in hairy-root cultures of *Astragalus boeticus* and *A. hamosus* as compared with non-transformed root cultures (*82*), and these levels were maintained on sub-culturing. Elimination of calcium in the medium reduced growth, but promoted digoxin synthesis in *Digitalis thapsi* (*83*).

Echinacea purpurea root extracts have been used as medicinals for numerous ailments. Transgenic root and callus cultures derived from *A. rhizogenes* have been examined for their production of alkamides, such as dodeca-2*E*,4*E*,8*Z*,10*E*/*Z*-tetraenoic acid isobutylamide and related compounds, which are believed to be some of the active ingredients (*84*). While transformed callus produced higher levels than natural roots, transgenic and normal root cultures produced the same amounts.

In transformed roots of *Beta vulgaris,* betalains (both betacyanins and betaxanthins) were found at levels comparable to those of intact roots (*85*) and were entirely retained within the roots, as might be expected. Betalain production begins somewhere behind the root tip, indicating that some degree of differentiation is necessary for pigment biosynthesis to commence.

Efforts have been made to scale-up production of chemicals by using fermenters. Hilton and Rhodes (*86*) solved the problem of damage incurred by impeller action by modifying the vessel segregating the impeller by covering it with a cage. Another solution is to use an air-lift system, whereby air is sparged through the medium (*87*). One difference between the hairy-root cultures and suspension cultures is that the fermenter system is not ameniable to sampling. Toivonen *et al.* (*87*) demonstrated, with hairy-root cultures of *Catharanthus roseus*, that root cultures behave differently from suspension cultures and the conductivity of the medium does not reflect root growth. That they found a direct correlation between sucrose consumption and dry-weight increase and therefore suggest that this is a better way of monitoring this type of culture system. There was no difference in chemical composition of the cultured roots grown in fermenters as opposed to shaker flasks (*87*).

Finally, it should be pointed out that no studies appear to have been made of plants regenerated from selected hairy-root cultures which produce high levels of specific phytochemicals. Such plants could be useful agriculturally. Shoots, regenerated from roots of *Glycine argyrea*, inoculated with *A. rhizogenes* carrying engineered plasmids, confirmed the stable regeneration and expression of the chimaeric kanamycin-resistance gene in transgenic tissues (*88*).

Elicitation of Secondary Metabolism

When plant cells or tissues are subjected to environmental stresses, there is an enhanced production of some of the usual secondary metabolites produced in the plant tissues (phytoalexins). This phenomenon is known as elicitation and there are two major types of elicitors. Biotic elicitors include preparations from pathogenic bacteria and fungi, whereas abiotic elicitors include the salts of heavy metals. Abiotic treatments will not be discussed here.

Tissue-culture systems lend themselves well to the use of elicitors to enhance the production of secondary metabolites. In early research, live pathogens were used in a variety of co-culture techniques. However, since in most cultures there may be little, if any, defense against the invading organism, the pathogenic species often overgrow the explant. Cell wall preparations from the organism are equally effective and, accordingly, are now generally employed.

Hairy-root cultures of *Ambrosia chamissonis*, elicited with a cell wall preparation of *Phytophthora megasperma,* did not yield higher concentrations of overall thiarubrines, but the chemical profile of these polyynes was altered (Ellis, S.M.; Towers, G. H. N.; unpublished data). Cultures generally yield higher concentrations of thiarubrine A relative to the more polar thiarubrines, but, on elicitation, the production of the vicinal diol was increased to levels approximately the same as those of thiarubrine A. Fungal elicitors prepared from *Phytophthora* and *Pythium* spp. have been shown to enhance the production of polyynes such as trideca-1,3,11-ene-5,7,9-yne in hairy-root cultures of *Bidens sulphureus* (*44*). Other species such as *Carthamus tinctorius*, *Bidens pilosus* and *Tagetes* spp. also showed enhanced levels of polyynes on elicitation. Mukundan and Hjortso (*89,90*) found that all pathogenic and non-pathogenic fungal elicitors tested enhanced thiophene production in hairy-root cultures of *Tagetes patula*. Most researchers have only employed cell-wall preparations of known pathogenic fungi. Robbins *et al.* (*91*) examined the production of the isoflavans, vestitol and sativan, in HR cultures of *Lotus corniculatus* elicited with glutathione. The activity of phenylalanine ammonia lyase (PAL) was enhanced at least three-fold over the elicitation period. Vestitol production was enhanced, whereas sativan was produced at low levels. The isoflavans accumulated both in the medium as well as in the tissues. They examined the response of different ages of tissue to elicitation and found that there were differing responses depending on age and on dosage of the elicitor.

There is great potential for the use of elicitors, although they do not always enhance the production of the complete spectrum of secondary metabolites of a given plant species. For example, abiotic elicitors increase the production of sesquiterpenoids rather than tropane alkaloids in *Datura* spp. (*92*). This aspect of the elicitation phenomenon has been insufficiently explored and deserves further attention.

Concluding Remarks

Plant-tissue cultures, transformed by *Agrobacterium tumefaciens* and *A. rhizogenes*, offer an interesting approach to the study of the patterns of secondary metabolism in relation to differentiation. Unfortunately, it is not possible to predict the phytochemical outcome of the transformations. In some cases, desired phytochemicals are produced in transgenic callus, but not in transgenic cell suspensions. In other cases, metabolite levels are increased above those of the parent plants, while in still others decreased, often depending on the selection of inocula, manipulation of media or other conditions of growth. Much study is still needed in order to understand the regulation of secondary metabolism under stress conditions so as to sustain and maximize yields of most desired classes of secondary compounds.

Transgenic-root cultures, however, because of their rapid growth rate relative to "normal" root cultures, offer the best material for large-scale production of phytochemicals, providing, of course, that they are normally biosynthesized in roots. Growth regulators are not required for their proliferation, as is the case for non-transgenic cultures. Commercial production of rare pharmaceuticals from transgenic-root cultures is in our opinion, eminently feasible and worth pursuing.

Literature Cited

1. Sakui, N.; Kuroyanagi, M.; Ishitobi, Y.; Sato, M.; Ueno, A. *Phytochemistry* 1992, *31*, 143-147.
2. Orihara, Y.; Saiki, K.; Furuya, T. *Phytochemistry* 1991, *30*, 3989-3992.
3. Tabata, M.; Mizukami, H.; Hiraoka, N.; Konoshima, M. *Phytochemistry* 1974, *13*, 927-932.

4. Tabata, M.; Mizukami, H.; Hiraoka, N.; Konoshima, M. *Abstr. 12th Phytochem. Sympos.* Kyoto, Japan, 1976 pp 1-8.
5. Eilert, U.; Kurz, W. G. W.; Constabel, F. *J. Plant Physiol.* 1985, *119*, 65-67.
6. Eilert, U.; Constabel, F.; Kurz, W. G. W. *J. Plant Physiol.* 1986, *126*, 11-22.
7. Bayliss, M. W. In *International Review of Cytology*, Vasil, I.K. (Ed.); Supplement IIA; Academic Press: New York, 1980; pp 113-133.
8. Karp, A.; Bright, S. W. J. *Oxford Surv. Plant Mol. Cell Biol.* .1985, *2*:, 199-234.
9. Binarova P.; Dolezel, J. *J. Plant Physiol.* 1988, *133*, 561-566.
10. Jha, S.; Sahu, N. P.; Sen, J.; Jha, T. B.; Mahato, S. B. *Phytochemistry* 1991, *30,* 3999-4003.
11. Furuya, T.; Yoshikawa, T.; Ushiyama, K.; Oda, H. *Experimentia* 1986, *42*, 193-194.
12. Chilton, M. D.; Drummond, M. H.; Merlo, D. J.; Sciaky, D.; Montoya, A. L.; Gordon, M. P.; Nester, E. W. *Cell* 1977, *11*, 263-271.
13. Chilton, M. D.; Tepfer, D. A.; Petit, A.; David, C.; Casse-Delbart. F.; Tempe, J. *Nature* 1982, *295*, 432-434.
14. Horsch, R. B.; Fraley, R. T.; Rogers, S. G.; Sanders, P. R.; Lloyd, A.; Hoffman, N. *Science* 1984, *223*, 557-564.
15. Douglas, C.J.; Staneloni, R.J.; Rubin, R.A.; Nester, E.W. *J. Bacteriol.* 1985, *161*, 850-860.
16. Spencer, P. A; Towers, G. H. N. *Phytochemistry* 1988, *27,* 2781-2785.
17. Stachel, S. E.; Messens, E.; Van Montagu, M.; Zambryski, P. *Nature* 1986, *318,* 624-629.
18. Spencer, P. A.; Towers, G. H. N. *Phytochemistry* 1991, *30*, 2933-2937.
19. Song, Y.-N.; Shibuya, M.; Ebizuka, Y.; Sankawa, U. *Chem. Pharm. Bull.* 1990, *38*, 2063-2065.
20. Mathews, H.; Bharathan, N.; Litz, R. E.; Narayanan, K. R.; Rao, P. S.; Bhatia, C. R. *J. Plant Physiol.* 1990, *136*, 404-409.
21. Owens, L. D.; Smigocki, A. C. *Plant Physiol.* 1988, *87*, 570-573.
22. Gould, J.; Devey, M.; Hasegawa, O.; Ulian, E. C.; Peterson, G.; Smith, R. H. *Plant Physiol.* 1991, *95*, 426-434.
23. Ooms, G.; Hooykaas, P. J. J.; Moolenar, G.; Schilperoort, R. A. *Gene* 1981, *14*, 33-50.
24. Vilaine, F.; Casse-Delbart, F. *Molec. Gen. Genet.* 1987, *206*, 17-23.

25. White, F. F.; Taylor, B. H.; Huffman, G. A.; Gordon, M. P.; Nester, E. W. *J. Bacteriol.* 1985, *164*, 33-44.
26. Slightom, J. L.; Durand-Tardif, M.; Jouanin, L.; Tepfer, D. *J. Biol. Chem.* 1986, *261*, 108-121.
27. Prinsen, E.; Van Dijck, R.; Spena, A.; Schmuelling, T.; De Greel, J.; Van Onckelen, H. *Physiol. Plant.* 1990, *79 (2 Part 2),* A41.
28. Cardarelli, M.; Spano, L.; Mariotti, D.; Mauro, M. L.; Constantino, P. *Mol. Gen. Genet.* 1987, *208*, 457-463.
29. Luza, J.; Sutter, E. G. *HortScience* 1990, *25*, 1155.
30. Powell, G. K.; Hommes, N. G.; Kuo, J.; Castle, L. A.; Morris, R. O. *Molec. Plant-Microbe Interact.* 1988, *1*, 35-242.
31. Penza, R.; Lunguin. P. F.; Fillipone, E. *J. Plant Physiol.* 1991, *138*, 39-43.
32. Philips, R.; Darrell, N. J. *J. Plant Physiol.* 1988, *133*, 447-451.
33 Moldenhauer, D.; Furst, B.; Dietrich, B.; Luckner, M. *Planta Med.* 1990, *56,* 435-438.
34. Sheerman, S.; Bevan, M. W. *Plant Cell Rep.* 1988 *7*, 13-16.
35. De Cleene, M. *J. Speculations Science Technol.* 1980, *3*, 353-356.
36. Kado, D. I. *Ann. Rev. Phytopath.* 1976 *14*, 265-305.
37. Kovacs, B. A.; Wakkary, J.A.; Goodfriend, L.; Rose, B. *Science* 1964 *44*, 285-296.
38. Norton, R. A. Ph. D. Thesis, University of British Columbia, Vancouver: Canada, 1984
39. Norton, R. A.; Towers, G.H.N. *J. Plant Physiol.* 1985, *120*, 273-283.
40. Norton, R. A.; Finlayson, R.J.; Towers, G.H.N. *Phytochemistry* 1985 *24*, 719-722.
41. Ichihara, K. I.; Noda, M. *Biochem. Biophys. Acta* 1977 *487,* 269-237.
42. Jente, R. *Tetrahedron* 1971, *27*, 4077-4083.
43. Ketel, D. H. *J. Exp. Bot.* 1987, *38,* 322-330.
44. Flores, H. E.; Pickard, J. J.; Hoy, M. W. In *Chemistry and Biology of Naturally Occurring Acetylenes and Related Compounds (NOARC)*; Lam, J.; Breteler, H.; Arnason, T.; Hansen, L. Eds. Elsevier: Amsterdam, 1988; pp 233-254.
45. Cosio, E. G.; Norton, R. A.; Towers, E.; Finlayson, A. J.; Rodriguez, E.; Towers, G. H. N. *J. Plant Physiol.* 1986, *124*, 155-164.
46. Lindsey, K.; Yeoman, M. M. *J. Exptl. Bot.* 1983, *34*, 1055-1065.
47. Radwan, S. S.; Kokate, C. K. *Planta* 1980, *147*, 340-344.
48. Skinner, S. E.; Walton, N. J.; Robins, R. J.; Rhodes, M. J. C. *Phytochemistry* 1987, *26*, 721-725.

49. Hartmann, T.; Ehmke, A.; Eilert, U.; von Borstel, K.; Theuring, C. *Planta* 1989, *177*, 98-107.
50. White, P. *Plant Physiol.* 1934, *9*, 585-600.
51. Mugnier, J. *Plant Cell Rep.* 1988, *7*, 9-12.
52. Norton, R. A.; Towers, G. H. N. *J. Plant Physiol.* 1986, *122*, 41-53.
53. Constabel, C. P.; Towers, G. H. N. *J. Plant Physiol.* 1988, *133*, 67-72.
54. Constabel, C. P.; Towers, G. H. N. *Phytochemistry* 1990, *28*, 93-95.
55. Mukundan, U.; Hjortso, M. *J. Exptl. Bot.* 1990, *41*, 1497-1501.
56. Parodi, F. L.; Fischer, N. H.; Flores, H. E. *J. Nat. Prod.* 1988, *51*, 594-595.
57. Croes, A. F.; van den Berg, A. J. R.; Bosveld, M.; Breteler, H.; Wullems, G. J. *Planta* 1989, *179*, 43-50.
58. Norton, R. A.; Finlayson, A. J.; Towers, G. H. N. *Phytochemistry* 1985, *24*, 356-357.
59. Ishimaru, K.; Yonemitsu, H.; Shimomura, K. *Phytochemistry* 1991, *30*, 2255-2257.
60. Hanisch ten Cate, C. H.; Ramulu, K. S.; Dikj, P.; de Groot, B. *Plant Sci.* 1987, *49*, 217-222.
61. Hamill, J. D.; Rhodes, M. J. C. *J. Plant Physiol.* 1988, *133*, 506-509.
62. Kamada, H.; Okamura, N.; Satake, M.; Harada, H.; Shimomura, K. *Plant Cell Rep.* 1986, *5*, 239-242.
63. Flores, H. E.; Filner, P. In *Primary and Secondary Metabolism of Plant Cell Cultures*; Newman, K.H.; Barz, W.; Reinhard, E.; Eds. Springer-Verlag, Heidelberg 1986, pp 174-186.
64. Payne, J.; Hamill, J. D.; Robins, R. J.; Rhodes, M. J. C., *Planta Med.* 1987, *53*, 474-478.
65. Yamada, Y.; Hashimoto, T. In *Applications of Plant Cell and Tissue Culture*; Bock, G.; Marsh, J.; Eds. 1988, pp 199-212.
66. Jaziri, M.; Legros, M.; Homes, J.; Vanhaelen, M. *Phytochemistry* 1988, *27*, 419-420.
67. Jung, G.; Tepfer, D. *Plant Sci.* 1987, *50,* 145-151.
68. Sauerwein, M.; Shimomura, K. *Phytochemistry* 1991, *30,* :3277-3280.
69. Lloyd, G. B.; McCown, B. H. *Internat. Plant Propag. Soc.* 1980 30, 421-427.
70. Shimomura, K.; Sauerwein, M.; Ishimaru. K. *Phytochemistry* 1991, *30*, 2275-2278.
71. Robins, R. J.; Hamill, J. D.; Parr, A. J.; Smith, K.; Walton, N. J.; Rhodes, M. J. C. *Plant Cell Rep.* 1987, *6,* 122-126.

72. Toivonen, L. J.; Balsevich, J.; Kurz, W. G. W. *Plant Cell, Tissue Organ Cult.* 1989, *18*, 79-84.
73. Davioud, E.; Kan, C.; Quirion, J.-C.; Das, B. C. *Phytochemistry* 1989, *28*, 1383-1387.
74. Parr, A. J.; Peerless, A. C. J.; Hamill, J. D.; Walton, N. J.; Robins, R. J.; Rhodes, M. J. C. *Plant Cell Rep* 1988, *7*:309-312.
75. Sauerwein, M.; Ishimaru, K.; Shimomura, K. *Phytochemistry* 1991, *30*, 2977-2978.
76. Reichling, J.; Thron, U. *Planta Med.* 1990, *56*:488-490.
77. Fujita, Y.; Maeda, Y.; Suga, C.; Morimoto, T. *Plant Cell Rep.* 1983, *2*:192-193.
78. Shimomura, K.; Sudo, H.; Saga, H.; Kamada, H. *Plant Cell Rep.* 1991, *10*:2, 82-285.
79. Sato, K.; Yamazaki, T.; Okuyama, E.; Yoshihira, K.; Shimomura, K. *Phytochemistry* 1991, *30*,1507-1509.
80. Furuya, T. In *The Somatic Cell Genetics of Plants. Vol. 5. Phytochemicals in plant cell cultures*; Constabel, F.; Vasil, I. K. Eds. Academic Press: New York 1988, pp 213-234.
81. Yoshikawa, T.; Furuya, T. *Plant Cell Rep.* 1987, *6*, 449-453.
82. Ionkova, I.; Alfermann, A. W. *Planta Med.* 1990, *56,* 634-635.
83. Corchete, M. P.; Jimenez, M. A.; Moran, M.; Cacho, M.; Fernandez-Tarrago, J. *Plant Cell Rep.* 1991, *10*, 394-396.
84. Trypsteen, M.; Van Lijsebettens, M.; Van Severen, R.; Van Monagu, M. *Plant Cell Rep.* 1991, *10*, 85-89.
85. Hamill, J. D.; Parr, A. J.; Robins, R. J.; Rhodes, M. J. C. *Plant Cell Rep* 1986, *5,* 111-114.
86. Hilton, M. G.; Rhodes, M. J. C. *Appl. Microbiol. Biotech.* 1990, *33*, 132-138.
87 Toivonen, L.; Ojala, M,; Kauppinen, V. *Biotech. Lett.* 1990, *12*, 519-524.
88. Kumar, V.; Jones, B.; Davey, M. R. *Plant Cell Rep.* 1991, *10*, 135-138.
89. Mukundan, U.; Hjortso, M. A. *Appl. Microbiol. Biotechnol.* 1990, *33*, 45-147.
90. Mukundan, U.; Hjortso, M. *Biotechnol. Lett.* 1990, *12*, 609-614.
91. Robbins, M.P.; Hartnoll, J.; Morris, P. *Plant Cell Rep.* 1991, *10,* 59-62.
92. Furze, J. M.; Rhodes, M. J. C.; Parr, A. J.; Robins, R. J.; Whitehead, I. M.; Threlfall, D. R. *Plant Cell Rep.* 1991, *10*, 111-114.

RECEIVED March 5, 1993

Anticancer and Cancer Chemopreventive Agents from Plants

Chapter 7

Role of Plants in the National Cancer Institute Drug Discovery and Development Program

Gordon M. Cragg, Michael R. Boyd, John H. Cardellina II, Michael R. Grever, Saul A. Schepartz, Kenneth M. Snader, and Matthew Suffness†

Developmental Therapeutics Program, Division of Cancer Treatment, National Cancer Institute, Bethesda, MD 20892

Over the past 30 years the National Cancer Institute (NCI) has developed a number of plant-derived drugs which have undergone clinical trials. Prior to 1986, plant collections were generally restricted to temperate areas of the world. Since then, collections have focused on tropical, primarily rainforest, regions, and to date over 28,000 plant samples have been collected from over 20 countries. Policies have been formulated aimed at establishing collaborations with scientists in these countries and providing long-term compensation to source countries for drugs which are developed as marketable products. The various facets of the NCI drug discovery and development program are discussed, including future plans for the large-scale production of biomass for the isolation of promising new anticancer and antiviral natural products.

The United States National Cancer Institute (NCI) was established in 1937, with its mission being "to provide for, foster, and aid in coordinating research related to cancer." During the early 1950s, there emerged a growing interest amongst the scientific and lay community for the need to develop a major program aimed at screening new agents for potential chemotherapeutic activity, and, in 1955, the NCI set up the Cancer Chemotherapy National Service Center (CCNSC) to coordinate a national voluntary cooperative cancer chemotherapy program. The original objectives involved the procurement of compounds, their screening, and preclinical studies and clinical evaluation of active agents. By 1958, however, the initial service nature had evolved into a large drug research and development operation with input from academic sources and massive participation by the pharmaceutical industry. The responsibility for drug discovery and preclinical development now rests with the Developmental Therapeutics Program (DTP), a major component of the Division of Cancer Treatment (DCT).

†Deceased

Thus, for the past 38 years, the NCI has provided a resource for the preclinical screening of compounds and materials submitted by the research community worldwide, and has played a major role in the discovery and development of many of the commercial and investigational anticancer agents (*1*). With the emergence of acquired immunodeficiency syndrome (AIDS) as a global epidemic in the 1980s, the DTP initiated a major new program within the NCI for the discovery and preclinical development of anti-HIV agents (*2*). A high-flux screen accommodating over 40,000 samples per year was developed, and since 1988, the DTP has provided a national resource which permits scientists from academic and industrial organizations worldwide to submit compounds for anti-HIV testing (*3*).

Plant Acquisition Program: 1960-1982

During the early years of the CCNSC, the screening of natural products was concerned mainly with the testing of fermentation products, and, prior to 1960, only about 1,500 plant extracts were screened for antitumor activity. Earlier work on the isolation of podophyllotoxin and other lignans exhibiting *in vivo* activity against the murine sarcoma 37 model from *Podophyllum peltatum* L. (*4*, *5*), and the discovery and development of vinblastine and vincristine from *Catharanthus roseus* (L.) G. Don (*6,7*), however, provided convincing evidence that plants could be sources of a variety of novel potential chemotherapeutic agents. A decision was made to explore plants more extensively as sources of agents with antitumor activity, and, in 1960, an interagency agreement was established with the United States Department of Agriculture (USDA) for the collection of plants for screening in the CCNSC program. Initially, collections were made in the United States and Mexico, but these were later expanded to about 60 countries through both field collections by USDA personnel and procurements from contract suppliers. Between 1960 and the termination of this collection program in 1982, about 114,000 extracts from an estimated 35,000 plant samples were screened against a range of tumor systems used as a primary screen, principally the L1210 and P388 mouse leukemias. The collection strategy, scope and achievements of this program have been discussed in earlier reports, and readers are referred to in-depth reviews by Perdue (*8*) and Suffness and Douros (*9*) for details.

Current Status of Plant-Derived Anticancer Agents

Currently, four plant-derived anticancer drugs are in regular clinical use. Even though none of these was discovered in the NCI program, the NCI played a substantial role in their development as clinical agents. The best known of these agents are the so-called Vinca alkaloids, vinblastine and vincristine (Figure 1), isolated from the Madagascan periwinkle, *Catharanthus roseus*. These alkaloids first became available in the 1960s, and are now used extensively, generally in combination with other agents, in the treatment of a wide variety of cancers (*10*). Long-term, disease-free survivals have been observed in the treatment of various lymphomas and leukemias, bladder cancer and testicular cancer, while significant palliative benefits have been seen in patients with breast cancer, melanoma, and small-cell lung cancer. The two other agents in regular clinical use are the lignan derivatives, etoposide (VP-26) and teniposide (VM-26) (Figure 2). These are semi-synthetic derivatives of epipodophyllotoxin, an epimer of podophyllotoxin isolated from *Podophyllum peltatum* or *P. emodii* Wallich. Etoposide shows activity against small-cell lung and testicular cancers, as well as lymphomas and leukemias (*11*), while teniposide is active against acute lymphocytic leukemia and neuroblastoma in children, and non-Hodgkin's lymphomas and brain tumors in adults (*12*).

With the inception of the CCNSC program in 1955, a number of *in vivo* tumor models were examined as potential screens (*1,13*), but major reliance was placed on the murine leukemias, L1210 and P388. From 1956-1975, the L1210 model served as the primary screen, but, in 1975, it was replaced by the more sensitive P388 model. After 1975, agents exhibiting significant activity against the P388 were tested further against a secondary tumor panel comprising four to eight other models, including animal tumors and some human tumor xenografts (*13*). Those agents showing significant, broad-spectrum activity in the secondary panel were generally given the highest priority for preclinical development and eventual advancement to clinical trials.

Bioassay-guided fractionation of active extracts led to the isolation and characterization of a large number of active agents belonging to a wide variety of chemical classes. The sources and structures of the agents have been reported in earlier reviews (*9,14*), and the present discussion will be limited to those agents which advanced to clinical development.

The most significant discovery to emerge from the NCI screening program to date has been the diterpenoid, taxol (Figure 3), isolated from the bark of *Taxus brevifolia* Nutt. and the needles of various *Taxus* species. Although its isolation and characterization were reported in 1971 (*15*), it was only the observation of significant activity against the B-16 melanoma system and several human tumor xenograft systems that led to its selection for preclinical development in 1977. The report of its unique mechanism of action in promoting tubulin polymerization and stabilizing microtubules against depolymerization (*16*) heightened interest, and Phase I trials were started in 1983. Taxol gained prominence in 1988 with the observation of significant activity in the treatment of patients with refractory ovarian cancer (*17*). Good clinical activity has also been observed against advanced breast cancer (*18*), and clinical trials are ongoing against a variety of other cancer disease-types, including small-cell lung and non-small-cell lung, colon and head and neck cancers. A detailed discussion of various aspects of taxol is presented in the chapter by Kingston in this volume.

Another product of significance in the cancer chemotherapy program is camptothecin (Figure 4), an alkaloid isolated from the Chinese ornamental tree, *Camptotheca acuminata* Decne. Clinical trials in China, using a fine suspension of microparticles, have reportedly shown responses in the treatment of patients with liver, gastric, head and neck, and bladder cancers, but clinical trials in the United States in the early 1970s, using the more soluble sodium salt, were terminated due to severe toxicity, despite observations of some responses in patients with gastrointestinal tumors (*19*). Several more soluble semi-synthetic derivatives of camptothecin having superior

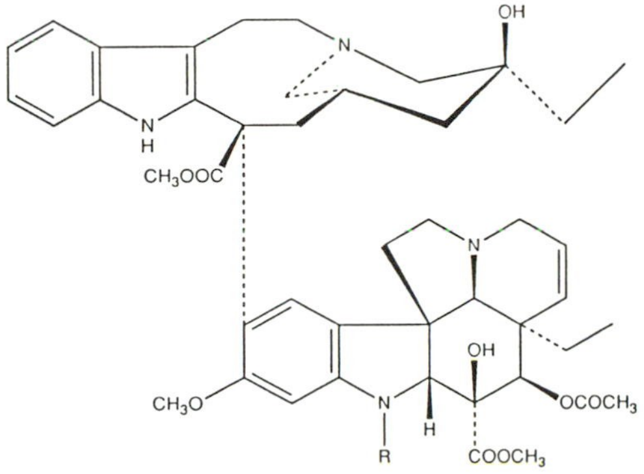

Vincristine : R = CHO

Vinblastine: R = CH_3

Figure 1 Structures of vincristine and vinblastine

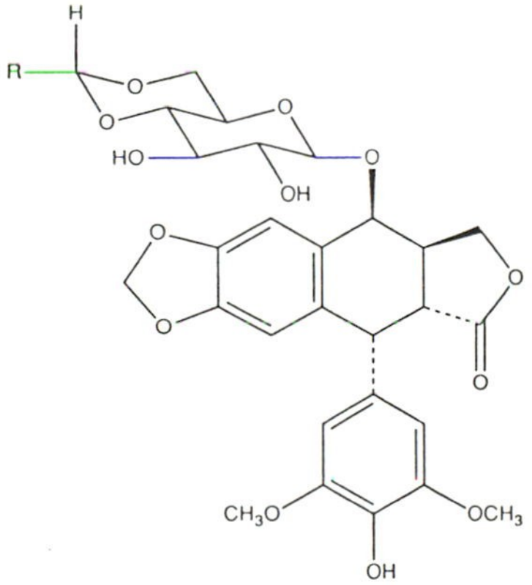

Etoposide : R = CH_3

Teniposide : R =

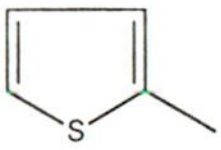

Figure 2 Structures of etoposide and teniposide

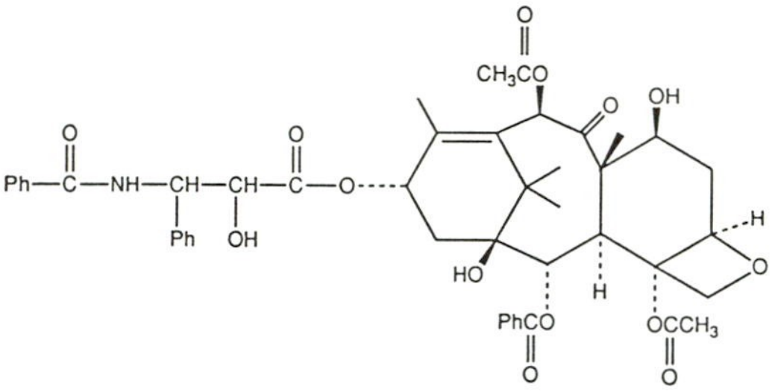

Taxol (Ph = Phenyl)

Figure 3 Structure of taxol

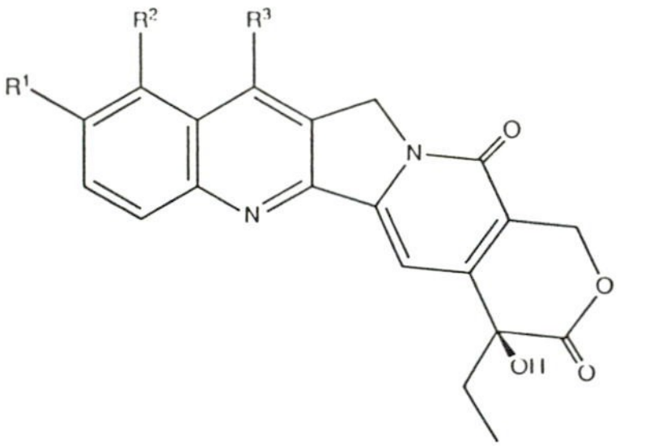

Camptothecin : $R^1 = R^2 = R^3 = H$

CPT–11: R^1 =

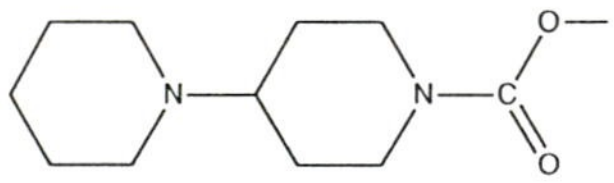

$R^2 = H$; $R^3 = CH_3CH_2$

Topotecan: $R^1 = OH$; $R^2 = CH_2N(CH_3)_2$; $R^3 = H$

Figure 4 Structures of camptothecin, CTP-11, and topotecan

activity and reduced toxicity in preclinical animal systems, have recently entered clinical trials. One such derivative, CPT-11 (Figure 4), has shown activity against refractory leukemia and lymphoma (*20*), while another derivative, topotecan (Figure 4), is showing activity against ovarian and various lung cancers (*21*). 9-Aminocamptothecin (Figure 4) is scheduled to enter Phase I clinical trials in the near future (*22*). Various aspects of these and other derivatives of camptothecin are discussed in the chapter by Wall and Wani in this volume.

Other plant-derived drugs currently undergoing clinical trials include homoharringtonine (Figure 5), an alkaloid isolated from the small Chinese evergreen tree, *Cephalotaxus harringtonia* var. *drupacea* (Sieb. & Zucc.) Koidzumi (*23*), and phyllanthoside (Figure 6), a terpene glycoside isolated from the Central American tree, *Phyllanthus acuminatus* Vahl (*24*). Homoharringtonine has shown activity against various leukemias, while phyllanthoside is in early clinical trials in the United Kingdom. 9-Hydroxy-2-methylellipticinium acetate (elliptinium; Figure 7), a semi-synthetic derivative of the alkaloid, ellipticine, is undergoing clinical trials in Europe, where activity has been reported against thyroid and renal cancers, and in the treatment of bone metastases resulting from advanced breast cancer (*25*). Ellipticine is isolated from species of the Apocynaceae family, including *Bleekeria vitensis A.C.Sm.*, *Aspidosperma subicanum Mart.*, and *Ochrosia* species (*26*). 4-Ipomeanol (Figure 8) is a pneumotoxic furan derivative produced by sweet potatoes (*Ipomoea batatas*) infected with the fungus, *Fusarium solani* (*27*). Ipomeanol, which has been shown to be activated by metabolism in lung cells of Clara origin and to exert selective cytotoxicity to human lung cancer cell lines (*28*), is in early clinical trials for the treatment of patients with lung cancer (*29*).

A number of agents, listed in Table I, were entered into clinical trials by the NCI, but the trials were terminated due to lack of efficacy or unacceptable toxicity. The protocols for the administration of agents to patients have evolved over the years from routine bolus injections to infusions of varying time-spans, dependent on the toxicity and bioavailability of the agents. It is interesting to speculate whether some of the agents listed in Table I might not prove to be more effective if administered as slow infusions according to more recent protocols.

Table I.
Plant-Derived Anticancer Agents for Which Clinical Trials Have Been Terminated

Agent	Source Plant	Reference
Acronycine	*Acronychia baueri* Schott	30
Bruceantin	*Brucea antidysenterica* J. F. Mill	31
Indicine *N*-Oxide	*Heliotropium indicum* L.	32
Lapachol	*Tabebuia* spp.	33
Maytansine	*Maytenus serrata* (Hochst. ex A. Rich.) R. Wilczek	34
Tetrandrine	*Stephania tetrandra* S. Moore	35
Thalicarpine	*Thalictrum dasycarpum* Fisch. et Lall	36
Tylocrebrine	*Tylophora crebriflora* S. T. Blake	37

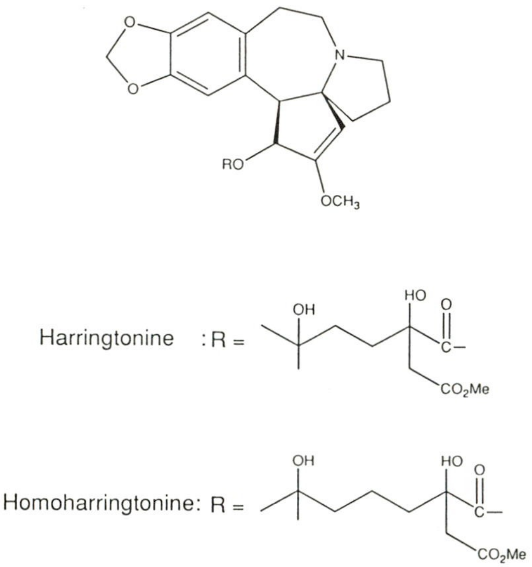

Harringtonine : R =

Homoharringtonine: R =

Figure 5 Structures of harringtonine and homoharringtonine

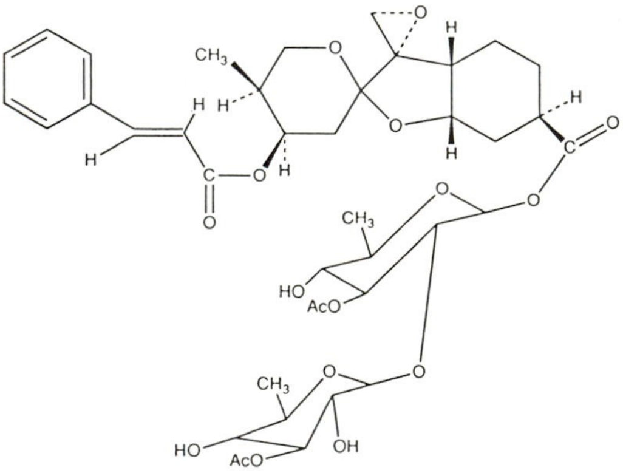

Phyllanthoside (Ac = CH_3CO)

Figure 6 Structure of phyllanthoside

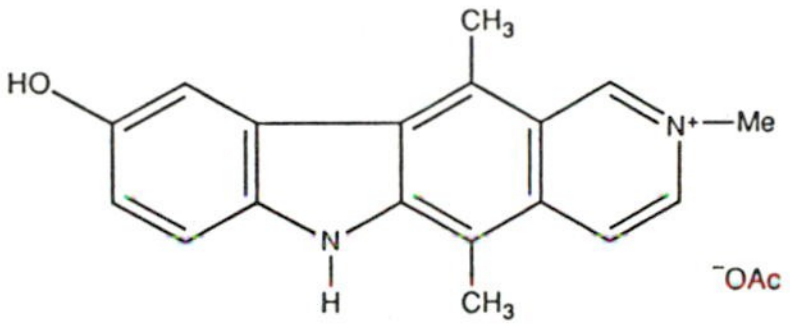

Figure 7 Structure of 9-hydroxy-2-methylellipticinium acetate

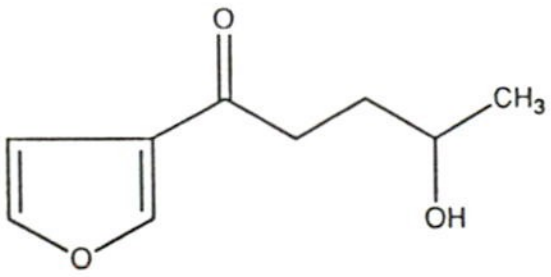

Figure 8 Structure of ipomeanol

Plant Acquisition Program: 1986 - Present

From 1960-1982, approximately 35,000 species of terrestrial plants were collected through the interagency agreement between the NCI and the USDA. As discussed earlier, a number of compounds discovered through this program advanced to clinical trials, and several of these, such as taxol and the camptothecin derivatives, possess great promise as novel chemotherapeutic agents. In the early 1980s, however, this collection program was discontinued since it was perceived that few novel active leads were being isolated from natural sources. This conclusion applied equally to plants, marine organisms and micro-organisms, and resulted in a general de-emphasis of natural products as potential sources of novel antitumor agents. Of particular concern was the failure to yield agents possessing activity against the resistant solid tumor disease-types. This apparent failure might, however, be attributed more to the nature of the primary screens being used at the time, rather than to a deficiency of Nature. Continued use of the primary P388 mouse leukemia screen appeared to be detecting only known active compounds or chemical structure-types having little or no activity against human solid tumors.

Suspicion of this potential weakness led, in 1985, to a revision of the antitumor screening strategy and development of a new *in vitro* human cancer cell line screen, incorporating panels of cell lines representing various solid tumor disease-types (*38*). This revision also was accompanied by the implementation of a new NCI natural products program involving new procurement, extraction and isolation components. The initiation, in 1987, of a major new program within the NCI for the discovery and development of anti-HIV agents (*2*) provided yet further impetus and resources for the revitalization of the NCI's focus upon natural products.

In September of 1986, three five-year contracts were awarded for the collection of plants in tropical and sub-tropical regions worldwide. Each contract called for the collection of 1,500 samples of 0.3-1.0 kg (dry weight) per year, with different plant parts (e.g., bark, roots, leaves, fruits, etc.) constituting discrete samples. It was specified that collections should encompass a broad taxonomic diversity, but that emphasis should be given to the collection of medicinal plants when reliable information on their use was available. Detailed documentation of each sample is required, including taxonomy, plant part, date and location of collection, habitat, hazards (e.g., thorny) and, when available, medicinal uses and methods of preparation used by the healer. Each sample is assigned a unique NCI collection number, expressed in the form of a barcode label, which is attached to the sample bag in the field. In addition, at least five voucher specimens of each plant species are prepared. One voucher is donated to the national herbarium in the country of collection, while another is deposited with the Botany Department of the Smithsonian Institution Museum of Natural History in Washington, D.C; this latter collection serves as the official national and NCI collection. The dried plant samples are shipped by air freight to the NCI Natural Products Repository (NPR) in Frederick, Maryland, where they are stored at -20°C for at least 48 hours in order to minimize the survival of plant pests and pathogens. This freezing period is a requirement of the import permit issued by the United States Department of Agriculture Animal and Plant Health Inspection Service, which has provided excellent support to the NCI in facilitating the importation of thousands of plant samples.

The initial contracts were awarded to the Missouri Botanical Garden, the New York Botanical Garden, and the University of Illinois at Chicago (assisted by the Arnold Arboretum at Harvard University and the Bishop Museum in Honolulu) for collections in Africa and Madagascar, Central and South America, and Southeast Asia, respectively. These contracts were subjected to competitive renewal in 1990, and five-year contracts were awarded to the same institutions in September, 1991. The logistics and problems associated with one of these collection programs are discussed in the chapter by Soejarto in this volume. As of December, 1991, 28,800 plant samples had been collected from over 20 countries.

Extraction and Screening of Plant Materials

Plant samples are sequentially extracted by room temperature percolation with a 1:1 mixture of methanol-dichloromethane and water to give organic solvent and aqueous extracts. Each extract is divided into five 100 milligram samples suitable for screening, while the remainder is kept as a bulk sample suitable for subsequent fractionation and isolation studies, if required. All the extract samples are assigned discrete NCI sample numbers, and are returned to the NPR for storage at -20°C until requested for screening or further investigation.

Extracts are tested *in vitro* for selective cytotoxicity against panels of human cancer cell lines representing major disease-types, including leukemia, lung, colon, central nervous system, melanoma, ovarian and renal cancers (*39*). *In vitro* anti-HIV activity is determined by measuring the survival rate of virus-infected human lymphoblastoid cells in the presence or absence of plant-extracts (*3*).

As of December, 1991, 16,650 samples had been extracted to yield 33,300 extracts. Over 14,000 of these extracts have been tested in the anti-HIV screen, and about 1,400 have exhibited some *in vitro* activity; of these, over 1,000 are aqueous extracts, and, in the majority of cases, the activity has been attributed to the presence of ubiquitous compound-types, such as polysaccharides or tannins. Such compounds are not considered to be viable candidates for drug development, and typically are eliminated early in the fractionation process. Several interesting *in vitro* active anti-HIV agents have been isolated. These include prostratin from the Samoan medicinal plant, *Homalanthus nutans* (Forster) Pax (*40*), the michellamines from a new species of the genus *Ancistrocladus* collected in Cameroon (*41*), and the calanolides from *Calophyllum lanigerum* miq. var. austrocoriaceum (T.C. Whitmore), collected in Sarawak (*42*). The isolation and structural elucidation of these compounds are reviewed in the chapter in this volume by Cardellina and colleagues. Of the close to 5,000 extracts tested in the *in vitro* human cancer cell line screen, about 2.7% have shown some degree of selective cytotoxicity; the majority of these are organic extracts. Interesting, novel patterns of differential cytotoxicity have been observed, some of which have been associated with known classes of compounds, such as cardenolides, cucurbitacins, lignans and quassinoids, while others appear to be new leads which are being investigated further. Progress in this area is reviewed by Cardellina and colleagues in their chapter in this volume.

Plant Drug Development and the Supply Issue

The initial plant collection sample (0.3-1.0 kg) will generally yield enough extract (10-40 g) to permit isolation of the pure, active constituent in sufficient (milligram) quantity for complete structural elucidation. Subsequent secondary testing and preclinical development, however, might require gram or even kilogram quantities, depending on the degree of activity and toxicity of the active agent (*1*).

In order to obtain sufficient quantities of an active agent for early preclinical development, recollections of 5-50 kg of dried plant material, preferably from the original collection location, might be necessary. Should the preclinical studies justify development of the agent towards clinical trials, considerably larger amounts of plant material would be required. The performance of large recollections necessitates surveys of the distribution and abundance of the plant, as well as determination of the variation of drug content in the various plant parts and the fluctuation of content with the time and season of harvesting. In addition, the potential for mass cultivation of the plant would need to be assessed. If problems are encountered due to scarcity of the wild plant or inability to adapt it to cultivation, a search for alternative sources would be necessary. Other species of the same genus, or closely related genera, can be analyzed for drug content, and techniques, such as plant tissue culture, can be investigated. While total synthesis must always be considered as a potential route for bulk production of the active agent, it is worth noting that the structures of most bioactive natural products are extremely complex, and bench-scale syntheses often are not readily adapted to large-scale economic production.

Agents exhibiting significant activity in primary and secondary screens are selected by the Decision Network Committee (DNC) of the NCI Division of Cancer Treatment for advanced preclinical and clinical development. The stages of development are defined in terms of Decision Network (DN) approval points. Approval at the DNIIA level signals initiation of studies directed at: (1) feasibility of various large-scale production methods; (2) development of suitable vehicles for *in vivo* studies; (3) pharmacological evaluation, involving the determination of half-lives and bioavailability, rates of clearance, excretion routes and metabolism; and (4) preliminary toxicological evaluation. Detailed toxicological studies directed towards submission of the Investigational New Drug Application (INDA) by the Food and Drug Administration (FDA) are performed after DNC approval at the DNIIB level, while DNIII approval clears the agent for INDA filing and entry into Phase I clinical trials.

Recent experience with the shortage of supplies of the promising anticancer agent, taxol, to meet a growing demand for the treatment of patients with ovarian, breast and other cancers (*43*), has highlighted the necessity for studying various methods of biomass production at an early stage of development of a new agent. To this end, the NCI is implementing a Master Agreement (MA) mechanism to promote the large-scale production of biomass for the isolation of promising new natural product agents. A Master Agreement mechanism involves establishing pools of qualified organizations with expertise in areas of particular interest to the NCI. In the case of production of biomass for the isolation of active plant-derived agents, areas of interest include large-scale recollection of source-plant materials, cultivation of the source plants, and source-plant tissue culture. The areas of cultivation and tissue culture are subdivided into two phases, one involving initiation of pilot-scale studies to explore the feasibility of the technique for production of the desired agent, and the second involving application of the method developed in phase one to large-scale production.

In establishing the pools of qualified organizations, a Master Agreement Announcement is issued by the NCI requesting proposals from organizations possessing expertise in the area of interest (e.g., plant tissue culture). The proposals, which elaborate on the organizations' experience and expertise in the area, are reviewed by a panel of external experts in the field, and those organizations considered to be technically qualified are awarded Master Agreements, and are placed in the pool of qualified organizations. When a plant-derived agent is approved by the DNC at the DNIIA level, Master Agreement Order (MAO) Requests for Proposals (RFPs) for a large-scale recollection of the source plant, and/or feasibility studies for the cultivation and/or tissue culture of the source plant, will be issued to the relevant pools of Master Agreement Holders. The MA Holders will submit technical and cost proposals addressing the particular RFP specifications, and an award will be made to the MA Holder whose proposal is considered best suited to the Government needs.

On DNC approval of the agent at the DNIII the level, MAO RFPs for large-scale production of the agent by the method (or methods) found to be most effective in phase one will be issued to the relevant pools of MA Holders. As before, technical and cost proposals will be submitted, and an award made to the MA Holder submitting the proposal best meeting Government requirements.

International Collaboration and Compensation

The recognition of the value of the natural resources (plant, marine and microbial) being investigated by the NCI, and the significant contributions being made by scientists and traditional healers in aiding the performance of the NCI collection programs, have led NCI to formulate policies aimed at facilitating collaboration with, and compensation of, countries participating in the NCI drug discovery program. Many of these countries are developing nations which have a real sensitivity to the possibility of exploitation of their natural resources by developed country organizations involved in drug discovery or other programs searching for natural bioactive agents (*44*).

The Letter of Intent formulated by the NCI contains both short-term and long-term measures aimed at assuring countries participating in NCI-funded collections of its intentions to deal with them in a fair and equitable manner (National Cancer Institute, unpublished). In the short term, the NCI periodically invites scientists from local organizations to visit the drug discovery facilities at the Frederick Cancer Research and Development Center (FCRDC) to discuss the goals of the program, and to explore the scope for collaboration in the drug discovery effort. When laboratory space and resources permit, suitably qualified scientists are invited to spend periods of up to 12 months working with scientists in NCI facilities on the bioassay-guided isolation and structure determination of active agents, preferably from organisms collected in their home countries. As of April, 1992, scientists from 17 countries have visited the FCRDC for periods of one to two weeks, while chemists and biologists from 10 countries have carried out, or are presently undertaking, collaborative research projects with scientists in NCI facilities.

As test data become available from the anticancer and anti-HIV screens, these are provided to the NCI collection contractors for dissemination to interested scientists in countries participating in the NCI collection programs. Each country receives only data obtained from extracts of organisms collected within their borders, and scientists are requested to keep data on active organisms confidential until the NCI has had sufficient time to assess the potential for the development of novel drugs from such organisms. Confidentiality is an important issue, since the NCI will apply for patents on agents showing particular promise; such patents may be licensed to pharmaceutical companies for development and eventual marketing of the drugs, and the NCI undertakes to make its best effort to ensure that a percentage of the royalties, accruing from sales of the drugs, will be paid by the licensee to the country of origin of the organism yielding the drug. Such compensation is regarded as a potential long-term benefit, since development of a drug to the stage of marketing can take from 10 to 20 years from its time of discovery.

Another potential benefit to the country of origin is the development of a large-scale cultivation program to supply sufficient raw material for bulk production of the drug. In licensing a patent on a new drug to a pharmaceutical company, the NCI will require the licensee to seek, as its first source of supply, the raw material produced in the country of origin. The policies concerning royalty payments and raw material supplies will also apply to inventions made by other organizations screening NCI extracts for activities against diseases related to the NCI mission. In addition, Master Agreement Holders investigating the large-scale production of biomass for the isolation of drugs of interest to NCI will be required to explore collaboration with the relevant source countries in the performance of their Master Agreement Order tasks.

The policies formulated in the Letter of Intent have received full support from all levels of the NCI and National Institutes of Health, and have also elicited support from members of the U.S. Congress and various non-governmental conservation organizations.

Conclusion

While a number of promising anticancer drugs, such as taxol and various camptothecin derivatives, have been developed as a result of the early NCI plant collection and screening program, the need to discover additional agents effective against resistant solid-tumor disease-types has given rise to new natural product acquisition and screening strategies at the NCI. The urgent need for drugs for the treatment of AIDS has intensified the search for novel natural product agents. Extensive NCI-funded plant collections in tropical and subtropical regions worldwide, and bioassay-guided fractionation of active extracts, have led to the discovery of several compounds showing promising *in vitro* anti-HIV activity, and these agents are now in various stages of preclinical development. Selective cytotoxicity against panels of human cancer cell lines has been observed in a number of plant-derived extracts, and these are now being further investigated by bioassay-guided fractionation. Through its Letter of Intent, the NCI is firmly committed to promoting international collaboration and fair compensation in the discovery and development of drugs from natural product sources in countries participating in its acquisition programs.

Literature Cited

1. Driscoll, J. S. *Cancer Treat. Rep.* **1984**, *68*, 63-76.
2. Boyd, M. R. In *AIDS: Etiology, Diagnosis, Treatment and Prevention*; DeVita, V. T.; Hellman, S.; Rosenberg, S. A., Eds.; J. B. Lippincott: Philadelphia, 1988; pp 305-319.
3. Weislow, O. S.; Kiser, R.; Fine, D. L.; Bader, J.; Shoemaker, R. H.; Boyd, M. R. *J. Natl. Cancer Inst.* **1989**, *81*, 577-586.
4. Hartwell, J. L.; Schrecker, A. W. In *Progress in the Chemistry of Organic Natural Products*; Zechmeister, L., Ed.; Springer-Verlag: Vienna, Austria, 1958, Vol. 15; pp 83-166.
5. Jardine, I. In *Anticancer Agents Based on Natural Product Models*; Cassady, J. M.; Douros, J. D., Eds.; Academic Press: New York, 1979; pp 319-351.
6. Johnson, I. S.; Wright, H. F.; Svoboda, G. H. *Proc. Am. Assoc. Cancer Res.* **1960**, *3*, 122.
7. Carter, S. K.; Livingston, R. B. *Cancer Treat. Rep.* **1976**, *60*, 1141-1156.
8. Perdue, Jr. R. E. *Cancer Treat. Rep.* **1976**, *60*, 987-998.
9. Suffness, M.; Douros, J. In *Methods in Cancer Research*; De Vita, V. T.; Bush, H., Eds.; Academic Press: New York, 1979, Vol. XVI; pp 73-125.
10. Neuss, N.; Neuss M. N. In *The Alkaloids. Antitumor Bisindole Alkaloids from Catharanthus Roseus (L.) G. Don*; Brossi, A.; Suffness, M., Eds.; Academic Press: New York, 1990, Vol. 37; pp 229-239.
11. O'Dwyer, P.; Leyland-Jones, B.; Alonso, M. T.; Marsoni, S.; Wittes, R. E. *New Engl. J. Med.* **1985**, *312*, 692-700.
12. O'Dwyer, P.; Alonso, M. T.; Leyland-Jones, B.; Marsoni, S. *Cancer Treat. Rep.* **1984**, *68*, 1455-1466.
13. Venditti, J. M.; Wesley, R. A.; Plowman, J. *Adv. Pharmacol. Chemother.* **1984**, *20*, 1-20.
14. Hartwell, J. L. *Cancer Treat. Rep.* **1976**, *60*, 1031-1067.
15. Wani, M. C.; Taylor, H. L.; Wall, M. E. *J. Amer. Chem Soc.* **1971**, *93*, 2325-2327.
16. Schiff, P.B.; Fant, J.; Horwitz, S. B. *Nature* **1979**, *277*, 665-667.
17. Rowinsky, E. K.; Cazenave, L. A.; Donehower R. C. *J. Natl. Cancer Inst.* **1990**, *82*, 1247-1259.

18. Holmes, F. A.; Walters, R. S.; Theriault, R. L.; Forman A. D.; Newton, L. K.; Raber, M. N.; Buzdar, A. U.; Frye, D. K.; Hortobagyi, G. N. *J. Natl. Cancer Inst.* **1991**, *83*, 1797-1805.
19. Suffness, M.; Cordell, G. A. In *The Alkaloids. Antitumor Alkaloids*; Brossi, A., Ed.; Academic Press: New York, Vol. 25; pp 73-89.
20. Ohno, R.; Okada, K.; Masaoka, T.; Kuramoto, A.; Arima, T.; Yoshida, Y.; Ariyoshi, H.; Ichimaru, M.; Sakai, Y.; Ogura, M.; Ito, Y.; Morishima, Y.; Yokomaku, S.; Ota, K. *J. Clin. Oncol.* **1990**, *8*, 1907-1912.
21. Kingsbury, W. D.; Boehm, J. C.; Jakas, D. R.; Holden, K. G.; Hecht, S. M.; Gallagher, G.; Caranfa, M. J.; McCabe, F. L.; Faucette, L. F.; Johnson, R. K.; Hertzberg, R. P. *J. Med. Chem.* **1991**, *34*, 98-107.
22. Wani, M. C.; Nicholas, A. W.; Wall, M. E. *J. Med. Chem.* **1986**, *29*, 2358-2362.
23. Suffness M.; Cordell, G. A. In *The Alkaloids. Antitumor Alkaloids*; Brossi, A., Ed.; Academic Press: New York, 1985, Vol. 25; pp 57-69.
24. Pettit, G. R.; Cragg, G. M.; Suffness, M.; Gust, D.; Boettner, F. E.; Williams, M.; Saenz-Renauld, J. A.; Brown, P.; Schmidt, J. M.; Ellis, P. D. *J. Org. Chem.* **1984**, *49*, 4258-4266.
25. Clarysse, A.; Brugarolas, A.; Siegenthaler, P.; Abele, R.; Cavalli, F.; De Jager, R.; Renard, G.; Rozencweig, M.; Hansen, H. H. *Eur. J. Clin. Oncol.* **1984**, *20*, 243-247.
26. Suffness, M; Cordell, G. A. In *The Alkaloids. Antitumor Alkaloids*; Brossi, A., Ed.; Academic Press: New York, 1985, Vol. 25; pp 89-142.
27. Boyd, M. R.; Burka, L. T.; Harris, T. M.; Wilson, B. J. *Biochem. Biophys. Acta* **1974**, *337*, 184-195.
28. McLemore, T. L.; Litterst, C. L.; Coudert, B. P.; Liu, M. C.; Hubbard, W. C.; Adelberg, S.; Czerwinski, M.; McMahon, N. A.; Eggleston, J. C.; Boyd, M. R. *J. Natl. Cancer Inst.* **1990**, *82*, 1420-1426.
29. Christian, M. C.; Wittes, R. E.; Leyland-Jones, B.; McLemore, T. L.; Smith, A. C.; Grieshaber, C. K.; Chabner, B. A.; Boyd, M. R. *J. Natl. Cancer Inst.* **1989**, *81*, 1133-1143.
30. Suffness, M.; Cordell, G. A. In *The Alkaloids. Antitumor Alkaloids*; Brossi, A., Ed.; Academic Press: New York, 1985, Vol. 25; pp 38-47.
31. Cragg, G.; Suffness, M. *Pharmacol. Therap.* **1988**, *37*, 425-426.
32. Cragg, G.; Suffness, M. *Pharmacol. Therap.* **1988**, *37*, 434-442.
33. Suffness, M.; Douros, J. D. In *Anticancer Agents Based on Natural Product Models*; Cassady, J. M.; Douros, J. D., Eds.; Academic Press: New York, 1979; p 474.
34. Suffness, M.; Cordell, G. A. In *The Alkaloids. Antitumor Alkaloids;* Brossi, A. Ed; Academic Press: New York, 1985, Vol. 25; pp. 146-155.
35. Suffness, M.; Cordell, G. A. In *The Alkaloids. Antitumor Alkaloids*; Brossi, A. Ed; Academic Press: New York, 1985, Vol. 25; pp. 164-171.
36. Suffness, M.; Cordell, G. A. In *The Alkaloids. Antitumor Alkaloids*; Brossi, A. Ed; Academic Press: New York, 1985, Vol. 25; pp. 171-178.

37. Suffness, M.; Cordell, G. A. In *The Alkaloids. Antitumor Alkaloids*; Brossi, A. Ed; Academic Press: New York, 1985, Vol. 25; pp. 156-163.
38. Boyd, M. R. In *Accomplishments in Oncology.* Cancer Therapy; Frei, E. J.; Freireich, E. J., Eds.; J. B. Lippincott: Philadelphia, 1986, Vol. 1; pp 68-76.
39. Boyd, M. R. *Principles and Practice of Oncology Updates* **1989**, *3*, 1-12.
40. Gustafson, K. R.; Cardellina II, J. H.; McMahon, J. B.; Gulakowski, R. J.; Ishitoya, J.; Szallasi, Z.; Lewin, N. E.; Blumberg, P. M.; Weislow, O. S.; Beutler, J. A.; Buckheit Jr., R. W.; Cragg, G. M.; Cox, P. A.; Bader, J. P.; Boyd, M. R. *J. Med. Chem.* **1992**, *35*, 1978-1986.
41. Manfredi, K. P.; Blunt, J. W.; Cardellina II, J. H.; McMahon, J. B.; Pannell, L. L.; Cragg, G. M.; Boyd, M. R. *J. Med. Chem.* **1991**, *34*, 3402-3405.
42. Kashman, Y.; Gustafson, K. R.; Fuller, R. W.; Cardellina II, J. H.; McMahon, J. B.; Currens, M. J.; Buckheit, R. W.; Hughes, S. H.; Cragg, G. M.; Boyd, M. R. *J. Med. Chem.* **1992**, *35*, 2735-2743.
43. Cragg, G. M.; Snader, K. M. *Cancer Cells* **1991**, *3*, 233-235.
44. *Anon. NAPRECA (Natural Products Research Network For Eastern and Central Africa, Addis Ababa University, Addis Ababa, Ethiopia) Newsletter* 1990, *7*, 2.

RECEIVED January 27, 1993

Chapter 8

Logistics and Politics in Plant Drug Discovery

The Other End of the Spectrum

Djaja Doel Soejarto

Program for Collaborative Research in the Pharmaceutical Sciences, College of Pharmacy, University of Illinois at Chicago, Chicago, IL 60612 and Department of Botany, Field Museum of Natural History, Chicago, IL 60605

Plant-drug discovery comprises a spectrum of activities, with the collection and identification of the plant materials to be tested on one end and the isolation and evaluation of biologically active compound(s) on the other. During the period of 1986-1991, a plant-collecting program was undertaken in the tropical rain forests of Southeast Asia as part of the National Cancer Institute's (NCI's) anticancer and anti-AIDS screening program. After five years of operation, the first cycle of the plant collection program was successfully completed, with the collection of more than 10,000 samples, comprising more than 2,500 angiosperm species, distributed in more than 200 families. In vitro anti-HIV test results provided by the NCI to date clearly demonstrate that logistics and politics have played an important role in this plant-drug discovery program.

Plant drug discovery and development comprise a spectrum of activities which start with field explorations to collect and identify the plant material to be investigated, followed by the extraction of the plant material, the testing of the extracts, the isolation of the active chemical constitutent(s), and the in-depth study of the purified compounds in preclinical and clinical evaluations. Whether it is an endeavor to discover drugs from plants of the temperate or tropical

0097–6156/93/0534–0096$06.00/0

forests, the field exploratory role is fundamental, since without the collected plant material, the remaining chain of events cannot be activated.

The logistical aspects of plant procurement in drug discovery has been discussed at length by Perdue (*1*), based on experiences from the U.S. National Cancer Institute's (NCI's) plant anticancer-screening program, during the period of 1960-1976. At that time, plant exploration and collection efforts were directed to a broad geographic area worldwide, though the major focus was on the temperate and subtropical regions of the world (*2*). In 1986, following a four-year hiatus (see chapter by Cragg et al. in this volume) the NCI's anticancer-terrestrial plant screening program was re-established (*3, 4*). This time, however, the plant exploration and collection efforts were targeted to the tropical rain forests of the world, which had received only token attention in the previous program.

Whether the goal is to collect plants from temperate or tropical rain forest regions, the basic logistical problems to be faced are the same. However, in tropical rain forest areas, the logistical challenges are compounded and include, among others, the high biodiversity of these forests, the inadequacy or lack of road communications in most forested areas, the various red tape issues and different cultures and language barriers that must be overcome in order to enter these forests, and the unique climatic conditions that present a challenge to the collection and drying of plant materials throughout major portion of the year.

It is the purpose of this chapter to examine some aspects of the plant-collection phase of the NCI's tropical rain forest plant drug discovery program, based on on experiences obtained in the Southeast Asian region during the past few years.

Plant Collecting Contracts

Since 1986, when the NCI re-established its terrestrial plant-screening program, the Program for Collaborative Research in the Pharmaceutical Sciences of the College of Pharmacy, University of Illinois at Chicago, has been under contract to explore the tropical rain forests of Southeast Asia to collect plants, originally to be screened against cancer cell lines, but later also against HIV, the causative agent of AIDS. The actual laboratory testing is performed at the NCI's Frederick Cancer Research and Development Center, Frederick, Maryland. Two other institutions have also been under contract, the New York Botanical Garden, for tropical American plants, and the Missouri Botanical Garden, for tropical African and Madagascaan plants. The goal of each of these contracts has been to collect a minimum of 1,500 different, vouchered and identified plant samples of high biological diversity per year (primarily flowering plants), each of 1/2-1 kg dry weight in quantity, as well as to recollect specific samples of interest that showed promiising activity in preliminary bioassay(s) in a somewhat larger quantity (10 kg dry weight). The University of Illinois at Chicago has carried out this mission with the support and collaboration of a number of U.S. botanical institutions. Among these are the Field Museum of

Natural History (Chicago), the Arnold Arboretum of Harvard University (Cambridge, Massachusetts) and the Botany Department of the Bishop Museum (Honolulu, Hawaii), as well as collaborating host institutions in Southeast Asian countries, in particular the Forest Herbarium of the Royal Forest Department (Bangkok, Thailand), the Department of Botany of the University of Malaya (Kuala Lumpur, Malaysia), the Heng-chun Tropical Botanical Garden (Kenting, Taiwan), the Philippine National Herbarium (Manila, Philippines), the Bogor Herbarium of the Biology Research Center (Bogor, Indonesia) and the Forest Research Institute (Lae, Papua New Guinea).

General Strategy

To undertake a plant procurement program of this magnitude, a number of factors must be considered. These include: (i) selection criteria, (ii) permits to conduct field research in and to export plant materials from the country of collection, and (iii) cooperation with a host institution(s) during the collecting operation.

The complete logistical aspects of the NCI-sponsored Southeast Asia plant collection program, in its broader context, is summarized in Figure 1.

Selection Criteria. An important consideration in any plant collecting operation in a drug discovery program is the selection of plant species to be collected for study. If the selection process is based on literature analysis, normally, a list of plants to be procured is used as a guide in the collection process. This approach is known as a selective-collecting program. If plants to be collected are selected randomly, without any specific criteria, this approach is called a random-collecting program. However, a truly random approach in which one blindly collects everything encountered, without giving consideration even to the possibility of duplication, is not practical. The approach used in most drug discovery programs falls between these two basic approaches. In the NCI-sponsored plant collection program, the ethnotaxonomic approach has been utilized. Plants to be collected are selected, first, with the primary goal of achieving the highest biodiversity of the samples as possible and, second, based on information of their medicinal uses. In order to achieve a high taxonomic diversity, a pre-collection documentation is necessary. This involves a literature survey and documentation on the floristic diversity of the area(s) targeted for collection, followed by a survey of the medicinal uses of plants from such area(s), as well as on-site field interviews with the populace or indigenous tribal groups on the uses of the plants for medicinal purposes. In the NCI plant-screening program, collection efforts have focussed primarily on the flowering plants, with field operations in a given region carried out in different months of the year.

Permits to Conduct Field Research in and to Export Plant Materials from the Country of Collection. All tropical rain forests of the world are located within

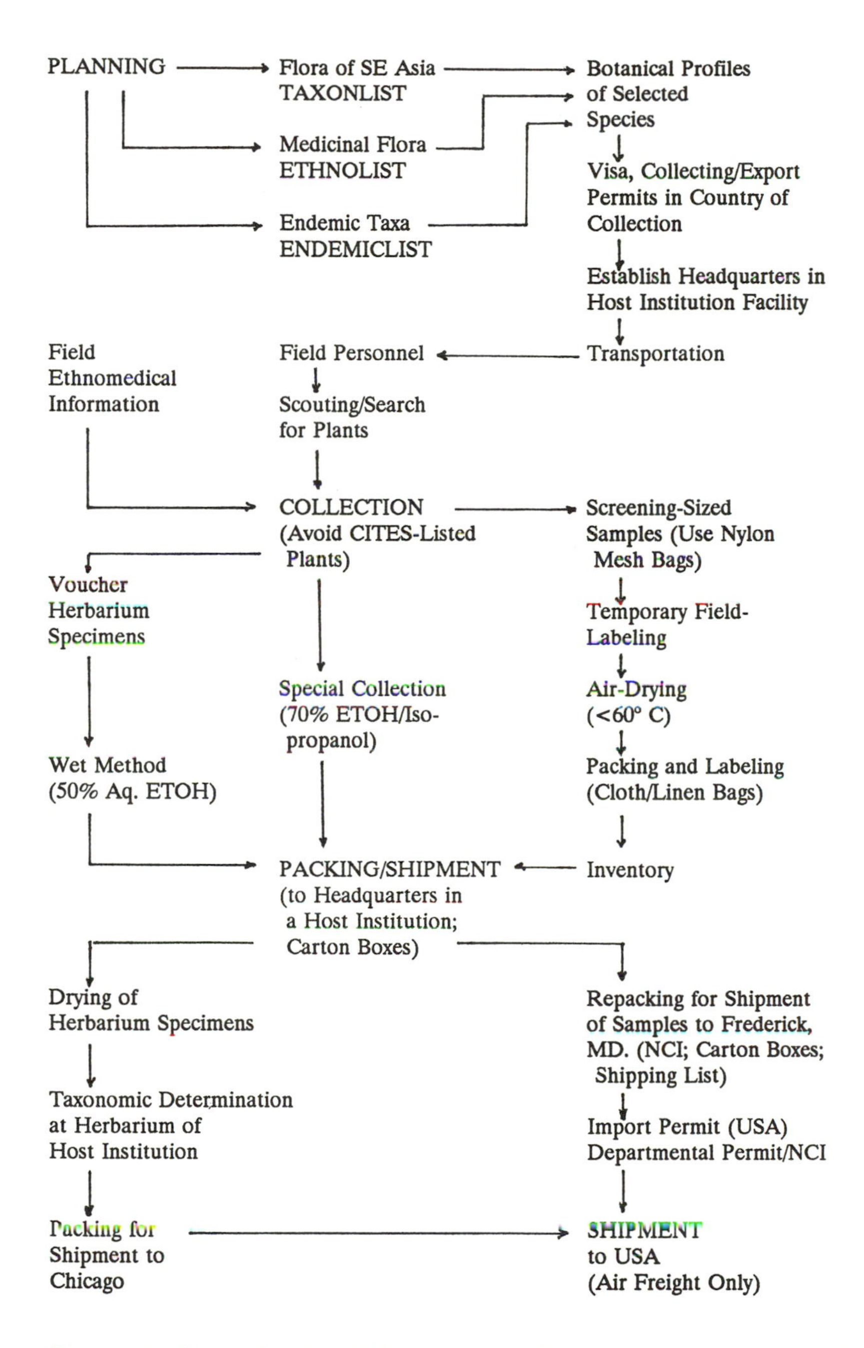

Figure 1. Steps in the NCI-sponsored Southeast Asia plant exploration/collection program.

the territory of different sovereign countries, in the tropical belt. Therefore, it must be recognized that biological materials (plant samples) are the property of the country in which they are found and that indigenous knowledge on the uses of plants (and other materials) in a particular country is the intellectual property of the people native to that country.

All countries, including those in the tropical belt, have rules and regulations for the collection and removal of biological materials from their territory. In order to explore the tropical rain forests and to collect and export plant materials from a particular country, permission must be secured from the respective governments, first, to enter the country legally, in the form of a visa, and second, to carry out field research in that country.

Cooperation with A Host Institution(s). Essential for the issuance of a research permit is the sponsorship of a collaborating host institution. Therefore, a close cooperation between the project leader and a host institution(s) greatly facilitates the undertaking of the field operation. In such a joint field work or plant-collecting expedition, the host institution provides a scientific counterpart to the U.S. project's personnel, plus technicians, while the project provides the financial support for the expedition.

In our collection program, the host collaborating institution has always been the primary headquarters for the field operation. It is in these institutions that all preparations for the field operation are finalized. It is also here that herbarium specimens collected are later dried, packed and shipped to the U.S.

Similarly, preliminary taxonomic determination is also carried out in the herbarium of each of the host botanical institutions or in a national herbarium in the country of collection. Overall, close cooperation with the host botanical institution has permitted the completion of all logistical and scientific formalities within the country of collection.

Logistical Aspects

Most collaborating host institutions are located in the capital city or near the capital city of the country of collection. In order to reach an expedition area, either airplanes or roadways (depending on the location and distance) would normally be used to transfer expedition technical personnel from the capital city to a town located near the collection site(s). In certain areas, passenger ships may be the only available mode of transportation.

Once the area of expedition is reached, local logistical problems include a series of activities directly related to the plant-collection process. First of all, this involves finding and setting up a field base. From such a base, collecting forays may be made into forest areas. Because of the nature of sample collecting (bulky material and voluminous loads) for bioassays and chemical analyses, the availability of such a semi-permanent base with access to transportation, either by road or waterway, is a primary consideration. In addition to the semi-permanent field base, in actual collecting sites, a temporary (2-5 day) field base may also need to be set up, in tents or huts, in the forests or at the edge of forests (Figure 2-a), where streams may be found. After a field base is established, the next step is finding a dependable mode of transportation that would provide mobility to the collecting party.

Reliable transportation is crucial, since forested areas are normally located at considerable distances from a field base. Transportation may be in the form of a motorized vehicle or boat, but one that should be big enough to transport expedition crew, expedition supplies, and the voluminous plant materials collected. When roads or waterways are not available, the use of porters to transport expedition supplies and plant materials collected may prove necessary.

Another basic component of a plant-collecting expedition that has to be organized is a labor force; together with the technical personnel, a crew of 5 to 12 or so, depending on local conditions, may have to be assembled. Such a crew will include the expedition leader and co-leader, herbarium technician(s) from the host institutions, tree climbers, general-purpose workers, and other helpers (local authority, guide, driver, cook). In addition, 5-10 porters to carry expedition supplies and plant materials collected may have to be employed temporarily, if vehicles cannot not penetrate deep into or near the collecting site(s). A large crew is essential if time is a constraint for the goal to be achieved [for example, to collect 200-400 samples within a period of one month).

Next comes the task of acquiring certain expedition supplies (twines, metal bar to dig roots, used newsprints to press voucher herbarium specimens, denatured alcohol to fix specimens, cutting tools, etc.) and provisions. Basic field equipment and supplies, such as a camera, altimeter, field binoculars, branch clippers, plastic bags, notebook, specimen tags, etc.) are normally brought from the headquarters.

One of the important considerations to be given in collecting plant samples for chemical and biological studies is to determine the type of drying method to be used in the field and to find a space to dry the plant materials collected. Drying either a large number of small-scale plant samples simultaneously (batches of 30-60 samples of approximately 1 kg dry weight each) or a large amount of only one type of plant sample (for example, 100 kg dry weight of leaves) is a real logistical challenge. The method widely used in the NCI-sponsored Southeast Asian plant collection program has been to use zippered, oversized nylon net or mesh bags of various colors (80 cm x 100 cm) (Figure 2-b) that would accomodate 1-2 kg fresh plant material in a loose manner, thus allowing the plant material being dried (in the sun) to be shaken occasionally for even drying (2-5 days, depending on the texture of the plant material). This method is clean, efficient and prevents possible sample contamination. These collecting/drying bags may be reused. Once plant samples are dry, they may then be transferred to special packing cloth bags (Figure 2-c).

At the collecting site, scouting (Figure 2-d) for plants to collect, and mobilizing the field crew to carry out individual tasks assigned (such as plucking herbaceous plants and shrubs, climbing trees, stripping tree bark, digging and cutting roots, chopping and bagging samples, numbering samples, etc.; Figures

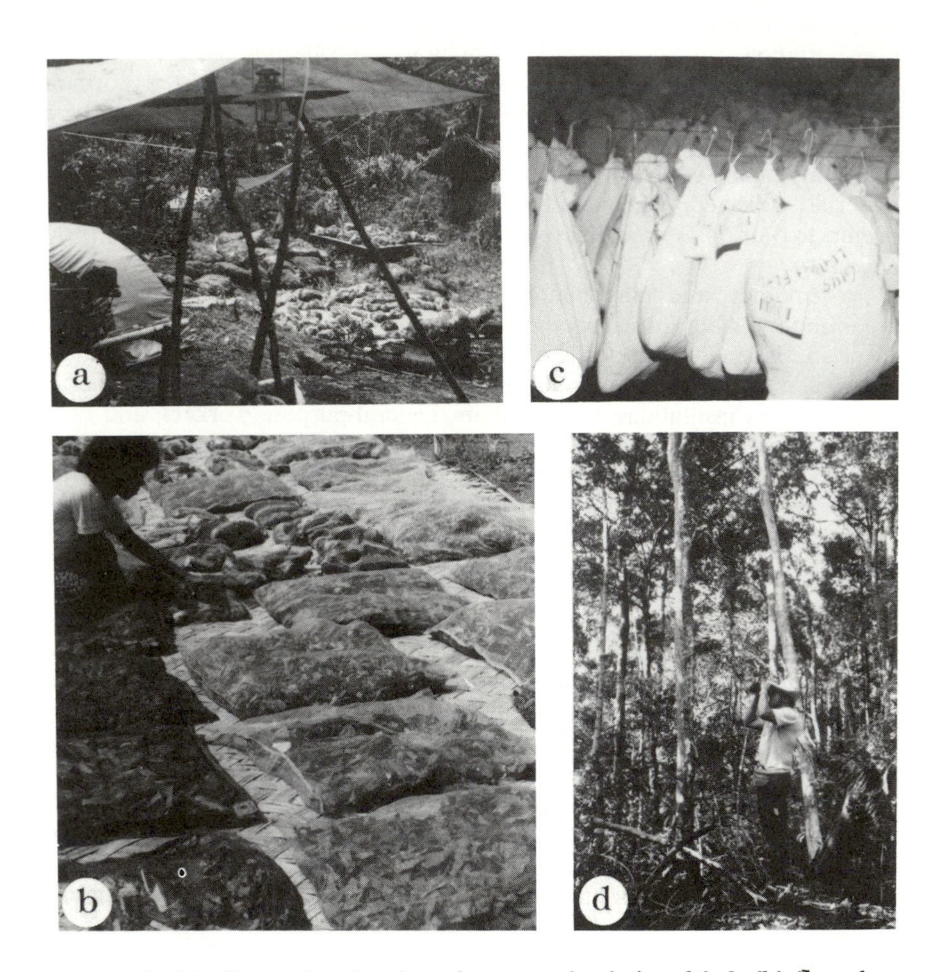

Figure 2. (a) Camp site showing plant samples being dried. (b) Sample drying method utilized in the Southeast Asian plant collection program, using oversized (80 cm x100 cm) nylon net or mesh bags; a clean and efficient method that prevents contamination of one sample by another. (c) Plant samples left hanging in a protected and well-ventilated location while the expedition is still ongoing. (d) The author stalking plants to collect in a Philippine forest. Palawan, Philippines. Photos by the author.

3-a, 3-b), are two routine, but important, activities. Nearing the end of an expedition, final labeling, packing, inventory (Figure 3-c) and shipping of the dried and packed plant material must be done. For shipment, dried plant samples are packed in large carton boxes for efficient use of space (cost is based on volume, not weight) and shipped by air cargo to the NCI's Frederick Cancer Research and Development Center in Frederick. Importation of samples is made through a special Departmental permit held by the NCI. In the shipping label, the following notice is written: "Keep Dry! Please Do Not Fumigate! Vacuum Sterilization Only!"

A final observation must be added. During the course of a field operation, it is the responsibility of the expedition leader to keep the crew happy, healthy and in good spirits, in order to ensure the success of the expedition. This means the crew must be well fed and prepared for any possible field hazards. Steps should be taken to prevent crew members from contracting infectious diseases (e.g., malaria, diarrhea), to anticipate accidents and minor casualties, to anticipate exposure to toxic or hazardous plants (especially, contact dermatitits, physical skin tears), and to anticipate encounters with poisonous animals (e.g., poisonous snakes, insect stings). In cases of political upheaval, attempts must be made to return rapidly to the headquarters and to avoid area(s) of high risk.

Documentation of Plants Collected

Post-collection documentation of the plants collected is basic and forms part of routine field activities. Most important of this aspect is the preparation of voucher herbarium specimens (Figure 3-d) according to standard botanical practice. In our operation, seven duplicate specimens are usually prepared, for distribution to collaborating botanical institutions, including one set for the host institution. Because of the large number of herbarium specimens collected, field drying is not considered; these specimens, therefore, are routinely fixed in denatured alcohol, by placing bundles of these specimens in extra newsprint in a strong (heavy duty) plastic bag in a vertical position and pouring an adequate quantity of aqueous alcohol (approximately 50:50 v/v) into the bag and through the specimens, until the material is well soaked. Such a bag may then be sealed and put into a second bag for added strength in transport and to lessen alcohol loss. These treated specimens will keep in good, fixed condition, without molding or decomposition, for a month or longer; bags must be opened, however, after arrival at the host collaborating institution and the specimens removed for drying.

At the time of collection of the herbarium specimens, field notes must be taken, including the characteristics of the plant, the geographic location of the collection site, the date of collection, and the habitat type. Field documentation also includes activities to mark trees by using numbered plastic ribbons or paint, a measure that would facilitate the relocation of a particular tree for recollection in case of positive test results; to perform field inquiries on the

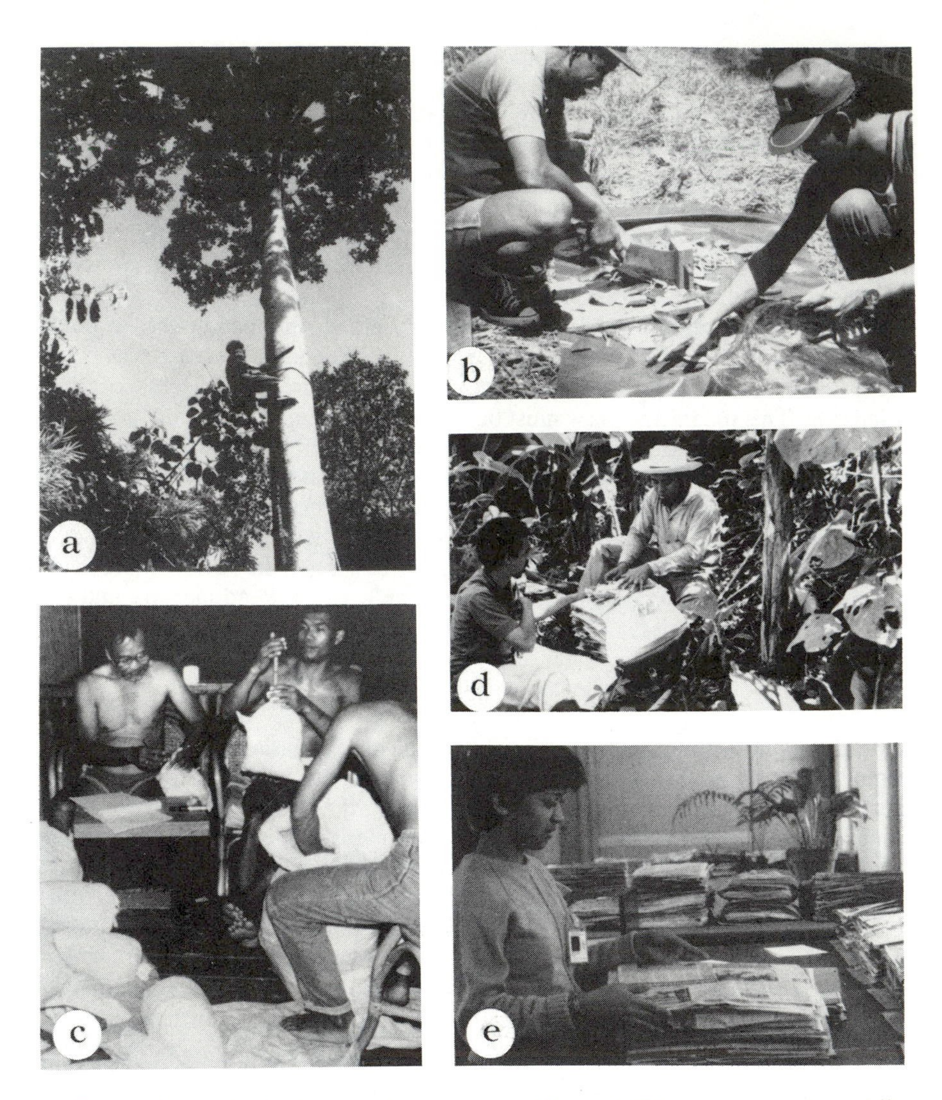

Figure 3. (a) A climber on the way up a 35-m tall Dipterocarpus hasseltii tree. (b) Plant samples are chopped to facilitate drying and packing. (c) Labeling, weighing, and packing dried samples for shipment to a host collaborating institution. (d) The author pressing voucher herbarium specimens. Palawan, Philippines. (e) Voucher herbarium specimens are labeled and distributed to collaborating herbarium institutions. Photos by the author.

medicinal uses of the plants collected and documenting such information; and to prepare photographic documentation.

Back at the herbarium of the host institution, post-collection documentation includes activities to dry herbarium specimens and to separate a set of duplicates of the voucher herbarium specimens to be deposited in the herbarium of the host collaborating institution (normally, a national herbarium), and preliminary taxonomic determinations. In our operation, the remaining six sets of herbarium specimens are shipped to the University of Illinois at Chicago, where they are then forwarded to the John G. Searle Herbarium, Field Museum of Natural History, Chicago, for further processing.

In Chicago, post-collection documentation comprises computer entry of field data, generation of printed field labels, distribution of duplicates of voucher herbarium specimens to various collaborating institutions (Figure 3-e), and, in some cases, taxonomic determination of the plants collected. Further determination is performed at the Arnold Arboretum of Harvard University and at the Rijksherbarium in Leiden (The Netherlands), as well as with the assistance of taxonomic specialists.

The conservation Challenge

The field work to collect small-scale samples for preliminary biological screening is only the first or initial phase of three exploratory stages in discovering new drugs from plants. When a sample is shown to be active in preliminary bioassays and is considered to be of interest for further development, a larger quantity of sample (10 kg or more, dry weight) will need to be recollected for further isolation and structure elucidation work, as well as for further bioassays. If the isolated compound turns out to be promising in preclinical studies, and synthetic efforts failed to provide the desired compound, an even larger quantity of plant material will be needed, probably up to quantities of several metric tons.

The logistical challenge in collecting such amounts of plant material lies in finding a large enough species population that will permit the collection of the required quantities, especially in the case of shrubs or herbaceous plants. Also, for rare tree species, the necessity of re-locating the original tree that gave the positive test result(s), to be used as a field reference, is another logistical challenge.

Another type of challenge to be faced is the performance of the collection work without endangering the environment and/or the species. Our plant collecting practices have been developed with the conservation challenge in mind; minimal disturbance to the forest habitat is given a high priority in the collection process. This is achieved primarily by using tree climbers to collect the aerial parts of a tree, by making a narrow longitudinal strip cut on one side of a tree in collecting the stembark, by tracing the root to a distance (2-3 m) from the tree in collecting the root, by leaving a collecting site in the same condition as it was found, by extinguishing all campfires completely, and by not harvesting all plants of a species from a single population.

Political Considerations

The major political consideration in a plant-collection program for drug discovery involves issues on the question of property rights and recompense to the people in the country of collection. In recent years, there has been an increasing awareness in many countries on issues of benefit return; namely, that some type of compensation should be provided in exchange for the collection of biological materials from a particular country. Such an awareness has been awakened, in part, by concerned scientists, who, through ethical considerations, feel justly that plant and animal resources of a country belong to the people of that country, and that indigenous knowledge on the medicinal uses of plants belongs to the people who originally discovered it, through eons of trial and error, and who are now providing such potentially valuable information, which may lead to the discovery and development of a new drug(s). Thus, a share of the monetary gain should be returned to the country of collection or to the people who gave the information responsible for the discovery. Proof of this concern is expressed by, among others, the Declaration of Belem, the Kunming Action Plan and the Hipolito Unanue Agreement (*5*), the Goteborg Resolution (*6*), the World Health Organization's Chiang Mai Declaration (*7*), the United States Agency of Ineternational Development-National Institute of Health-National Science Foundation joint Commentary (statement) (*8*), and articles such as "Folklore, Tradition, or Know-How" (*9*) and "No Hunting" (*10*).

Among the measures that have been implemented in the Southeast Asia plant- collecting program to respond to such issues are the training of technicians and junior scientists in field and herbarium techniques; the deposit of a set of duplicates of voucher herbarium specimens in the host (normally, national) herbarium institution; the provision of relevant test results for plants collected in a particular country to appropriate governmental agencies [primarily the host institution] in that country; the provision of curatorial support to host botanical institutions; and the sponsorship of travel for scientists from the country of collection to visit the NCI laboratories and other facilities of U.S. institutions that have connection with the NCI-sponsored program.

Other measures include assurance to the country of collection of a share of future financial benefit, should a drug derived from a plant collected in that particular country be developed and marketed, within the guidelines as provided by the NCI. A provision is also in force for a selected junior scientist (primarily natural products scientist) in the country of collection to undertake special training in the NCI's laboratories at Frederick or in other laboratories in the U.S. under the NCI's sponsorship.

Starting with the second cycle of the new plant collection program (1991-1996), the NCI, additionally, provides a Letter of Intent, if requested, which represents a legal document to strengthen any pledge to return share of possible future monetary benefits to the country of collection.

Collection Results (1986-1991)

Through the adoption of the measures which have been outlined above, our group completed the first plant exploration cycle (1986-1991) successfully. During this period, a total of more than 10,000 plant samples were collected and delivered to the NCI Frederick Cancer Research Center. This number of samples represents 2,500-3,000 species, mostly flowering plants, distributed in about 1,000 genera and 214 families.

Fifty-one percent of the samples collected consisted of leaves plus twigs, 21% of stems, 17% of reproductive parts (flowers, fruits, seeds), 9% of underground parts, while the remainder (2%) consists of entire plants. The low number of entire plants (herbaceous species) and underground parts sampled reflects the difficulty of obtaining these parts in an adequate quantity required for screening.

A large number of these plant samples have been tested at the NCI's laboratories against HIV. Although these *in vitro* results are considered preliminary,
an analysis of these data may provide an insight into the validity of the approach utilized in the collection program and may serve to point out the prospects for the discovery of potential drug candidates.

In this analysis (Table I), only species scoring "A" (for active) in the NCI screen were considered. Any "M" (for moderate), "W" (for weak), "G" (for glimmer) test results were considered inactive. A total of 3,003 different extracts (giving 170 "A" test results) were scored, using the following scoring system. When two or more different extracts belonging to the same species receive an "A" test result, only one count for the species was scored; when one of two or more different samples of a species receive an "A" test result (the others may be "M", "W", -, or others), one count for the species was scored; when none of the different samples belonging to a species has an "A" test result, no active count was scored. Only fully and definitively identified species were scored.

A total of 776 fully identified species (766 species of angiosperms, 4 gymnosperms, 5 pteridophytes and 1 lichen) were scored. Taking only the angiosperm species only into consideration, 106 species (of 766) (13.8%) were active, while 660 (86.2%) inactive. Of the 106 active angiosperm species, 62 had a history of medicinal use, while 44 did not. Of the 660 inactive species, 289 had a history of medicinal use, while 371 did not (Table I). Ethnomedical data were obtained from three sources: Perry and Metzger (*11*) (an extensive literature review published in 1980 on the medicinal uses of East and Southeast Asian plants), the NAPRALERT database (which summarizes world literature on ethnomedicine, phytochemistry and plant bioactivity; *12*) and field observations made during our plant collecting expeditions as recorded on the herbarium labels.

A chi-square test shows that the proportion of plants with medicinal uses that scored "A" is greater than would be expected if anti-HIV test activity and medicinal use had no interrelationship (chi-square 8.76, $p<0.02$). In other words, ethnomedical use can, to some extent, predict activity. Nevertheless, the large number of active species that have no history of medicinal use (44) is noteworthy.

The chemotaxonomic distribution of the medicinal *vs.* non-medicinal plant species does not appear to explain the difference in the "A" test results. The 106 species that gave an "A" test result belong to 68 different plant families, none of which is known for its high tannin content, such as the Fagaceae. (High tannin-containing plants often give positive anti-HIV test results.) No single family has an overwhelmingly high number of active species (Table II). The highest number of species scoring "A" in given families were Leguminosae *sensu latu*, 7 species, followed by Araliaceae, 6 species, Asteraceae, 5 species, Arecaceae, Cyperaceae, Verbenaceae, 4 species each, and Annonaceae, Rubiaceae, Urticaceae, Vitaceae, 3 species each. The greater number of the families scored have only one species showing an "A" test result.

Table I. Results of Anti-HIV Tests of Angiosperm Species Collected in Southeast Asia[a]

Test Results[b]	Species with a history of medicinal use	Species without a history of medicinal use	Subtotal
Active ("A")	62	44	106
Inactive ("M", "W", "G", etc.)	289	371	660
Subtotal	351	415	766

[a]A total of 766 fully and definitively identified species were included in these statistics; "cf.", "aff.", "sp. nov.", etc. taxonomic determinations were excluded.
[b]"A" = active; "M" = moderately active; "W" = weak activity; "G" = glimmer (very weak activity).

Summary and Conclusions

In its broader context, the role of field-work logistics consists of overcoming problems that relate to: (i) the implementation of an appropriate collection strategy (ii) the availability of dependable transport, (iii) the availability of local labor, (iv) the availability of adequate field supplies and equipment, (v) the availability of local facilities to dry plant material, and (vi) the completion of post-collection documentation. The role of politics, on the other hand, consists

Table II. Taxonomic distribution of "A" species[a]

Family	Number of "A" Species	Family	Number of "A" Species
Acanthaceae	1	Icacinaceae	1
Alangiaceae	1	Lamiaceae	2
Amaranthaceae	1	Lauraceae	1
Anacardiaceae	1	Lecythidaceae	1
Annonaceae	3	Limnocharitaceae	1
Apiaceae	1	Linaceae	1
Apocynaceae	2	Loganiaceae	1
Aquifoliaceae	1	Malvaceae	2
Araceae	1	Marantaceae	1
Araliaceae	6	Melastomaceae	1
Arecaceae	5	Meliaceae	2
Asteraceae	5	Moraceae	1
Balsaminaceae	1	Myrtaceae	1
Barringtoniaceae	1	Nyctaginaceae	1
Bignoniaceae	1	Olacaceae	1
Bombacaceae	1	Pandanaceae	1
Boraginaceae	1	Piperaceae	1
Burseraceae	2	Poaceae	1
Caesalpiniaceae	3	Rhizophoraceae	1
Caricaceae	1	Rubiaceae	3
Clusiaceae	1	Rutaceae	1
Commelinaceae	1	Sambucaceae	1
Convolvulaceae	1	Sapindaceae	2
Costaceae	1	Sapotaceae	1
Cucurbitaceae	1	Sterculiaceae	1
Cyperaceae	4	Stilaginaceae	1
Dipterocarpaceae	1	Symplocaceae	1
Ehretiaceae	2	Theaceae	1
Erythropalaceae	1	Thymelaeaceae	2
Euphorbiaceae	1	Urticaceae	3
Fabaceae	4	Verbenaceae	4
Flacourtiaceae	1	Vitaceae	3
Gesneriaceae	1		
Hydrocotylaceae	1		

[a]Family concept as used in ref. 13.

of overcoming problems that relate to: (i) the acquisition of appropriate permits to enter into, collect in and export plant materials from the country of collection, (ii) the establishment of a close working relationship and cooperation with a host-institution(s) in the country of collection, (iii) the implementation of appropriate measures of recompense, and (iv) addressing the issues and challenges of biological conservation. In the NCI-sponsored Southeast Asian plant-collection effort, logistics and politics have clearly played an important role in the successful completion of the first-cycle of the new plant-collection program. Such a role must be recognized as indispensable and as of fundamental importance in any plant drug discovery endeavor.

Based on the analysis of the data from the anti-HIV screening performed to date, ethnotaxonomic selection criteria appear to represent a valid plant collection approach. Although ethnomedical information did predict anti-HIV activity, the high number of species (44) without a history of medicinal use that tested active is also noteworthy.

The significance of this information lies in the fact that the depth of the ethnomedical reservoir for drug discovery is rather limited; only about 20,000 species of angiosperms have a history of medicinal use (*14*). On the other hand, the depth of taxonomic diversity is much greater: 250,000 species of angiosperms are known to occur on our planet; close to half of these are found in the tropical rain forests (*15, 16*).

These tropical rain forest species are sensitive to changes in their habitat through deforestation; they may also be rare, threatened, or in danger of extinction. In view of the fact that: (i) a relatively large proportion of such species showed active anti-HIV test results, from which a clinically useful drug(s) to treat AIDS may eventually be derived, and (ii) that a great number of these species may become extinct by the year 2000 (*17*), it is imperative that efforts to study whatever species currently remain in these forests be given very high priority. With the loss of these species through human-induced extinction, gone also will be, forever, the prospect of finding any new drug(s) to treat AIDS from tropical rain forest species.

Acknowledgments

Plant explorations in Southeast Asia were sponsored by the United States National Cancer Institute under Contract NO1-CM-67925. Anti-HIV test results used for data analysis were provided by the NCI and are gratefully acknowledged. The cooperation of all scientists and individuals, who have performed and assisted in the plant collection work in Southeast Asia for the project, is gratefully acknowledged. Special thanks are due to Dr. Charlotte Gyllenhaal of the Program for Collaborative Research in the Pharmaceutical Sciences, who kindly performed the data analysis for anti-HIV tests.

Literature Cited

1. Perdue, Jr., R.E. *Cancer Treat. Rep.* **1976**, *68*, 987-998.

2. Spjut, R.W. *Econ. Bot.* **1985**, *39*, 266-288.
3. Booth, W. *Science (Washington, D.C.)* **1987**, *237*, 969-970.
4. Suffness, M. In *Biologically Active Natural Products*; Hostettmann, K; Lea, P.J., Eds.; Clarendon Press: Oxford, U.K., 1987; p 100.
5. Anon. *Internatl. Trad. Med. Newslett.* **1992**, *4*(2), 1-2, 4.
6. Eisner, T.; Meinwald, J. *Chemoecol.* **1990**, *1*, 38-40.
7. Akerele, O; Heywood, V.; Synge, H., Eds. *The Conservation of Medicinal Plants*; Cambridge University Press: Cambridge, U.K., 1991; p ix.
8. Schweitzer, J.; Handley, F.G.; Edwards, J.; Harris, W.F.; Grever, M.R.; Schepartz, S.A.; Cragg, G.M.; Snader, K.; Bhat, A. *J. Natl. Cancer Inst.* **1991**, *83*, 1294-1298.
9. Elisabetsky, E. *Cultur. Surviv. Quart.* **1991**, Summer, 9-13.
10. Kloppenburg, Jr., J. *Cultur. Surviv. Quart.* **1991**, Summer, 14-18.
11. Perry, L.M.; Metzger, J. *Medicinal Plants of East and Southeast Asia*; MIT Press: Cambridge, MA, 1980.
12. Loub, W.; Farnsworth, N.R.; Soejarto, D.D.; Quinn, M.L. *J. Chem. Inform. Comp. Sci.* **1985**, *25*, 99-103.
13. Willis, J.C.; Airy Shaw, H.K. *Dictionary of Flowering Plants and Ferns;* Cambridge University Press: Cambridge, U.K., 1980.
14. Livingston, R.; Zamora, J. *Unasylva* **1983**, *35* (no. 140), 7-10.
15. Myers, N. *The Primary Source*; Norton: New York, 1984.
16. Myers, N. In *Conservation Biology*; Soule, E.M., Ed.; Sinauer: Sunderland, MA, 1986.
17. Raven, P. In *Biodiversity*; Wilson, E.O.; Peter, F.M., Eds; National Academy Press: Washington, DC, 1988; p 119-122.

RECEIVED March 5, 1993

Chapter 9

Simple Bench-Top Bioassays (Brine Shrimp and Potato Discs) for the Discovery of Plant Antitumor Compounds

Review of Recent Progress

Jerry L. McLaughlin, Ching-jer Chang, and David L. Smith

Department of Medicinal Chemistry and Pharmacognosy, School of Pharmacy and Pharmacal Sciences, Purdue University, West Lafayette, IN 47907

Advances in separation technology and spectral methodology have greatly simplified the science of phytochemistry. New challenges lie in the isolation of nature's trace phytochemicals which exert various bioactivities and, thus, present potential usefulness to mankind. To help meet these challenges in discovering novel plant antitumor compounds, we have developed two simple bench-top bioassays which conveniently expedite screening and activity-directed fractionation of active plant extracts. These are lethality to brine shrimp and the inhibition of crown gall tumors on discs of potato tubers. Progress in this laboratory over the past three years is reviewed here. Our most exciting leads are the Annonaceous acetogenins of which bullatacin offers the best potential for antitumor drug development; the potent pesticidal effects of these new agents, which act via mitochondrial inhibition, also offer economic potential.

In 1992, about 1,130,000 people in the United States were diagnosed as having cancer (excluding skin cancers); also during this year about 520,000 people (1,400 people/day) in the U.S. died of cancer; there has been a steady rise in our cancer mortality rate in the last half-century in spite of the fact that 51% of the cancer patients (excluding those with skin cancer), through medical and chemotherapeutic intervention, can now expect to survive with their disease for at least five years following diagnosis (*1*). Thus, the search for new anticancer agents, especially those effective against breast, prostate, ovarian, and cervical cancers, has been given a top priority at the National Cancer Institute (NCI) (Broder, S.; letter to NIH grantees, December 12, 1991).

To help in this search, NCI has, from 1957 through 1991, screened, for antitumor activities, more than 156,000 plant extracts from over 53,000 plant species (*2*, *3*; McCloud, T.G., NCI/FCRF personal communication, 1992). Through the combination of efforts at NCI and various pharmaceutical companies, several chemically useful antitumor drugs have been developed and are marketed from the *Catharanthus* alkaloids and the lignans of *Podophyllum* species (*4*, *5*). Current hopes are high that camptothecin and/or its analogues (*6*) and taxol and/or its congeners (*7*, *8*) will also survive the rigors of clinical testing and become routinely

0097–6156/93/0534–0112$07.50/0

useful. Additional plant products such as phyllanthoside and homoharringtonine (*9, 10*) have clinical activity and are in drug development. These plant products often contribute a novel mechanism of action and favorably attest to the fact that plants are a good source of lead antitumor compounds.

Plant Antitumor Screening at the National Cancer Institute

Bioassay methods for the screening and testing of the plant extracts at NCI have varied over the years, but by 1980 they had evolved primarily into *in vivo* testing against 3PS (P388) (methylcholanthrene-induced) leukemia in mice and *in vitro* screening for cytotoxicities in 9PS or 9KB [human nasopharyngeal carcinoma or, perhaps, Hela cells (*11*)] (*12, 13*). Often cytotoxicity does not translate well into *in vivo* activity. Thus, for years the 3PS *in vivo* system was recognized as the most satisfactory known primary screen; however, it was especially predictive of potential antileukemic (soft tumor) agents and has fallen into disfavor because it, perhaps, overlooked agents that might be more effective against solid tumors (*13*).

Throughout the past 35 years, the NCI plant antitumor screening methods have been the subject of continual debate and evolution. During the 1980's, the Developmental Therapeutics program of NCI scuttled most of their *in vivo* screening capacity to implement a primary *in vitro* screening system which comprises a panel of 60 human cancer cell lines, including those of lung, colon, breast, melanoma, and other refractory (solid tumor) disease types. Housed at the Frederick Cancer Research Facility (FCRF), the system is specifically directed at identifying agents having cell-type selective toxicity; a screening rate of 400 compounds or extracts per week has been consistently achieved (*14*).

Pure compounds showing activity in this cell-panel prescreen were to be classified either as nonselective (e.g., cytotoxic in many cell lines across the panels) or selective (e.g., cytotoxic in one or a few specific cell lines). Selective compounds would then be subjected to subsequent *in vivo* tumor panels for further evaluation, especially against the cell type in which they were active *in vitro* (Boyd, M.; NCI, personnal communication, 1984). Adriamycin (NSC 123127) can be considered; it gave an average molar log LC_{50} 5.01, (delta 0.91, range 1.3). The range is a good indicator of selectivity, and this range of 1.3 shows only a ten to one-hundred fold difference in LC_{50} values among the cell lines. (Care must be taken when inspecting the range data to be sure that dilutions have been made to give accurate LC_{50} values for all cell lines). Some selectivity is suggested for adriamycin against melanomas, but the data is not as impressive as one might think.

Inspection of the mean graphs generated recently through the NCI cell panel (with 48 hour cell-drug exposures), for over 100 compounds known to be *in vivo* active, is, likewise, quite alarming. Tamoxifen (an antiestrogen) lacks cytotoxic activity, and there are evidently no breast cell lines or other estrogen-dependent cells available in the current panel. Taxol, etoposide (VP-16), cis-platinum, and several other agents known to be clinically useful produce mean graphs which are not at all distinctive. Some other clinically useful agents, such as busulfan and 5-fluorouracil, give only marginal LC_{50} values, and show no selectivity, whatsoever. It is doubtful if these important agents would ever have been detected and developed if this cell panel were the only primary screen in use. One questions whether the 3PS *in vivo* screen might not have been so bad after all.

In addition, one is compelled to question: How might crude extracts behave in this selective system? We have observed that a single plant often yields a mixture of several cytotoxic compounds, representing several chemical classes and several possible modes of action, each with its unique cytotoxicity pattern; crude extracts may, therefore, only produce a pattern of general cytotoxicity with the mixtures obscuring the selectivity of the individual active components. Thus, we argue that the high cost and high technology needs for all of these cytotoxicity assays, at the crude extract stage, are neither necessary nor advisable for the detection and isolation of new natural antitumor agents. Inexpensive prescreens that are proven as statistically equal or superior alternatives to cytotoxicity should be considered, at least as a supplement or even as a prerequisite, for the expensive and often over-burdened NCI cell panel. Certainly the cell panel would be less burdened if it were not being used as the primary screen of the thousands of plant and marine organism extracts coming from the NCI's large intramural screening program.

The need for simple, rapid, reliable, and inexpensive antitumor prescreens has long been recognized. Several models have been presented (*15, 16*), and the NCI has considered a number of alternatives (2). However, during the 1980's, the drug-discovery programs of the American pharmaceutical industry moved rapidly toward mechanism-based screening assays, with target receptors and enzymes, and away from cell culture and animal models (*17*). It was argued that a) cytotoxic compounds are quite common, b) only a small portion of cytotoxic compounds demonstrate any degree of *in vivo* activity (as previously noted), and c) only a small portion of *in vivo* actives demonstrate a spectrum and selection of activity that warrants clinical trials. The debate goes on and on with the NCI's major plea being that it is too early to judge (*18, 19*).

Mechanism-Based Antitumor Assays

Mechanism-based assays detect potential anticancer drugs that interfere with processes relevant to neoplastic growth (e.g., protein kinase C and protein tyrosine kinases), or they focus on target receptors that were discovered as the mechanism of action or drug resistance of various prototype drugs (e.g., camptothecin and topoisomerase I; etoposide and topoisomerase II; vincristine and tubulin denaturation; bleomycin and DNA snipping; taxol and tubulin stabilization; adriamycin-resistant cells; etc.). They often utilize radioligand-binding and are very sensitive, requiring only small amounts of sample; thus, in the absence of financial constraints, they can be automated to screen hundreds of samples per week.

Yet, for independent investigators, a battery of mechanism-based assays would be expensive to create and maintain, and they can, at best, operate only a few of these assays in their discovery efforts (*20*). These assays are, furthermore, highly specific and, therein, lies their major disadvantage. Libraries of compounds and extracts must be maintained and re-screened, over and over again, every time a new mechanism-based assay comes on line. Some of us would prefer to screen materials just once, in a broad-based, general bioassay system, and be done with the inactives which take up the bulk of the screening capacity. In addition, some of us would hope to discover some novel compounds which have, as yet, undiscovered mechanisms of action. A simple, general, and affordable alternative, which offers the advantage of detecting a broad range of both novel and known antitumor effects, is the inhibition of crown gall tumors.

Crown Gall Tumors on Potato Discs

Crown gall is a neoplastic disease of plants induced by specific strains of the Gram-negative bacterium, *Agrobacterium tumefaciens* (*21*). The bacteria contain large Ti (tumor-inducing) plasmids which carry genetic information (T-DNA) that transforms normal, wounded, plant cells into autonomous tumor cells (*22*). The T-DNA encodes enzymes that participate directly in phytohormone biosynthesis, thus, promoting mitosis (*23*). This genetic infection by *Agrobacterium* is a verified example of natural genetic exchange between prokaryotic and eukaryotic species (*24*), and these natural gene vectors are highly studied as tools for plant genetic engineering (*25*).

Thus, certain tumorigenesis mechanisms, in both plants and animals, may have in common the intracellular incorporation of extraneous nucleic acids containing oncogenes (*26*, *27*). Considering their many biochemical similarities, antitumor drugs might well inhibit tumor growth in both plant and animal systems. The development of a simple antitumor prescreen, using these convenient and inexpensive plant tumor systems, thus, offers numerous advantages in the search for new anticancer drugs.

Galsky *et al.* (*28*) first demonstrated that inhibition of crown gall tumor initiation on discs of potato tubers showed good correlation with compounds and plant extracts known to be active in the 3PS *in vivo* antitumor assay. Inhibition of the growth of the tumors, in addition to the inhibition of tumor initiation, correlates well with 3PS activity (*29*). The antitumor effect with crown gall is independent of antibiosis (*28*), although antibiotics, if present in sufficient concentration and active against the Gram-negative *Agrobacterium*, could confuse the assay by giving false positives. However, several antibiotics do have useful antitumor activity, and antitumor compounds often have antibiotic activities. Thus, as a prescreen, antibiosis is not really a problem. Nonetheless, antibiotic activity can be easily detected in the active extracts by simple screens such as the agar plate dilution method (*30*).

In preliminary studies, our group 1) modified Galsky's potato disc methods and 2) tested the effectiveness, with appropriate statistical evaluation, of the modified assay as an antitumor prescreen for 3PS activity in crude plant extracts (*31*). The screening results of plant extracts in the potato disc assay were strongly associated with the 3PS results (p=0.002), while the cytotoxicity assays, 9PS (p=0.114) and 9KB (p=0.140), were much less reliable as predictors of *in vivo* 3PS activity. The calculated kappa values for potato disc, 9PS, and 9KB were 0.64, 0.24, and 0.40, respectively; a kappa value of 0.75 is excellent; less than 0.40 is poor (*32*). Thus, this data substantiated the predicted usefulness of the potato disc bioassay and, furthermore, questioned the value of 9KB and 9PS cytotoxicities as statistically reliable prescreens of *in vivo* activity. Additional comparisons of testing data, for both extracts and pure compounds, between 3PS and potato discs gave even more impressive statistical agreements: $p=2\times10^{-6}$ and kappa=0.69 (*33*).

We concluded that, with these modified procedures (*31*), crown gall tumors on potato discs could be employed routinely as comparatively rapid, inexpensive, safe, and statistically reliable prescreens for *in vivo* antitumor activity. The methodology is simple, and the assay can be performed in-house, with minimal technical training and equipment. The assay also circumvents mouse toxicity, which is an inherent disadvantage of the 3PS assay; and it offers a reliable alternative to extensive antitumor screening with small animals for models. With the demise of 3PS testing, the uncertainty of any useful *in vivo* predictive value of the NCI tumor panel results, and the impracticality of establishing a battery of mechanism-based assays on a small scale, the need for such an *in vitro* alternative, that is statistically validated to an *in vivo* system, is readily apparent.

Brine Shrimp Lethality as an Antitumor Prescreen

Antitumor compounds are almost always toxic in high doses. Thus, *in vivo* lethality in a simple zoologic organism might be used as a rapid and simple prescreen and convenient monitor during the fractionation of known antitumor plant extracts. The eggs of brine shrimp, *Artemia salina*, are readily available in pet shops at low cost and can remain viable for years in the dry state. Upon being placed in a brine solution, the eggs hatch within 48 hours, providing large numbers of larvae (nauplii). Desiring a rapid, inexpensive, in-house, bioassay to supplement cytotoxicities for screening and fractionation monitoring, we developed simple bench-top procedures for using this tiny crustacean as a bioassay tool (*34*). The lethality data are then processed through a personal computer program to estimate LC_{50} values with 95% confidence intervals, for statistically significant comparisons of potencies and reproducibility. Cell culture results give no such statistical data to analyze variance.

The biology and physiology of brine shrimp have been extensively studied (*35-37*), but, for screening and fractionation monitoring of active materials from higher plants, they apparently had not been used before our work was initiated. In screening a number of compounds with both brine shrimp and 9KB, we found a positive correlation between brine shrimp toxicity and 9KB cytotoxicity ($p=0.036$ and $kappa=0.56$) (*33*).

Thus, for extracts of antitumor plant species, it is possible to prescreen and monitor fractionation using principally the more rapid and less expensive brine shrimp bioassay rather than the more tedious and expensive *in vivo* and *in vitro* assays. Confirmation of the prescreen and occasional cross-monitoring in more specific antitumor systems, e.g., on potato discs and in our in-house cell-culture panel, is performed to ensure that the brine shrimp activities continue to correlate. Isolated active compounds can then be run expeditiously through a battery of mechanism-based assays at cooperating pharmaceutical companies (e.g., Bristol and Glaxo) in order to search out their mechanisms of action, and selected *in vitro* active compounds can be suggested for testing *in vivo* at NCI or at a cooperating company such as Abbott or Eli Lilly. The brine shrimp assay has advantages of being rapid (24 hours following introduction of shrimp), inexpensive, and simple (e.g., no aseptic techniques are required). It easily utilizes a large number of organisms for statistical considerations and requires no special equipment and a relatively small amount of sample (20 mg for screening initial extracts at 1000 ppm, lesser amounts for active fractions). It has the added advantage, over cytotoxicity assays, of not requiring higher animal serum (*38*). We have received no criticism, so far, from animal rights advocates for killing brine shrimp. On the negative side, as a more general bioassay, it, like cytotoxicity, gives a number of false positives; thus, it has to be used in conjunction with the more specific potato disc assay to eliminate finding toxic compounds which are not antitumor compounds.

In a recent publication (*39*), the brine shrimp and potato disc bioassays were evaluated for their accuracies at detecting a series of known *in vivo* active antitumor agents which were supplied (blind) by the National Cancer Institute. In every case, the potato disc assay correctly differentiated between the *in vivo* actives and inactives ($p=0.008$), and the brine shrimp assay ($p=0.033$) proved to be superior, or equally as accurate at predicting *in vivo* activity as cytotoxicities in a series of human solid tumor cell lines ($p=0.033$-0.334).

Demonstrations and Acceptance of these Bench-Top Bioassays

For ten years, since the development and statistical validation of the potato disc and brine shrimp bioassays, our work has been directed toward the improvement and utilization of these two simple systems in plant antitumor fractionation problems. With grant support from NCI, a series of convincing demonstrations, validating the utility of these assays in the isolation of new antitumor compounds, is in progress, and we hope to attract the interest of other workers, many of whom are simply phytochemists and have never had their compounds bioassayed, to this drug discovery area. Attention given to our publications has resulted in invited seminars and symposium presentations worldwide. Week-long workshops have been conducted, teaching these methodologies, in ten different countries. We continue to have a number of visiting scholars and visiting professors who come to Purdue to learn these bioassays; their efforts have helped to make our progress more successful; in addition, they return to their countries to teach and use the bioassays on their own plants. Several laboratories have now adapted these simple bioassays or modified them to their needs (*40-42*; *inter alia*).

A recent computer search on our original brine shrimp paper (*34*) uncovered 116 citations (41 of which were in papers from our own laboratory, and 75 of which were in papers from other laboratories). A similar search showed that the potato disc assay paper (*31*) has not been quite as popular (45 citations of which 20 were in our own papers), but, nevertheless, the method is being used more and more. Two recent book chapters (*43, 44*) and an invited review in a Central American journal (*45*) have updated the methodology and summarized our earlier progress.

Progress in the Past Three Years

During the past three-years, over 40 papers from our laboratory have been published dealing with this research. Thus, a good level of productivity in this area is quite possible with a small group of dedicated scientists using these workable bioassays. To illustrate the success of this simple bioassay approach, a summary of some of our progress, during this three year period, with several different plant species, follows. A diversity of structural types, among the dozens of compounds presented, has been detected and characterized. Significant progress is especially evident with the discovery of numerous new Annonaceous acetogenins, several of which are extremely potent (some show over a billion times the cytotoxic potency of adriamycin). The protein tyrosine kinase activity of piceatannol (**1**) and the inhibition of mitochondrial electron transport at complex I by the acetogenins, such as bullatacin (**3**), offer modes of antitumor action for which no clinical agents are currently available. For example, the ATP dependence of the P-170 glycoprotein pump in multiple drug-resistant cells suggests a new type of combination therapy using the acetogenins with traditional chemotherapeutic agents to treat resistant tumors. In addition, a new type of rescue therapy is now suggested with succinate should the acetogenins be given in too toxic a dose.

***Euphorbia lagascae* Spreng. (Euphorbiaceae).** In the NCI tumor cell panel, piceatannol (**1**) (*43*) showed selectivities for certain colon (including HT-29), leukemia, and melanoma cell lines. We have now supplied **1** to a dozen or so laboratories as a standard protein tyrosine kinase (PTK) inhibitor (*46*). It effectively inhibits the PTK that mediates mast cell degranulation and may be useful in allergies as well as in cancer chemotherapy. The synthesis of over fifty derivatives and structural analogues failed to improve its PTK inhibiting activity (*47,48*). The combretastatins are similar stilbenes from *Combretum caffrum* (Combretaceae), but they act via tubulin inhibition and not through PTK (*49*).

Piceatannol (**1**) and related hydroxylated *trans*-stilbenes are now being recognized as phytoalexins produced in response to fungal infections (*50, 51*) and very likely act in plants via inhibition of PTK. A retrospective literature search revealed that crown gall tumors on potato discs do contain protein kinases (*52*). Erbstatin, an actinomycete-derived PTK inhibitor, has recently shown impressive effects against human breast (MCF-7) and esophageal tumor xenografts in athymic mice (*53*).

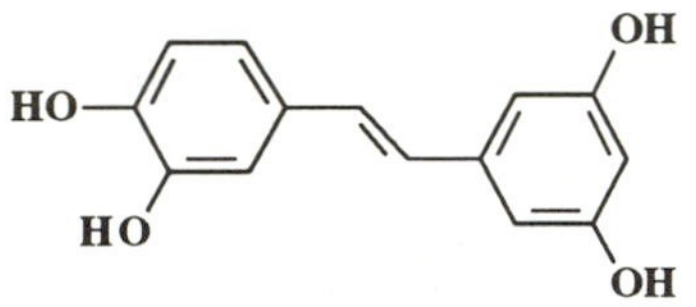

HO, HO, OH, OH

1

Piceatannol (**1**) (NSC 365798): BS LC_{50} 278 ppm; 9KB ED_{50} <10 μg/ml; 9PS ED_{50} 2.5 μg/ml; 3PS 131% T/C at 25 mg/kg; B16 153% T/C at 25 mg/kg; 3 MBG5 (mamm. carcin.) 123% T/C at 37.5 mg/kg; A-549 ED_{50} >10 μg/ml; MCF-7 ED_{50} >10 μg/ml; HT-29 3.89 μg/ml; potato disc average of 38% tumor inhibition; p40 protein tyrosine kinase, Ki=15mM.; selective in the NCI tumor panel, ave. molar log IC_{50} 4.5, delta 0.87, range 1.38.

***Asimina triloba* Dunal. (Annonaceae).** Asimicin (**2**), the first 4-hydroxylated linear acetogenin, was isolated earlier as a major bioactive component of the seeds and bark of the common paw paw, *Asimina triloba* (*54*). The potent pesticidal actions of the acetogenins were discovered and patented (*55*), and a divisional patent was awarded on the structure of asimicin (*56*). A number of firms are now considering the crude acetogenin mixture from paw paw (F020) and other species for licensing and development as a new class of natural pesticides (*57*). F020 shows pesticidal synergisms with pyrethrins and neem extract at AgriDyne. The twigs of tetraploid varieties of paw paw (supplied by A.E. Kehr) and diploids showed no significant differences in bioactivities. Evaluation of the various parts of the paw paw tree determined that the twigs and small branches could be harvested to provide a renewable yield of the biomass as a source of acetogenins (*58*). In the fall of 1991, we collected 2,500 pounds (dry weight) of this biomass from 600 trees growing at the University of Maryland. A guinea pig maximization test demonstrated that asimicin and F020 are not serious skin allergens (*59*), and Ames testing of F020 showed no significant mutagenicity.

At the University of Ottawa, F020 and asimicin from our laboratory were demonstrated to be potent inhibitors of insect respiration by blocking complex I of the mitochondrial electron transport system (*60*). Workers at the Bayer Company in Germany simultaneously reached the same conclusion with the "annonins", a group of acetogenins from *Annona squamosa* (*61*). Thus, the acetogenins act as inhibitors of mitochondrial ATP production. Rapidly growing cells, including cancer cells, are, evidently, selectively poisoned because they have larger bioenergetic demands.

Further fractionation of F020 (F005) from the bark, gave six more bioactive compounds including the known acetogenins bullatacin (**3**) and bullatacinone (**4**) (*62*), a new acetogenin, trilobacin (**5**), and three known compounds (*N-p*-coumaroyltyramine, *N-trans*-feruloyltyramine, and (+)-syringaresinol) all of which are cytotoxic (*63*). In further cytotoxicity tests at Abbott Labs and in the NCI tumor panel, trilobacin (**5**) is more potent or equipotent to bullatacin (**3**) (*62*). Bullatacin (**3**) remains as our most promising compound because trilobacin (**5**) is relatively rare.

In work yet to be published two new adjacent bis-THF-acetogenins, asiminacin (**6**) and isoasiminacin (**7**), along with squamocin (**8**) (*64*) were isolated. In addition, the extracts have yielded two new mono-THF acetogenins, annonacin-A-one (**9**) and asimininone (**10**) along with annonacin (**11**) and isoannonacin (**12**) (*65*). At least ten additional acetogenins are being pursued in further bioactive fractions of these extracts. By comparing ^{1}H-nmr spectra of the acetogenin acetates with a series of synthetic diacetylated dibutyl bis-THF's (*66, 67*) we were able to predict certain relative stereochemical features. A recent paper, which involved considerable work here and in Hoye's lab at the University of Minnesota, describes, for the first time, certain absolute stereochemistries of the acetogenins such as **2, 3,** and **11**; Mosher esters were prepared and analyzed by ^{19}F and ^{1}H nmr (*68*).

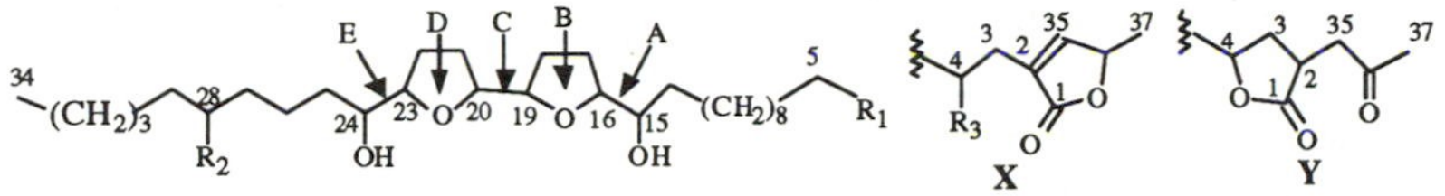

34 (CH2)3 28 R2 24 OH 23 O 20 19 O 16 15 OH (CH2)8 5 R1 E D C B A; X: 4 3 2 35 37 1 O O R3; Y: 4 3 35 37 1 2 O O O

	A	B	C	D
2	threo	trans	threo	trans
3	threo	trans	threo	trans
4	threo	trans	threo	trans
5	threo	trans	erythro	trans
6	threo	trans	threo	trans
7	threo	trans	threo	trans
8	threo	trans	threo	trans
31	threo	trans	threo	trans
	E	**R_1**	**R_2**	**R_3**
2	threo	X	H	OH
3	erythro	X	H	OH
4	erythro	Y	H	--
5	threo	X	H	OH
6	threo	X	R-OH	H
7	threo	X	S-OH	H
8	erythro	X	OH	H
31	threo	X	H	H

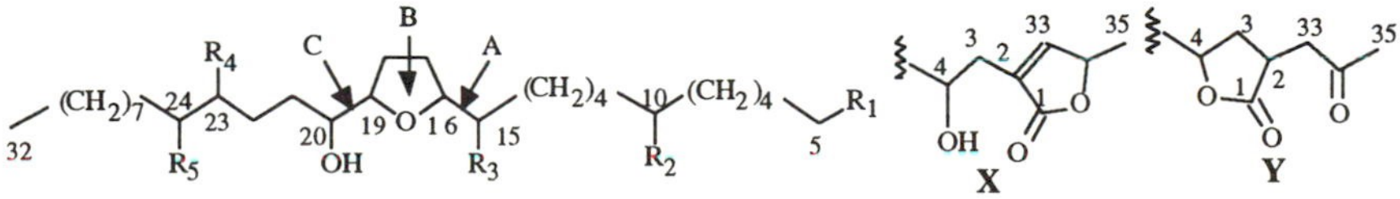

32 (CH2)7 24 R4 23 R5 20 OH 19 O 16 15 R3 C B A (CH2)4 10 (CH2)4 R2 5 R1; X: 4 3 2 33 35 1 O OH O; Y: 4 3 33 35 1 2 O O O

	A	B	C	R_1
9	threo	trans	erythro	Y
10	-	trans	threo	Y
11	threo	trans	threo	X
12	threo	trans	threo	Y
	R_2	**R_3**	**R_4**	**R_5**
9	OH	OH	H	H
10	H	H	OH	OH
11	OH	OH	H	H
12	OH	OH	H	H

Asimicin (**2**) (NSC 609700): BS LC_{50} 0.03 ppm; potato disc 70% tumor inhibition; 9 KB ED_{50} $<10^{-5}$ μg/ml; 9PS ED_{50} $< 10^{-7}$ μg/ml; 9ASK cytotoxic at 10 μg/ml; 3PS toxic at 0.22 mg/kg, active T/C 124% at 0.025 mg/kg; has potent antimalarial, insecticidal, nematocidal, and thymocytocidal activities. NCI tumor panel (7 days) ave. molar log IC_{50} 6.0, delta 2.3, range 3.6. *In vivo* active in L1210 leukemia T/C 131% at 200 μg/kg (Upjohn). NCI tumor panel (3 days) mean molar log GI_{50} 5.88, delta 2.12, range 2.85 with some selectivity.

Trilobacin (**5**) (NSC 643353): BS LC_{50} 0.0087 ppm; potato disc 46% tumor inhibition; at Purdue A-549 ED_{50} 8.02×10^{-4} μg/ml, MCF-7 ED_{50} 3.29×10^{-1} μg/ml, HT-29 $<10^{-15}$ μg/ml; at Abbott: A-549 ED_{50} 1.7×10^{-6} μg/ml, HT-29 ED_{50} 2.4 μg/ml, B16F10 ED_{50} 7.9×10^{-1} μg/ml, 9PS ED_{50} 2.2×10^{-6} μg/ml; essentially inactive in antimicrobial and other Abbott screens; NCI tumor panel (3 days) mean molar log GI_{50} 6.04, delta 1.90, range 3.02 with some selectivities.

Asiminacin (**6**): BS LC_{50} 0.041 ppm; (cell culture data pending).

Isoasiminacin (**7**): BS LC_{50} 0.049 ppm; (cell culture data pending).

Annonacin-A-one (**9**): BS LC_{50} 0.288 ppm; A-549 ED_{50} 3.39×10^{-2} μg/ml; MCF-7 ED_{50} 1.44 μg/ml; HT-29 2.82×10^{-2} μg/ml.

Asimininone (**10**): BS LC_{50} 0.133 ppm; (cell culture data pending).

***Asimina parviflora* (Michx.) Dunal. (Annonaceae).** This dwarf paw paw is indigenous to the southeastern U.S. Early column fractions of the bioactive twig extract (F005) produced asimicin (**2**) and a series of new acetogenins which are still being pursued. Later column fractions appeared to be devoid of acetogenins but were still toxic to the brine shrimp; these were fractionated further to yield liriodenine (**11**) and the known lignan, (+)-syringaresinol (**12**) (**11** and **12** are not illustrated), and two new compounds: asimicilone (**13**) (a 2-quinolone alkaloid) and docosenamide (**14**) (a new amide of a long chain fatty acid). Both **13** and **14** are only moderately cytotoxic, but **14** was surprisingly active against HT-29 (colon). HMBC and HMQC correlations were very helpful in identifying the structure of **13** (*69*).

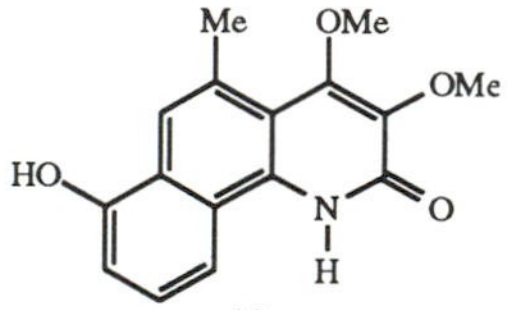

Me OMe OMe HO N H O

13

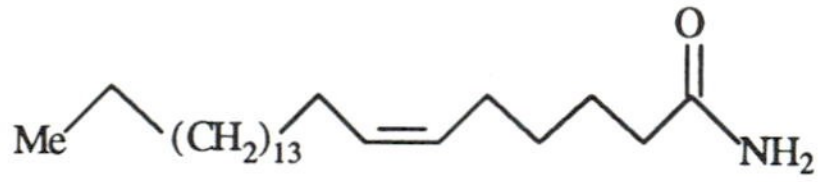

O Me $(CH_2)_{13}$ NH_2

14

Asimicilone (**13**): BS LC_{50} 2.15 ppm; A-549 ED_{50} 2.14 μg/ml, MCF-7 ED_{50} 7.22 μg/ml, HT-29 ED_{50} 3.25 μg/ml.

Docosenamide (**14**): BS LC_{50} 2.45 ppm; A-549 ED_{50} 3.16 μg/ml, MCF-7 ED_{50} 2.91 μg/ml, HT-29 ED_{50} $<10^{-2}$ μg/ml.

***Goniothalamus giganteus* Hook. f. and Thomas (Annonaceae).** Our continuing work on the stem bark of this Thai species is producing additional compounds in two major phytochemical groups. The styryllactones are formed biogenetically only from cinnamic acid and two acetates, yet a surprising number of novel ring combinations and structures result; six such compounds were reported previously (*70-72*). The three acetogenins previously found in this species are of the mono-THF (*73*) and the nonadjacent bis-THF groups (*74*).

During the past three years, six additional styryllactones in this species were isolated and described: 9-deoxygoniopypyrone (**15**), 7-epi-goniofufurone (**16**), and goniodiol (**17**) (*75*) and goniobutenolide A (**18**), goniobutenolide B (**19**) and goniofupyrone (**20**) (*76*). While they all show cytotoxicities, they are not very potent, with the best ones (**17** and **20**) showing their best selective activities at one-tenth the potency of adriamycin. Compounds **15**, **16**, and **19** are related to more active compounds previously reported. X-Ray crystallography was used to help solve the relative stereochemistries of **15-17**. In addition, two novel 8-membered lactone ring compounds from this group are currently being characterized. These compounds continue to attract the attention of the synthetic chemists (*77*, Sharpless, K.B., Scripps Research Institute; personal communication, 1992).

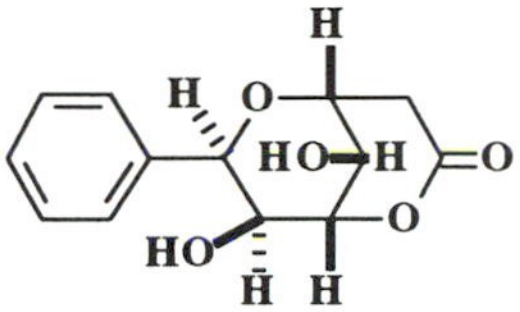

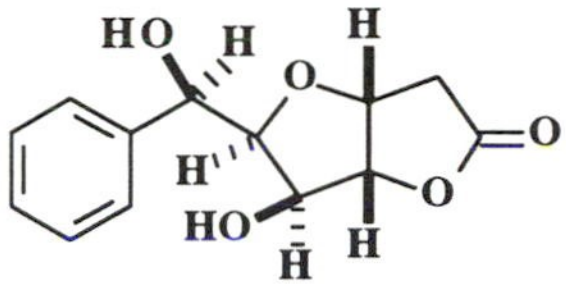

9-Deoxygoniopypyrone (**15**): BS LC_{50} >500 ppm; A-549 ED_{50} 27 μg/ml, MCF-7 ED_{50} 25 μg/ml, HT-29 ED_{50} 7.4 μg/ml.

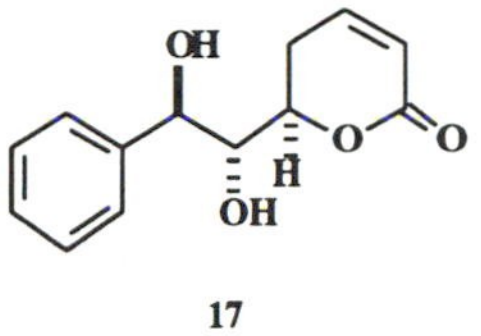

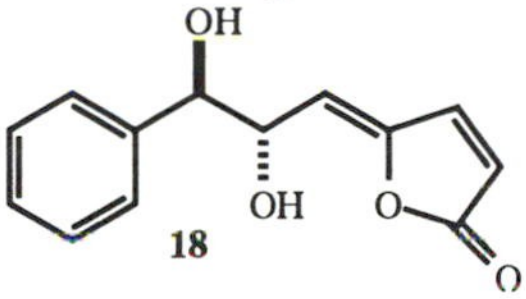

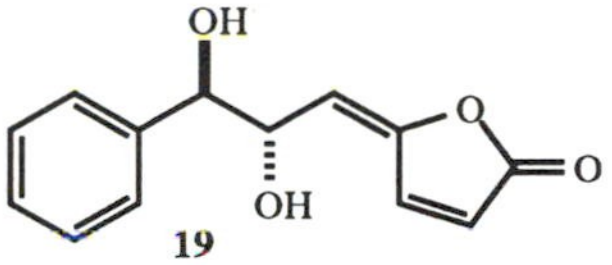

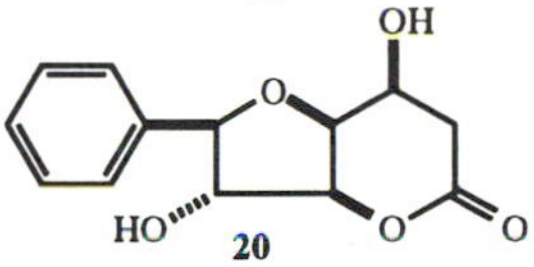

7-epi-Goniofufurone (**16**): BS LC_{50} 475 ppm; A-549 ED_{50} 86 μg/ml, MCF-7 ED_{50} 49 μg/ml, HT-29 >100 μg/ml.

Goniodiol (**17**): BS LC_{50} >500 μg/ml; A-549 ED_{50} 1.22×10^{-1} μg/ml, MCF-7 ED_{50} 8.27 μg/lm, HT-29 ED_{50} 2.45 μg/ml.

Goniobutenolide A (**18**): BS LC_{50} 53 ppm; A-549 ED_{50} 3.73 μg/ml; MCF-7 ED_{50} 7.76 μg/ml; HT-29 ED_{50} 3.41 μg/ml.

Goniobutenolide B (**19**): BS LC_{50} 60 ppm; A-549 ED_{50} 9.05×10^{-1} μg/ml; MCF-7 ED_{50} 19.85 μg/ml; HT-29 ED_{50} 2.67 μg/ml.

Goniofupyrone (**20**): BS LC_{50} >500 ppm; A-549 ED_{50} 56.36 μg/ml, MCF-7 ED_{50} >100 μg/ml, HT-29 ED_{50} 38.02 μg/ml.

However, the most bioactive chromatographic fractions continue to contain additional acetogenins. Gigantetrocin (**21**) and gigantriocin (**22**) are novel mono-THF acetogenins with vicinyl hydroxyl groups (*78*); these are equivalent to adriamycin in cytotoxic potency and show interesting selectivities in the NCI tumor cell panel. Giganenin (**23**) is a very potent, novel, mono-THF acetogenin with a double bond along the hydrocarbon chain between the lactone and THF rings, and 4-deoxygigantecin (**24**) is less active than gigantecin, which we have previously reported (*79*). Gigantetronenin (**25**) and gigantrionenin (**26**) are two new mono-THF compounds each with a double bond along the terminal hydrocarbon chain. Annomonatacin (**27**), a known compound (*80*), was also isolated (*81*). All of these compounds fit into a nice biogenetic scheme which likely proceeds from double bonds, to epoxides, to diols, to THF rings as we proposed in our earlier review on acetogenins (*82*). Our most recent discovery here is giganin (**28**) which has no THF ring but has a double bond which sets it up perfectly as a percursor of goniothalamicin (**38**) (*73*) and gigantetrocin (**21**); **28** shows unexpectedly good cytotoxicities which are quite comparable in potency to reference adriamycin in the same runs. At Abbott, gigantetrocin (**21**) inhibited a fungal topisomerase, but we suspect that these substances are all mitochondrial inhibitors like asimicin (**2**).

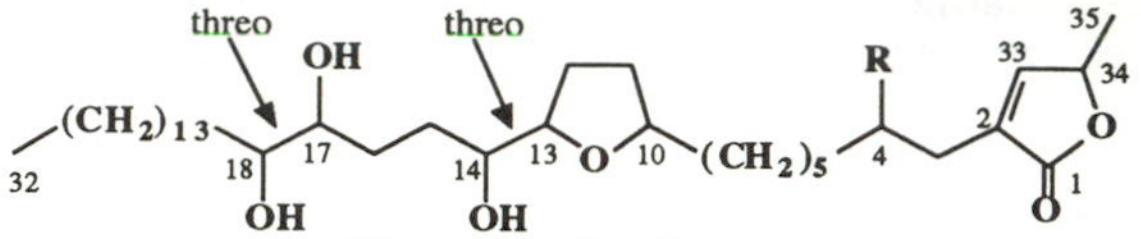

21 R = OH **22** R = H

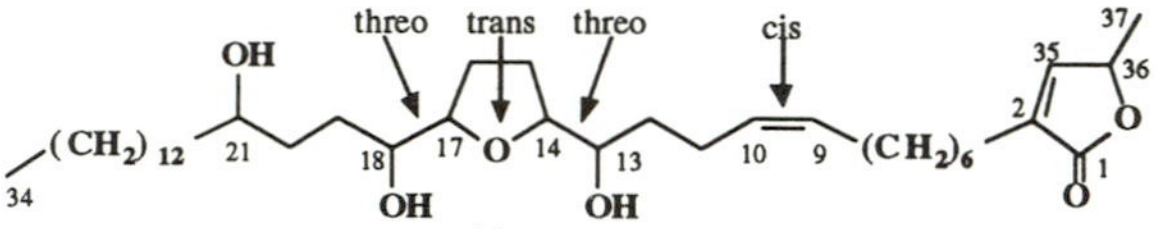

23

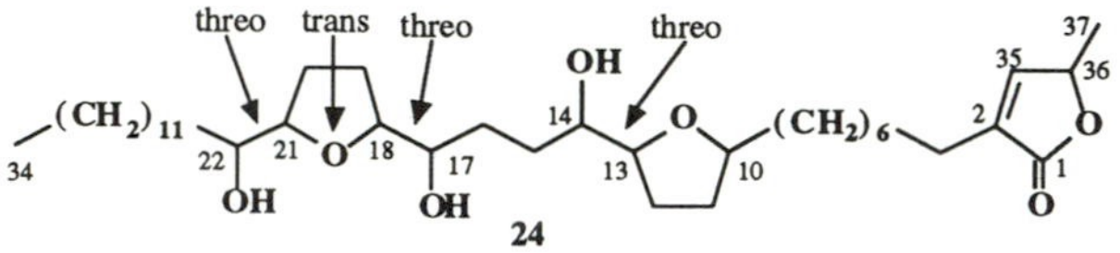

24

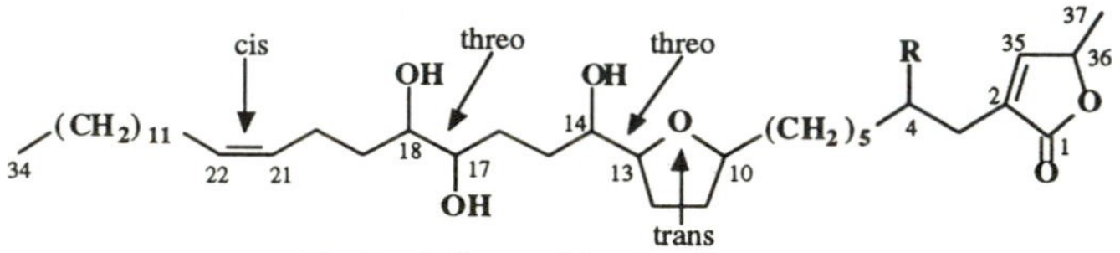

25 R = OH **26** R = H

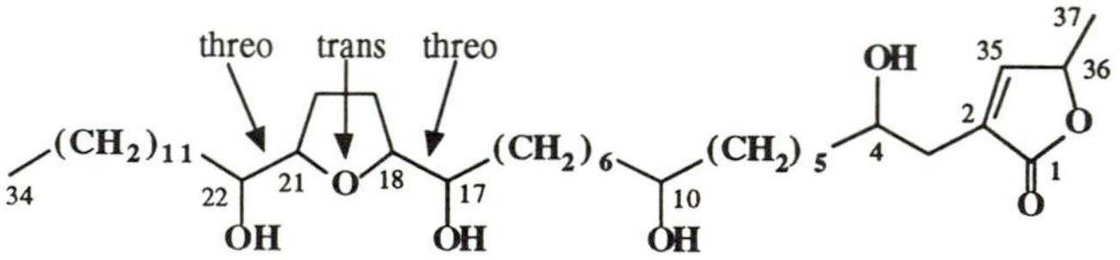

27

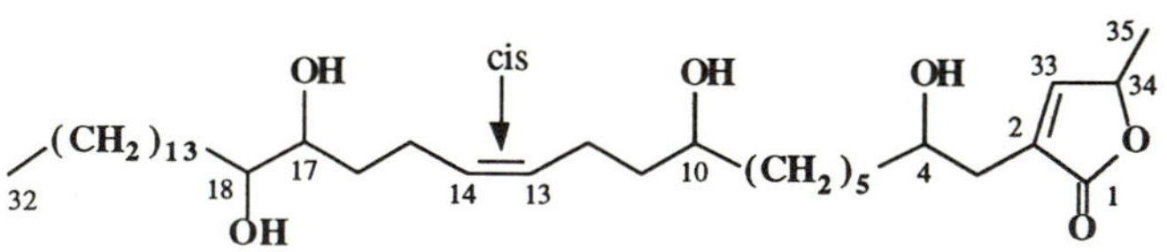

28

Gigantetrocin (**21**) (NSC 643354): BS LC_{50} 0.60 ppm; potato disc 66% tumor inhibition; at Purdue A-549 ED_{50} 3.48×10^{-3} μg/ml, MCF-7 6.49×10^{-3} μg/ml, HT-29 1.24 μg/ml; at Abbott A-549 ED_{50} 1.7×10^{-3} μg/ml, HT-29 ED_{50} 1.8 μg/ml, B16F10 ED_{50} 5.7×10^{-1} μg/ml, 9PS ED_{50} 2.0×10^{-3} μg/ml;

NCI tumor panel selectivity with mean molar log GI_{50} 6.60, delta 1.94, range 3.04; at Abbott inhibits a fungal topoisomerase.

Gigantriocin (**22**) (NSC 643355): BS LC_{50} 5.6 ppm; potato disc 53% tumor inhibition; at Purdue A-549 ED_{50} $2.69x10^{-2}$ μg/ml, MCF-7 ED_{50} 1.47 μg/ml, HT-29 ED_{50} 2.18 μg/ml; at Abbott A-549 ED_{50} $2.9x10^{-3}$ μg/ml, HT-29 ED_{50} 2.03 μg/ml, B126F10 ED_{50} $6.3x10^{-1}$ μg/ml, 9PS ED_{50} $2.1x10^{-3}$ μg/ml; NCI tumor panel selectivity with mean molar log GI_{50} 5.58, delta 2.42, range 3.84.

Giganenin (**23**): BS LC_{50} 89.02 ppm; A-549 ED_{50} $6.97x10^{-7}$ μg/ml, MCF-7 ED_{50} $2.59x10^{-2}$ μg/ml, HT-29 ED_{50} $5.80x10^{-8}$ μg/ml.

4-Deoxygigantecin (**24**): BS LC_{50} 29.46 ppm; A-549 ED_{50} $2.51x10^{-2}$ μg/ml; MCF-7 ED_{50} 5.3 μg/ml, HT-29 ED_{50} 3.58 μl/ml..

Gigantetronenin (**25**): BS LC_{50} 9.4 ppm; A-549 ED_{50} $4.71x10^{-3}$ μg/ml, MCF-7 ED_{50} $6.03x10^{-1}$ μg/ml, HT-29 ED_{50} $5.37x10^{-2}$ μg/ml.

Gigantrionenin (**26**): BS LC_{50} 13.9 ppm; A-549 ED_{50} $3.9x10^{-3}$ μg/ml, MCF-7 ED_{50} $1.65x10^{-1}$ μg/ml, HT-29 ED_{50} $2.58x10^{-3}$ μg/ml.

Annomonatacin (**27**): BS LC_{50} 12.5 ppm; A-549 ED_{50} $3.9x10^{-3}$ μg/ml, MCF-7 ED_{50} $1.65x10^{-1}$ μg/ml, HT-29 $2.58x10^{-3}$ μg/ml.

Giganin (**28**): BS LC_{50} 890 ppm; A-549 ED_{50} $1.88x10^{-2}$ μg/ml; MCF-7 ED_{50} 1.03 μg/ml; HT-29 ED_{50} $6.88x10^{-3}$ μg/ml.

***Annona bullata* Rich. (Annonaceae).** During the screening of almost 50 species of Annonaceae, this Cuban species gave extracts which are very potent (F005, BS LC_{50} 0.003 ppm, potato disc 78% tumor inhibition). Previously, we have reported bullatacin (**3**) and its ketolactone, bullatacinone (**4**) (*62*), which are very potent adjacent bis-THF acetogenins; a patent application on these has now been allowed. We have also reported bullatalicin (**29**) as a potent nonadjacent bis-THF compound (*83*). New *in vivo* and *in vitro* testing data are now available on these compounds from Upjohn, NCI, Abbott, and Glaxo.

At Upjohn in L-1210 leukemia *in vivo*, bullatacin (**3**) gave 138% T/C at 50 μg/kg and bullatacinone (**4**) gave 144% T/C at 400 μg/kg . (As a comparison, taxol gives 139% T/C at 15 mg/kg; thus, **2** and **3** were 300 times and 30 times, respectively, as potent as taxol.) In the NCI tumor panel (7 days), these compounds showed selective activities against A2780 human ovarian cancer; Upjohn verified this *in vitro* activity and conducted expensive xenograft studies in athymic mice; the results are impressive with bullatacin (**3**) giving 67% tumor growth inhibition (TGI) at 50 μg/kg, bullatacinone (**4**) giving 52% TGI at 125 μg/kg, bullatalicin (**29**) giving 75% TGI at 1 mg/kg, and cis-platinum (positive control) giving 78% TGI at 5 mg/kg (day one only). In preliminary tests at NCI in athymic mice, bullatacin (**3**) caused a 41% TGI against SK-MEL-5 (melanoma) at 81 μg/kg and an 18% increase in life span at 112 μg/kg against human HL-60 promyelocytic leukemia. At Abbott, a panel of *in vitro* cytotoxicity studies on the eleven acetogenins submitted showed as much as 10,000 times the potency of adriamycin. At Glaxo, a battery of thirty mechanism-based studies failed to determine the mode of action of bullatacin (**2**). However, at Michigan State University (MSU), bullatacin (**3**) and bullatacinone (**4**), like asimicin (**2**), inhibited mitochondrial electron transport (complex I) at subnanomolar concentrations (Hollingworth, R.M.; MSU; personal communication, 1991). Inhibition of coenzyme Q and bioenergetics has been long proposed but little explored as a site for attacking cancer cells (*84*).

NCI recently ran the COMPARE computer program (*85*) using **3** as the seed at the GI_{50} level; 21 compounds (of the thousands in the data bank) showed good correlation, and these included 10 of the 14 acetogenins that we had entered; it seems that not all acetogenins work the same way, but some are similar. Compound **3** has shown potent effects on the NADH oxidase system of plasma membranes (*86*); thus, sites of action for the acetogenins, other than mitochondria, may exist.

Additional *in vivo* tests are now underway at NCI, Abbott, Lilly, and UCLA, and we have supplied bullatacin (**3**) to Clinical Reference Laboratory, Bristol and Degussa for additional tests. We have about 1.2 g of bullatacin (**3**) still in hand, having found it in five other Annonaceous species, including paw paw. It is interesting that *Agrobacterium* produces high levels of coenzyme Q_{10} (*87*).

Bullatacin (**3**) (NSC 615484): BS LC_{50} 0.0016 ppm; potato disc 51% tumor inhibition; at Purdue 9PS ED_{50} 10^{-15}-10^{-16} μg/ml; 9KB ED_{50} 6.12×10^{-14} μg/ml, A-549 ED_{50} 1.3×10^{-13} μg/ml, MCF-7 ED_{50} 9.7×10^{-18} μg/ml, HT-29 ED_{50} 1×10^{-12} μg/ml, SK-MEL-5 ED_{50} 1×10^{-11} μg/ml, MALME-3M ED_{50} 4.5×10^{-4} μg/ml; NCI (7 day) tumor panel mean (μg/ml) ave. log IC_{50} 1.5 , delta 3.53, range 5.98; NCI (3 day) tumor panel selectivity with mean molar log GI_{50} 5.72, delta 2.28, range 4.0; at Abbott A-549 ED_{50} 3.7×10^{-6} μg/ml, HT-29 ED_{50} 3.5 μg/ml, B16F10 ED_{50} > 1.0 μg/ml, 9PS ED_{50} 3.4×10^{-6} μg/ml; insecticidal (Dow Elanco); mitochondrial ETS inhibitor (MSU); L-1210 *in vivo* 138% T/C at 50 μg/kg (Upjohn); A-2780 *in vivo* in athymic mice 67% TGI at 50 μg/kg (Upjohn).

Bullatacinone (**4**) (NSC 620595): BS LC_{50} 0.003 ppm; potato disc 15% tumor inhibition; at Purdue 9PS ED_{50} 4.23×10^{-3} μg/ml, 9KB ED_{50} $<10^{-12}$ μg/ml, A-549 ED_{50} 10^{-3} μg/ml, MCF-7 ED_{50} 5.4×10^{-18} μg/ml, HT-29 5×10^{-12} μg/ml; NCI (7 day) tumor panel mean (μg/ml) log IC_{50} 1.3, delta 2, range 2.2; NCI (3 day) tumor panel selectivity with mean molar log GI_{50} 5.50, delta 2.0, range 3.41; at Abbott A-549 ED_{50} 3.4×10^{-4} μg/ml, HT-29 ED_{50} >10.0 μg/ml, B16F10 ED_{50} 0.97 μg/ml, 9PS ED_{50} 2.0×10^{-4} μg/ml; nematocidal (Upjohn); inactive as an insecticide (Lilly); mitochondrial ETS inhibitor (MSU); L-1210 *in vivo* 144% T/C at 400 μg/kg (Upjohn); A-2780 *in vivo* in athymic mice 52% TGI at 125 μg/kg (Upjohn).

Bullatalicin (**29**) (NSC 634802): BS LC_{50} 0.154 ppm; potato disc 63% tumor inhibition; at Purdue 9KB ED_{50} >10 μg/ml, A-549 ED_{50} 2.34×10^{-7} μg/ml, MCF-7 ED_{50} 2.34 μg/ml, HT-29 ED_{50} 8.8×10^{-6} μg/ml, SK-MEL-5 ED_{50} 7.1×10^{-9} μg/ml, MALME-3M ED_{50} $>10^{-5}$ μg/ml, 9ASK cytotoxic with slight astrocyte reversal at 100 μg/ml; NCI (3 day) tumor panel selectivity with mean molar log GI_{50} 5.22, delta 2.78, range 4.00; at Abbott A-549 ED_{50} 1.0×10^{-5} μg/ml, HT-29 ED_{50} 3.9 μg/ml, B16F10 ED_{50} 1.9 μg/ml, 9PS ED_{50} 3.4×10^{-5} μg/ml,A-2780 *in vivo* in athymic mice 75% TGI at 1 μg/kg (Upjohn).

In the past three years we have isolated seven more acetogenins, as well as an inseparable mixture of four more from these extracts: bullatalicinone (**29**) was identified as the ketolactone of **28** along with the known squamocin (**8**) (*88*). Bullatencin (**30**) is a new mono-THF acetogenin with a double bond in the terminal part of the hydrocarbon chain; it was found with 4-deoxyasimicin (**31**) (structure not illustrated), a mixture of the mono-THF compounds uvarimicins I-IV (**32**), and the known alkaloid, atherospermidine (**33**) (*89*). Just recently we have found two new nonadjacent bis-THF compounds, bullatanocin (**34**) and bullatanocinone (**35**), which appear to be diastereomers of bullatalicin (**28**) and bullatacinone (**29**), along with desacetyluvaricin (**36**), a known compound. The potent bioactivities of **34** and **35** are equivalent to those of **28** and **29**. A Japanese patent on this series of anticancer compounds has just been issued (*90*) but seems to ignore our papers as prior art.

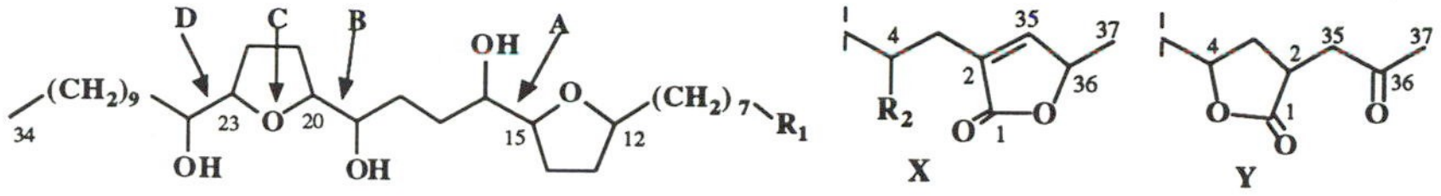

	A	B	C	D	R_1	R_2
29	threo	threo	threo	erythro	Y	-
34	threo	threo	threo	threo	X	OH
35	threo	threo	threo	threo	Y	OH

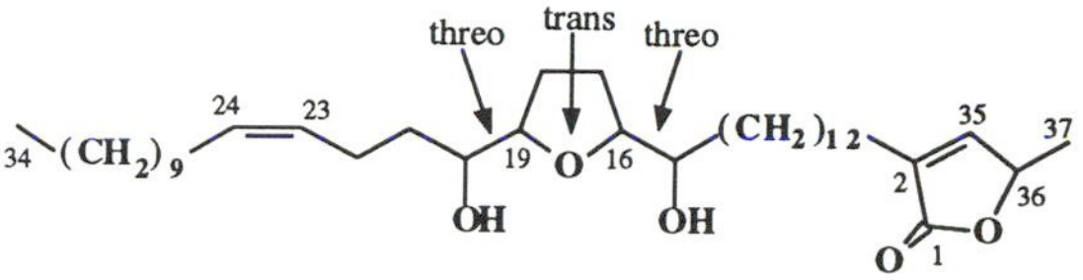

30

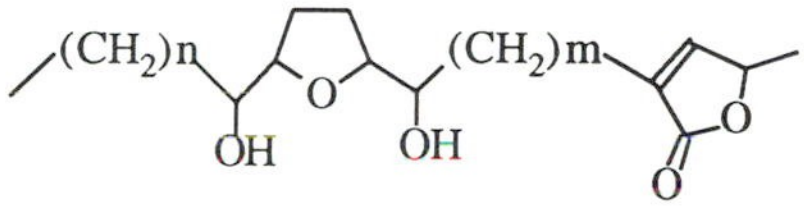

	m	n
I	12	13
II	14	11
III	16	9
IV	10	15

32 (Uvarimicin: **I-IV** mixture), **40** = **II**

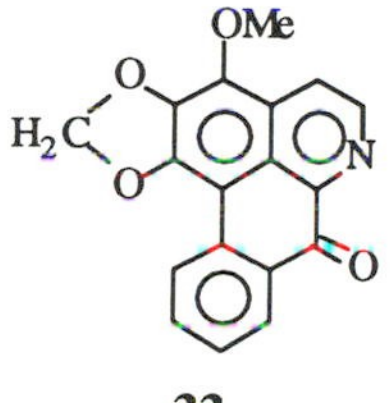

33

Bullatalicinone (**29**) (NSC 643359): BS LC_{50} 46 ppm, potato disc 56% tumor inhibition; at Purdue A-549 ED_{50} 1.6×10^{-2} µg/ml, MCF-7 ED_{50} 8.5×10^{-4} µg/ml, HT-29 ED_{50} 5.0×10^{-5} µg/ml, SK-MEL-5 ED_{50} 3.1×10^{-4} µg/ml, MALME-3M ED_{50} 5.0×10^{-4} µg/ml; NCI tumor panel selectivity with mean molar log GI_{50} 5.47, delta 2.53, range 3.46; at Abbott A-549 ED_{50} 1.3×10^{-3} µg/ml, HT-29 ED_{50} 4.1 µg/ml, B16F10 ED_{50} >1.0 µg/ml, 9PS ED_{50} 1.0×10^{-4} µg/ml; mitochondrial ETS inhibitor (MSU).

Squamocin (**8**) (NSC 643356): BS LC_{50} 2×10^{-2} ppm; potato disc 82% tumor inhibition; at Purdue A-549 ED_{50} 1.9×10^{-9} µg/ml, MCF-7 ED_{50} 3.2×10^{-2} µg/ml, HT-29 ED_{50} >10 µg/ml, SK-MEL-5 ED_{50} 5.7×10^{-12} µg/ml, MALME-3M ED_{50} >10 µg/ml; NCI tumor panel selectivity with mean molar log GI_{50} 5.15, delta 2.65, range 3.80.

Bullatencin (**30**): At Purdue A-549 ED_{50} 9.6×10^{-2} µg/ml, MCF-7 ED_{50} 8.1×10^{-1} µg/ml, HT-29 ED_{50} 1.7×10^{-2} µg/ml, SK-MEL-5 ED_{50} 7.4×10^{-2} µg/ml, MALME-3M ED_{50} 6.4×10^{-2} µg/ml.

4-Deoxyasimicin (**31**) (NSC 643352): BS LC_{50} 5.4×10^{-1} ppm; potato disc 49% tumor inhibition; at Purdue A-549 ED_{50} 3.3×10^{-2} µg/ml, MCF-7 ED_{50} 2.5×10^{-3} µg/ml, HT-29 ED_{50} 2.1 µg/ml, SK-MEL-5 ED_{50} 3.6×10^{-3} µg/ml, MALME-3M ED_{50} 3.3 µg/ml; NCI tumor panel selectivity with mean molar log GI_{50} 5.87, delta 2.13, range 3.07.

Uvarimicins I-IV (mixture) (**32**): BS LC_{50} 31 ppm; potato disc 24% tumor inhibition; at Purdue A-549 ED_{50} $2.8x10^{-1}$ μg/ml, MCF-7 ED_{50} 3.63 μg/ml, HT-29 ED_{50} $8.0x10^{-3}$ μg/ml, SK-MEL-5 ED_{50} $2.1x10^{-1}$ μg/ml, MALME-3M ED_{50} >10 μg/ml.

Atherospermidine (**33**): BS LC_{50} 3.9 ppm; at Purdue A-549 ED_{50} 2.05 μg/ml, MCF-7 ED_{50} >10 μg/ml, HT-29 ED_{50} 4.43 μg/ml.

Bullatanocin (**34**): BS LC_{50} $4.33x10^{-1}$ μg/ml; at Purdue A-549 ED_{50} $<10^{-8}$ μg/ml, MCF-7 ED_{50} $6.09x10^{-1}$ μg/ml, HT-29 ED_{50} $<10^{-8}$ μg/ml.

Bullatanocinone (**35**): BS LC_{50} 1.24 ppm; at Purdue A-549 ED_{50} $<10^{-8}$ μg/ml, MCF-7 ED_{50} $3.22x10^{-1}$ μg/ml, HT-29 ED_{50} $<10^{-8}$ μg/ml.

***Annona muricata* L. (Annonaceae).** The fruits of this tropical tree are called "sour sop" or "guanabana" and are grown extensively as the basis of a tropical juice industry. Thus, the seeds are an abundant by-product as tens of millions of pounds per year are currently wasted. We are working on seeds obtained from the Dominican Republic, Costa Rica, and Indonesia. These are rich in annonacin (**11**), isoannonacin (**12**), annonacin-10-one (**36**), isoannonacin-10-one (**37**), goniothalamicin (**38**), and gigantetrocin (**21**) (*91*). In addition, six new acetogenins have been isolated, bioassayed, and almost completely characterized. Early fractions gave muricatacin (**39**) which we proposed is a degradation product of annonacin (**11**) (*92*); **39** has already been synthesized (*93*; Sharpless, K.B, Scripps Research Institute, personal communication, 1992). The F005 fraction of the seeds was strongly pesticidal in tests at AgriDyne. The bark contains bullatacin (**3**) and other bis-THF acetogenins, but the seeds, so far, have only yielded mono-THF acetogenins. Heinstein in our department has succeeded in producing the acetogenins in tissue cultures of these seeds (Heinstein, P.F., Purdue University, personal communication, 1991).

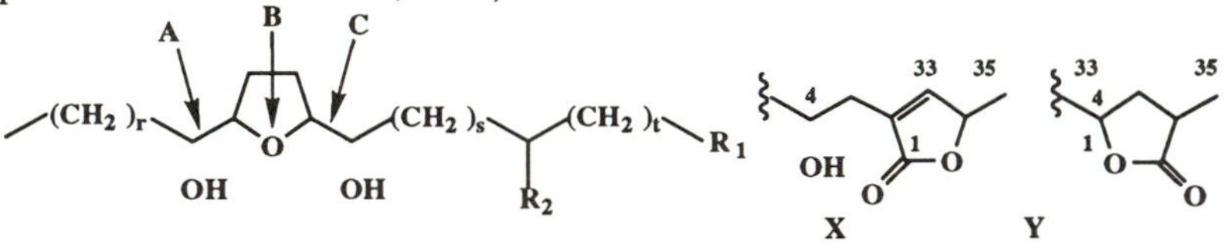

	A	B	C	R_1	R_2	r	s	t
11	threo	trans	threo	X	OH	11	4	5
12	threo	trans	threo	Y	OH	11	4	5
36	threo	trans	threo	X	=O	11	4	5
37	threo	trans	threo	Y	=O	11	4	5
38	threo	trans	threo	X	OH	13	2	5

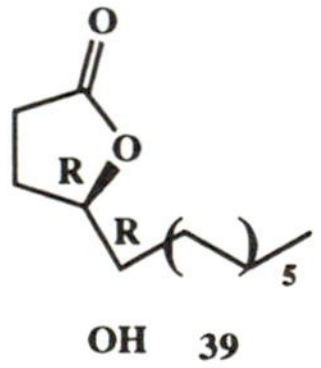

Annonacin (**11**) (NSC 606194): BS LC_{50} 0.3 ppm; potato disc 72% tumor inhibition; 3PS *in vivo* 124% T/C at 0.95 μg/kg, 9ASK 15-30% reversal at 100 μg/ml; insecticidal (Dow Elanco); antimalarial (Water Reed); NCI tumor panel (7 day) ave. IC_{50} 0.75, delta 2, range 3.7; at Purdue A-549 ED_{50} $<10^{-3}$ μg/ml, MCF-7 ED_{50} $1.0x10^{-2}$ μg/ml, HT-29 ED_{50} $<10^{-3}$ μg/ml; at Abbott A-549 ED_{50} $4.8x10^{-3}$ μg/ml, HT-29 ED_{50} 2.9 μg/ml, B16F10 ED_{50} $6.4x10^{-1}$ μg/ml, 9PS ED_{50} $1.7x10^{-3}$ μg/ml; NCI tumor panel (3 day) selectivity with mean molar log GI_{50} 6.09, delta 1.91, range 2.55.

Isoannonacin (**12**) (NSC 618571): BS LC_{50} 2.3 ppm; at Purdue A-547 ED_{50} $9.6x10^{-3}$ μg/ml, MCF-7 ED_{50} $1.1x10^{-2}$ μg/ml, HT-29 ED_{50} $<10^{-3}$ μg/ml;

at Abbott A-549 ED_{50} 2.7×10^{-2} µg/ml, HT-29 ED_{50} 3.4 µg/ml, B16F10 ED_{50} 4.9×10^{-1} µg/ml, 9PS ED_{50} 1.1×10^{-2} µg/ml; NCI tumor panel (3 day) selectivity with mean molar log GI_{50} 5.87, delta 2.13, range 3.03.

Annonacin-10-one (**36**) (NCI 643357): BS LC_{50} 1.4 ppm; at Purdue A-549 ED_{50} 7.6×10^{-2} µg/ml, MCF-7 ED_{50} 1.1×10^{-2} µg/ml, HT-29 ED_{50} $<10^{-3}$ µg/ml; at Abbott A-549 ED_{50} 1.5×10^{-2} µg/ml, HT-29 ED_{50} 8.3 µg/ml, B16F10 ED_{50} 4.6×10^{-1} µg/ml, 9PS ED_{50} 1.19×10^{-2} µg/ml; NCI tumor panel (3 day) selectivity with mean molar log GI_{50} 5.81, delta 1.55, range 2.41.

Isoannonacin-10-one (**37**) (NCI 643358): BS LC_{50} 3.4 ppm; at Purdue A-549 ED_{50} 9.7×10^{-3} µg/ml, MCF-7 ED_{50} 1.4×10^{-2} µg/ml, HT-29 ED_{50} 1.8×10^{-3} µg/ml; NCI tumor panel (3 day) selectivity with mean molar log GI_{50} 5.52, delta 1.59, range 2.41.

Goniothalamicin (**38**) (NCI 617594): BS LC_{50} 24 ppm; potato disc 68% tumor inhibition; at Purdue 9KB ED_{50} $<10^{-2}$ µg/ml, 9PS ED_{50} $<10^{-1}$ µg/ml, A-549 ED_{50} 8.0×10^{-3} µg/ml, MCF-7 ED_{50} 5.7×10^{-2} µg/ml, HT-29 1.1×10^{-3} µg/ml; NCI tumor panel (7 day) ave. molar log IC_{50} -0.9, delta 1.7, range 3.5; NCI tumor panel (3 day) selectivity with mean molar log GI_{50} 6.03, delta 1.97, range 2.91; insecticidal (Dow Elanco); antimalarial (Walter Reed).

Muricatacin (**39**): BS LC_{50} 2.86 ppm; at Purdue A-549 ED_{50} 23.3 µg/ml, MCF-7 ED_{50} 9.8 µg/ml HT-29 ED_{50} 14.0 µg/ml.

***Annona reticulata* L. (Annonaceae).** This is another tropical fruit tree popularly known as "bullock's heart". Previous phytochemical work had identified an acetogenin for which we revised the structure (*82*), and a sterically unidentified adjacent bis-THF acetogenin had been briefly described without experimental details (*94*). The bark gave an extract (F005) which is quite active. In our first efforts (*95*), we reported bullatacin (**3**), liriodenine, two known diterpenes, and a new mono-THF acetogenin, reticulatacin (**40**), which may be identical to uvarimicin II with a *threo*, *trans*, *threo*-configuration from C17-C22. Subsequently, bullatacinone (**4**), asimicin (**2**), and *trans*-isoannonacin-10-one (**41**) have been isolated; the ketolactones like isoannonacin-10-one (**37**) are usually *cis* and *trans* mixtures at C2, and tedious hplc resolved these. Work is continuing with this plant with some additional novel acetogenins on the way.

Reticulatacin (**40**): BS LC_{50} 27.32 ppm; at Purdue A-549 ED_{50} 3.49 µg/ml, MCF-7 ED_{50} 2.91 µg/ml, HT-29 ED_{50} 4.66 µg/ml.

trans-Isoannonacin-10-one (**41**) (see structure **37**): BS LC_{50} 3.38; at Purdue A-549 ED_{50} 9.7×10^{-3} µg/ml, MCF-7 ED_{50} 1.4×10^{-2} µg/ml, HT-29 ED_{50} 1.8×10^{-3} µg/ml.

***Melodorum fruticosum* Lour. (Annonaceae).** In our previous review (*44*), six known bioactive compounds (*96*) and three new heptenes were described in these extracts from Thailand (*97*). Of these, the triterpene polycarpol (NSC 637387), recently showed promising selectivity in the NCI tumor panel against non-small cell lung and a single melanoma cell line (mean molar GI_{50} 5.33, delta 1.71, range 2.46). Four additional new bioactive heptenes, melodorinol (**42**), homomelodienone (**43**), 7-hydroxy-6-hydromelodienone (**44**), and homoisomelodienone (**45**), have been isolated and reported (*98*). The most cytotoxic of these were submitted to NCI for the cell panel and AIDS antiviral screening, but no results have been reported to us as yet. These 13 compounds seem to represent all of the bioactivity detected in this species, and it does not contain acetogenins even though the plant is in the Annonaceae. The heptenes are a new class of bioactive compounds. Six of the same compounds were subsequently reported from the same species but by using conventional cytotoxicities to direct the work (*99*). Compounds **42-45** were not tested in our simple bioassays because of their short supply.

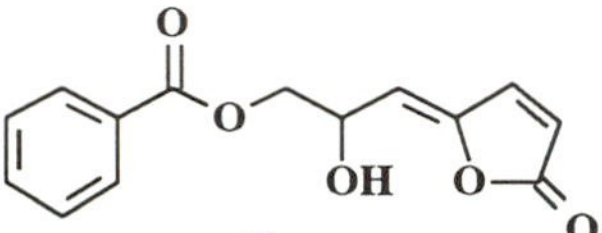

42 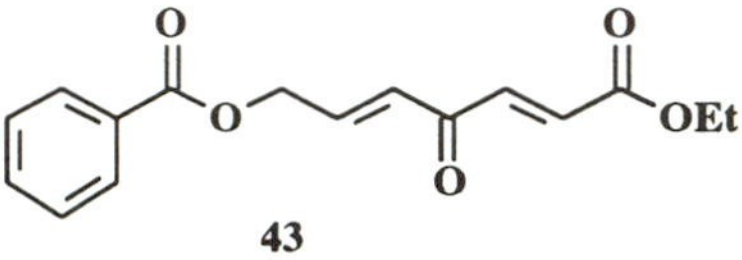 **43**

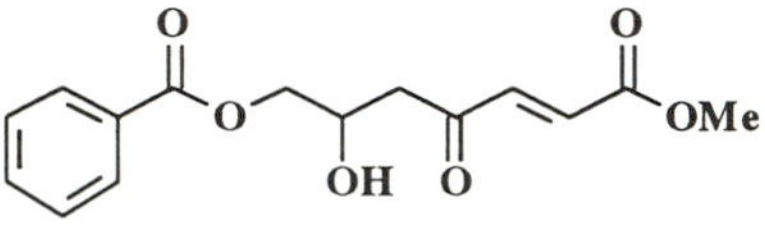

44 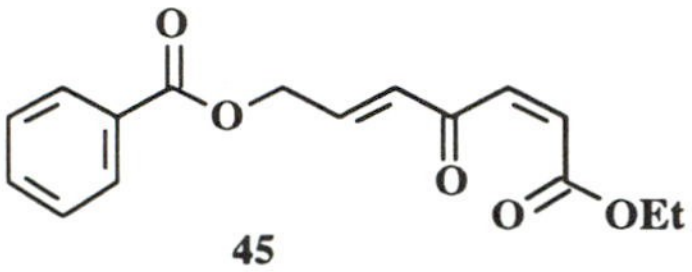 **45**

Melodorinol (**42**) (NSC 637382): At Purdue A-549 ED_{50} 5.89 μg/ml, MCF-7 ED_{50} 1.99 μg/ml, HT-29 2.87 μg/ml, SK-MEL-5 3.75 μg/ml, MALME-3M 3.32 μg/ml.
Homomelodienone (**43**): At Purdue A-549 ED_{50} 36.91 μg/ml, MCF-7 ED_{50} 25.61 μg/ml, HT-29 ED_{50} 37.88 μg/ml.
7-Hydroxy-6-hydromelodienone (**44**) (NSC 637385): At Purdue A-549 ED_{50} 3.28 μg/ml MCF-7 ED_{50} 1.92 μg/ml, HT-29 ED_{50} 2.53×10^{-1} μg/ml, SK-MEL-5 ED_{50} 1.04 μg/ml, MALME-3M ED_{50} 1.10 μg/ml.
Homoisomelodienone (**45**): At Purdue A-549 ED_{50} 6.70 μg/ml, MCF-7 ED_{50} 4.26 μg/ml, HT-29 ED_{50} 36.32 μg/ml.

***Polyalthia longifolia* Thw. (Annonaceae).** This is a popular ornamental tree in India. Extracts of the bark were bioactive, and the brine shrimp test guided the fractionation to three cytotoxic clerodane diterpenes: 16α-hydroxy-cleroda-3,13(14)*Z*-dien-15,16-olide (**46**), kolavenic acid (**47**), and polyalthialdoic acid (**48**); **48** is a novel natural product and was quite significantly cytotoxic, but all are apparently nonselective (*100*). No bioactive acetogenins were found in this Annonaceae species, and bioactive diterpenes were not expected. The methyl esters of **47** and **48** were cytotoxic but less active.

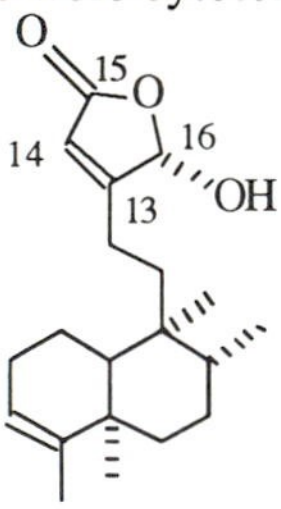

46

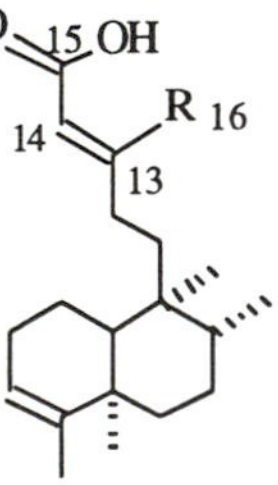

47 R=CH_3
48 R=CHO

16α-Hydroxy-cleroda-3,13(14)Z-dien-15,16-olide (**46**): BS LC_{50} 4.9 ppm; potato disc 83% tumor inhibition; at Purdue A-549 ED_{50} 2.97 μg/ml, MCF-7 ED_{50} 2.15 μg/ml, HT-29 ED_{50} 1.91 μg/ml.

Kolavenic acid (**47**): BS LC_{50} 9.1 ppm; potato disc 82% tumor inhibition; at Purdue A-549 ED_{50} 3.34 μg/ml, MCF-7 ED_{50} 1.81, HT-29 ED_{50} 1.39 μg/ml.

Polyalthialdoic acid (**48**): BS LC_{50} 0.6 ppm; potato disc 95% tumor inhibition; at Purdue A-549 ED_{50} $6.84x10^{-1}$ μg/ml, MCF-7 $5.5x10^{-1}$ μg/ml HT-29 ED_{50} $7.53x10^{-1}$ μg/ml.

***Persea major* Mill. (Lauraceae).** Previously two novel bioactive δ-lactones, majorynolide (**49**, NSC 644647) and majorenolide (**50**, NSC 644648), were isolated from the bark extracts of this Brazilian avocado species (*101*). Now a third compound, majoranolide (**51**) has been isolated (*102*); the absolute stereochemistry at C5 was predicted by Horeau's method and cd studies. Isolate **51** was quite bioactive but not selective; **50** was quite insecticidal against melon aphids (80% mortality at 10 ppm) at Dow Elanco. No cytotoxicity results have been received yet from NCI regarding the testing of **49** and **51**, but both are inactive in the AIDS screen.

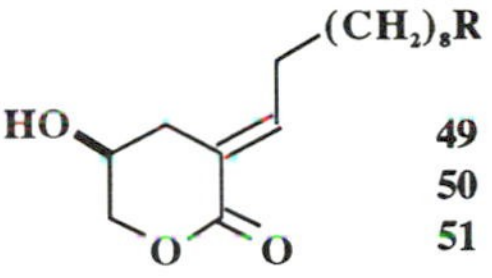

$(CH_2)_8R$

HO O O

49 R = CCH
50 R = $CHCH_2$
51 R = $(CH_2)_3CH_3$

Majoranolide (**51**): BS LC_{50} 63 ppm; potato disc 84% tumor inhibition; at Purdue A-549 ED_{50} $4.98x10^{-1}$ μg/ml, MCF-7 ED_{50} $5.04x10^{-1}$ μg/ml, HT-29 ED_{50} $3.11x10^{-1}$ μg/ml.

***Cryptocarya crassinervia* Miq. (Lauraceae).** The bark of this species from Malaysia gave extracts positive in the brine shrimp test. Column chromatography quickly resulted in crystals of (-)-grandisin (**52**), a known lignan, as the bioactive component; X-ray crystallography was used to determine the relative stereochemistry (*103*). Substance **52** was not significantly cytotoxic, but these lignans have been patented by Merck and Co. as antagonists of platelet-activating factor.

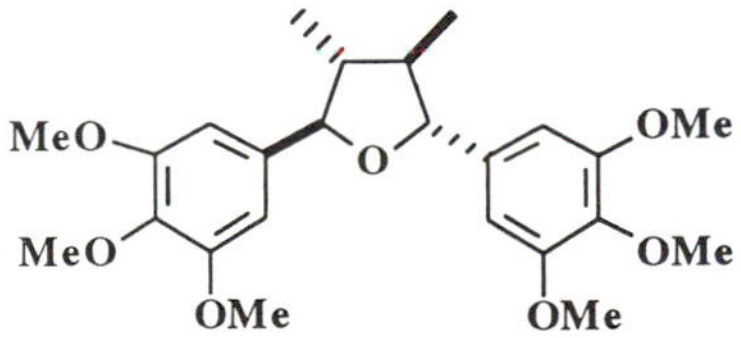

MeO MeO OMe O OMe OMe OMe

52

(-)-Grandisin (**52**): BS LC_{50} 21 ppm; at Purdue A-549 LC_{50} >10 μg/ml, MCF-7 LC_{50} >10 μg/ml, HT-29 >10 μg/ml.

***Endlicheria dysodantha* Miq. (Lauraceae).** Root extracts of this Peruvian laurel were active in our simple bioassays. Activity-directed fractionation, using brine shrimp lethality, initially led to four benzylbenzoates (**53-56**) which, however, failed to explain the good potato disc and cell culture activities of the original extract (*104*). Further fractionation of some fractions that were less toxic to the shrimp led to five new dysodanthins which are hexahydrobenzofuranoid neolignans (**57-61**) and three known neolignans (**62-64**) (*105*; *106*). Two of the known neolignans, megaphone acetate (**63**) and megaphyllone acetate (**64**), are significantly bioactive, and tumor panel tests results have just been received from NCI on **63**.

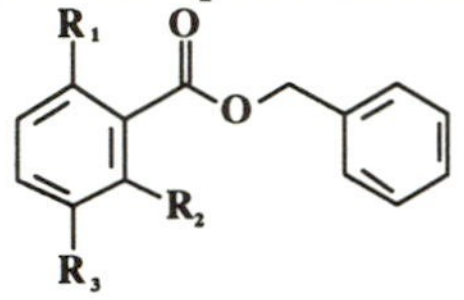

	R1	R2	R3
53	OH	H	H
54	OH	OMe	H
55	OMe	OMe	H
56	OMe	H	OMe

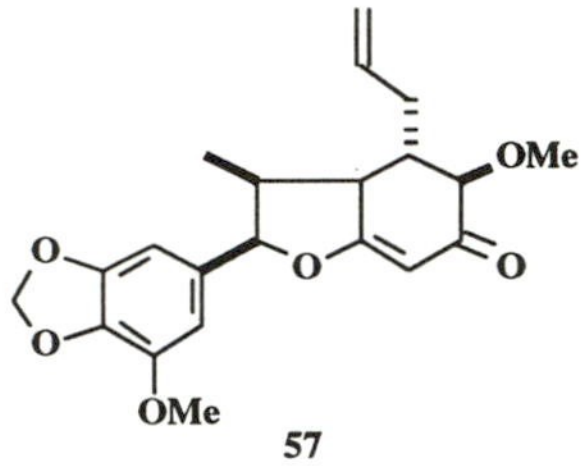

57 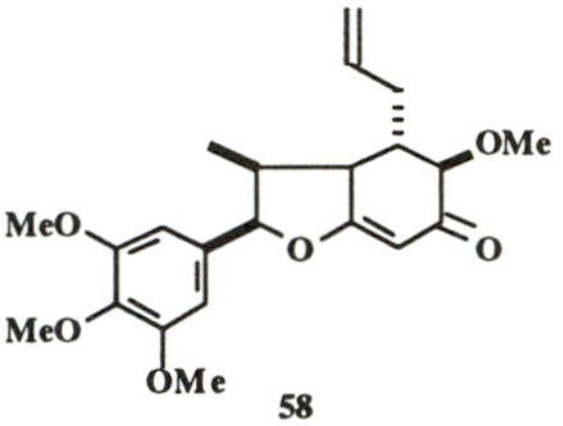 **58** 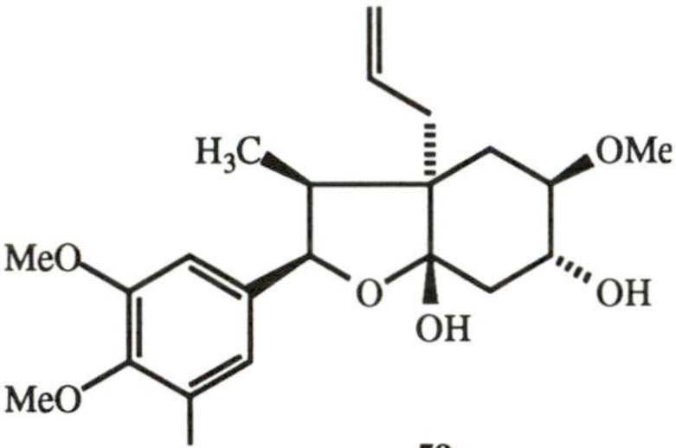 **59** 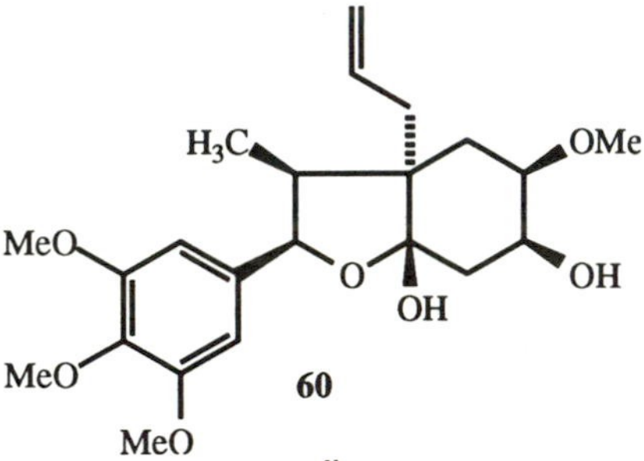 **60** 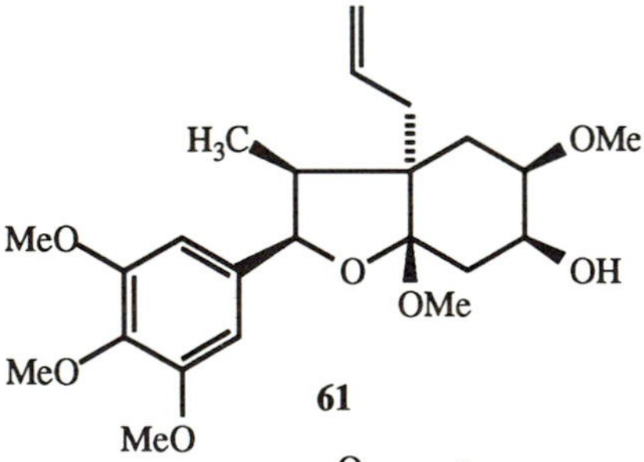 **61** 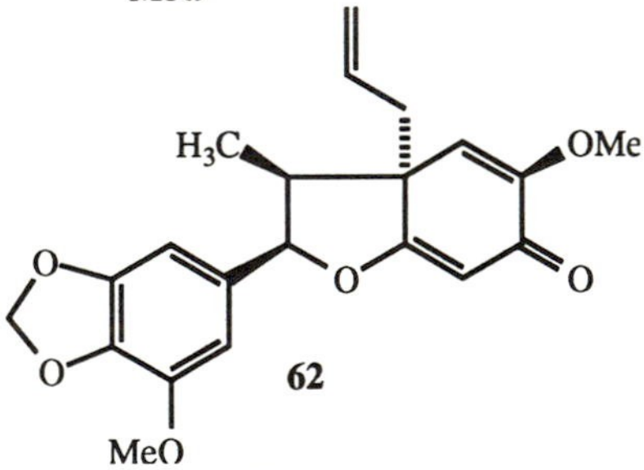 **62** 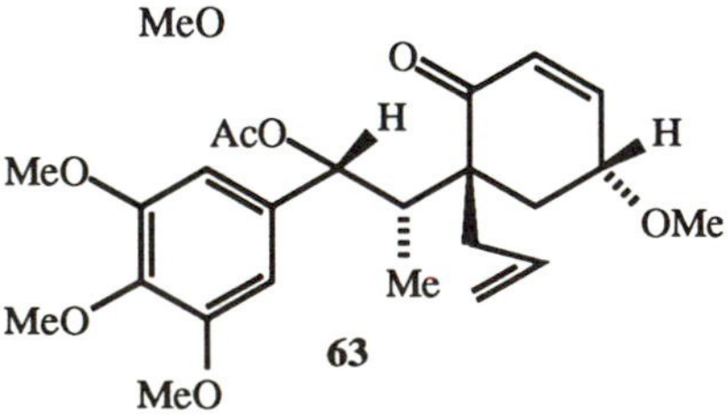 **63** 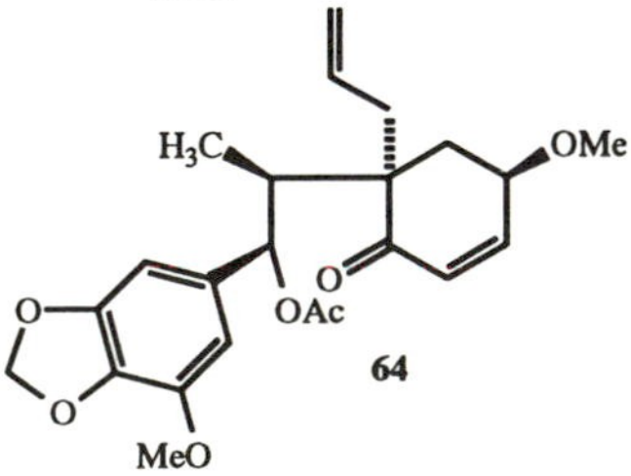 **64**

Benzyl 2-hydroxybenzoate (**53**): BS LC_{50} 0.34 ppm; potato disc 38% tumor inhibition; at Purdue A-549 ED_{50} 75.0 μg/ml, MCF-7 38.4 mg/ml, HT-29 ED_{50} 27.7 μg/ml.

Benzyl 2-hydroxy-6-methoxybenzoate (**54**): BS LC_{50} 3.60 ppm; potato disc 35% tumor inhibition; at Purdue A-549 ED_{50} >100 μg/ml, MCF-7 ED_{50} 53.3 μg/ml, HT-29 ED_{50} 33.0 μg/ml.

Benzyl 2,6-dimethoxybenzoate (**55**): BS LC_{50} 1.83 ppm; potato disc 39% tumor inhibition; at Purdue A-549 ED_{50} 33.3 μg/ml, MCF-7 ED_{50} 45.4 μg/ml, HT-29 ED_{50} 31.3 μg/ml.

Benzyl 2,5-dimethoxybenzoate (**56**): BS LC50 1.45 ppm; potato disc 6% tumor inhibition; at Purdue A-549 ED_{50} >100 μg/ml, MCF-7 ED_{50} 38.8 μg/ml, HT-29 ED_{50} 23.6 μg/ml.

Dysodanthin A (**57**): BS LC_{50} 47.5 ppm; potato disc 32% tumor inhibition; at Purdue A-549 ED_{50} 4.23 μg/ml, MCF-7 ED_{50} 5.51 μg/ml, HT-29 ED_{50} 2.39 μg/ml.

Dysodanthin B (**58**): BS LC_{50} 334 ppm; potato disc 52% tumor inhibition; at Purdue A-549 ED_{50} 3.68 μg/ml, MCF-7 ED_{50} 4.09 μg/ml, HT-29 ED_{50} 2.76 μg/ml.

Dysodanthin D (**59**): BS LC_{50} >1000 ppm; potato disc 11% tumor inhibition; at Purdue A-549 ED_{50} 41.8 μg/ml, MCF-7 ED_{50} 52.6 μg/ml, HT-29 ED_{50} 32.1 μg/ml.

Dysodanthin E (**60**): BS LC_{50} >1000 ppm; potato disc 8% tumor inhibition; at Purdue A-549 ED_{50} 9.86 μg/ml, MCF-7 ED_{50} 3.32 μg/ml, HT-29 ED_{50} 4.14 μg/ml.

Dysodanthin F (**61**): BS LC_{50} >1000 ppm; potato disc 13% tumor inhibition; at Purdue A-549 ED_{50} 54.6 μg/ml, MCF-7 ED_{50} 53.5 μg/ml, HT-29 ED_{50} 38.1 μg/ml.

Dysodanthin C (bruchellin analogue)(**62**): BS LC_{50} 162 ppm; potato disc inactive; at Purdue A-549 ED_{50} >100 μg/ml, MCF-7 ED_{50} 79.0 μg/ml, HT-29 ED_{50} 96.5 μg/ml.

Megaphone acetate (**63**) (NSC 644649): BS LC_{50} 0.34 ppm; potato disc 48% tumor inhibition; A-549 ED_{50} 4.35×10^{-2} μg/ml, MCF-7 ED_{50} 2.16 μg/ml, HT-29 ED_{50} 6.08×10^{-2} μg/ml; NCI tumor panel (3 day) mean GI_{50} 4.74, delta 0.74, range 1.48.

Megaphyllone acetate (**64**): BS LC_{50} 0.34 ppm; potato disc 42% tumor inhibition; at Purdue A-549 ED_{50} 2.26×10^{-1} μg/ml, MCF-7 ED_{50} 4.02×10^{-1} μg/ml, HT-29 ED_{50} 4.10×10^{-2} μg/ml.

***Lindera benzoin* (L.) Blume (Lauraceae).** This species is called spicebush and is native to Indiana. We initially tested the leaves because they repel mosquitoes and Gypsy moths. However, the ripe berries were more bioactive and became the subject of a major project (*107*). Brine shrimp lethality-directed fractionation of F005 led to the isolation of three new C_{21} alkane-alkene γ-lactones (**65-67**) as well as the known series of C_{17} and C_{19} obtusilactones (**68-71**) previously isolated from *Lindera obtusiloba* (*108*). The novel methylketoalkenes (**72,73**) and the known sesquiterpene (**74**) were also isolated as bioactive constituents. With the exception of **73** and **74**, all of these compounds showed moderate, but significant, cytotoxicities in one or more of our cell lines; **67** showed good selectivity for the lung cells (A-549).

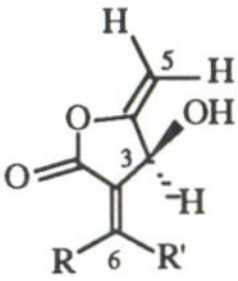

65 R=H, R'= (21)

66 R=H, R'= (21)

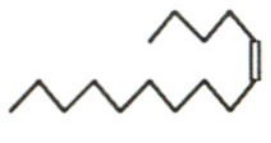

67 R= (21), R'=H

68 R=H, R'= (19)

69 R= (19), R'=H

70 R=H, R'= (17)

71 R= (17), R'=H

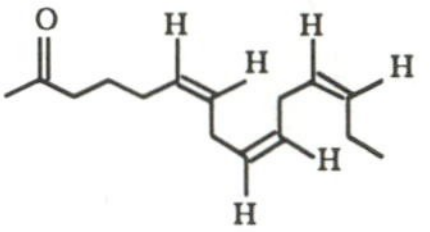

72

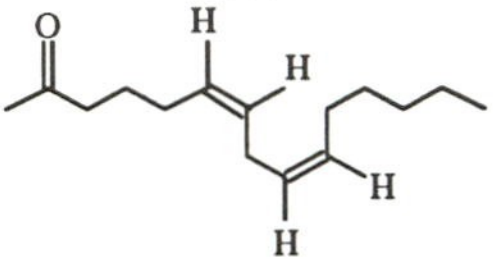

73

74

Isolinderanolide (**65**): BS LC_{50} 0.96 ppm; potato disc 45% tumor inhibition; at Purdue A-549 ED_{50} 4.57 µg/ml, MCF-7 ED_{50} 3.12 µg/ml, HT-29 3.07 µg/ml; insecticidal against corn root worm (20% mortality at 12 ppm).

Isolinderenolide (**66**): BS LC_{50} 0.52 ppm; potato disc 42% tumor inhibition; at Purdue A-549 ED_{50} 3.01 µg/ml, MCF-7 ED_{50} 2.58 µg/ml, HT-29 ED_{50} 1.90 µg/ml.

Linderanolide (**67**): BS LC_{50} 5.23 ppm; at Purdue A-549 9.28×10^{-1} µg/ml, MCF-7 ED_{50} 73.6 µg/ml, HT-29 ED_{50} 30.5 µg/ml.

Isoobtusilactone A (**68**): BS LC_{50} 0.72 ppm; at Purdue A-549 ED_{50} 3.75 µg/ml, MCF-7 ED_{50} 2.31 µg/ml, HT-29 ED_{50} 2.48 µg/ml.

Obtusilactone A (**69**): BS LC_{50} 0.89 ppm; at Purdue A-549 ED_{50} >10 µg/ml, MCF-7 ED_{50} 5.12 µg/ml, HT-29 ED_{50} 2.93 µg/ml.

Obtusilactone (**71**): BS LC_{50} 0.035 ppm; potato disc 64% tumor inhibition; at Purdue A-549 ED_{50} 3.32 µg/ml, MCF-7 ED_{50} 4.58 µg/ml, HT-29 ED_{50} 3.26 µg/ml.

(6Z,9Z,12Z)-pentadecatrien-2-one (**72**): BS LC_{50} 3.0 ppm; at Purdue A-549 ED_{50} >10 µg/ml, MCF-7 ED_{50} 5.15 µg/ml, HT-29 ED_{50} 3.01 µg/ml.

(6Z,9Z)-pentadecadien-2-one (**73**): BS LC_{50} 2.9 ppm; A-549 ED_{50} >10 µg/ml, MCF-7 ED_{50} >10 µg/ml, HT-29 ED_{50} >10 µg/ml.

(+)-(*Z*)-Nerolidol (**74**): BS LC_{50} 5.75 ppm; A-549 ED_{50} >10 µg/ml, MCF-7 ED_{50} >10 µg/ml, HT-29 ED_{50} >10 µg/ml.

***Aralia spinosa* L. (Araliaceae).** This spiny species is called "devil's walking stick" and is abundant in southern Indiana. In the fall it bears huge umbels of blue-colored berries which birds and other animals seem to shun. The extracts of the berries were moderately toxic to brine shrimp and fractionation yielded free petroselinic acid (**75**) which showed slight cytotoxicity (*109*).

$Me(CH_2)_9$ — 7 — 6 — $(CH_2)_4$ — 1 — C(=O)OH

75

Petroselinic acid (**75**): BS LC_{50} 18.32 ppm; at Purdue A-549 ED_{50} 11.9 µg/ml, MCF-7 ED_{50} 33.76 µg/ml, HT-29 ED_{50} 27.91 µg/ml.

Summary

In the past three years, simple bioassays, using brine shrimp and potato discs, have quickly led us to several dozen new cytotoxic compounds in higher plant extracts. These convenient bench-top methods detect a wide array of bioactive chemical classes representing a diversity of antitumor mechanisms (*39*). *In vivo* antitumor studies have revealed that our most promising leads are the Annonaceous acetogenins (*82, 110*); these compounds are powerful inhibitors of ATP production via blockage of complex I (NADH: ubiquinone oxidoreductase) in mitochondrial electron transport systems (*61*). No current chemotherapeutic agents act by this mechanism.

Acknowledgments

We are indebted to all of the students, postdoctorates, colleagues, and visiting scholars whose names as co-authors are listed on many of the references cited below. Major financial support for the past three years has come from R01 Grant No. CA 30909-07-09 from the NCI/NIH. Special thanks are due to the Upjohn Company, Abbott Laboratories, Glaxo Inc., Dow Elanco, the National Cancer Institute, Eli Lilly and Company, AgriDyne, Inc., D.J. Morré, and the Purdue Cell Culture Laboratories for bioassay results. Nelson R. Ferrigni and Brian N. Meyer deserve special thanks for their perserverance in developing these convenient bench-top methods.

Literature Cited

1. Cancer Facts and Figures-1992; American Cancer Society: Atlanta, 1992; pp 1-7.
2. Suffness, M.; Douros, J. *J. Nat. Prod.* **1982**, *45*, 1-14.
3. Cassady, J.M.; Chang, C.-J.; Cooks, R.G. In *Natural Products Chemistry*, Atta-ur-Rahman; LeQuesne, P.W., Eds.; Springer-Verlag: Heidelberg;1988, Vol. 3; pp 291-304.
4. *The Catharanthus Alkaloids*; Taylor, W.I.; Farnsworth, N.R. Eds.; Marcel Dekker: New York, 1975, pp 1-323.
5. *Etoposide (VP-16): Current Status and New Developments*; Issell, B.F., Muggia, F.M.; Carter, S.K., Eds.; Academic Press, Orlando, Florida, 1984, pp 1-355.
6. Giovanella, B.C., Stehlin, J.S.; Wall, M.E.; Wani, M.C.; Nicholas, A.W.; Liu, L.F.; Silber, R.; Potmesil, M. *Science* **1989**, *246*, 1046-1047 .
7. Rowinsky, E.K.; Arzenane, L.A.; Donehower, R.C. *J. Natl. Cancer Inst.* **1990**, *82*, 1247-1259.

8. Kingston, D.G.I.; Samaranayake, G.; Ivey, C.A. *J. Nat. Prod.* **1990**, *53*, 1-12; and references cited therein.
9. Suffness, M. *Gann Monogr. Cancer Res.*, **1989**, *36*, 21-44; and references cited therein.
10. Lomax, N.R.; Narayanan, V.L. *Chemical Structures of Interest to the Division of Cancer Treatment*, National Cancer Institute, Bethesda, Maryland, August, 1988; Vol. 6.
11. Shoemaker, R.H.,; Abbott, B.J.; MacDonald, M.M.; Mayo, J.G.; Venditti, J.M.; Wolpert-DeFillippes, M.K. *Cancer Treat. Rep.* **1983**, *67*, (no. 1), 97.
12. Geran, R.I.; Greenberg, N.H.; MacDonald, M.M.; Schumacher, A.A.; Abbott, B.J. *Cancer Chemother. Rep.* **1972**, *3*, part 3, 1-103.
13. Suffness, M.; Douros, J. In *Methods in Cancer Research*; DeVita, V.T.; Busch, H.; Eds.; Academic Press; New York, 1979; Vol. 16; pp 73-126.
14. Monks, A.; Scudiero, D.; Skelhan, P.; Shoemaker, R.; Paull, K.; Vistica, D.; Hose, C.; Langley, J.; Cronice, P.; Vaigio-Wolff, A.; Gray-Goodrich, M.; Cambell, H.; Mayo, J.; Boyd, M. *J. Natl. Cancer Inst.* **1991**, *83*, 757-766.
15. White, R.J., *Ann. Rev. Microbiol.* **1982**, *36*, 415-433.
16. Otani, T.T.; Briley, M.R.; Geran, R.I. *J. Pharm. Sci.* **1984**, *73*, 264-265.
17. Johnson, R.K.; Hertzberg, R.P. *Ann. Rep. Med. Chem.* **1990**, *25*, 129-140.
18. Johnson, R.K., *J. Natl. Cancer Inst.* **1990**, *82*, 1082-1083.
19. Chabner, B.A., *J. Natl. Cancer Inst.* **1990**, *82*, 1083-1084.
20. Chang, C.-j.; Ashendel, C.L.; Chan, T.; Geahlen, R.L.; McLaughlin, J.L. In *Advances in New Drug Development,* Kim, B.-K.; Lee, E.-B.; Kim, C.-K.; Han, Y.-N., Eds.; The Pharmaceutical Society of Korea: Seoul, 1991; pp 448-457.
21. Hooyokaas, P.J.J.; Melchjers, L.S.; Rodenburg, K.W.; Turk, S.C.H. In *Plant Molecular Biology*; Herrmann, R.G.; Larkins, B., Eds.; Plenum Press; New York, 1991; Vol. 2; pp 193-204.
22. Chilton, M.-D.; Saihi, R.K.; Yadow, N.; Gordon, M.P.; Quetier, F. *Proc. Natl. Acad. Sci. USA*, *77*, 4060-4064.
23. Nester, E.W.; Gordon, M.P.; Amasino, R.M.; Yanofsky, M.F. *Ann. Rev. Plant Physiol.* **1984**, *35*, 387-413.
24. Cangelosi, G.A.; Best, E.A.; Martinetti, G.; Nester, E.W. In *Methods in Enzymology*, Miller, J.H., Ed.; Academic Press: San Diego, California, 1991; Vol. 204; pp 384-397.
25. Saito, K.; Yamazaki, M.; Murakoshi, I. *J. Nat. Prod.* **1992**, *55*, 149-162.
26. Karpas, A., *Amer. Sci.* **1982**, *70*, 277-285.
27. Binns, A.N.; Thomashow, M.F. *Ann. Rev. Microbiol.* **1988**, *42*, 575-606.
28. Galsky, A.G.; Wilsey, J.P.; Powell, R.G. *Plant Physiol.* **1980**, *65*, 184-185.
29. Galsky, A.G.; Kozimor, R.; Piotrowski, D.; Powell, R.G. *J. Natl. Cancer Inst.* **1981**, *67*, 689-692.
30. Mitscher, L.A.; Len, R.-P.; Bathala, M.S.; Wu, W.-W.; Beal, J.L. *Lloydia* **1972**, *35*, 157-166.
31. Ferrigni, N.R.; Putnam, J.E.; Anderson, B.; Jacobsen, L.B.; Nichols, D.E.; Moore, D.S.; McLaughlin, J.L. *J. Nat. Prod.* **1982**, *45*, 679-686.
32. Landis, J.R.; Koch, G.G. *Biometrics* **1977**, *33*, 159-174.
33. McLaughlin, J.L.; Ferrigni, N.R. *Proceedings of Symposium on Discovery and Development of Naturally Occurring Antitumor Agents*, NCI/FCRF, Frederick, Maryland, June 27-29, 1983, pp 9-12.
34. Meyer, B.N.; Ferrigni, N.R.; Putnam, J.E.; Jacobsen, L.B.; Nichols, D.E.; McLaughlin, J.L. *Planta Med.* **1982**, *45*, 31-34.

35. *Fundamental and Applied Research on the Brine Shrimp, Artemia salina (L.) in Belgium;* Jaspers, E. Ed.; European Mariculture Society, special public. no. 2, Breden, Belgium, 1977.
36. Persoone, G., *Proceedings of the International Symposium on Brine Shrimp, Artemia salina,*; Universa Press, Witteren, Belgium, 1980; Vols. 1-3.
37. Sleet, R., *Lab Animal* **1992**, *21*, 26-36.
38. Barnes, D., *BioTechniques* **1987**, *5*, 534-542.
39. Anderson, J.E.; Goetz, C.M.; McLaughlin, J.L.; Suffness, M. *Phytochem. Anal.* **1991**, *2*, 107-111.
40. MacRae, W.D.; Hudson, J.B.; Towers, G.H.N. *J. Ethnopharmacol.* **1988**, *22*, 143-172.
41. Mata, R., *Recent Advances in Phytochemistry* (submitted for publication) (1993).
42. Hamburger, M.; Hostettmann, K. *Phytochemistry* **1991**, *30*, 3864-3874.
43. McLaughlin, J.L. In *Methods in Plant Biochemistry*, Hostettmann, K., Ed.; Academic Press, London, **1991**, Vol. 6; pp 1-31.
44. McLaughlin, J.L.; Chang, C.-J.; Smith, D.L. In *Studies in Natural Products Chemistry*, Atta-ur-Rahman, Ed. Elsevier: Amsterdam, 1991; Vol. 9; pp 383-409.
45. McLaughlin, J.L. *Brenesia* **1991**, *34*, 1-14.
46. Geahlen, R.L.; McLaughlin, J.L. *Biochem. Biophys. Res. Commun.* **1989**, *165*, 241-245.
47. Cushman, M.; Nagarathnan, D.; Gopal, D.; Geahlen, R.L. *Bioorg. Med. Chem. Lett.* **1991**, *1*, 211-214.
48. Cushman, M.; Nagarathnan, D.; Gopal, D.; Geahlen, R.L. *Bioorg. Med. Chem. Lett.* **1991**, *1*, 215-218.
49. Pettit, G.R.; Singh, S.B.; Hamel, E.; Lin, C.M.; Alberts, D.S.; Garcia-Kendall, D. *Experientia* **1989**, *45*, 209-211; and references cited therein.
50. Pont, V.; Pezet, R. *J. Phytopathol.*, **1990**, *130*, 1-8.
51. Brinker, A.M.; Seigler, D.S. *Phytochemistry* **1991**, *30*, 3229-3232.
52. Kahl, G.; Schaefer, W. *Plant Cell Physiol.* **1984**, *25*, 1187-1196.
53. Toi, M.; Mukaida, H.; Wada, T.; Hirabayashi, N.; Toge, T.; Hori, T.; Umezawa, K. *Eur. J. Cancer,* **1990**, *26*, 722-724.
54. Rupprecht, J.K.; Chang, C.-J.; Cassady, J.M.; McLaughlin, J.L.; Mikolajczak, K.L.; Weisleder, D. *Heterocycles* **1986**, *24*, 1197-1201.
55. Mikolajczak, K.L.; McLaughlin, J.L.; Rupprecht, J.K. U.S. Patent No. 4,721,727; issued January 26, 1988.
56. Mikolajczak, K.L.; McLaughlin, J.L.; Rupprecht, J.K. U.S. Patent No. 4,855,319; issued August 8, 1989.
57. Alkofahi, A.; Rupprecht, J.K.; Anderson, J.E.; McLaughlin, J.L.; Mikolajczak, K.L.; Scott, B.A. In *Insecticides of Plant Origin*, Arnason, J.T.; Philogene, B.J.R.; Morand, P., Eds.; ACS Symposium Series 387, American Chemical Society: Washington, DC, 1989; pp 25-43.
59. Avalos, J.; Rupprecht, J.K.; McLaughlin, J.L.; Rodriguez, E. *Contact Dermatitis* **1992**, in press.
60. Lewis, M.A.; Arnason, J.T.; Philogene, B.J.R.; Rupprecht, J.K.; McLaughlin, J.L. *Pesticide Biochem. Physiol.* **1993**, in press.
61. Londershausen, M.; Leicht, W.; Lieb, F.; Moeschler, H. *Pesticide Science* **1991**, *33*, 427-438.
62. Hui, Y.-H.; Rupprecht, J.K.; Liu, Y.-M.; Anderson, J.E.; Smith, D.L.; Chang, C.-J.; McLaughlin, J.L. *J. Nat. Prod.* **1989**, *52*, 463-477.

63. Zhao, G.-X.; Hui, Y.-H.; Rupprecht, J.K.; McLaughlin, J.L.; Wood, K.V. *J. Nat. Prod.* **1992**, *55*, 347-356.
64. Fujimoto, Y.; Eguchi, T.; Kakinuma, K.; Ikekawa, N.; Sahai, M.; Gupta, Y.K. *Chem. Pharm. Bull.* **1988**, *36*, 4802-4806.
65. Xu, L.; Chang, C.-J.; Yu, J.-G.; Cassady, J.M.; *J. Org. Chem.* **1989**, *54*, 5418-5421.
66. Hoye, T.R. ; Shuhadolnik, J.C. *J. Am. Chem. Soc.* **1987**, *109*, 4402-4403.
68. Rieser, M.J.; Hui, Y.-H.; Rupprecht, J.K.; Kozlowski, J.F.; Wood, K.V.; McLaughlin, J.L.; Hoye, T.R.; Hanson, P.R.; Zhuang, Z.-P. *J. Am. Chem. Soc.*, **1992**, *114*, 10203-10213.
69. Ratnayake, S.; Fang, X.-P.; Anderson, J.E.; McLaughlin, J.L.; Evert, D.R. *J. Nat. Prod.* **1992**, *55*, 1462-1467.
70. Ebrahim El-Zayat, A.A.; Ferrigni, N.R.; McCloud, T.G.; McKenzie, A.T.; Byrn, S.R.; Cassady, J.M.; Chang, C.-j.; McLaughlin, J.L. *Tetrahedron Lett.*, **1985**, *26*, 955- 956.
71. Alkofahi, A.; Ma, W.-W.; McKenzie, A.T.; Byrn, S.R.; McLaughlin, J.L. *J. Nat. Prod.* **1989**, *52*, 1371-1373.
72. Fang, X.-P.; Anderson, J.E.; Chang, C.-J.; Fanwick, P.E.; McLaughlin, J.L. *J. Chem. Soc., Perkin Trans. I*, **1990**, 1655-1661.
73. Alkofahi, A.; Rupprecht, J.K.; Smith, D.L.; Chang, C.-J.; McLaughlin, J.L. *Experientia*, **1988**, *44*, 83-85.
74. Alkofahi, A.; Rupprecht, J.K.; Liu, Y.-M.; Chang, C.-J.; Smith, D.L.; McLaughlin, J.L. *Experientia*, **1990**, *46*, 539-541.
75. Fang, X.-P.; Anderson, J.E.; Chang, C.-J.; McLaughlin, J.L.; Fanwick, P.E. *J. Nat. Prod.* **1991**, *54*, 1034-1043.
76. Fang, X.-P.; Anderson, J.E.; Chang, C.-J.; McLaughlin, J.L. *Tetrahedron*, **1991**, *47*, 9751-9758.
77. Gracza, T.; Jaeger, V. *Synthet. Commun.*, in preparation (1993).
78. Fang, X.-P.; Rupprecht, J.K.; Alkofahi, A.; Hui, Y.-H.; Liu, Y.-M.; Smith, D.L.; Wood, K.V.; McLaughlin, J.L. *Heterocycles* **1991**, *32*, 11-17.
79. Fang, X.-P.; Anderson, J.E.; Smith, D.L.; Wood, K.V.; McLaughlin, J.L. *Heterocycles* **1991**, *34*, 1045-1083.
80. Jossang, A.; Bubaele, B.; Cave, A.; Bartoli, M.-H.; Beriel, H. *J. Nat. Prod.* **1991**, *54*, 967-971.
81. Fang, X.-P.; Anderson, J.E.; Smith, D.L.; Wood, K.V.; McLaughlin, J.L. *Nat. Prod.* **1992**, *55*, 1655-1663.
82. Rupprecht, J.K.; Hui, Y.-H.; McLaughlin, J.L. *J. Nat. Prod.* **1990**, *53*, 237-278.
83. Hui, Y.H.; Rupprecht, J.K.; Anderson, J.E.; Lui, Y.M.; Smith, D.L.; Chang, C.-J.; McLaughlin, J.L. *Tetrahedron* **1989**, *45*, 6941-6948.
84. Folkers, K., *Cancer Chemother. Rep.* **1974**, *4*, part 2, 19-22.
85. Paull, K.D.; Shoemaker, R.H.; Hodes, L.; Monks, A.; Scudiero, D.A.; Rabinstein, L.; Plowman, J.; Boyd, M.R. *J. Natl. Cancer Inst.* **1989**, *81*, 1088-1092.
86. Morré, D.J.; Brightman, A.O. *J. Bioenerg. Biomemb.* **1991**, *23*, 469-489.
87. Kwatsu, Y.; Sakurai, M.; Hagino, H.; Inuzuka, K. *Agric. Biol. Chem.* **1984**, *48*, 1997-2002.
88. Hui, Y.-H.; Rupprecht, J.K.; Anderson, J.E.; Wood, K.V.; McLaughlin, J.L. *Phytother. Res.* **1991**, *5*, 124-129.
89. Hui, Y.-H.; Wood, K.V.; McLaughlin, J.L. *Natural Toxins* **1992**, *1*, 4-14.
90. Ikegawa, N.; Fujimoto, G.; Ikegawa, T.; Japanese Patent no. Hesei 3-41076, issued February 21, 1991.

91. Rieser, M.J.; Fang, X.-P.; Rupprecht, J.K.; Hui, Y.-H.; Smith, D.L.; McLaughlin, J.L. *Planta Med.* **1993**, in press.
92. Rieser, M.J.; Kozlowski, J.F.; Wood, K.V.; McLaughlin, J.L. *Tetrahedron Lett.* **1991**, *32*, 1137-1140.
93. Scholtz, G.; Tochtermann, W. *Tetrahedron Lett.* **1991**, *32*, 5535-5538.
94. Cassady, J.M.; Baird, W.M.; Chang, C.-J. *J. Nat. Prod.* **1990**, *53*, 23-41.
95. Saad, J.M.; Hui, Y.-H.; Rupprecht, J.K.; Anderson, J.E.; Kozlowski, J.F.; Zhao, G.-X.; Wood, K.V.; McLaughlin, J.L. *Tetrahedron* **1991**, *47*, 2751-2756.
96. Jung, J.H.; Pummangura, S.; Chaichantipyuth, C.; Patarapanich, C.; McLaughlin, J.L. *Phytochemistry* **1990**, *29*, 1667-1670.
97. Jung, J.H.; Pummangura, S.; Chaichantipyuth, C.; Patarapanich, C.; Fanwick, P.E.; Chang, C.-J.; McLaughlin, J.L. *Tetrahedron* **1990**, *46*, 5043-5054.
98. Jung, J.H.; Chang, C.-J.; Smith, D.L.; McLaughlin, J.L.; Pummangura, S.; Chaichantipyuth, C.; Patarapanich, C. *J. Nat. Prod.* **1991**, *54*, 500-505.
99. Tuchinda, P.; Udchachon, J.; Rentrakul, V.; Santisuk, T.; Taylor, W.C.; Farnsworth, N.R.; Pezutto, J.M.; Kinghorn, A.D. *Phytochemistry* **1991**, *30*, 2685-2689.
100. Zhao, G.-X.; Jung, J.H.; Smith, D.L.; Wood, K.V.; McLaughlin, J.L. *Planta Med.* **1991**, *57*, 380-383.
101. Ma, W.-W.; Anderson, J.E.; Chang, C-.J.; Smith, D.L.; McLaughlin, J.L. *J. Nat. Prod.* **1989**, *52*, 1263-1266.
102. Ma, W.-W.; Anderson, J.E.; Chang, C.-J.; Smith, D.L.; McLaughlin, J.L. *Phytochemistry* **1990**, *29*, 2698-2699.
103. Saad, J.M.; Soepadamo, E.; Fang, X.-P.; McLaughlin, J.L.; Fanwick, P.E. *J. Nat. Prod.* **1991**, *54*, 1681-1683.
104. Ma, W.-W.; Anderson, J.E.; McLaughlin, J.L. *Internat. J. Pharmacog.* **1991**, *29*, 237-239.
105. Ma, W.-W.; Kozlowski, J.F.; McLaughlin, J.L. *J. Nat. Prod.* **1991**, *54*, 1153-1158.
106. Ma, W.-W.; Anderson, J.E.; McLaughlin, J.L.; Wood, K.V. *Heterocycles* **1992**, *34*, 5-11.
107. Anderson, J.E.; Ma, W.-W.; Smith, D.L.; Chang, C.-J.; McLaughlin, J.L. *J. Nat. Prod.* **1992**, *55*, 71-83
108. Niwa, M.; Iguchi, M.; Yamamura, S., *Chem. Lett.* **1977**, 581.
109. Ratnayake, S.; McLaughlin, J.L. *Int. J. Pharmacog.* **1993**, (in press).
110. Fang, X.-P.; Rieser, M.J.; Gu, Z.-M.; Zhao, G.-X.; McLaughlin, J.L. *Phytochem. Anal.* **1993**, *4*, 27-48 (plus appendices in press).

RECEIVED April 2, 1993

Chapter 10

Taxol, an Exciting Anticancer Drug from *Taxus brevifolia*

An Overview

David G. I. Kingston

Department of Chemistry, Virginia Polytechnic Institute and State University, Blacksburg, VA 24061–0212

The taxane diterpenoid taxol was first reported in 1971, but it has only recently been recognized as a highly effective anticancer drug. The history of taxol's development is reviewed with an emphasis on the problems that almost prevented the discovery of its clinical activity, and on the key factors that kept it under investigation. Recent research on the structure-activity relationships and the synthesis of taxol is also reviewed.

The toxic properties of yew have been known for at least two thousand years. Thus, Julius Caesar recorded that Catuvolcus, king of the Eburenes, poisoned himself with yew rather than face capture by Caesar's legions (*1*), and many other early accounts of yew poisoning have been reviewed (*2*). Because of this, initial chemical studies of the constituents of yew (either the English yew, *Taxus baccata,* or the Japanese yew, *Taxus cuspidata*) concentrated primarily on its toxic principles, culminating in the structure elucidation of the first taxane diterpenoids in the 1960's. It is thus ironic that the yew, long known as a tree of death, should become the source of one of the most promising and important new anticancer drugs of the last twenty years.

Discovery, Isolation, and Structure Elucidation of Taxol

By 1960, the National Cancer Institute (NCI) had recognized the importance of natural products as potential anticancer drugs, largely through the leadership of the late Jonathan Hartwell and his work on the constituents of *Podophyllum peltatum.* Under his guidance, a contract program was established to collect plant materials, screen them for biological activity, and fractionate them to obtain the pure active compounds. The plant collection was conducted by a team at the U.S. Department of Agriculture (USDA) under Dr. Robert Perdue, Jr., and one of the fractionation contracts was awarded to a team at the newly established Research Triangle Institute in North Carolina under Dr. Monroe Wall.

One day in 1964, Dr. Wall received a sample of the bark of the western yew, *Taxus brevifolia,* from NCI, collected by the USDA botanists. This sample was one of hundreds received that year, but this one was unusual in that its extracts showed a high cytotoxicity in the KB cell culture assay, although this activity was not exceptional. However, *T. brevifolia* bark was selected for fractionation, and a large recollection of bark was received in 1966. Fractionation of the bark extract proved

0097–6156/93/0534–0138$06.00/0

difficult because it turned out that the active material was present in very low abundance, and it was not until 1969 that adequate quantities of the active compound became available for structural work.

The structure elucidation of the active compound, named taxol because of its botanical origin and the fact that it had a hydroxyl group, proved to be a challenging task. Initial studies were complicated by the difficulty of obtaining reliable mass spectrometric data, and by the failure of taxol to crystallize in a form suitable for X-ray analysis. The structure was finally solved when Dr. Mansukh Wani at RTI discovered that it was possible to cleave taxol into two portions by Zemplen methanolysis. He was then able to obtain crystalline derivatives of each fragment. The structures of these two derivatives were elucidated by means of X-ray crystallography by Dr. Andrew McPhail at Duke University as the *p*-bromobenzoate derivative, **1**, and the bisiodoacetate **2**. The structure of taxol was then determined by oxidation studies with maganese dioxide, which established that the side chain

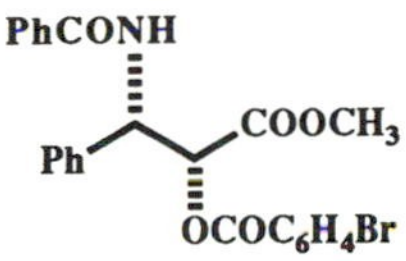

1 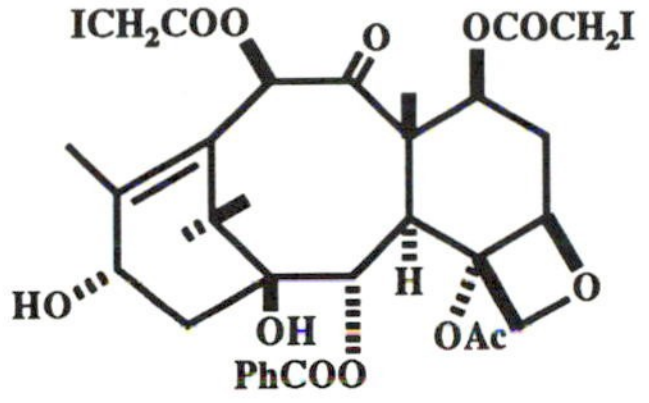 **2**

esterifies the taxane diterpenoid unit at the allylic hydroxyl group. The structure of taxol, first published in 1971, is thus **3** (*3*). It belongs to the class of taxane

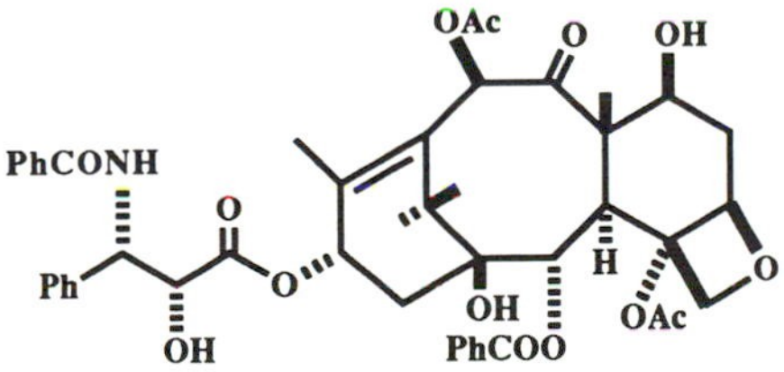

3

diterpenoids, or taxoids, and is structurally related to the toxic constituents of yew such as taxine B (**4**).

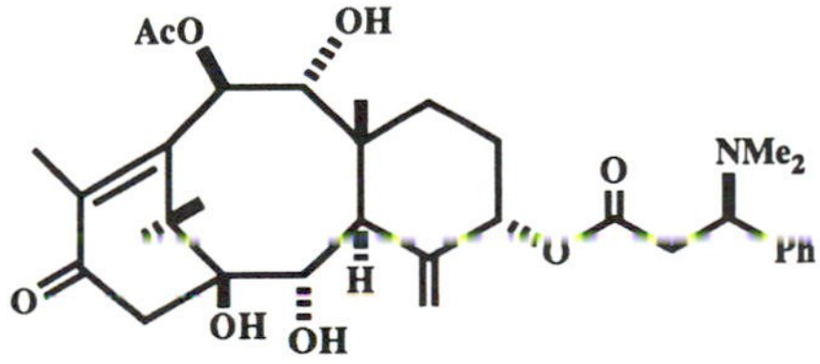

4

Development of Taxol as an Antitumor Agent

At the time of its discovery, taxol was but one of a number of promising natural product leads that were being investigated. The Wall group had earlier isolated camptothecin (*4*), and his and other research groups were reporting new active compounds on a regular basis. Taxol showed clear but modest *in vivo* activity in the P-388 and L-1210 leukemia assays, and its activity in these assays was no better than that of various other compounds competing for limited development resources, and activity against leukemias was not in and of itself sufficient cause for development. In addition, taxol had two significant disadvantages.

The first disadvantage was that of *supply*. Taxol was obtained from *T. brevifolia* bark by a laborious isolation procedure in an overall yield of only 0.02% w/w (*3*). Although *T. brevifolia* is not a rare tree, and was until recently treated as trash and burned by timber harvesters in the Pacific Northwest, it was nevertheless clear from the very first studies on taxol that its use as an anticancer drug would present a massive supply problem. To put this into perspective, the bark of about three trees is required to produce one gram of taxol. An early unpublished survey carried out by the National Cancer Institute (Dr. M. Suffness, personal communication) showed that the bark of *T. brevifolia* was the best source of biological activity, and thus there was little hope that a more abundant plant part might yield adequate amounts of taxol.

The second major disadvantage was *solubility*. Because of their narrow therapeutic index, anticancer drugs are almost all administered by intravenous infusion, and taxol is inactive orally against mouse tumors. A water-soluble formulation is thus necessary. Taxol is, however, almost completely insoluble in water, and it was not immediately obvious that a satisfactory water-soluble formulation could be achieved. In due course, a solution to this problem was eventually found, as described below, but at the cost of very nearly ending taxol's development.

Because of these two major disadvantages, the development of taxol as an anticancer drug was not pursued aggressively during the early 1970's. However, various new *in vivo* assay systems were developed during this time, including the use of athymic mice as hosts for human tumor xenografts. The NCI was thus able to test promising compounds against several new assays such as mammary, lung, and colon xenografts, and the B16 mouse melanoma assay. Among the compounds tested in this way was taxol, and the results were highly encouraging; taxol showed clear and convincing activity against various human solid tumors, including the MX-1 mammary xenograft, and against the B16 mouse melanoma (Table 1) (*5*). Based on these assay results, the decision was made in 1977 to begin the development of taxol as an anticancer agent.

Interest in taxol received a significant boost from the discovery by Susan Horwitz in 1979 that it promoted the assembly of tubulin into stable microtubules (*6*). Tubulin is a ubiquitous cellular protein that is intimately involved in the mitotic process. During mitosis, tubulin, which exists in α- and β-forms, reversibly assembles to form hollow microtubules, and the chromosomes separate with the assistance of these microtubules. After mitosis, the microtubules disassemble to regenerate tubulin. Several drugs, including the important anticancer drugs vinblastine and vincristine, are believed to exert their effects by preventing the assembly of tubulin into microtubules (*7*). Taxol, however, promotes the assembly of tubulin into heat- and calcium-stable microtubules, and this presumably prevents cellular division and facilitates cell death. Taxol binds stoichiometrically and non-covalently to tubulin, but the binding site is on the assembled microtubule rather than the tubulin sub-unit (*8*).

Table 1. Selected Antitumor Activity Data for Taxol

Tumor system	*Route tumor/drug*	*Regimen*	*Optimal dose (mg/kg/inj)*	*T/C (%) at OD*	*Evaluation*
B16 melanoma	ip/ip	Daily x 9	5	283	++
P1534 leukemia	ip/ip	Daily x 10	3.75	300	++
P388 leukemia	ip/ip	q4d x 3	43	170	+
L1210 leukemia	ip/ip	Daily x 15	20	131	+
Colon 26	ip/ip	q4d x 2	30	161	+
MX-1 mammary xenograft	src/sc	Daily x 10	200	(-67)	++
LX-1 lung xenograft	src/sc	Daily x 10	200	(13)	+
CX-1 colon xenograft	src/sc	Daily x 10	400	(12)	+

During the period 1978-1982, various preclinical studies were carried out on taxol; the taxol required for these studies was obtained by large-scale isolation from *T. brevifolia* bark. As noted above, formulation of taxol was a difficult problem, and eventually a formulation was developed using Cremophor EL, a polyethoxylated castor oil, and absolute ethanol, the whole being diluted with 5% dextrose in water or normal saline before use (*9*). Toxicology studies were also performed, and LD_{50} values ranged from 34 mg/kg for rats to 9 mg/kg for beagle dogs (*5*).

Phase I clinical trials were initiated in 1983, and these very nearly proved disastrous. Taxol must be given at relatively high doses [a typical course of treatment uses a dose of 250 mg/m^2 (*10*)], and thus relatively large levels of Cremophor adjuvant must be administered. Some severe allergic reactions were observed during the Phase I clinical trials, including at least one death (*11*). These problems were probably due to the Cremophor adjuvant, since allergic reactions have been observed with this compound in other cases (*12*), but they very nearly halted further studies with taxol. Fortunately, the novelty of taxol's mechanism of action encouraged the clinicians to persevere, and the allergic reactions were minimized by premedication with glucocorticoids and antihistamines, and by lengthening the period of infusion. Phase I studies were then successfully completed, and in one study partial responses were observed in four of 12 patients with melanoma (*13*). The dose-limiting toxicity of taxol was found to be leukopenia, with other toxicities being neurotoxicity, nausea and vomiting, various allergic reactions, cardiotoxicity, and stomatitis.

Phase II clinical trials were initiated in 1985. Although the extent of these trials has been limited by the availability of taxol, the results to date have been excellent, especially in comparison with other anticancer drugs. At the time of this writing, complete reports have appeared of results in ovarian and breast cancers. In ovarian cancer, a response rate of 30% was observed in a group of 40 patients, with one complete response (*14*). Particularly encouraging was the fact that these responses were noted in heavily pre-treated patients, many of whom were resistant to cisplatin. In breast cancer, with a group of 25 patients who had only one prior chemotherapy regimen, a response rate of 56% was observed, with 12% complete and 44% partial responses (*10*).

These results, together with other as yet unpublished studies, clearly demonstrate the clinical effectiveness of taxol as an anticancer drug, and presage a bright future for it once the twin problems of supply and solubility have been overcome.

Enhancement of the Taxol Supply

Because of the importance of increasing the taxol supply, various approaches have been adopted in addressing this problem. Each will be discussed briefly.

Isolation from *T. brevifolia* Bark. This approach is the most expeditious one for the production of taxol for clinical use, since it has already been approved by the Food and Drug Administration. Currently (1992) all taxol used for clinical purposes is obtained by this route, but the limited supply of *T. brevifolia* bark and the destruction of the tree that necessarily accompanies collection of it prevent this from being a viable long-term source of taxol.

Isolation from Other Parts of Various Taxus Species. Several surveys of the taxoid content of various *Taxus* species have appeared; since the taxoids cephalomannine (**5**) and baccatin III (**6**) can be converted to taxol by procedures described below, their occurrence is of almost as much interest as is that of taxol. The major finding to emerge from this work is that the taxol content of *Taxus* leaves

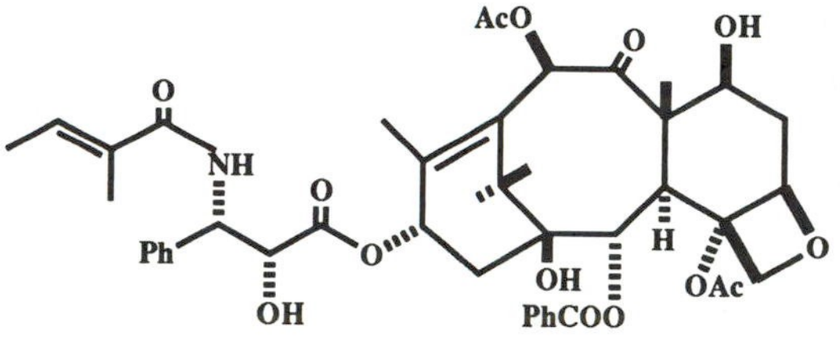

5

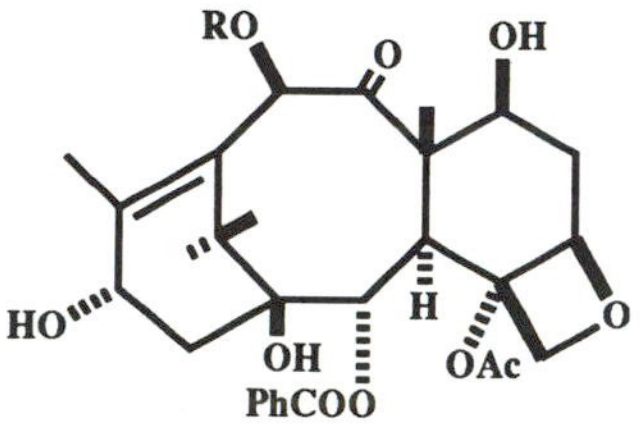

6 R = Ac
7 R = H

(or needles) is at least as great as that reported from bark. Thus Wheeler et al., found taxol contents up to 0.033% w/w in *T. brevifolia* shoots (needles plus twigs), and a maximum of 0.010% w/w in the bark (*17*). Significant variations in taxol content are seen depending on factors such as the season when collected, handling procedures, geographical location, and population. The content of other taxoids can be even greater; thus baccatin III was observed at levels up to 0.2% w/w in *T. brevifolia* (*17*), and 10-deacetylbaccatin III (**7**) can be obtained in yields of 0.1% w/w from fresh leaves of *T. baccata* (*18*).

Partial Synthesis from Baccatin III. Since baccatin III and 10-deacetylbaccatin III both occur in good yield in the needles of *Taxus* spp., methods to convert them to taxol would be of significant importance in the overall approach to improving the taxol supply. Although the desired conversion is simply that of acylation at the C-13 hydroxyl group of baccatin III with an appropriately substituted acid, this conversion is in fact difficult to carry out because of the very hindered nature of the C-13 hydroxyl group. Several approaches have, however, been devised to overcome this problem.

The first approach, developed by Potier and his collaborators, involved acylation of a suitably protected baccatin III with cinnamic acid. The cinnamoyl group was then functionalized by the Sharpless hydroxyamination procedure to yield taxol and various stereo- and regio-isomers (*19*). This approach is of limited utility because of the formation of these isomers, but it did have the important fringe benefit of leading to the synthesis of the taxol analogue taxotere (**8**) from 10-deacetylbaccatin III. This analogue shows a better activity than taxol in some assays (*20*), and is currently in clinical trials in France.

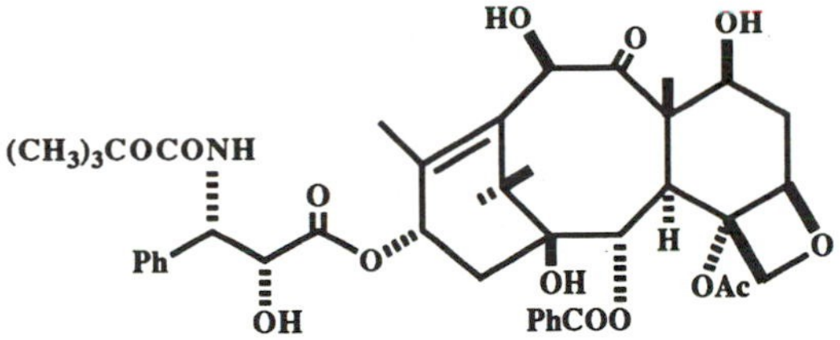

8

The second approach, developed by Potier and Greene, involve a direct acylation of a protected baccatin III or 10-deacetylbaccatin III with a complete protected taxol side chain (*18*). The key reaction in this process proceeds in 80% yield at 50% conversion, and taxol can be obtained in 38% overall yield from 10-deacetylbaccatin III, assuming no recycling.

The third general approach has been developed by Holton, who has shown that a suitably derivatized β-lactam (**9**) will couple to a protected baccatin III (**10**) in excellent yield, as shown below (Holton, R.A., personal communication, 1990).

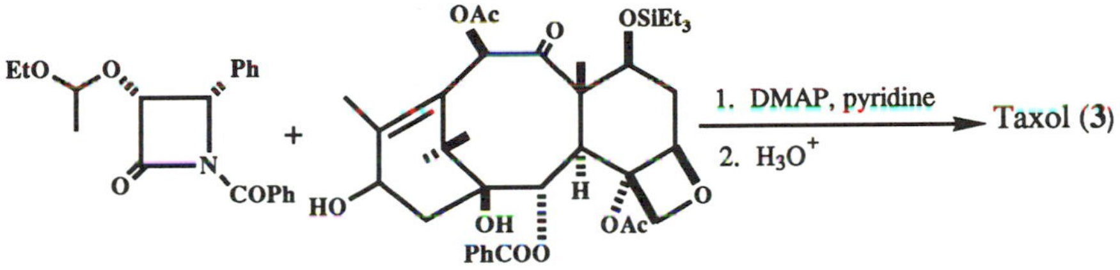

9 + **10** → 1. DMAP, pyridine; 2. H_3O^+ → Taxol (**3**)

The partial synthesis route to taxol is attractive as compared with direct isolation of taxol for two reasons. In the first place, the availability of baccatin III or 10-deacetylbaccatin III appears to be greater than that of taxol; based on published data it would seem that the best yield of baccatin III could be up to 6 times that of the best yield of taxol. Secondly, the partial synthesis approach allows great flexibility in the nature of the side chain, and the synthesis of taxotere is a good example of the benefits of this approach. It is very likely that other taxol analogues will be prepared in the future with even greater activity.

Total Synthesis. The total synthesis of a molecule as complex as taxol is a formidable challenge, but many research groups are working on this problem and significant progress has been made. The scope of this review is too limited to allow a detailed discussion of the synthetic approaches, but mention should be made of the synthesis of taxusin from patchouli alcohol by Holton and his collaborators (*21*). In addition, an elegant approach to taxol has been disclosed by Wender and his group (*22*), and it is possible that this will be completed before this review is published.

Chemistry of Taxol

Studies of the chemistry of taxol have proved very valuable in providing various analogues with improved solubility properties or with structural modifications that lead to our understanding of structure-activity relationships. The work described below is from our own group; a review of the chemistry of taxol that summarizes work from all groups has recently appeared (*23*).

Preparation of Prodrugs. As noted earlier, one of the major problems with taxol is that it is very insoluble in water, and thus must be administered as an emulsion in Cremophor EL. To address this problem, we investigated the acylation chemistry of taxol. Acetylation of taxol under mild conditions yields its 2'-acetate (**11**), while more vigorous conditions lead to the 2',7-diacetate (**12**). Mild hydrolysis then yields the 7-acetate (**13**) (*24*). The 2'-acetate, although inactive in a tubulin-assembly assay, was almost as cytotoxic as taxol. This finding indicated that the 2'-acetate was undergoing hydrolysis to taxol in cell culture, a conclusion which was supported by the ready chemical hydrolysis of the 2',7-diacetate to the 7-acetate. The 2'-position

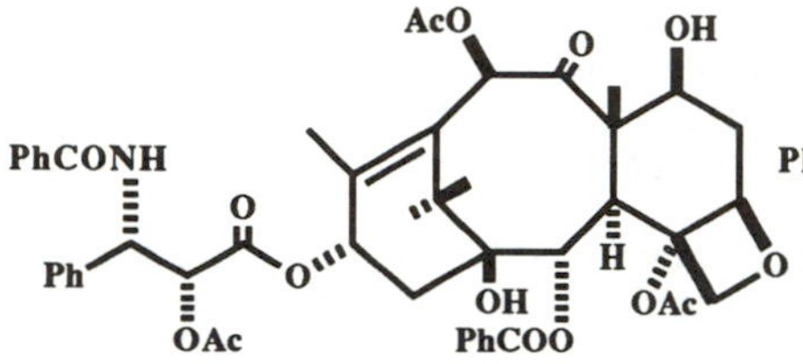

11

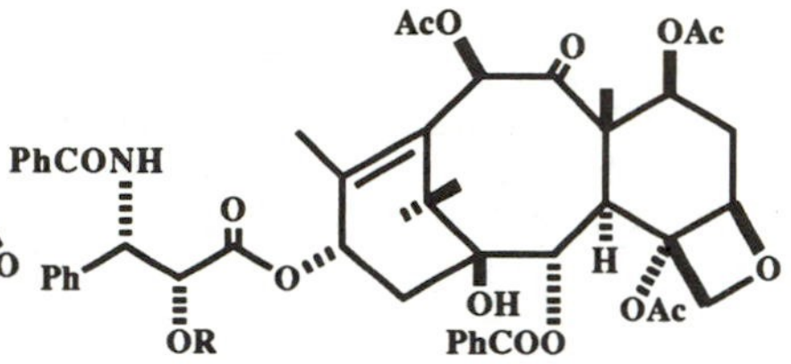

12 R = Ac
13 R = H

was thus suitable for esterification to form a prodrug, and therefore both we (*25, 26*) and others (*27, 28*) have prepared prodrugs of taxol. Among those prepared in our laboratory are 2'-succinyltaxol and its sodium salt (*14*) and various sulfonic acid derivatives such as **15**.

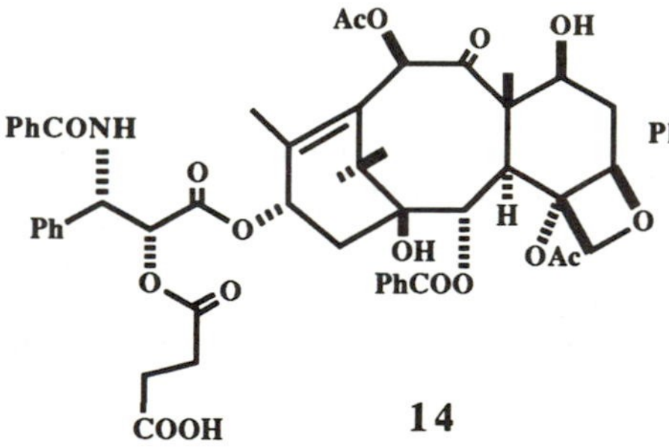

14

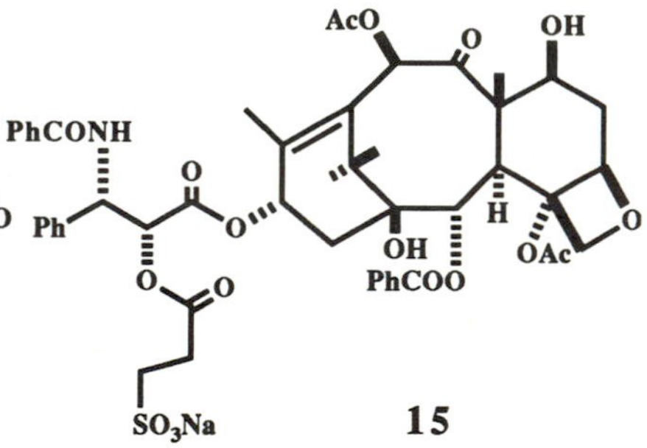

15

The activities of the prodrugs that have been made are either comparable to or less than that of taxol, with the actual activity presumably dependent on the rate of bioconversion back to taxol. It is thus reasonably certain that an effective prodrug form of taxol will be developed in the near future.

Conversion of Taxol to Baccatin III. Although the conversion of taxol to baccatin III might appear a retrograde step, it is nevertheless a very useful reaction in many situations. In particular, since crude plant extracts typically contain mixtures of taxoids such as taxol, cephalomannine, baccatin III, and 10-deacetylbaccatin III, a method to convert these to a common intermediate would enable isolation to proceed more readily and in higher overall yield.

Initial hydrolysis studies were carried out on cephalomannine (*29*) and showed that cleavage of the side chain was accompanied by deacetylation at C-10 and epimerization at C-7, leading to a complex mixture of products. However, we found that reductive cleavage of the side chain could be effected selectively and in high yield by treatment of taxol (**3**) with tetrabutylammonium borohydride in dichloromethane (*30*) to yield baccatin III (**6**), and this reaction has found wide utility. The reductive nature of the cleavage was confirmed by the isolation of the diol, **16**; the reaction occurs selectively because the borohydride anion can complex with the 2'-hydroxyl group and deliver a hydride to the ester carbonyl group selectively. As proof of this, taxol derivatives in which the 2'-hydroxyl group is protected are not reduced.

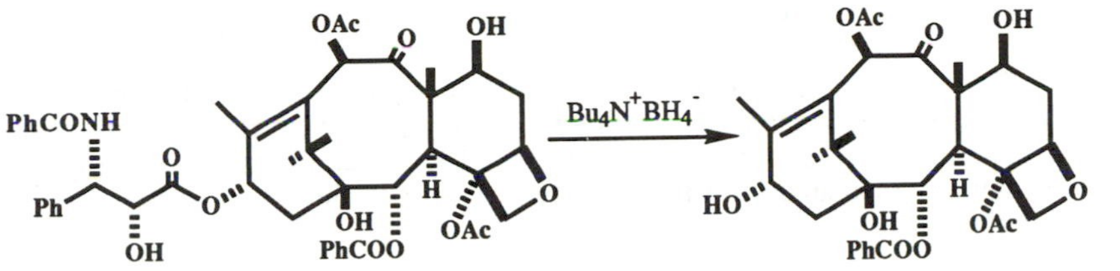

3 **6**

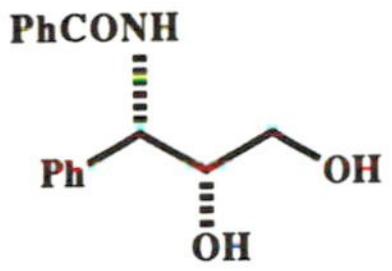

16

Reactions Leading to Opening of the Oxetane Ring. The oxetane ring is an unusual structural unit in natural products, and it was of significant interest to develop methods to open it selectively, so as to determine its effect on the activity of taxol.

The first method we developed involved oxidation of taxol. Mild oxidation of taxol converted it selectively into its 7-oxo derivative, **17**, and treatment of this with mild base resulted in opening of the oxetane ring to give the D-secotaxol **18** (*31*).

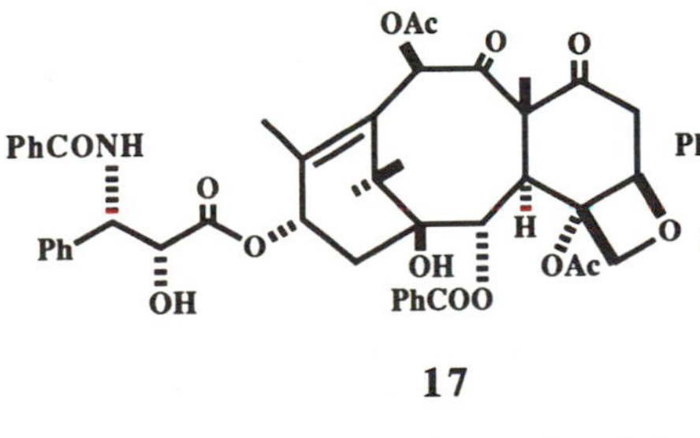

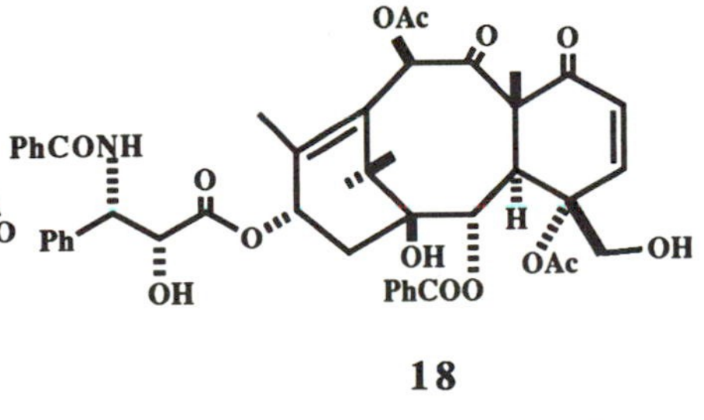

17 **18**

A second method was developed when we discovered fortuitously that reaction of taxol with Meerwein's reagent (triethyloxonium tetrafluoroborate) yielded the ring-opened product, **19** (*32*). A similar ring-opening, but this time in conjunction with a contraction of the A-ring, was observed after vigorous treatment with acetyl chloride to yield the acetate, **20** (*32*).

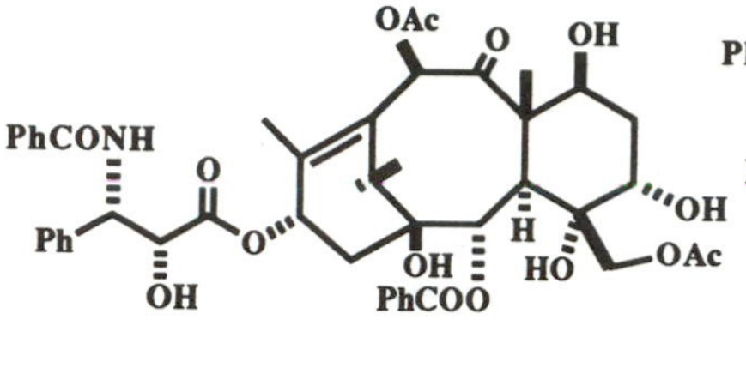

19

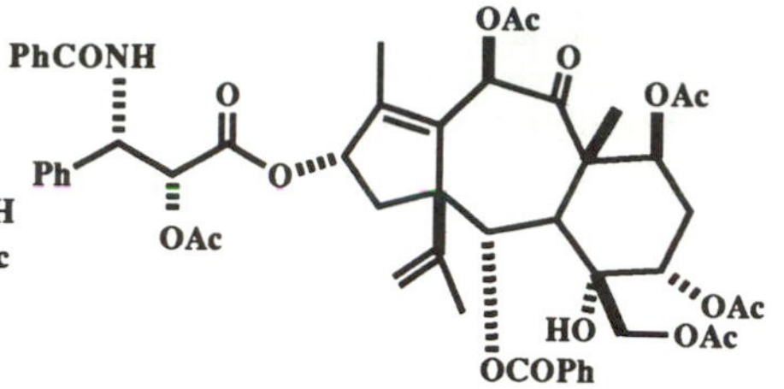

20

The cytotoxic activities of the D-secotaxols, **18**, **19**, and **20**, were all very much less than that of taxol, indicating that an intact oxetane ring or some similar structural unit is necessary for taxol's activity. The reason for this is not fully clear, but it may be associated with the rigidity imparted to the taxane skeleton by the oxetane ring.

Rearranged Taxol. Protection of taxol as its 2',7-di(triethylsilyl) derivative, followed by mesylation at C-1 and then deprotection, led to the formation of the A-nortaxol, **21**. This compound is interesting in that molecular models show it to have

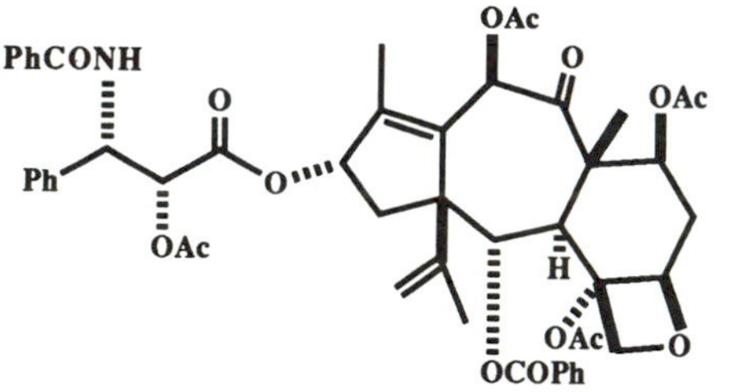

21

a very similar overall shape to that of taxol. It also has a comparable activity to taxol in a tubulin-assembly assay, although it is not particularly cytotoxic (*32*). This compound is thus an interesting lead compound for the development of future generations of taxol analogues.

Summary and Conclusions

The story of taxol is an instructive one in that its present success as an antitumor agent could so easily have been missed. If Drs. Wall and Wani had been less persistent in following up a difficult lead, taxol would not have been isolated when it was. If scientists at NCI had not had faith in taxol, it would not have been selected for clinical trials. If Dr. Horwitz had not discovered its unusual mechanism of action, clinicians might have been less willing to take a chance on a novel drug, and especially one that showed initial toxicity. That taxol did survive these and other problems and emerge as a clear winner can be attributed to the faith and persistence of the scientists who have worked with it, and perhaps also to the hand of Providence. Its success points the way to the development of other natural products as medicinal agents.

Acknowledgments

The work from my laboratory was carried out with the help of an able and dedicated group of students and research associates, whose names are indicated in the references. This review is dedicated to them and to all others who have helped to take taxol from a crude active plant extract to its present place as one of the most promising anticancer drugs of the last twenty years. The author is grateful to the American Cancer Society, the National Cancer Institute, and Bristol-Myers Squibb for financial support, and to Dr. M. Suffness for a critical reading of the manuscript.

Literature Cited

1. Caesar, J. *The Battle for Gaul*, Book 6, Section 31; Wiseman, A.; Wiseman, P., Translators; Chatto and Windus: London, UK, 1980; p 126.
2. Bryan-Brown, T. *Quart. J. Pharm. Pharmacol.* **1932**, *5*, 205-219.
3. Wani, M. C.; Taylor, H. L.; Wall, M. E.; Coggon, P.; McPhail, A. T. *J. Am. Chem. Soc.* **1971**, *93*, 2325-2327.
4. Wall, M. E.; Wani, M. C.; Cook, C. E.; Palmer, K. H.; McPhail, A. T.; Sim, G. A. *J. Am. Chem. Soc.* **1966**, *88*, 3888-3890.
5. Suffness, M.; Cordell, G. A. In *The Alkaloids*; Vol. 25; Brossi, A., Ed.; Academic Press: New York, NY, 1985, pp 1-369.
6. Schiff, P. B.; Fant, J.; Horwitz, S. B. *Nature* **1979**, *277*, 665-667.
7. Neuss, N.; Neuss, M. N. In *The Alkaloids*; Vol. 37; Brossi, A.; Suffness M., Eds.; Academic Press: New York, NY, 1990, pp 229-239.
8. Parness, J.; Horwitz, S. B. *J. Cell. Biol.* **1981**, *91*, 479-487.
9. National Cancer Institute, Division of Cancer Treatment: Taxol (IND 22850, NSC 125973). Annu. Rep. to FDA, March 1991.
10. Holmes, F. A.; Walters, R. S.; Theriault, R. L.; Forman, A. D.; Newton, L. K.; Raber, M. N.; Buzdar, A. U.; Frye, D. K.; Hortobagyi, G. N. *J. Natl. Cancer Inst.* **1991**, *83*, 1797-1805.
11. Kris, M. S.; O'Connell, J. P.; Gralla, R. J.; Wertheim, M. S.; Parente, R. M.; Schiff, P. B.; Young, C. W. *Cancer Treat. Rep.* **1986**, *70*, 605-607.
12. Huttel, M. S.; Olesen, A. S.; Stoffersen, E. *Br. J. Anaesth.* **1980**, *52*, 77-79.
13. Wiernik, P. H.; Schwartz, E. L.; Einzig, A.; Strauman, J. J.; Lipton, R. B.; Dutcher, J. P. *J. Clin. Oncol.* **1987**, *5*, 1232-1239.
14. McGuire, W. P.; Rowinsky, E. K.; Rosenshein, N. B.; Grumbine, F. C.; Ettinger, D. S.; Armstrong, D. K.; Donehower, R. C. *Ann. Intern. Med.* **1989**, *111*, 273-279.
15. Witherup, K. M.; Look, S. A.; Stasko, M. W.; Ghiorzi, T. J.; Muschik, G. M.; Cragg, G. M. *J. Nat. Prod.* **1990**, *53*, 1249-1255.
16. Vindensek, N.; Lim, P.; Campbell, A.; Carlson, C. *J. Nat. Prod.* **1990**, *53*, 1609-1610.
17. Wheeler, N. C.; Jech, K.; Masters, S.; Brobst, S. W.; Alvarado, A. B.; Hoover, A. J.; Snader, K. M. *J. Nat. Prod.* **1992**, *55*, 432-440.
18. Denis, J.-N.; Greene, A. E.; Guénard, D.; Guéritte-Voegelein, F.; Mangatal, L.; Potier, P. *J. Am. Chem. Soc.* **1988**, *110*, 5917-5919.
19. Mangatal, L.; Adeline, M.-T.; A. E.; Guénard, D.; Guéritte-Voegelein, F.; Potier, P. *Tetrahedron* **1989**, *45*, 4177-4190.
20. Guéritte-Voegelein, F.; Guénard, D.; Lavelle, F.; LeGoff, M.-T.; Mangatal, L.; Potier, P. *J. Med. Chem.* **1991**, *34*, 992-998.
21. Holton, R. A.; Juo, R. R.; Kim, H. B.; Williams, A. D.; Harusawa, S.; Lowenthal, R. E.; Yogai, S. *J. Am. Chem. Soc.* **1988**, *110*, 6558-6560.

22. Wender, P. A.; Mucciaro, T. P. *J. Am. Chem. Soc.* **1992**, *114* , 5878-5879.
23. Kingston, D. G. I. *Pharmac. Therap.* **1991**, *52*, 1-34.
24. Mellado, W.; Magri, N. F.; Kingston, D. G. I.; Garcia-Arenas, R.; Orr, G. A.; Horwitz, S. G. *Biochem. Biophys. Res. Commun.* **1984**, *124*, 329-335.
25. Magri, N. F.; Kingston, D. G. I. *J. Nat. Prod.* **1988**, *51*, 298-306.
26. Zhao, Z.; Kingston, D. G. I. *J. Nat. Prod.* **1991**, *54*, 1607-1611.
27. Deutsch, H. M.; Glinski, J. A.; Hernandez, M.; Haugwitz, R. D.; Narayanan, V. L.; Suffness, M.; Zalkow, L. H. *J. Med. Chem.* **1989**, *32*, 788-792.
28. Mathew, A. E.; Mejillano, M. R.; Nath, J. P.; Himes, R. H.; Stella, V. J. *J. Med. Chem.* **1992**, *35*, 145-151.
29. Miller, R. W.; Powell, R. G.; Smith, C. R., Jr.; Arnold, R.; Clardy, J. *J. Org. Chem.* **1981**, *46*, 1469-1474.
30. Magri, N. F.; Kingston, D. G. I.; Jitrangsri, C.; Piccariello, T. *J. Org. Chem* **1986**, *51*, 3239-3242.
31. Magri, N. F.; Kingston, D. G. I. *J. Org. Chem.* **1986**, *51*, 797-802.
32. Samaranayake, G.; Magri, N. F.; Jitrangsri, C.; Kingston, D. G. I. *J. Org. Chem.* **1991**, *56*, 5114-5119.

RECEIVED January 27, 1993

Chapter 11

Camptothecin and Analogues

Synthesis, Biological In Vitro and In Vivo Activities, and Clinical Possibilities

Monroe E. Wall and Mansukh C. Wani

Research Triangle Institute, P.O. Box 12194, Research Triangle Park, NC 27709–2194

A large number of camptothecin (CPT) analogs have been prepared in the 20(S), 20(RS) and 20(R) configurations with a number of ring A substituents. Topoisomerase 1 (T-1) inhibition data (IC_{50}) has been obtained by standard procedures. In general, substitution at the 9 or 10 positions with amino, halogeno or hydroxyl groups in compounds with 20(S) configuration results in compounds with enhanced T-1 inhibition. Compounds in the 20(RS) configuration were less active in vitro and in vivo and those in the 20(R) configuration were inactive. Compounds with 10,11-methylenedioxy substitution in ring A displayed a marked increase in potency in T-1 inhibition assay. The activities of some of the analogs were determined in vivo L-1210 mouse leukemia assay and in general were in accord with T-1 inhibition. A number of water soluble analogs such as 20-glycinate esters, 9-glycinamides, or hydrolyzed lactone salts were prepared and tested in vitro and in vivo assays. In general, these compounds were less active than CPT both in terms of T-1 inhibition and life prolongation in the L-1210 assay. However, certain 20-glycinate esters showed good in vivo activity after IV administration.

This chapter will emphasize the relationship of the structure of camptothecin (CPT) 1 (*1*) and analogs to topoisomerase 1 (T-1) inhibition and other in vitro and in vivo antitumor assays. The structures of this parent compound and its analogs to be discussed in this chapter are shown in Figure 1. As a background for these studies, structural features of 1 (Figure 1) early SAR studies, and syntheses of 1 and analogs will be briefly reviewed prior to a discussion of the T-1 inhibition activity of 1 and analogs and the relationship to in vivo antitumor activity.

0097–6156/93/0534–0149$06.25/0

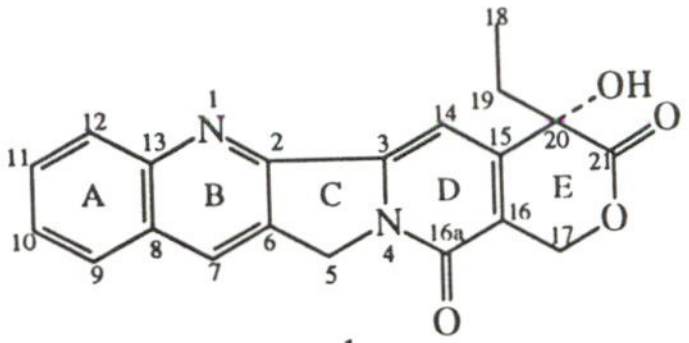

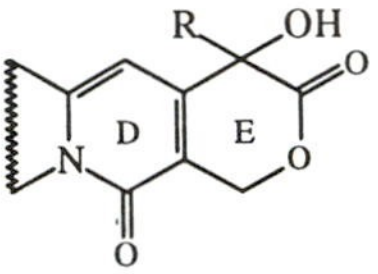

41, R = CH_3, 20(RS)
42, R = CH_2CH_2OH, 20(RS)
43, R = CH_2CH_2Br, 20(RS)

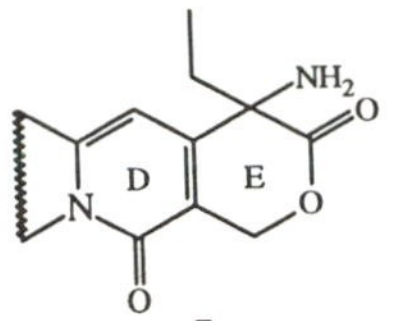

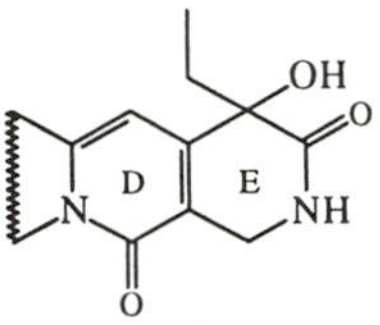

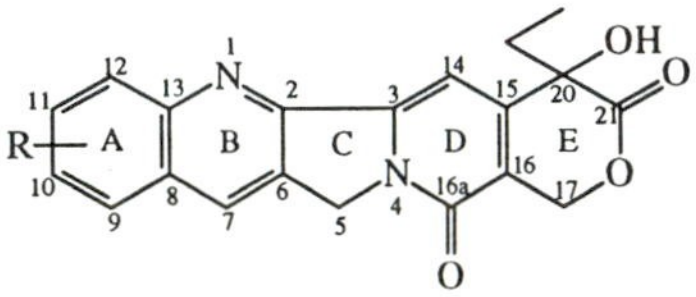

3, R = 10-OH, 20(S)
6, R = 9-NH_2-10,11-OCH_2O-, 20(S)
9, R = 9-NH_2, 20(S)
10, R = 10-NH_2, 20(S)
11, R = 9-NO_2, 20(S)
12, R = 10-NO_2, 20(RS)
13, R = 10-NO_2, 20(S)
14, R = 9-OH, 20(RS)
15, R = 9-OH, 20(S)
16, R = 9-Cl, 20(S)
17, R = 10-Cl, 20(S)
18, R = 9-CH_3, 20(S)
19, R = 11-CN, 20(RS)
20, R = 11-OH, 20(RS)
21, R = 11-NH_2, 20(RS)
22, R = 12-NH_2, 20(S)
23, 9-NO_2-10-OH, 20(S)
24, 9-$NHCOCH_3$-10-OH, 20(S)
25, R = 9-$CH_2)_2NCH_2$-10-OH•HCl
26, R = 10,11-OCH_3, 20(RS)
27, R = 10,11-OCH_2O-, 20(RS)
28, R = 10,11-OCH_2O-, 20(S)
29, R = 9-NH_2-10,11-OCH_2O-, 20(RS)
30, R = 9-Cl-10,11-OCH_2O-, 20(S)
31, R = 10,11-$O(CH_2)_2O$-, 20(RS)
32, R = 9,10-OCH_2O-, 20(RS)

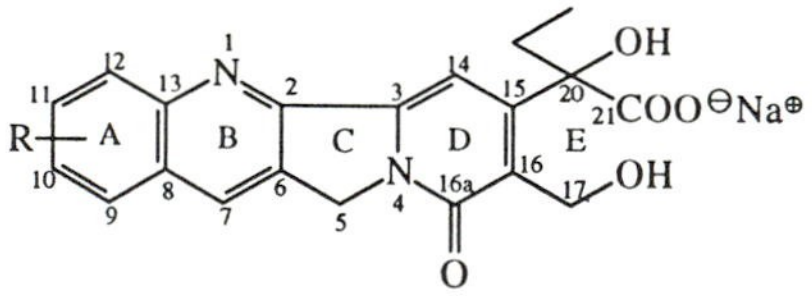

2, R = H, 20(S)
33, R = 9-NH_2-10,11-OCH_2O-, 20(S)
34, R = 10-ONa, 20(S)
35, R = 9-NH_2, 20(RS)
36, R = 10,11-OCH_2O-, 20(RS)

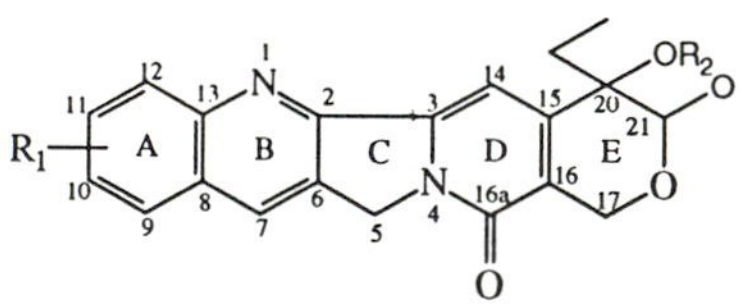

37, R_1 = 9-$NHCOCH_2NH_2$•HCl, R_2 = H, 20(RS)
38, R_1 = 9-NH_2, R_2 = $COCH_2NH_2$•HCl, 20(RS)
39, R_1 = 10-NH_2, R_2 = $COCH_2NH_2$•HCl, 20(RS)
40, R_1 = 10,11-OCH_2O-, R_2 = $COCH_2NH_2$•HCl, 20(RS)

Figure 1. Structures of camptothecin and its analogs discussed in this chapter.

Structural Features of CPT 1

The potent antitumor activity of CPT was first discovered somewhat serendipitously in 1958 in extracts of the fruit of Camptotheca acuminata Decaisne (Nyssaceae) (*2*). The compound was isolated by M. E. Wall and co-workers at Research Triangle Institute and its structure elucidated in 1966 (*1*). The taxonomy, occurrence, and the interesting manner in which the seeds of this Chinese tree were collected from Szechwan Province and sent to the United States many years ago has been reported (*3*). It will be noted that CPT has a pentacyclic ring structure with only one asymmetric center in ring E with the 20(S) configuration. Other notable features are the presence of the α-hydroxy lactone system in ring E, the pyridone ring D moiety, and the conjugated system which links rings A through D. The water-soluble sodium salt 2 was prepared from 1 (*1*) and several other CPT analogs, including 10-hydroxy-20(S)-CPT 3, were discovered somewhat later (*4*). As we will discuss in another section, water-soluble analogs of 3 are in clinical trial. Because of its noteworthy life-prolongation activity in mice bearing murine L1210 leukemia, 1 was of great interest from the time of its initial isolation (*1*). The compound demonstrated this antitumor activity at 0.25 mg to 3.0 mg/kg with T/C values frequently in excess of 200% (T/C = (survival time of treated animals ÷ survival time of control animals) x 100). In addition, 1 has great activity in inhibition of the growth of solid tumors in rodents. In the form of the water soluble sodium salt 2 (Figure 2) the compound received early clinical trial but was not effective (*5-7*). Some years later 1, salt 2, 10-hydroxy analog 3 and several other CPT analogs were subjected to the P-388 mouse life prolongation assay at the same time under identical experimental protocols (*8*). Under these conditions salt 2 was found to have only one-tenth the potency of 1 (*9*). More recently, 1 was found to be unique as a potent inhibitor of T-1 (*10,11*). As a result, interest in this class of compounds is currently at a high level. T-1 activity of CPT analogs will be reviewed in detail in the section entitled "Effect of CPT and Analogs on T-1."

Initial Structure-Activity Relationship Studies

Initial studies revealed some interesting structure-activity relationships (SAR) (*12*) (cf. Figure 2), and several important features were noted. Hydroxylation of ring A of CPT, 1 is compatible with activity and, indeed, the 10-hydroxy analog 3 has greater activity than 1 (*9,12*). CPT reacts readily with bases or amines to give water-soluble salts or amides, formed by reaction of the lactone moiety (pH 6.0). Under acidic conditions both the sodium salt and amide were readily relactonized (*9, 12, 13*). Reaction of the lactone moiety with nucleophilic components of DNA and/or enzymes analogous to amide formation may possibly be involved in the antitumor activity and potency of 1 and certain of its analogs in the inhibition of T-1 (*13*).

Major reduction of antineoplastic activity was noted as a result of reactions involving the C20 hydroxyl or C21 lactone moiety in ring E (cf. Figure 2) (*12*). After acetylation of CPT, the resultant C20 acetate was nearly inactive. Other reactions focused at modifying the hydroxyl at C20 gave inactive analogs which also point to the absolute requirement of the C20 α-hydroxyl group. After

Camptothecin 1
L1210, T/C 220 at 2 mg/kg
PS, T/C 197 at 4 mg/kg

Camptothecin Sodium 2
L1210, T/C 209 at 3 mg/kg
PS, T/C 212 at 40 mg/kg

L1210, T/C 172 at 3.5 mg/kg

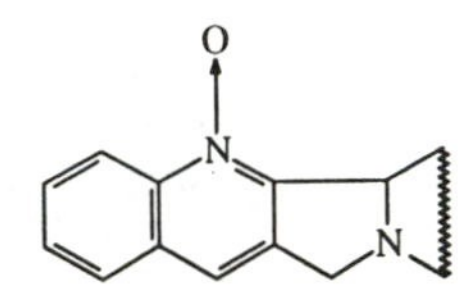

L1210, T/C 144 at 2 mg/kg

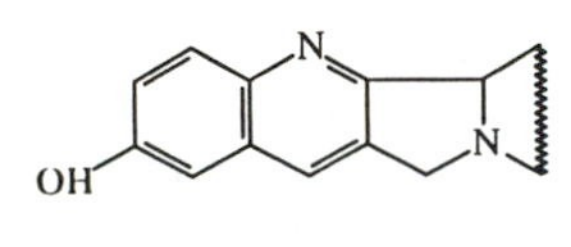

10-Hydroxycamptothecin 3
L1210, T/C 230 at 0.5 mg/kg
PS, T/C 314 at 4 mg/kg

Slightly Active at 25 mg/kg
L1210, T/C 125 mg/kg

a, R = H
b, R = Cl
Inactive
L1210

Inactive
L1210

Figure 2. Structure activity relationships in the camptothecin series.

replacement of this group by chlorine, the activity in L1210 was lost. After hydrogenolysis of the chloro group, the resulting 20-deoxy-CPT analog was inactive. Reduction of the lactone under mild conditions to give the lactol also resulted in complete loss of activity. These early studies have thus provided strong evidence that the presence of the α-hydroxy lactone moiety is one of several essential structural requirements for activity of CPT and analogs.

Analog Synthesis and Effect of Substituents on Biological Activities

After our report on the structure of 1, many novel total syntheses of 1 were reported. However, none of these numerous early syntheses of 1, including one from our laboratory, provided 1 in adequate yields, nor were they sufficiently versatile to permit analog synthesis (cf. reference 14 for a comprehensive review).

Friedlander Synthesis. Improved procedures for the total synthesis of CPT and its analogs developed at RTI involving the Friedlander reaction of a properly substituted o-aminobenzaldehyde with a tricyclic synthon have permitted studies of the effects of modifications in rings A, D and E. Figure 3 presents the application of this synthesis to a wide variety of A-ring substituted CPT analogs, to the synthesis of a tetracyclic CPT analog 4 and to the synthesis of a ring D benzo analog 5 (*9*, *13*, *15-17*), (Wall, M. E. and co-workers, submitted for publication in J. Med. Chem.).

Stereochemistry at C20. As discussed previously, the C20 α-hydroxyl moiety is required for in vitro and in vivo activity of CPT. Recent studies have provided even greater information on the specificity required in this system. Resolution of the 20(RS) tricyclic synthon (Figure 4) into the 20(S) and 20(R) components enabled us to prepare 20(S)- 1 and the corresponding 20(R)-CPT (*18*). The stereochemistry of the naturally occurring alkaloid is 20(S). The 20(R) form is inactive in the in vivo L1210 mouse leukemia assay and in vitro assays (*18*). Thus not only is the
C20-α-hydroxyl moiety required, but correct specific stereochemistry at C20 is also an absolute necessity for in vitro cytotoxicity, inhibition of T-1 activity, and in vivo antitumor activity (*18-20*). Figure 5 shows the synthesis of a highly active compound, 9-amino-10,11-methylenedioxy-20(S)-CPT, 6, by the Friedlander synthesis. This compound is exceptionally active at very low dose levels (<1.5 mg/kg) in mouse L1210 leukemia, shows high T-1 inhibition in vitro, and recently has shown strong inhibition in vitro of resistant human chronic lymphocytic leukemia cells (Wall, M. E. and co-workers, submitted for publication to J. Med. Chem.).

Effect of Substitution of Nitrogen for Oxygen in Ring E. Recent studies from our laboratory have shown that not only must the correct C20 stereochemistry be present for activity, but that the oxygen of the C20 hydroxyl moiety and the oxygen in the lactone ring E cannot be replaced by nitrogen. Thus 20(RS)-amino-CPT 7 and the 20(S)-lactam analog 8 are inactive both in vitro and in vivo (*13*).

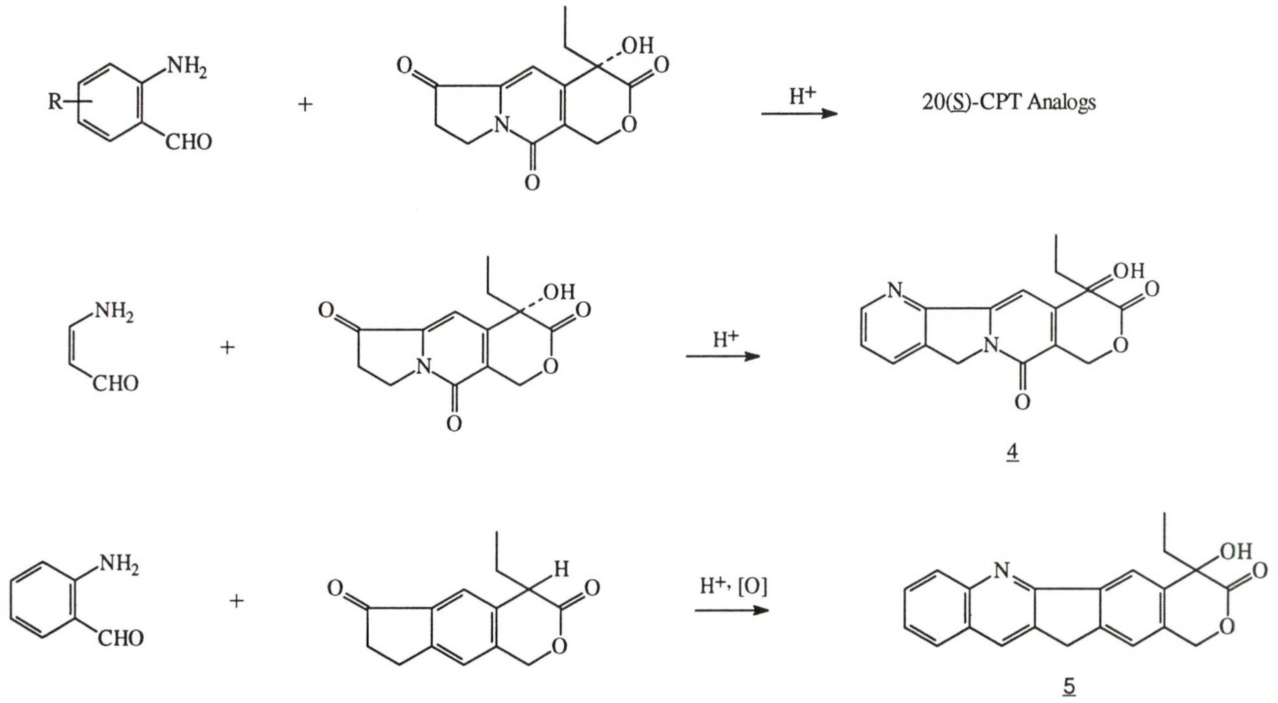

Figure 3. Friedlander condensation of an appropriately substituted o-aminobenzaldehyde with a tricyclic synthon.

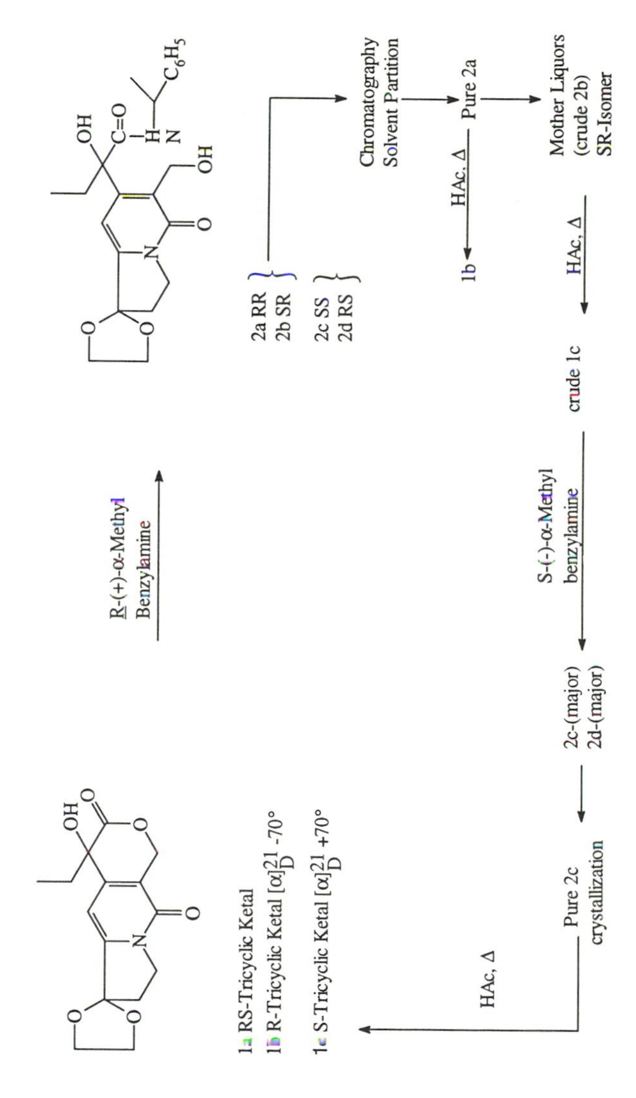

Figure 4. Resolution of 20(RS)-tricyclic synthon.

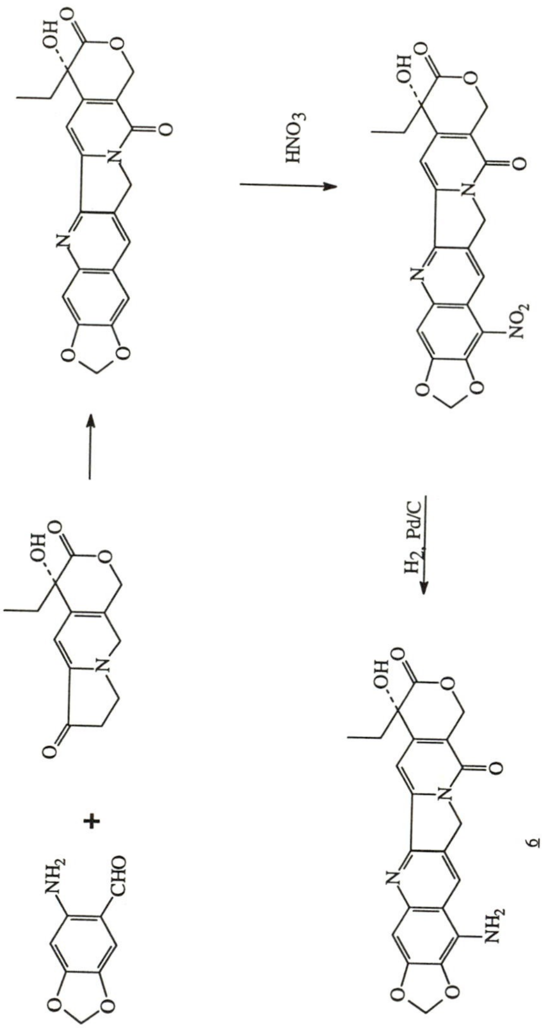

Figure 5. Synthesis of 9-amino-10,11-methylenedioxy-20(S)-camptothecin.

The lack of activity of the 20-amino analog may be due to the fact that hydrogen bonding to the C21 carbonyl moiety in ring E is weaker than is the case when the C20 hydroxyl moiety is present. As a consequence, the C21 carbonyl in the amino analog 7 is less positively charged and hence is less susceptible to nucleophilic attack. The inactivity of the lactam 8 is readily explicable since lactams are much less reactive than lactones. It seems increasingly apparent that one of the major features of the biological activity of the CPT molecule is the chemical reactivity of the α-hydroxy-lactone moiety towards nucleophiles. For example, 1 readily forms carboxylic salts in the presence of bases, or forms amides by reaction with amines (*1, 12*). The reactions are reversible, and hence one can postulate a readily reversible reaction involving the lactone carbonyl and a nucleo-philic group or an enzyme or enzyme-DNA complex.

Ring A Analogs. The synthetic scheme shown in Figure 3 for the synthesis of CPT has also permitted the preparation of many ring A analogs by suitable manipulation of the o-aminobenzaldehyde synthon. Initially, many of the compounds prepared had the 20(RS) stereochemistry at C20 (*9, 15-17*). Potencies of these compounds were lower than those of the corresponding 20(S) analogs. Thus, for example, in the mouse P-388 leukemia life prolongation assay, 20(RS)- 1 was found to have a T/C of 200 at 8 mg/kg whereas in the case of 20(S) 1, the same activity was found at a dose of 4 mg/kg (*9*). The availability of the 20(S)- and 20(R)- tricyclic synthons (*18*) have permitted the preparation of a number of 20(S)- and 20(R)- analogs of 1 (Wall, M. E. and co-workers, submitted for publication to J. Med. Chem.). As will be shown subsequently, 20(S)-CPT analogs always exhibit greater potency than the corresponding 20(RS) compounds in vitro and in vivo.

Modification of ring A has yielded many compounds with a wide range of in vitro and in vivo activity, from exceptionally active to inactive (*15-17*) (cf. Tables I,II). As we will show in subsequent sections, inactivity or loss of activity may result from several factors.

Effect of Substitution at C9 and C10. Substitution at C9 or C10 by an amino group leads to compounds with considerably greater in vivo activity than is found with the corresponding parent compound 1 (*16, 17*). Thus both 9-amino-20(S)-CPT 9 and the corresponding 10-amino-20(S) analog 10 are very active compounds in the in vivo life prolongation assay for murine L1210 leukemia (Table I). The 10-amino-20(RS) analog also shows high activity . All of the 9-nitro-20(S) analog 11, the corresponding 10-nitro-20(RS) analog 12 and the 10-nitro-20(S) analog 13 also have excellent activity in vivo, cf. Table I (albeit at a considerably higher dose level is required). Although at present there is no experimental proof available that reduction of the nitro substituent to the corresponding amino group occurs in vivo, the nitro compounds 11 and 13 may be pro-drugs for the corresponding amino analogs 9 and 10.

We have shown previously that 10-hydroxy-20(S)-CPT 3 has excellent life prolongation activity in both L1210 and P-388 murine leukemia (*9, 16, 17*). Rather surprisingly, in view of the high activity in vivo (L-1210) and in vitro (T-1 inhibition) shown by the 9- and 10-amino analogs 9 and 10, and by the 10-hydroxy analog 3, both the 20(RS) and 20(S) 9-hydroxy analogs 14 and 15 respectively, are weakly active in vitro and in vivo, (Wall, M. E. and co-workers, submitted for publication, J. Med. Chem.) (cf. Tables I and II).

Active CPT analogs were noted with other substituents in the 9- and 10- positions. For example, the lipophilic 9-chloro-20(S)-CPT 16 and the corresponding 10-chloro-20(RS) analogs have shown high activity in inhibition of L-1210 mouse leukemia and the 9- and 10-chloro-20(S) analogs 16 and 17 respectively have shown high activity in the T-1 inhibition assay, Table II. The 10-chloro-20(RS) analog is also potent in vivo, Table I. The 9-methyl-20(S) analog 18 has potent activity both in the T-1 and L-1210 in vivo assay (Wall, M. E. and co-workers, submitted for publication to J. Med. Chem.). Compounds 16, 17 and 18 are chloroform soluble whereas 1 has low solubility in all solvents except DMSO.

Effect of Substitution at C11 and C12. In general, the effect of monosubstitution at C11 and C12 is to lower the activity and/or potency. Of a large number of compounds synthesized with C11 substituents (*15-17*) only the 11-cyano-20(RS) 19 and 11-hydroxy-20(RS) 20 analogs exhibited appreciable life prolongation in the murine L1210 leukemia assay, and both compounds had low potency (*15,17*). In contrast to the enhancement of activity by amino groups at C9 and C10, the 11-amino-20(RS) analog 21 had low activity and the 12-amino-20(S) analog 22 was inactive both in vivo and in vitro (*17, 19*).

Effect of Two or More Ring A Substituents. The effects of two or more substituents in the A ring on in vivo and/or in vitro activity is complex and variable. Moreover, this area still requires considerably more study.

Analogs Substituted at Both 9 and 10 Positions. A combination of a 10-hydroxy substituent with substituents at C9 results in formation of compounds that are less active than the parent 10-hydroxy-CPT 3 in vivo and in vitro (T-1 inhibition). Compounds of this type prepared to date are also less active than their mono 9- substituted CPT analogs. Thus, for example, 9-nitro-10-hydroxy-20(S)-CPT 23 is much less active than either 9-nitro-20(S)-CPT 11 or 10-hydroxy-20(S)-CPT 3 (16). Similarly, 9-acetamido-10-hydroxy-20(S)-CPT 24 is less active than the corresponding mono-substituted 10-hydroxy analog 3 or 9-acetamido-20(S)-CPT. 9-Dimethylaminomethyl-10-hydroxy-20(S)-CPT•HCl 25 (Topotecan, SmithKline Beecham, compound currently in clinical trial) is considerably less active than 10-hydroxy-CPT 3 in T-1 inhibition (cf. Table II, compounds 3, 9, 25, and section entitled "Effect of CPT and Analogs on T-1).

Analogs Substituted at Both 10 and 11 Positions. Disub-stitution at the 10 and 11 positions, with one notable exception, frequently leads either to inactive or less active compounds. We have previously described the inactivity of 10,11-dimethoxy-20(RS)-CPT 26 (*9*). The corresponding 10,11-dihydroxy-20(RS)-CPT, while showing modest activity in L1210 mouse leukemia life prolongation assay, i.e., T/C 157% at a dose of 40 mg/kg, is more active than the dimethoxy analog 26, which is inactive, but also much less active than either the corresponding mono-10- or 11-hydroxy analogs of 1.

Table I

L1210 Mouse Leukemia Life Prolongation by Water-Insoluble CPT Analogs[a]

Compound	Dose Regimen[b]	Highest Active Dose, mg/Kg (% T/C)[c]	Active Dose Range (mg/kg)	KE[d]	Cures	Toxic Dose mg/kg
20(S)-CPT 1	Q04DX02	12(250)	5.3-12	4.8	1/6	10
9-Amino-20(S)-CPT 9	Q04DX02	5(361)	0.6-5	5.97	3/6	10
10-Amino-20(S)-CPT 10[e]	Q04DX02	4.5(565)	2-4.5	>5.99	6/6	*
9-Nitro-20(S)-CPT 11	Q04DX02	10(348)	1.25-20	5.97	5/6	40
9-Amino-10,11-methylenedioxy-20(S)-CPT 6[f]	Q04DX02	1.5(536)	0.3-1.5	5.8	4/6	*
9-Chloro-20(S)-CPT 16	Q04DX02	12(150)	2.37-12.00	0	0/6	*
9-Methyl-20(S)-CPT 18	Q04DX02	8(274)	2.37-8.00	5.6	4/6	12
9-Hydroxy-20(RS)-CPT 14	Q04DX02	20(134)	10-40	*	0/6	*
10-Nitro-20(RS)-CPT 12	Q04DX02	7.25(233)	3.63-15.5	5.97	0/6	31
10-Amino-20(RS)-CPT	Q04DX02	3.75(365)	1.35-3.74	5.97	3/6	6.25
10-Chloro-20(RS)-CPT	Q04DX02	10(280)	5-20	5.97	2/6	40
10,11-Methylenedioxy-20(RS)-CPT 27	Q04DX02	2(325)	2-4	5.97	2/6	8.0

a IP implants.
b Q04DX02 = Drug dosing IP on days 1 and 5.
c % T/C = (median survival time of treated/median survival time of control animals) x 100.
d $\log_{10}$ of initial tumor cell population minus $\log_{10}$ of tumor cell population at the end of treatment.
e 100% long-term (day 45) tumor free survivors.
f 4/6 long-term (day 60) tumor free survivors.
* Data not available.

Table II

Topoisomerase I Inhibition from Cleavable Complex Formation by CPT Analogs

Compound	IC-50 (μM)[a]	SE of IC-50[a]
10,11-Methylenedioxy-20(S)-CPT 28	0.027	0.005
9-Methyl-20(S)-CPT 18	0.038	0.013
9-Amino-10,11-methylenedioxy-20(S)-CPT 6	0.048	0.016
9-Chloro-10,11-methylenedioxy-20(S)-CPT 30	0.061	0.019
9-Amino-10,11-methylenedioxy-20(RS)-CPT 29	0.076	0.030
9-Chloro-20(S)-CPT 16	0.086	0.054
10-Hydroxy-20(S)-CPT 3	0.106	0.031
9-Amino-20(S)-CPT 9	0.111	0.024
10-Amino-20(S)-CPT 10	0.140	0.022
10-Chloro-20(S)-CPT 17	0.141	0.027
10,11-Ethylenedioxy-20(RS)-CPT 31	0.468	0.140
10-Nitro-20(S)-CPT 13	0.635	0.116
20(S)-CPT 1	0.677	0.215
10,11-Methylenedioxy-20(RS)-CPT, Na^+Salt 36	0.843	0.681
9-Hydroxy-20(S)-CPT 15	0.873	0.302
9-Dimethylaminomethyl-10-hydroxy-20(S)-CPT•HCl 25	1.110	–
20(RS)-CPT	1.436	0.261
10-Hydroxy-20(S)-CPT, di Na^+Salt 34	3.202	0.603
9,10-Methylenedioxy-20(RS)-CPT 32	3.815	0.633
9-Amino-10,11-methylenedioxy-20(S)-CPT, Na^+Salt 33	4.303	2.512
20(S)-CPT, Na^+Salt 2	11.586	5.852
9-Amino-20(RS)-CPT. Na^+Salt 35	13.875	5.681
9-Amino-10,11-methylenedioxy-20(R)-CPT	>30	
10,11-Methylenedioxy-20(R)-CPT	>30	
20(R)-CPT	>30	

[a] IC-50 = the minimum drug concentration (μM) that inhibited the cleavable complex formation by 50%.

Effects of Substitution of Methylenedioxy and Ethylenedioxy Groups on In Vitro and In Vivo Activity. Substitution of the 10,11-methyl-enedioxy group in ring A of CPT results in a remarkable enhancement of both in vivo (L1210) and in vitro activity (T-1 inhibition). Both the 20(RS) 27 and 20(S) 28 analogs are among the most active T-1 inhibition agents (cf. Table II). The 20(RS) analog 27 has also shown high activity in the L1210 mouse leukemia assay (*16*, *17*, *19*, *20*) and in the inhibition of human xenograft cancer lines in nude mice (*21*, *22*). Recently, we have shown that 9-amino-10,11-methylenedioxy-CPT in either the 20(RS) 29 or 20(S) 6 configurations are very active in inhibition of T-1 (Tables II, III.) The 20(S) analog 6 is very active and potent in the L1210 mouse leukemia assay (cf. Table I). Recently, in vitro studies have been conducted on the effects of various CPT analogs on B-lymphocytic human leukemia cells which were resistant to standard treatment with chlorambucil. Compound 6 had greater cytotoxicity than any other CPT analog tested including the parent 10,11-methylenedioxy analog 28 and was much superior to chlorambucil and fludarabine (*23*). The 9-chloro-10,11-methylenedioxy-20(S) analog 30 is also a potent T-1 inhibitor, cf. Table II.

The 10,11-ethylenedioxy-20(RS) analog 31 is less potent than the corresponding 10,11-methylenedioxy analog 27, but is a more active inhibitor of T-1 than 1 (cf. Table II). 9,10-Methylenedioxy-20(RS)-CPT 32 exhibited considerably reduced T-1 inhibition activity (Table II), (Wall, M.E. and co-workers, submitted for publication in J. Med. Chem.). The 10,11-methylenedioxy group is coplanar with ring A and hence does not cause the steric hindrance observed with other 10,11-disubstituted groups. The 10,11-ethylenedioxy moiety is not completely coplanar with ring A and hence compound 31, because of possible steric interaction, lacks the enhanced activity of the 10,11-methylenedioxy analog. Recent molecular modeling studies have shown that all the CPT analogs with 10,11-methyl-enedioxy substituents have much higher max/min values (Kcal mol^{-1}) than CPT and many other CPT analogs (Private communication; A. P. Bowen; Computational Center for Molecular Structure and Design, University of Georgia).

It is conceivable that CPT may bind to an enzyme or enzyme-DNA complex on the face proximal to the C11 and, particularly, the C12 region. Hence groups substituted in these positions may cause unfavorable steric or stereoelectronic interactions. Substituents at positions C9 and C10 are more distant from this region, and substitution of certain groups at these locations is of less steric consequence. On the other hand, the bulky 9,10-methylenedioxy analog 32, although coplanar to ring A, may bind less well to the T-1-DNA cleavable complex.

Water Soluble CPT Analogs. The majority of the water-soluble analogs prepared by us were less potent in vitro, T-1 and L1210 in vivo (cf. Tables I, II, and IV). We have prepared three types of water soluble analogs, cf. Table IV. Typical examples are: (1) sodium salts of the carboxylic acid obtained by hydrolytic cleavage of the ring E lactone, (*1*), (2) 20-Glycinate ester-hydrochloride salts, (*24*), and (3) 9-glycinamido-hydrochloride salts (Wall, M. E. and co-workers, submitted for publication to J. Med. Chem.). The objective was to determine if any of these compounds were active per se or if the glycinate ester and 9-glycinamido

Table III

Comparison of Topoisomerase I Inhibition of 20(S), 20 (RS), and 20(R)-CPT Compounds

Compound	IC_{50} (μM)[a]
20(S)-CPT 1	0.68
20(RS)-CPT	1.44
20(R)-CPT	>30
9-Amino-20(S)-CPT 9	0.11
9-Amino-20(RS)-CPT	0.50
10,11-Methylenedioxy-20(S)-CPT 28	0.03
10,11-Methylenedioxy-20(RS)-CPT 27	0.08
10,11-Methylenedioxy-20(R)-CPT	>30
9-Amino-10,11-methylenedioxy-20(S)-CPT 6	0.05
9-Amino-10,11-methylenedioxy-20(RS)-CPT 29	0.07
9-Amino-10,11-methylenedioxy-20(R)-CPT	>30

[a] See footnote [a] (Table I).

Table IV

L1210 Life Prolongation[a] by Water-Soluble CPT Analogs

Compound	Dose Regimen	Route	Highest Active Dose, mg/kg (% T/C)[c]	Active Dose Range (mg/kg)	KE[d]	Cures	Toxic Dose mg/kg
9-Amino-20-glycinate Ester-20(RS)-CPT•HCl	Q04DX02[b]	IP	10(132)	10	-1.00	0	NT at 10
9-Amino-20-glycinate Ester-20(RS)-CPT•HCl	Q04HX02[e]	IV	5(168)	2.5-5.0	1.67	1/6	NT at 5
10-Amino-20-glycinate Ester-20(RS)-CPT•HCl	Q04HX02	IV	20(225)	1.25-20	>5.97	0	NT at 20
10,11-Methylenedioxy-20-glycinate Ester-20(RS)-CPT• HCl 39	Q04HX02	IV	10(236)	1.25-20	5.97	3/6	NT at 20
9-Glycinamido-20(RS)-CPT•HCl 37	Q04HX02	IV	20(180)	1.25-20	2.95	0	NT at 20
10,11-Methylenedioxy-20(RS)-CPT Sodium 36	Q04HX02	IV	10(157)	2.85-10	4.79	0	NT at 20
10-Hydroxy-20(S)-CPT Disodium 34	Q04HX02	IV	20(184)	2.5-20	3.24	0	NT at 20

For footnotes [a], [b], [c], and [d] see Table I

[e] Q04HX02 = Drug Dosing IV, Two injections on day 1 and two on day 5.

hydrochloride salts would be hydrolyzed by esterases and amidases known to be present in human plasma and tissues or whether facile ring-closure of sodium salts would occur in vivo. The results shown in Tables II and IV are typical of a much larger body of data (Wall, M. E. and co-workers, submitted for publication in J. Med. Chem.).

Sodium Salts. As shown in Table II the sodium salts 2, and 33 - 36 exhibit very weak activity in T-1 inhibition, whereas the corresponding water insoluble compounds are potent T-1 inhibitors. In vivo, compounds 2, 34, and 36 show some activity in L1210 mouse leukemia, albeit of a modest order, cf. (*9*) and Table IV, probably due to some degree of relactonization under physiological conditions. It is evident however, that extensive relactonization did not occur.

9-Glycinamido Hydrochlorides. We have not studied extensively water soluble compounds of this type. As shown in Table IV, the 9-glycin-amide 37 has modest activity in L1210 mouse leukemia at a relatively high dose by IV administration. T-1 data for this structure was not obtained. Because the parent 9-amino-20(RS)-CPT and the corresponding 20(S) analog 9 are very active in L1210 mouse assays, T/C >300% at 3-6 mg/kg, IP, it is evident that significant proportions of 37 were not obtained by hydrolytic cleavage under IV administration conditions.

20-Glycinate Esters. Although much more testing would be required for a definitive conclusion, it would appear that certain water-soluble 20-glycinate esters as exemplified by 39 and 40 when administered IV at 10-20 mg/kg, afforded substantial life prolongation in the L1210 mouse leukemia assay (cf. Table IV). It should be noted that compounds 39 and 40 are 20(RS) analogs. Even more active compounds might be anticipated from the corresponding 20(S) analogs.

New Structural Types. Certain new structural types of CPT analogs which in some cases have been briefly mentioned in earlier sections of this report will be reviewed in terms of biological activity.

Tetracyclic CPT Analog 4. In order to determine whether the pentacyclic ring structure of CPT was required for maximal activity, bicyclic ring DE and tricyclic ring CDE compounds with the α-hydroxy-lactone group were prepared and were found to be inactive (*12*). A hexacyclic analog was made and found to be of the same order of potency as 1 (*9*). A prediction was made some years ago that the tetracyclic ring de-A analog 4 (cf. Figure 3) might be active (*25*). Compound 4 was prepared with considerable difficulty and was found to be inactive in T-1,
IC_{50}=>30 μm, (Wall, M. E. and co-workers, submitted to J. Med. Chem.). It is apparent therefore that the pentacyclic ring structure of CPT is required for activity.

Ring D Benzo Analog 5. The role of the Ring D pyridone moiety has never been clearly understood. The cyclic unsaturated lactam seemed to be relatively unreactive. Synthesis of the Ring D benzo CPT analog 5 was therefore undertaken (*13*). Molecular modelling indicated that the shape of the pentacyclic Ring D benzo CPT analog was similar to that of 1. However, 5 was found to be almost inactive in the T-1 assay, and much less active than 1 in other cytotoxic assays, such as 9KB and 9PS. Hence the ring D pyridone moiety is also essential for the in vitro and in vivo activity of 1.

Analogs with Variation of C20 Ethyl Substituent. Some years ago it was noted that the C20 ethyl group could be replaced by allylic groups with no loss of in vivo activity (*26*). Several new CPT analogs were pre-pared in which the C20-ethyl group was replaced by CH_3, CH_2CH_2OH, and CH_2CH_2Br, e.g. analogs 41, 42, and 43 respectively. All of the compounds were inactive in T-1 inhibition.

Effect of CPT and Analogs on T-1

Interest in the potential clinical utilization of CPT 1 or its analogs has been cyclic in intensity. After an initial period of clinical study of the water soluble sodium salt 2, interest lagged due to the inactivity and/or toxicity of this compound (*5-7*).

Mechanism of Inhibition. However, interest greatly increased when it was found that 1 interferes with the DNA breakage-reunion reaction, a biological function exerted by DNA-T-1, by trapping the enzyme-DNA intermediate termed the "cleavable complex" (*10, 11*). In this complex, T-1 is linked to the 3' phosphoryl end of the broken DNA backbone. The event occurs in cell free systems as well as in mammalian cells. Collision of the cleavable complex with the DNA replicative machinery stops fork elongation. This may represent the first step of a process that irreversibly inhibits DNA synthesis and ultimately leads to cell death (*10, 11*).

Relation Between In Vivo Activity and T-1 Inhibition. There seems to be a reasonable correlation between in vivo activity and activity of CPT analogs in the inhibition of T-1 (cf. Tables I, II, IV). The relationship is not exact because in vivo activity depends not only on the intrinsic structure of the CPT analog, but also on many other factors: cell penetration, overall toxicity, plasma binding, solubility, etc. But, in general, inspection of Tables I or II shows that compounds that have high T-1 inhibition activity also have, for example, excellent life prolongation in the L1210 mouse leukemia assay. Compounds with 20(S) stereochemistry are at least twice as potent as their analogs with the racemic 20(RS) configurations and compounds with the 20(R) configuration are always inactive in vitro and in vivo, 18 - 20, cf. also Tables II, III. Some excellent examples of the correlation of in vivo activity (L1210) and T-1 inhibition have been shown in previous research by our group (*19*). Thus the 9-amino-20(S) analog 9 is potent in both assays, the 11-amino-20(RS) analog 21 is much less active, and the 12-amino-20(S) analog 22 was inactive in both assays. A similar relationship is shown for nitro analogs 11 and 12, the nitro analog 11 being potent in vivo with progressively decreasing activity in the 11- and 12-nitro series, (cf. Table I and references 16, 17, 19). The high potency in both assays of both the 20(RS)- and 20(S)- of

10,11-methylenedioxy analogs 27 and 28 is demonstrated by comparison of data in Table I, cf. also refer-ence (*19*).

The correlations of in vivo assays and T-1 inhibition activities are so good that, in general, initial assessment of the potential in vivo activity of a new CPT analog can be carried out rapidly by use of the T-1 inhibition assay (*19*). The case of prodrugs which might be weakly active in T-1 inhibition but which can be bioactively converted to active compounds are exceptions. This is the case with the Japanese compound CPT-11, a water- soluble derivative of 7-ethyl-10-hydroxy-20(S)-CPT which is converted in vivo to the active 7-ethyl-10-hydroxy compound (*27,28*).

Since levels of T-1 have been found to be higher in certain cancerous cells than in normal tissues, e.g., cancerous colorectal tissue (*22*), specific targets for CPT and its analogs may be available. As a consequence, the synthesis of DNA in tumor cells may be inhibited. It has been suggested that DNA-T-1 is an intracellular target for 1 and analogs (*21,22*). The induction of T-1 linked DNA breaks and the reversibility of the DNA damage in CPT treated cells provides strong evidence that 1 interferes with the breakage-reunion reaction of cellular T-1.

T-1 Inhibition of CPT Analogs. As a consequence of the successful resolution of the 20(RS) tricyclic synthon, the chiral 20(S) and 20(R) synthons are now readily available (*18*), cf. Figure 3. Friedlander reactions of the chiral tricyclic synthons with appropriately substituted ortho-aminobenzaldehydes have yielded for the first time chiral 20(S) and 20(R)-CPT analogs. Table I presents T-1 inhibition IC_{50} values for a number of 20(S) and a few 20(RS) and 20(R) analogs substituted for the most part in the 9, 10, or 10,11- positions. The T-1 inhibition data are compared directly in Table III for selected CPT analogs.

The CPT analogs are listed in Table I in order of decreasing potency against purified T-1 (as assessed by formation of the cleavable complex) and the discussion will relate in this case to the IC_{50} potency data. Four distinct classes of potency can be more or less arbitrarily assigned: Class 1, compounds with IC_{50} values between 0.01 and 0.10 µM; Class 2, compounds with values between 0.1-1.0 µM; Class 3, compounds with values between 1.0-10.0 µM; and Class 4, compounds with values
>10.0 µM.

Data in Table III shows that the 20(S) form of a particular CPT analog is always much more potent than the 20(R) form and is approximately twice as potent as the 20(RS) form (range 1.5-4.0 fold).

The compounds with maximal IC_{50} values in general are water insoluble 20(S) and 20(RS) analogs substituted in the 9 or 10 position. CPT analogs 9-amino, 9, 10-hydroxy, 3, 9-chloro-10,11-methylenedioxy, 30 or 9-methyl, 18 show in general high T-1 inhibition activity. The relatively low activity of 9-hydroxy-20(S)-CPT 15 is an exception. The presence of a 10,11-methylenedioxy substituent, for example in analogs 27 and 28, greatly enhances the basic CPT activity, either when present singly or when 9-amino or 9-chloro substitutents are also present as in analogs 6 or 30. Planarity of the substituents in ring A appears crucial for enhancing the biological activity of the various CPT analogs. For example, the 10,11-ethylenedioxy-20(RS) analog 31 is only about one-seventh as potent in T-1 inhibition as the corresponding 10,11-methylenedioxy-20(RS) analog 27. Moreover the location of the planar substituent in ring A also appears to be crucial for activity. Thus the planar analog, 9,10-ethyl-enedioxy-20(RS)-CPT 32 has only one-fiftieth the activity of 27.

Water soluble analogs which cannot undergo bioactive transfor-mation in the T-1 assay, or which interact poorly with the DNA-T-1 cleavable complex (10,11) show poor to modest activity in T-1 inhibition. Compare in Table II for example, 27 with the corresponding sodium salt 36, and 10-hydroxy-20(S)-CPT 3 with the corresponding water soluble analog 34.

As shown in Tables II and III all the 20(R) analogs are essentially inactive in T-1 inhibition. 20(R)-CPT has been shown to be inactive also in 9KB cytotoxicity and in the in vivo L1210 mouse leukemia assay (*18*), (*19*).

Clinical Prospects

At present CPT (1) and three CPT analogs, 9-amino-20(S)-CPT 9, 9-dimethylaminomethyl-10-hydroxy-20(S)-CPT•HCl (Topotecan, 25) and CPT-11, a water-soluble analog of 7-ethyl-10-hydroxy-20(S)-CPT are in clinical trial. Water insoluble 1 is in Phase I clinical trial administered orally (personal communication; B. C. Giovanella, Stehlin Foundation, Houston, Texas). As a consequence of research which showed that water insoluble 9 exhibited high in vivo antitumor activity in the L1210 assay (*16,17,29*), remarkable activity in human xenograft colorectal tumors (*21*) in conjunction with high T-1 inhibition (*19*), and because of the development of a parenteral infusion procedure by the National Cancer Institute, water insoluble 9 is now receiving Phase I clinical trial. Water-soluble drugs Topotecan and CPT-11 have been in clinical trial for several years and some results are now available.

Data from initial clinical trials of 1 and 9 are as yet not available. Initial Phase I and Phase II trials with CPT-11 have shown that objective responses have been obtained in treatment of lung, colorectal, ovarian and cervical cancers (30,31). Administration of Topotecan has also resulted in responses to treatment of lung, ovarian and colorectal cancer (*32,33*).

Several other CPT analogs may eventually receive consideration for clinical trial. These include the 10-amino analog 10 and the 9-amino-10,11-methylenedioxy analog 6, both highly active in L1210 mouse leukemia assay (cf. Table I). Compound 6 has recently been shown to have considerable cytotoxicity toward B-cell chronic lymphocytic leukemia (CLL) cells with activity much greater than chlorambucil, a drug currently in clinical use for CLL patients, or than other CPT analogs (*23*).

Acknowledgements

We thank Dr. Jeffrey Besterman, Glaxo, Inc. for T-1 inhibition data. Many of the compounds reported in this chapter have been the result of skilled experimental research by Drs. Allan Nicholas and Govindarajan Manikumar. The research reported in this chapter has been supported by NIH-NCI CA-38996-01-06 and CA50529.

Literature Cited

1. Wall, M. E.; Wani, M. C.; Cook, C. E.; Palmer, K. H.; McPhail, A. T.; Sim, G. A. J. Am. Chem. Soc. **1966**, *88*, 3888-3890.
2. Wall, M. E. In Chronicles of Drug Discovery; Vol. 3; Lednicer, D., Ed.; American Chemical Society; Washington, D.C., in press.
3. Perdue, R. E., Jr.; Smith, R. L.; Wall, M. E.; Hartwell, J. L.; Abbott, B. J. U. S. Department of Agriculture, Agric. Res. Serv. Tech. Bull., **1970**, No. 1415, 1-26.
4. Wani, M. C.; Wall, M. E. J. Org. Chem. **1969**, *34*, 1364-1367.
5. Moertel, C. G.; Schutt, H. J.; Reitmerer, R. J.; and Hahn, R. G., Cancer Chemother. Rep., **1972**, 56, 95-101.
6. Gottleib, J. A.; Guarino, A. M.; Call, J. B.; Oliverio, V. T.; and Block, J. B. Cancer Chemother. Rep., **1970**, 54, 461-470.
7. Muggia, F. M., Creaven, P. J., Jansen, H. A., Cohen, M. H., Selawry, O. S. Cancer Chemother. Rep. **1972**, 56, 515-521.
8. Geran, R. I., Greenberg, N. H., MacDonald, M. M., Schumacher, A. M., Abbott, B. J. Cancer Chemother. Rep., **1972**, 3(2), 1-63.
9. Wani, M. C.; Ronman, P. E.; Lindley, J. T.; Wall, M. E., J. Med. Chem. **1980**, 23, 554-560.
10. Hsiang, Y. H.; Hertzberg, R.; Hecht, S.; Liu, L. F. J. Biol. Chem. **1985**, 260, 14873-14878.
11. Hsiang, Y. H. and Liu, L. F. Cancer Res. **1988**, 48, 1722-1726.
12. Wall, M. E. In Biochemistry and Physiology of Alkaloids; Mothes, K.; Schreiber, K.; Schutte, H. R., Eds.; Akademie-Verlag; Berlin, 1969; pp 77-87.
13. Nicholas, A. W.; Wani, M. C.; Manikumar, G.; Wall, M. E.; Kohn, K. W.; Pommier, Y. J. Med. Chem. **1990**, 33, 972-978.
14. Cai, J. C.; Hutchinson, C. R. In Alkaloids, vol. 21; Brossi, A., Ed.; Academic Press: New York, 1983; pp 102-136.
15. Wall, M. E.; Wani, M. C.; Natschke, S. M.; Nicholas, A. W. J. Med. Chem. **1986**, 29, 1553-1555.
16. Wani, M. C.; Nicholas, A. W.; Wall, M. E. J. Med. Chem. **1986**, 29, 2358-2363.
17. Wani, M. C.; Nicholas, A. W.; Manikumar, G.; Wall, M. E. J. Med. Chem. **1987**, 30, 1774-1779.
18. Wani, M. C.; Nicholas, A. W.; Wall, M. E. J. Med. Chem. **1987**, 30, 2317-2319.
19. Jaxel, C.; Kohn, K. W.; Wani, M. C.; Wall, M. E.; Pommier, Y. Cancer Res. **1989**, 49, 1465-1469.
20. Wall, M. E.; Wani, M. C. In Economic and Medicinal Plant Research, Vol 5; Wagner, H.; Farnsworth, N. R.; Eds.; Academic Press: New York, 1991; pp 111-127.
21. Giovanella, B. C.; Stehlin, J. S.; Wall, M. E.; Wani, M. C.; Nicholas, A. W.; Liu, L. F.; Silber, R.; Potmesil, M. Science **1989**, 246, 1046-1048.

22. Potmesil, M.; Giovanella, B. C.; Liu, L. F.; Wall, M. E.; Silber, R.; Stehlin, J. S.; Hsiang, Y.-H.; Wani, M. C. In DNA Topoisomerases in Cancer; Potmesil, M.; Kohn, K. W., Eds.; Oxford University Press: New York, 1991; pp 299-311.
23. Costin, D.; Potmesil, M.; Morse, L.; Mani, M.; Canellakis, Z. W.; Silber, R. Abstracts of the Fourth Conference on DNA Topoisomerases in Therapy, New York, 1992; No. 62, p. 53.
24. Vishnuvajjala, B. R., Cradock, J. C., Garzon-Aburbek, A., Pharm. Res., **1986**, 3, 22S.
25. Flurry, Jr., R. L.; Howland, J. C.; Abstracts of Papers, 162nd National Meeting of the American Chemical Society, Washington, DC, 1971, MEDI, 30.
26. Sugasawa, T.; Toyoda, T.; Uchida, J.; Yamaguchi, K.; J. Med. Chem., **1976**, 19, 675-679.
27. Kingsbury, W. D.; Boehm, J. C.; Dalia, R. J.; Holden, K. G.; Hecht, S. M.; Gallagher, G.; Caranea, M. J.; McCabe, F. L.; Faucette, C. F.; Johnson, R. K.; Herzberg, R. P. J. Med. Chem. **1991**, 34, 98-107.
28. Sawada, S.; Okayima, S.; Aiyama, R.; Ken-ichiro, N.; Furuta, T.; Yokokura, T.; Sugino, E.; Yamaguchi, K.; Miyasaka, T. Chem. Pharm. Bull. **1991**, 39, 1446-1454.
29. Wall, M. E.; Wani, M. C. In DNA Topoisomerases in Cancer; Potmesil, M.; Kohn, K. W., Eds.; Oxford University Press: New York, 1991; pp 93-102.
30. Taguchi, T. Abstracts of the Fourth Conference on DNA Topoisomerases in Therapy, New York, 1992, No. 30, p. 31.
31. Rothenberg, M. L.; Rowinsky, E.; Kuhn, J. G.; Burrisil, H. A., Van Hoff, D. Abstracts of the Fourth Conference on DNA Topoisomerases in Therapy, New York, 1992, No. 31, p. 31.
32. Verweij, J. Abstracts of the Fourth Conference on DNA Topoisomerases in Therapy, New York, 1992, No. 34, p. 32.
33. Hochster, H. Abstracts of the Fourth Conference on DNA Topoisomerases in Therapy, New York, 1992, No. 33, p. 32

RECEIVED April 14, 1993

Chapter 12

Antineoplastic Agents and Their Analogues from Chinese Traditional Medicine

Kuo-Hsiung Lee

Natural Products Laboratory, Division of Medicinal Chemistry and Natural Products, School of Pharmacy, University of North Carolina, Chapel Hill, NC 27599

Cytotoxic antitumor principles from Chinese traditional medicines, plant materials, and their semi-synthetic analogs are reviewed with emphasis on those discovered in the author's laboratory. The active compounds include sesquiterpene lactones, diterpenes, quassinoids, triterpenes, alkaloids, quinones, diamides, coumarins, flavonoids, lignans, macrolides, polyacetylenes, polyphenols, and styrylpyrones, as well as their derivatives and analogs. The structure-activity relationships and mechanism of action studies among these compounds are discussed wherever feasible.

Chinese traditional medicines have been used in the treatment of cancer for centuries, but only recently has a systematic effort been made to isolate and characterize the active principles from their antitumor-active extracts. The progress made during the past two decades in the discovery and development of numerous such cytotoxic antitumor agents and their analogs has been reviewed (*1-8*). Studies on Chinese plant-derived antineoplastic agents and their analogs are proceeding in many laboratories. However, to limit the scope of the present discussion, this review will deal mainly with the work carried out in the author's laboratory, especially on those compounds discovered recently.

Chinese Plant-Derived Antineoplastic Agents and Their Analogs in Clinical Use, Clinical Trials, and Under Development

Several clinically useful anticancer drugs and their analogs discovered and developed initially in the U.S. and in other countries are currently also used in China as the sources of these drugs are readily available there. These include those used in the U.S., such as the *Catharanthus* alkaloids, vinblastine (**1**) and vincristine (**2**), and their analog, vindesine (**3**) (*3, 9-12*), and the podophyllotoxin-derived lignan glycosides, etoposide (**4**) and teniposide (**5**) (*3, 13-15*).

Other anticancer drugs used in China include 10-hydroxycamptothecin (**6**) (*3, 5, 16, 17*), homoharringtonine (**7**) (*3, 18-20*), monocrotaline (**8**) (*3, 7, 21*), lycobetaine (**9**) (*3, 5, 8*), colchicinamide (**10**) (*3, 22*), *d*-tetrandrine (**11**) (*3, 22-24*), (-)-sophocarpine (**12**) (*3, 8, 25*), indirubin (**13**) and its derivative *N,N*'-dimethylindirubin (**14**), and *N*-methylindirubin oxime (**15**) (*3, 5, 7, 26*), curzerenone (**16**) (*3, 4, 6, 27*), curcumol (**17**) (*4, 7*), curdione (**18**) (*4, 7*), oridonin (**19**) (*3, 4, 7, 28, 29*), and gossypol (**20**) (*3, 4, 7, 8*). Compounds **6**, **7**, **8**, **11**, **12**, and **16-20** were initially discovered in the U.S. (**6**, **7**, **11**, and **20**),

0097–6156/93/0534–0170$06.25/0

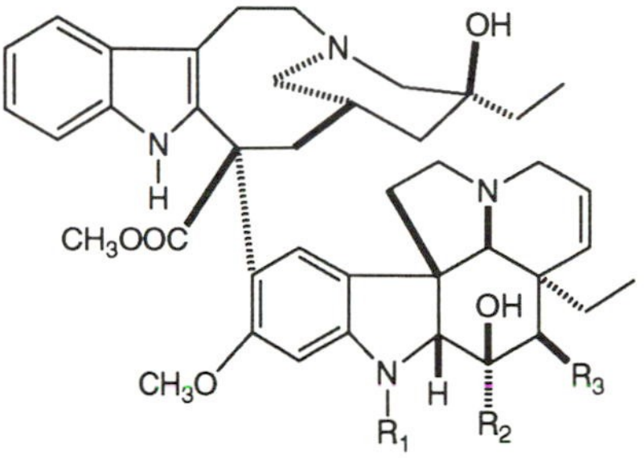

1, Vinblastine (Velban): $R_1 = CH_3$, $R_2 = COOCH_3$, $R_3 = OAc$

2, Vincristine (Oncovin): $R_1 = CHO$, $R_2 = COOCH_3$, $R_3 = OAc$

3, Vindesine: $R_1 = CH_3$, $R_2 = CONH_2$, $R_3 = OH$

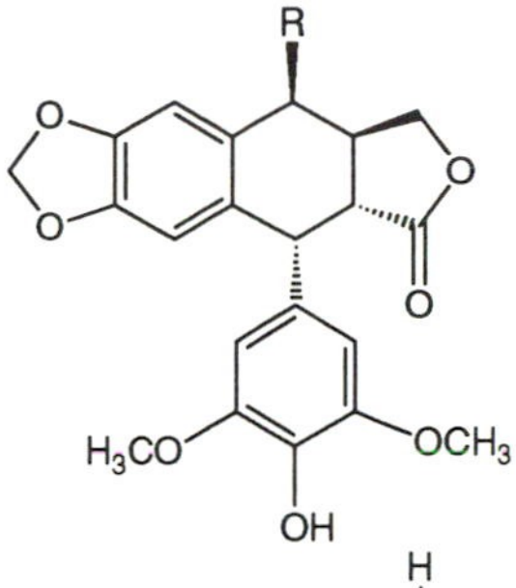

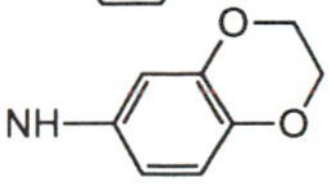

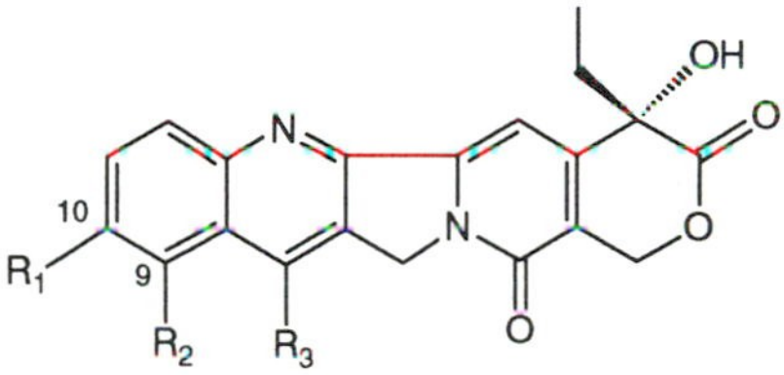

6, 10-Hydroxycamptothecin: $R_1 = OH$, $R_2 = R_3 = H$

21, Camptothecin: $R_1 = R_2 = R_3 = H$

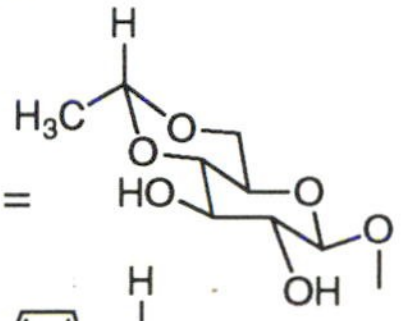

22, Hycamptamine: $R_1 = OH$, $R_2 = CH_2N(CH_3)_2$, $R_3 = H$

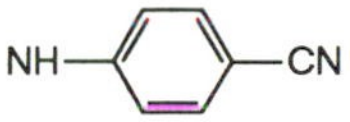

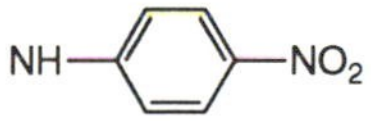

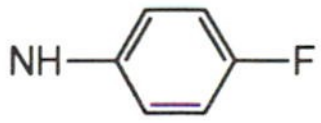

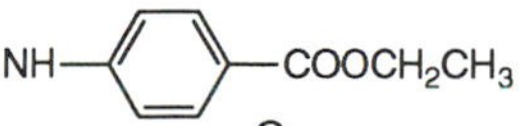

23, 9-Aminocamptothecin: $R_1 = OH$, $R_2 = NH_2$, $R_3 = H$

24, CPT-11: R_1 = OCO-N(piperidinyl)-N(piperidine) • HCl

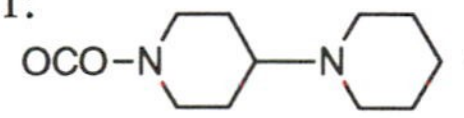

$R_2 = H$, $R_3 = CH_2CH_3$

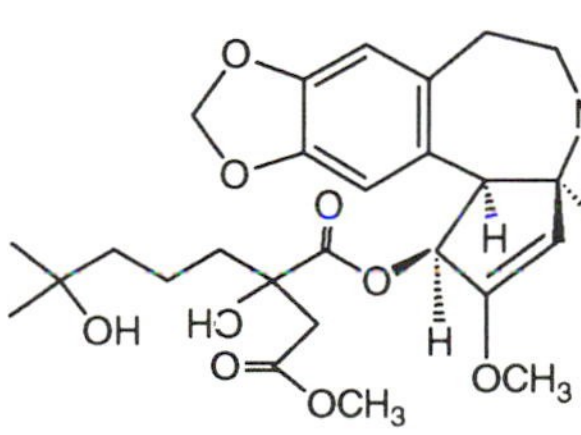

7, Homoharringtonine

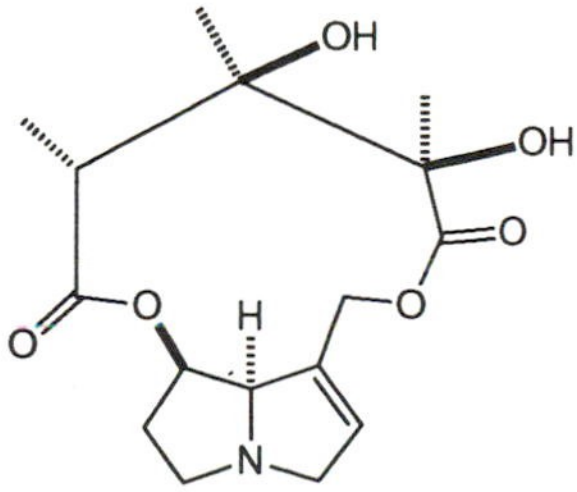

4, Etoposide: R =

5, Teniposide: R =

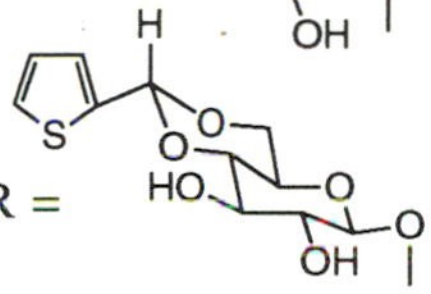

25, R = NH–C_6H_4–NH_2 • HCl

26, R = NH–C_6H_4–CN

27, R = NH–C_6H_4–NO_2

28, R = NH–C_6H_4–F

29, R = NH–C_6H_4–$COOCH_2CH_3$

30, R = NH–

8, Monocrotaline

Australia (**8**), and Japan (**11, 12, 18**, and **19**), respectively, but were first introduced into the clinic as anticancer drugs in China. Compounds **9, 10, 13, 14**, and **15** originated in China. A discussion on the major uses of these compounds has been included in a recent review (*3*) and is briefly summarized in Table I.

Modification and analog development of camptothecin (**21**) and etoposide (**4**) aimed at producing improved drug candidates are of current interest.

Derivatives of Camptothecin. With the recent discovery of **21** and **6** as potent inhibitors of DNA topoisomerase I (*30, 31*), their derivatives, such as hycamptamine (**22**) (*32*), 9-aminocamptothecin (**23**) (*33*), and CPT-11 (**24**) (*34*), are being tested clinically as anticancer drugs (*35*) against colon and other cancers in Europe, the U.S.A., and Japan, respectively.

Analogs of Etoposide. The podophyllotoxin-derived etoposide (**4**) and teniposide (**5**) are effective anticancer drugs for the treatment of small-cell lung and testicular cancers, as well as lymphoma and leukemia (*14, 15*). There is evidence to suggest that these drugs block the catalytic activity of DNA topoisomerase II by stabilizing an enzyme-DNA complex in which the DNA is cleaved and linked covalently to the enzyme (*36-38*). It is possible to associate the cytotoxicity of **4** and **5** with the phenoxy free radical and its resulting *ortho*-quinone species formed by the biological oxidation of these drugs (*39, 40*). The *ortho*-quinone could cleave DNA if it forms metal complexes with Ca^{2+} and Fe^{3+} ions (*41*). Metabolic activation of **4** is believed to produce the phenoxy free radical and *ortho*-quinone derivatives of **4**. This is because **4** as an alkylating species could bind to critically important cellular macromolecules causing dysfunction and, subsequently, cell death (*42, 43*). Recent studies in the author's laboratory involving the replacement of the 4β-sugar moiety with 4β-arylamino groups have yielded several compounds (**25-30**) which were 5- to 10-fold more potent than **4** as inhibitors of DNA topoisomerase II *in vitro*. All of these compounds (**25-30**) could generate the same amount or more protein-linked DNA breaks in cells than **4** at 1-20 μM. In addition, these new compounds were cytotoxic not only to KB cells but also to their **4**-resistant and vincristine(**2**)-resistant variants which showed decreased cellular uptake of **4** and a decrease in DNA topoisomerase II content or overexpression of the MDR1 phenotype (*44-46*). Further development of these compounds for clinical trials as anticancer drug candidates is in progress.

Chinese Plant-Derived Antineoplastic Agents and Their Analogs Discovered in the Author's Laboratory

Bioassay-directed isolation and characterization of potent cytotoxic antitumor agents and analogs from Chinese medicinal plants has been one of our research objectives over the past two decades. This program has led to the identification of numerous new leads for further development as anticancer drugs. Recently, we have discovered more than 50 such compounds which are of interest to the U.S. National Cancer Institute (NCI). They are currently undergoing NCI's detailed evaluation as potential new anticancer agents. The key to the success for new drug discovery is the proper selection of a bioassay method. Previously in the 1970's and early 1980's, we used NCI's *in-vitro* KB and *in-vivo* P-388 as the prescreen methods (*47*) to detect potential cytotoxic antileukemic agents. We have recently also used NCI's expanded protocol (*48, 49*) to include several disease-oriented human cancer cell lines, such as A-549 (lung carcinoma), HCT-8 (ileocecal adenocarcinoma), MCF-7 (breast adenocarcinoma), and RPMI-7951 (melanoma) in our routine screening program aimed at discovering agents which might be active against slowly

Table I: Major Uses of Compounds 1-21

Compounds	*Major Uses*
1 (Vinblastine)	Hodgkin's disease
2 (Vincristine)	Acute childhood leukemia
3 (Vindesine)	Adult nonlymphocytic leukemia, acute childhood lymphocytic leukemia, Hodgkin's disease, and malignant melanoma
4 (Etoposide)	Small cell lung cancer, testicular cancer, lymphoma, and leukemia
5 (Teniposide)	Acute lymphocytic leukemia, childhood neroblastoma, adult non-Hodgkin's lymphoma and brain tumors
6 (10-hydroxycamptothecin)	Cancers of liver, head, and neck
7 (Homoharringtonine)	Acute myeloblastic and monocytic leukemias, and erythroleukemia
8 (Monocrotaline)	External treatment of skin cancer
9 (Lycobetaine)	Ovarian carcinoma and gastric cancer
10 (Colchicinamide)	Mammary carcinoma
11 (*d*-tetrandrine)	Lung cancer
12 ([-]-sophocarpine)	Trophocytic tumor, chorion-epothelioma and leukemia
13 (Indirubin)	Chronic myelocytic leukemia
14 (*N,N'*-Dimethylindirubin)	Chronic myelocytic leukemia
15 (*N*-methylindirubin oxime)	Chronic myelocytic leukemia
16 (Curzerenone)	Uterine cervix and skin carcinomas
17 (Curcumol)	Early stage cervix cancer
18 (Curdione)	Early stage cervix cancer
19 (Oridonin)	Last stage of cancer of esophagus
20 (Gossypol)	Stomach, esophageal, liver, mammary, and bladder cancers
21 (Camptothecin)	Gastric, rectal, colonic, and bladder cancers

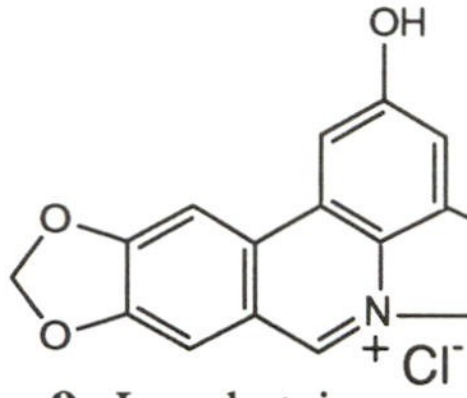

9, Lycobetaine

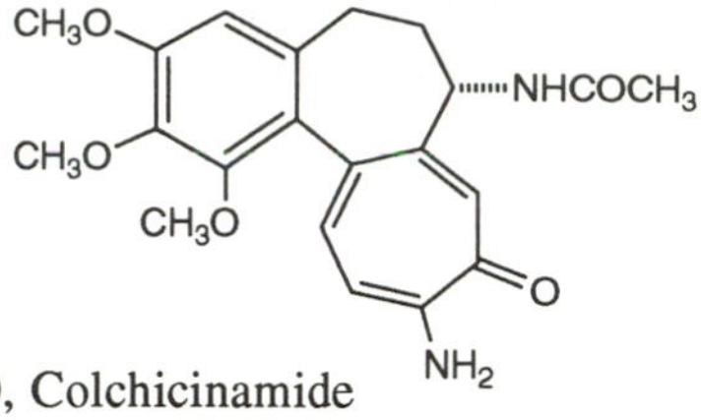

10, Colchicinamide

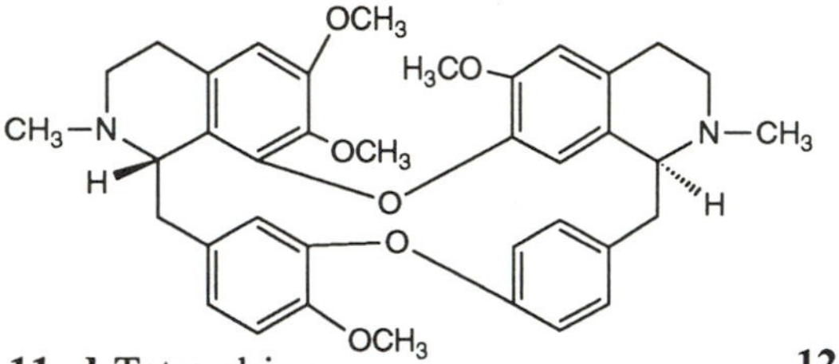

11, **d**-Tetrandrine

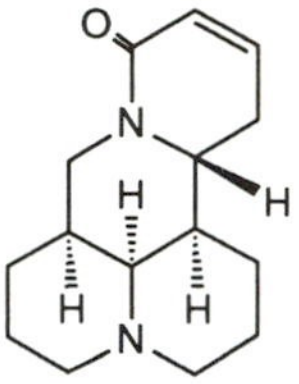

12, (-)-Sophocarpine

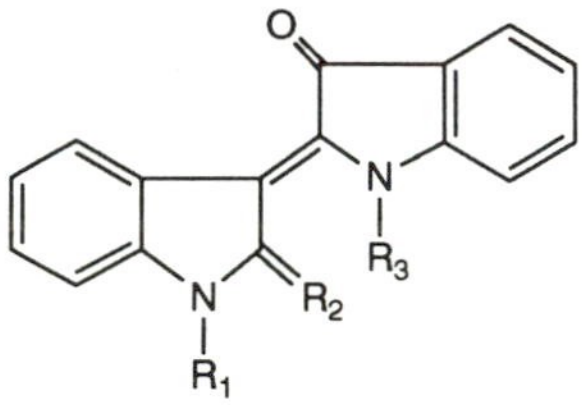

13, Indirubin: $R_1 = R_3 = H, R_2 = O$
14, **N**,**N**'-Dimethylindirubin:
$R_1 = R_3 = CH_3,\ R_2 = O$
15, **N**-Methylindirubin oxime:
$R_1 = CH_3, R_2 = NOH, R_3 = H$

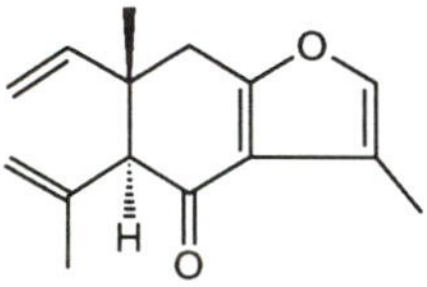

16, Curzerenone

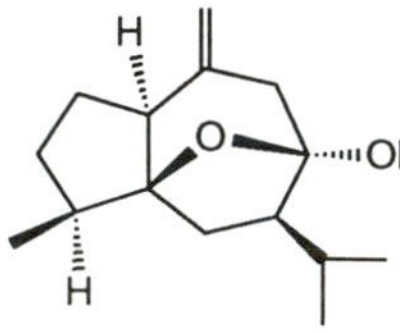

17, Curcumol

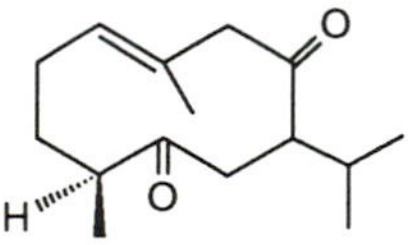

18, Curdione

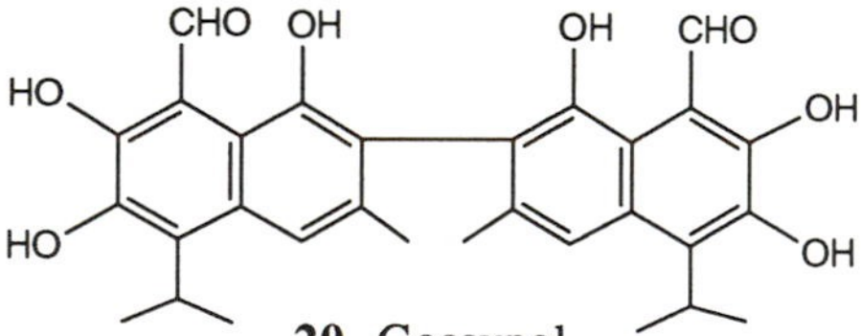

20, Gossypol

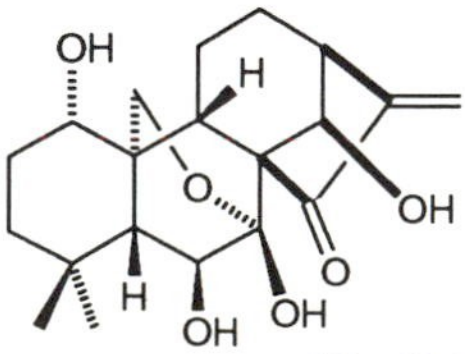

19, Oridonin

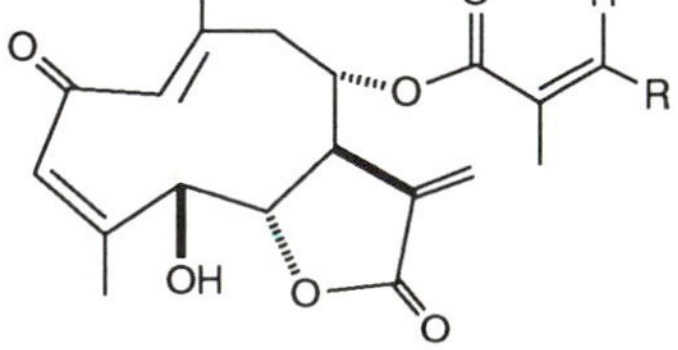

31, Molephantin: R=H
32, Molephantinin: $R = CH_3$

growing solid tumors. On the basis of this approach, many novel cytotoxic antitumor agents have been isolated and characterized from Chinese medicinal materials, and some of their representative examples are very briefly discussed below.

Sesquiterpene Lactones. Several potent cytotoxic antitumor germacranolides have been isolated. These include molephantin (**31**) (*51*), molephantinin (**32**) (*52*), and phantomolin (**33**) (*53*) from *Elephantopus mollis* (*Bei Deng Shu Wu*) (Compositae) as well as eupaformonin (**34**) (*54*) and eupaformosanin (**35**) (*55*) from *Eupaforium formosanum* (*Shan Che Lan*) (Compositae). Compound **35** showed remarkable antitumor activity against the fast-growing solid tumor Walker 256 carcinosarcoma with a T/C of 471% at 2.5 mg/kg (Table I) and deserves further development as a potentially useful anticancer drug.

The known pseudoguaianolide, helenalin (**36**), was isolated from *Anaphalis morrisonicola* (*Yu Shan Shu Chu Tsao*) (Compositae) (*50, 56*). An extensive study on the structure-activity relationships among **36**-related sesquiterpene lactones and derivatives by this laboratory (*57-59*) has established that an enone O=C-CH=CH system, either present in the form of a β-unsubstituted cyclopentenone or an exocyclic α-methylene-γ-lactone, as seen in **36**, is regarded as an alkylating center and is directly responsible for its enhanced cytotoxic antineoplastic activity, possibly via a rapid Michael-type addition of sulfhydryl group-bearing macromolecules in key regulatory enzymes of nucleic acid and cellular metabolism.

The bis-helenalinyl esters including bis-helenalinyl malonate (**37**), which was designed based on the fact that both an enone O=C-CH=CH system and lipophilicity could contribute to enhanced cytotoxic antitumor activity (*60*), showed potent antileukemic activity (P-388-UNC) with T/C=261% at 15 mg/kg (*61*). Both **36** and **37** are potent inhibitors of IMP dehydrogenase (*62*) and of protein synthesis by preventing the formation of the 48S initiation complex specifically inactivating eIF-3 (*63, 64*).

Diterpenes. Two antileukemic diterpenes from *Daphne genkwa* (*Yuan Hua*) (Thymelaeaceae) are the new genkwadaphnin (**38**) and the known yuanhuacin (**39**), which was previously reported as odoracin or gnidilatidin (*65*). Compounds **38** and **39** demonstrated potent antileukemic (P-388) activity in low doses (Table I). Compound **39** is used clinically as an abortifacient in China (*66*).

Other novel antileukemic diterpene esters from *Euphorbia kansui* (*Kan Sui*) (Euphorbiaceae) include kansuiphorin-A (**40**) and -B (**41**) (*67*). Compound **40** is also selectively cytotoxic to certain human cancer cell lines, such as leukemia, non-small-cell lung cancer, colon cancer, melanoma, and renal cancer cells (Table II). While *Kan Sui* is one of the commonly prescribed anti-cancer herbs in China, further development of **40** and **41** as potentially useful antitumor agents is needed.

Pseudolaric acid-A (**42**) and -B (**43**), the novel diterpene acids isolated from the root bark of *Pseudolaris kaempferi* (*Tu Jin Pi*) (Pinaceae) showed potent cytotoxicity. Compound **42** is more effective against leukemia P-388, CNS cancer U-251, and melanoma SK-MEL-5, while **43** inhibits preferentially leukemias HL-60TB and P-388, colon cancer SW-620, CNS cancer TE 671, melanoma SK-MEL-2, and ovarian cancer A-2780 (*68*).

Quassinoids. The fruit of *Brucea javanica* (Simaroubaceae), known as *Ya Tan Tzu* in Chinese folklore, has been used as a herbal remedy for human cancer, amebiasis, and malaria in traditional Chinese medicine. Bioassay-directed fractionation of the antitumor-active extract of *B. javanica* in this laboratory led to the isolation of bruceoside-A (**44**) and -B (**45**), the first two novel quassinoid

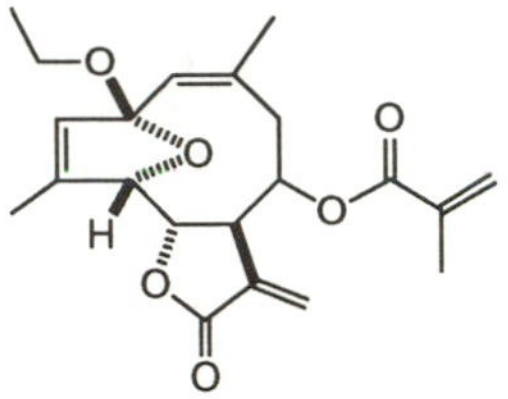

33, Phantomolin

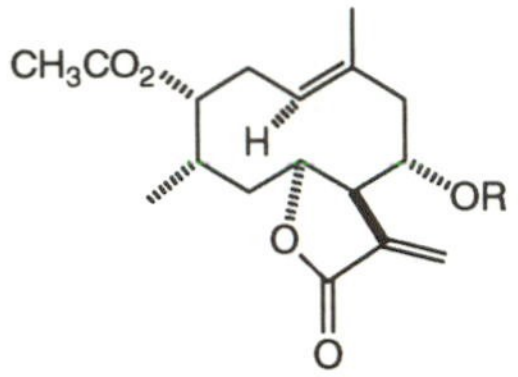

34, Eupaformonin: R= H
35, Eupaformosanin: R= —CO(C(CH$_2$OH)=CH(CH$_2$OH))

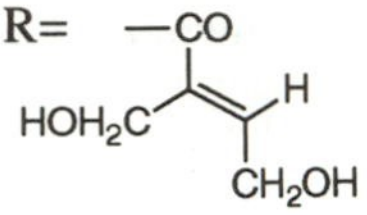

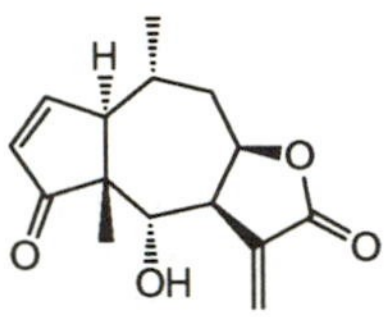

36, Helenalin

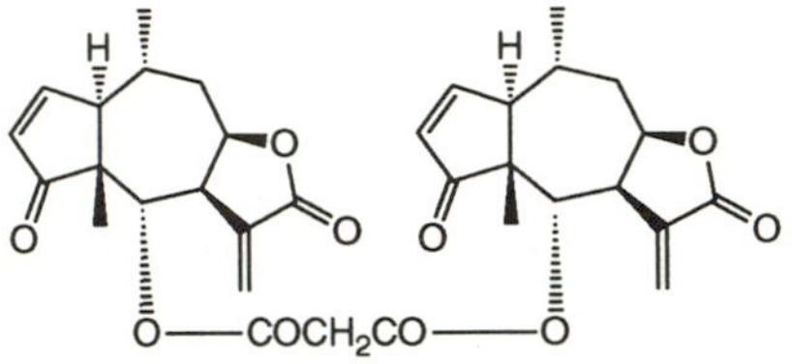

37, Bis-helenalinyl Malonate

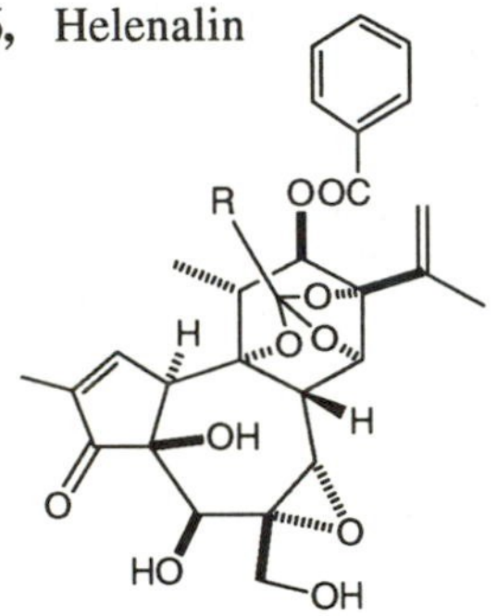

38, Genkwadaphnin: R= C_6H_5
39, Yuanhuacin: R= -CH=CH-CH=CH(CH$_2$)$_4$CH$_3$

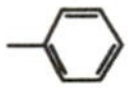

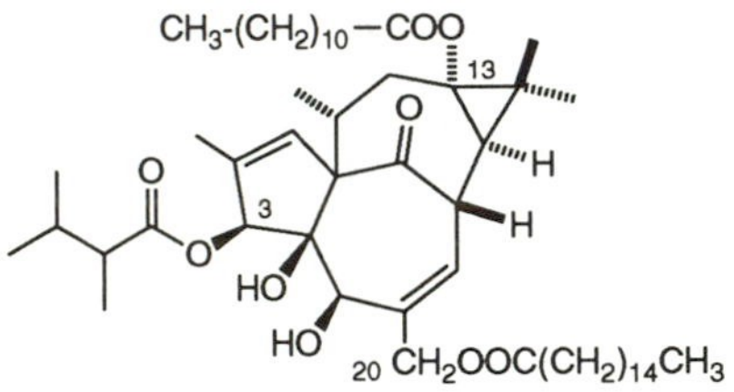

40, Kansuiphorin-A

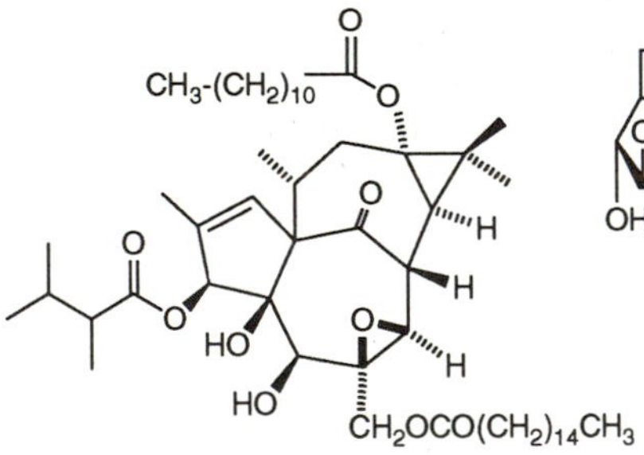

41, Kansuiphorin-B

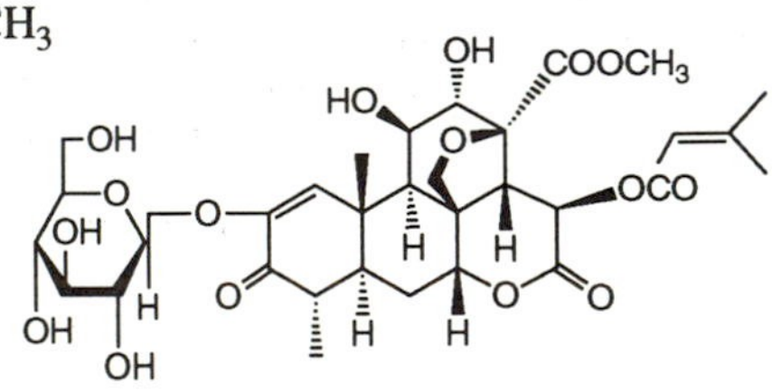

44, Bruceoside-A

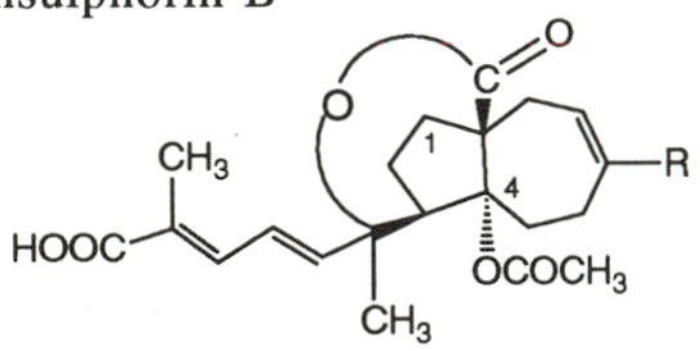

42, Pseudolaric Acid-A: R = CH$_3$
43, Pseudolaric Acid-B: R = COOCH$_3$

Table II. Cytotoxic and Antineoplastic Activity of Agents Derived from Chinese Medicinal Plants and Their Analogs

Compd.	ED_{50} (µg/ml)							%T/C (mg/kg)		
	KB[a]	P-388[b]	L-1210[b]	A-549[c]	HCT-8[d]	RPMI-7951[e]	MISC.-1	W-256[f]	P-388[g]	MISC.-2
31	1.05							149(2.5)		
32	0.90							397(2.5)	146(25)	
33	4.00							278(2.5)		
34							<1.00(H.EP.-2[h])			
35								471(2.5)	147(25)	
36	0.19						0.08(H.EP.-2[h])	316(2.5)	168(8)	142(25) (LL[i])
37									261(15[j])	
38									175(0.8)	
39									145(0.03)	
40						0.05	0.03(HL-60-TB[b]) 0.06(HOP-62[c]); 0.08(SW-620[d]); 0.02(A-498[k])		>176(0.1)	
41									>177(0.5)	
42	0.33	0.16	0.46	0.66	0.60		0.04(U-251[l]) 0.22(SK-MEL-5[e])			
43	0.22	0.04	0.42	0.66	0.61		0.03(HL-60-TB[b]); 0.01(SW-620[d]); 0.05(TE-671[l]); 0.47(SK-MEL-5[e]); 0.03(A-2780[m])			
44									156(6.0)	
45									132(1.5)	
46									158(0.1)	
47	0.59								204(1.0)	168(0.5) (B-16[n])
48									270(0.6)	
49									165(1.5)	178(3.0) (B-16[n]) 142(6.0) (L-1210[o])
50									148(6.25)	
51									120(10)	

Continued on next page

Table II Continued

Compd.	KB[a]	P-388[b]	L-1210[b]	A-549[c]	HCT-8[d]	RPMI-7951[e]	MISC.-1	W-256[f]	P-388[g]	MISC.-2
52	3.30									
53	4.20									
54	2.70									
55		3.18	4.10	4.00	4.50					
56	<0.08									
57	0.014			0.01	0.03					
58				0.04	0.01					
59	0.37									
60	0.20									
61	1.00	0.57	2.33	0.72	0.70					
62	2.50									
63	1.50	3.75			4.30					
64	3.30									
65	3.30									
66	4.00									
67	0.40									
68		2.90			4.60					
69		0.90								
70		1.85							129(10)	
71	4.00									
72	0.62									
73	0.09									
74	2.60								150(10)	
75	0.16									
76	3.00									
77	0.40									
78	0.40									
79	0.30									
80	0.60									
81	3.00									
82		0.59			4.90					

83							136(5)	
84								97% inhib.(3) (EAP)
85		0.40						
86							174(12.5)	
87							130(12.5)	
88	2.00							
89							154(16)	
90	<1.00							
91	<1.00							
92							>140(0.2)	
93	1.00		3.30					
94	0.50	3.30	2.90	0.98				
95	<0.10	<0.10			<0.10	<0.10(TE-671[l])		

[a]KB (human epidermoid carcinoma of nasopharynx).
[b]P-388 and L-1210 (lymphocytic leukemias).
[c]A-549 and HOP-62 (non-small-cell lung cancers).
[d]HCT-8 and SW-620 (colon cancers).
[e]RPMI-7951 and SK-MEL-5 (melanomas).
[f]W-256 (Walker-256 carcinosarcoma in Sprague-Dawley rat).
[g]P-388 (murine P-388 lymphocytic leukemia).
[h]H.EP.-2 (human epidermoid carcinoma of larynx).
[i]LL (murine Lewis lung carcinoma).
[j]P-388-UMC.
[k]A-498 (renal cancer).
[l]U-251 and TE-671 (CNS cancers).
[m]A-2780 (ovarian cancer).
[n]B-16 (murine B-16 melanoma).

glucosides, and brusatol (**46**) as the antileukemic principles (Table I) (*69*). Recently, Sasaki *et al.* (*70*) reported the isolation of yadaziosides-A, -B, -C, -F, -I, -J, -K, -L, -M, and -O as new quassinoid glucosides from this same plant species. However, all of these compounds showed antileukemic (P-388) activity of the same magnitude as **44**.

Brusatol (**46**) is structurally identical to bruceantin (**47**) (*71*) except for a slight difference in their C-15 ester side chains. Compound **47** was used by NCI in Phase II clinical trials as an anticancer drug candidate (*72*). Compound **47** was synthesized from either **44**, **45** or **46** in our laboratory (*4, 73, 74*).

Bis-brusatolyl malonate (**48**) and bis-bruceantinyl malonate (**49**), designed based on a systematic structure-activity relationship study, were found to demonstrate potent antileukemic [T/C (P-388-UNC) = 270% at 0.6 mg/kg] and anti-melanoma [T/C (B-16) = 178% at 3.0 mg/kg] activities, respectively (*4, 75*).

Compounds **44**, **46**, **47**, and **48** suppressed DNA, RNA, and protein biosynthesis, with **48** being the most potent, followed by **46** and **47** (*76*). Protein biosynthesis of P-388 cells was blocked by these compounds at the elongation step by interfering with peptide bond formation (*77*). The same mechanism was also observed with **46** in a rabbit reticulocyte system (*78, 79*). The **46**-related quassinoids deserve further study as useful anticancer agents.

Triterpenes. Two new antileukemic triterpenes are maytenfolic acid (**50**) and maytenfoliol (**51**). These were isolated from *Maytenus diversifolia* (*Pei Chung* or *Tsu Lou Shih*) (Celastraceae) (*80*).

Another novel cytotoxic triterpene is radermasinin (**52**), isolated from *Radermachia sinica* (*Shan Tsai Tou*) (Bignoniaceae) (*81*).

The extract of *Hyptis capitata* (*Bai Mou Ku Shiao*) (Labiatae) yielded as cytotoxic principles the new hyptatic acid-A (**53**) and 2α-hydroxyursolic acid (**54**) (*82*) as well as the known ursolic acid (**55**) (*83*).

Others include the new bryophyllin-B (**56**) and the known bryotoxin-C (=bryophyllin-A) (**57**) and bersaldegenin-3-acetate (**58**) from *Bryophyllum pinnatum* (*Luo Di Sheng Ken*) (Crassulaceae) (*84, 85*). Compounds **56-58** demonstrated remarkable cytotoxicity for KB, A-549, and HCT-8 tumor cells.

Pseudolarolide-I (**59**), a novel cytotoxic peroxytriterpene dilactone, was isolated from *Pseudolarix kaempferi* (*Tu Jin Pi*) (*86*).

Alkaloids. Kuafumine (**60**) is a new cytotoxic alkaloid from *Fissistigma glaucescens* (*Bei Yeh Kua Fu Mu*) (Annonaceae) (*87*). The stem and stem bark of *Artabotrys uncinatus* (*Ying Zhao Hua*) afforded two cytotoxic aporphine alkaloids, liriodenine (**61**) and atherospermidine (**62**) (*88*). Compound **61** and the cytotoxic (+)-thalifarazine (**63**) were also isolated from *Thalictrum sessile* (*Yu Shan Tang Suong Tsao*) (Ranunculaceae) (*89*). (-)-Norannuradhapurine (**64**), isolated from *Fissistigma oldhamii* (*Mao Kua Fu Mu*) (Annonaceae), is cytotoxic to HCT-8 tumor cells (*90*).

A novel skeletal indole-naphthoquinone alkaloid and cytotoxic principle, murrapanine (**65**), was isolated from *Murraya paniculata* var. *omphalocarpa* (*Chang Kuo Yueh Chu*) (Rutaceae). The synthesis of **65** from indole-3-aldehyde was also accomplished (*91*).

Emarginatine-A (**66**) and -B(**67**), two novel cytotoxic sesquiterpene pyridine alkaloids, along with maytansine, were isolated from *Maytenus emarginata* (*Lan Yu Lou Shih*) (Celastraceae) (*92, 93*).

The cytotoxic principles of *Securinega virosa* (*Bai Huan Shu*) (Euphorbiaceae) are virosecurinine (**68**) and viroallosecurinine (**69**). A comparison of the cytotoxicity of **68** and several of its derivatives indicates that the α, β- and γ, δ-

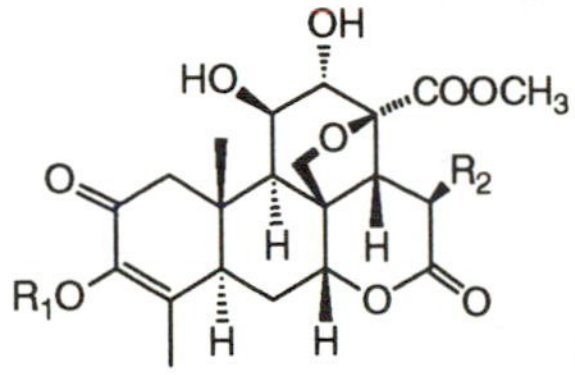

45, Bruceoside-B: R_1 = (glucosyl) , R_2 = $\overset{}{CO}$–CH=C(CH_3)$_2$

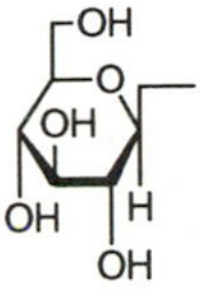

46, Brusatol: R_1 = H, R_2 = CO–CH=C(CH_3)$_2$

47, Bruceantin: R_1 = H, R_2 = CO–CH=C(CH_3)–CH(CH_3)$_2$

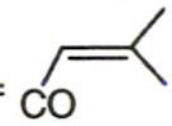

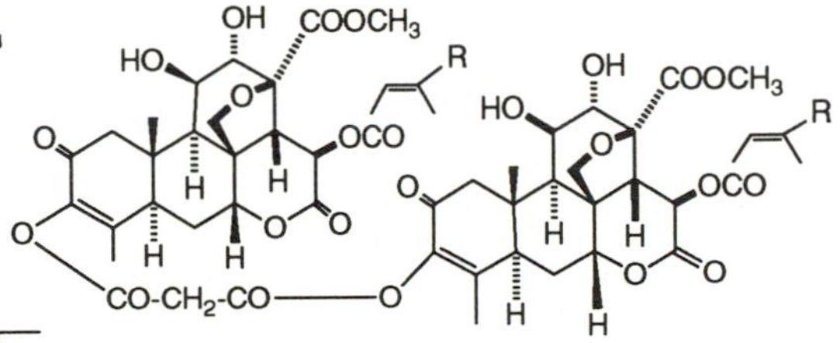

48, Bis-brusatolyl malonate: R = H

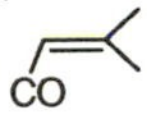

49, Bis-bruceantinyl malonate: R = $CH(CH_3)_2$

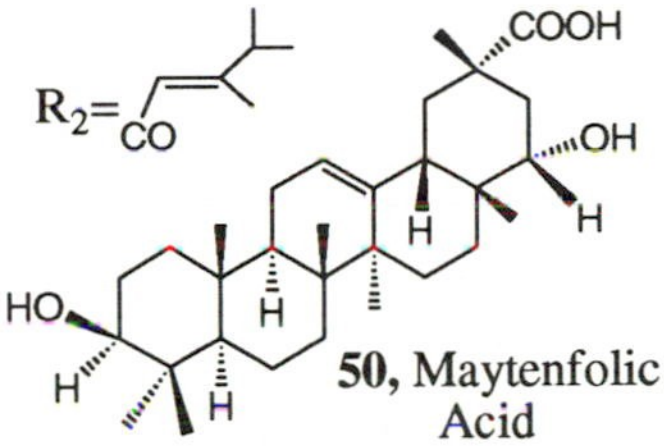

50, Maytenfolic Acid

51, Maytenfoliol

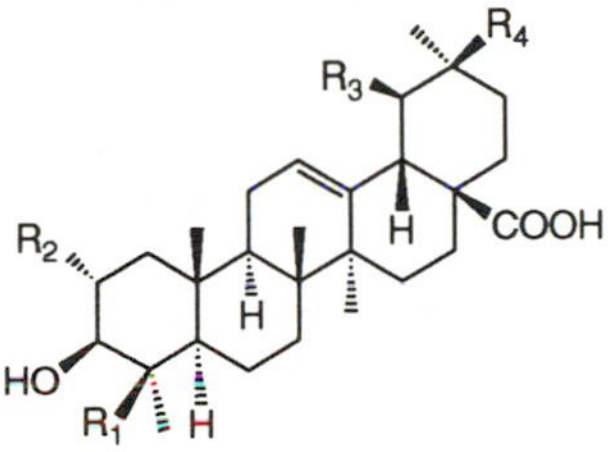

53, Hyptatic Acid-A: R_1 = CH_2OH, R_2 = OH, R_3 = H, R_4 = CH_3

54, 2α-Hydroxyursolic acid: R_1 = CH_3, R_2 = OH, R_3 = CH_3, R_4 = H

55, Ursolic acid: R_1 = CH_3, R_2 = H, R_3 = CH_3, R_4 = H

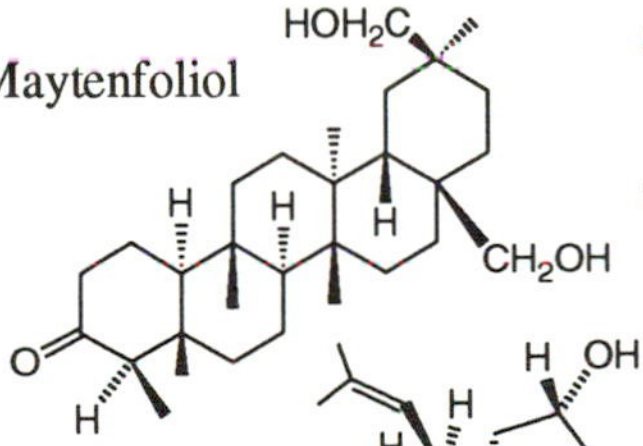

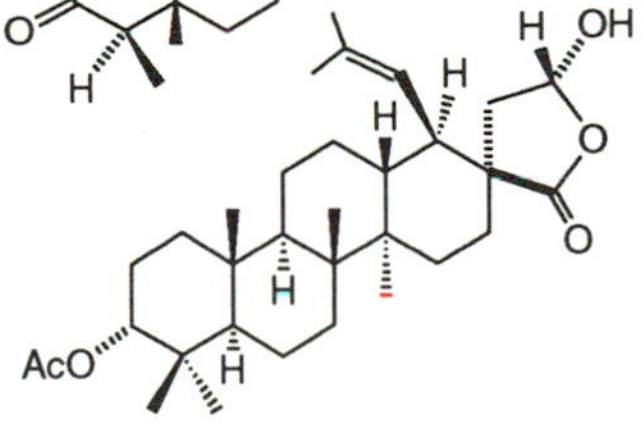

52, Radermasinin

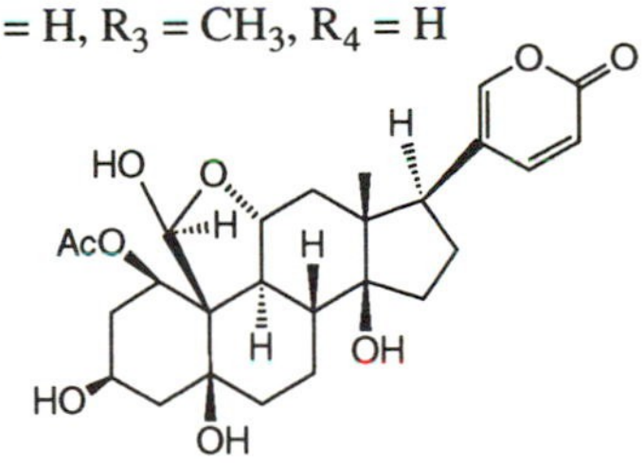

56, Bryophyllin-B

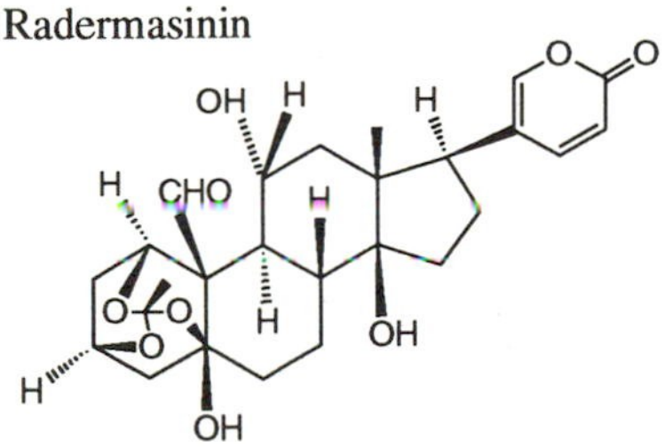

57, Bryotoxin-C (= Bryophyllin-A)

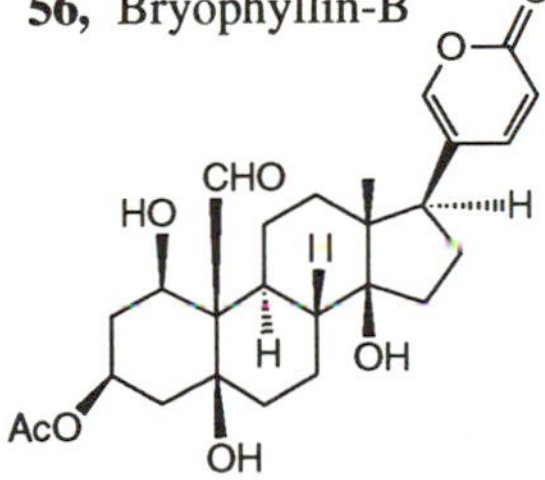

58, Bersaldegenin 3-acetate

unsaturated lactone located in a strained ring system, such as rings -B, -C, and -D of **68**, is structurally required for significant cytotoxicity (*94*).

Quinones. Five cytotoxic antileukemic anthraquinones were isolated from *Morinda parvifolia (Hong Chu Teng)* (Rubiaceae). These include the new morindaparvin-A (**70**) and -B (**71**) as well as the known 1,3 dihydroxy-2-methoxy-methylanthraquinone (**72**), digiferruginol (**73**), and 2-hydroxymethylanthraquinone (**74**) (*95, 96*). The total synthesis of **71** and **73** has been achieved (*97*).

A new phenanthrene-1,4-quinone, annoquinone-A (**75**), was isolated from *Annona montana* (*Bai Hsi Fan Li Shih*) (Annonaceae) (*98*).

The alcoholic extract of *Psychotria rubra* (*Chiou Chie Mu*) (Rubiaceae) furnished psychorubrin (**76**), a new cytotoxic naphthoquinone. Naphthoquinone derivatives (**77-80**) were prepared and exhibited superior cytotoxic activity to that of **76** (*99*).

Rhinacanthin-B (**81**), a new cytotoxic naphthoquinone, was isolated from *Rhinacanthus nasutus* (*Bai Her Ling Zhi*) (Acanthaceae) (*100*).

Another naphthoquinone isolated from *Diospyros morrisiana* (*Shan Hung Shih*) (Ebenaceae) is the known isodiospyrin (**82**). However, its cytotoxicity against HCT-8 and P-388 tumor cells was demonstrated for the first time (*101*).

Diamides (-)-Odorinol (**83**) was isolated as an antileukemic diamide from *Aglaia odorata* (*Shu Lan*) (Meliaceae) (*102*).

Coumarins. Daphnoretin (**84**), a known phenolic dicoumaryl ether, was isolated from *Wikstroemia indica* (*Nan Ling Jao Hua*) (Thymelaeaceae) and showed potent antitumor activity *in vivo* against Ehrlich ascites carcinoma growth in mice (*103*). Compound **84** is an inhibitor of DNA and protein synthesis in Ehrlich ascites tumor cells (*104, 105*).

The chloroform extract of *Brucea javanica* yielded the coumarino lignan, cleomiscosin-A (**85**), which showed potent cytotoxicity against P-388 leukemia cells for the first time (*106*).

Flavonoids. Two antileukemic flavonoids, tricin (**86**) and kaempferol-3-*O*-β-D-glucopyranoside (**87**), were isolated from *Wikstroemia indica*. The strong antileukemic activity demonstrated by **86** is noteworthy, as the cytotoxic flavonoids seldom show significant activity *in vivo* against P-388 (*103*).

Hinokiflavone (**88**) was isolated as the cytotoxic principle from the drupes of *Rhus succedanea* (Anacadiaceae). A comparison of the cytotoxicity of **88** and other related biflavonoids indicates that an ether linkage between two units of apigenin as seen in **88** is structurally required for significant cytotoxicity (*107*).

Lignans. *Wikstroemia indica* also yielded an antileukemic lignan, (+)-nortrachelogenin (=wikstromol) (**89**) (*103*).

Justicidin-A (**90**) and diphyllin (**91**) were isolated from *Justicia procumbens* (*Chu Wei Hung*) (Acanthaceae) as cytotoxic compounds for the first time. A comparison of the cytotoxicity of **90** and **91** and related compounds indicates that the γ-lactone ring with the carbonyl group α to C-3 (instead of C-2) is required for potent cytotoxicity (*108*).

Macrolides. The resorcylate macrolide, lasiodiplodin (**92**), was first isolated from *Euphorbia splendens* (*Chi Lin Hua*) (Euphorbiaceae) as a potent antileukemic agent (*109*).

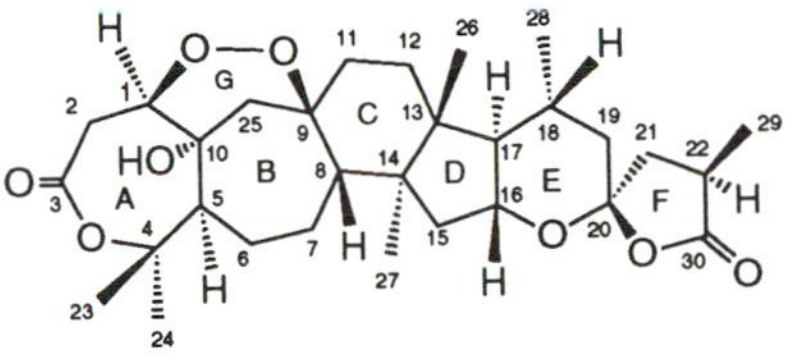

59, Pseudolarolide-I

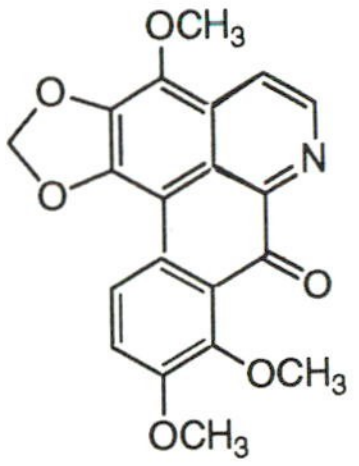

60, Kuafumine

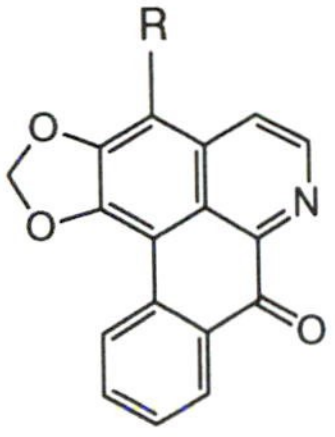

61, Liriodenine: R = H
62, Atherospermidine: R = OCH_3

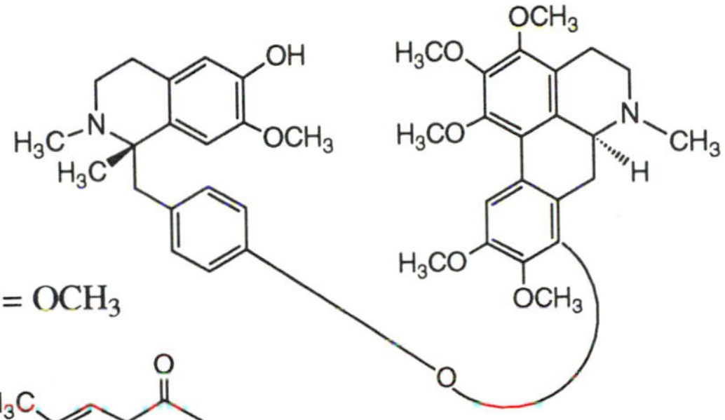

63, (+)-Thalifarazine

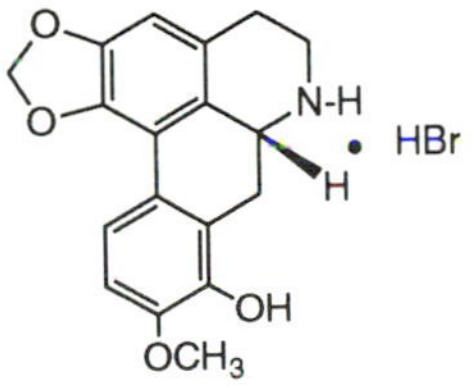

64, (-)-Norannuradhapurine • HBr

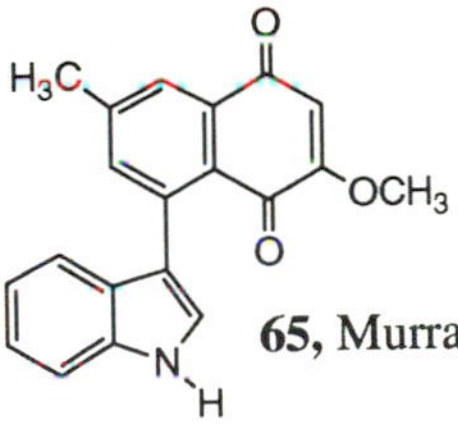

65, Murrapanine

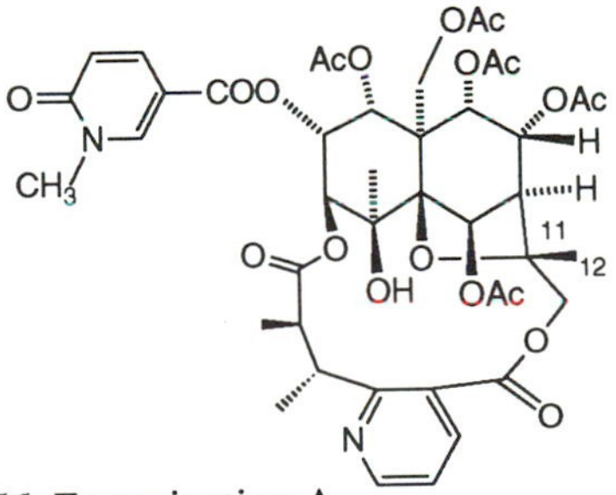

66, Emarginatine-A

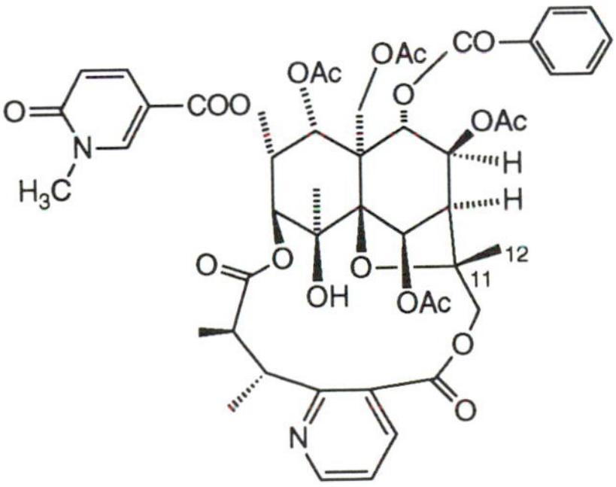

67, Emarginatine-B

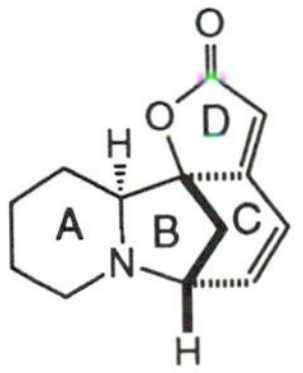

68, Virosecurinine

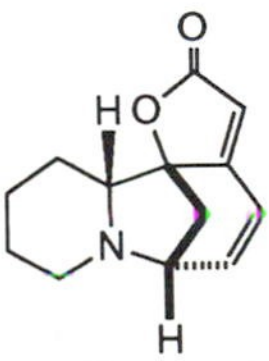

69, Viroallosecurinine

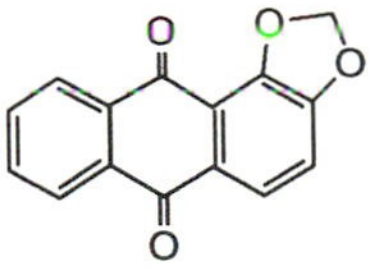

70, Morindaparvin-A

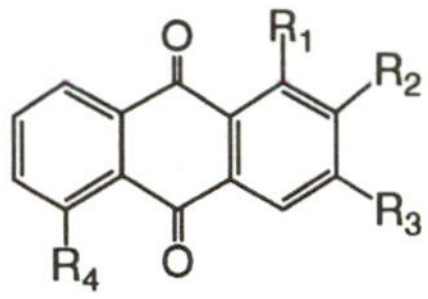

71, Morindaparvin-B:
$R_1 = R_4 = OH, R_2 = CH_2OH, R_3 = H$
72, 1,3-Dihydroxy-2-methoxymethylanthraquinone:
$R_1 = R_3 = OH, R_2 = CH_2OCH_3, R_4 = H$
73, Digiferruginol:
$R_1 = OH, R_2 = CH_2OH, R_3 = R_4 = H$
74, 2-Hydroxymethylanthraquinone:
$R_1 = R_3 = R_4 = H, R_2 = CH_2OH$

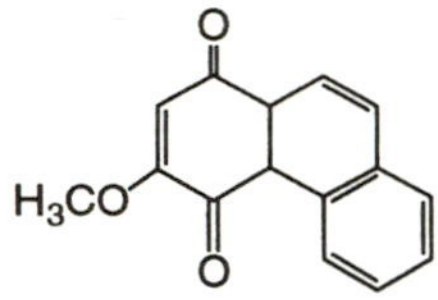

75, Annoquinone-A

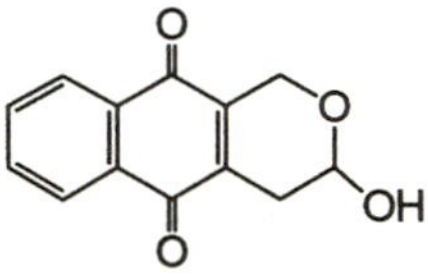

76, Psychorubrin

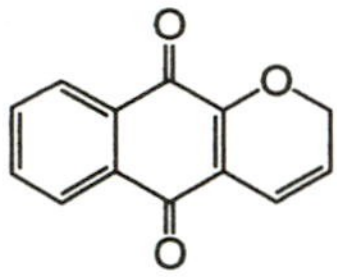

78

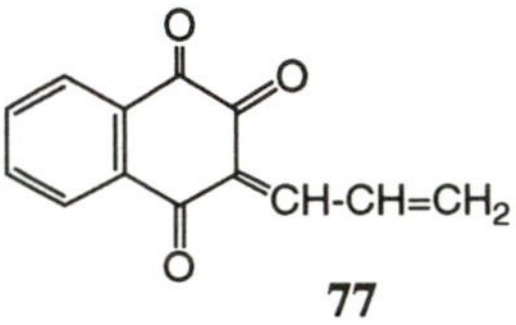

77

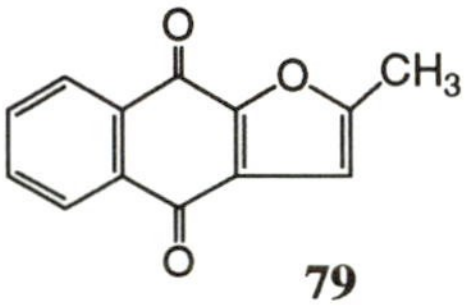

79

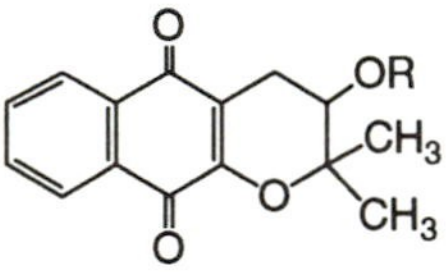

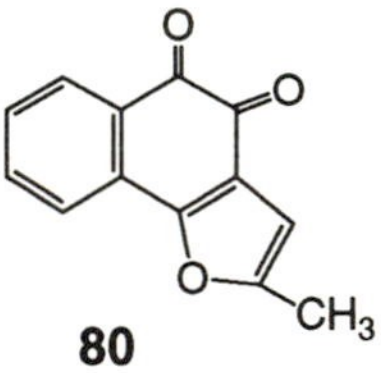

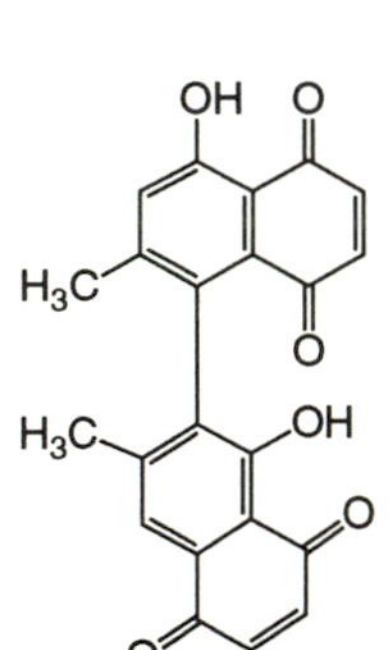

80

81, Rhinacanthin-B:
$R = -CO-C(CH_3)=CH-CH_2-CH_2-C(CH_3)=CH-CH_3$

82, Isodiospyrin

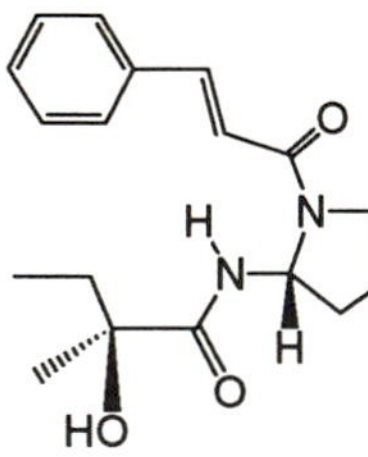

83, (-)-Odorinol

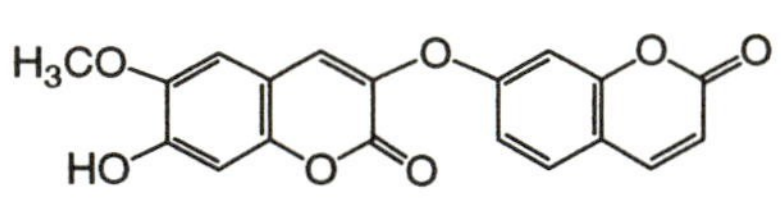

84, Daphnoretin

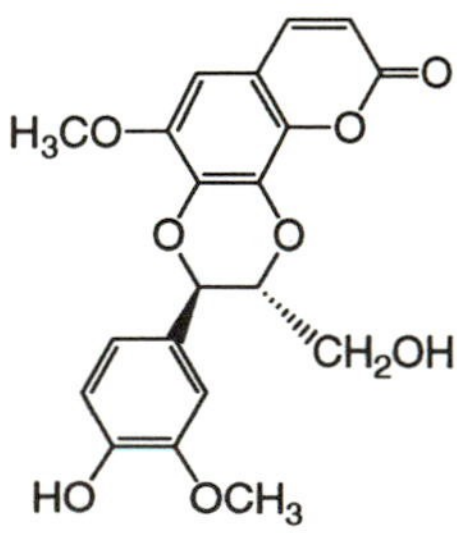

85, Cleomiscosin-A

86, Tricin

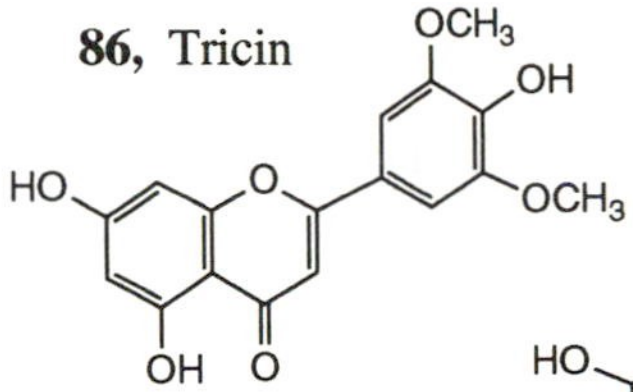

88, Hinokiflavone

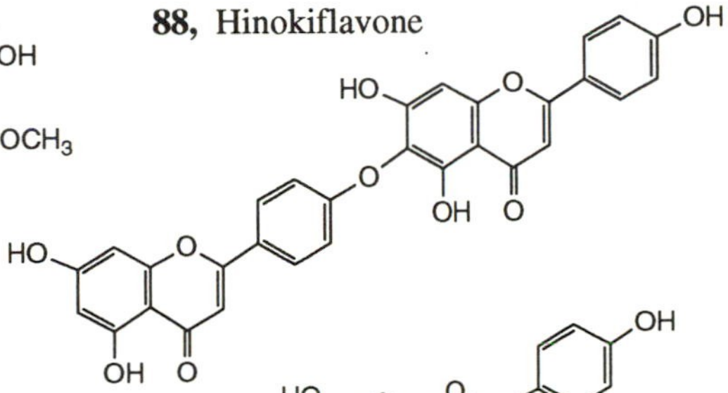

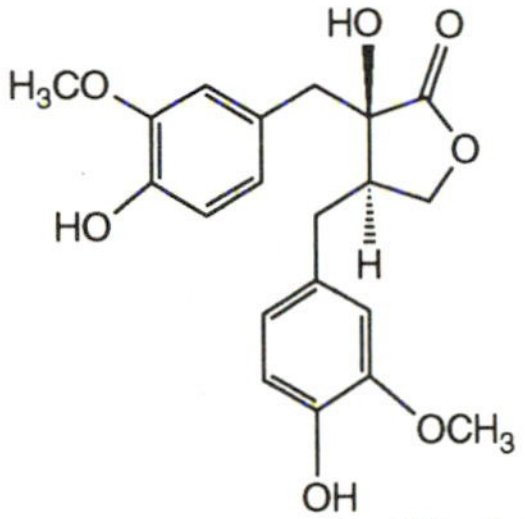

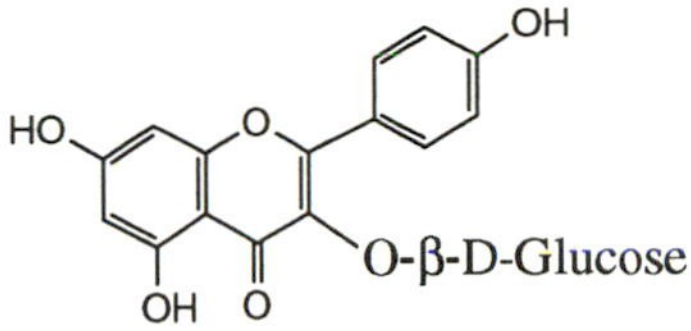

87, Kaempferol-3-*O*-β-D-glucopyranoside

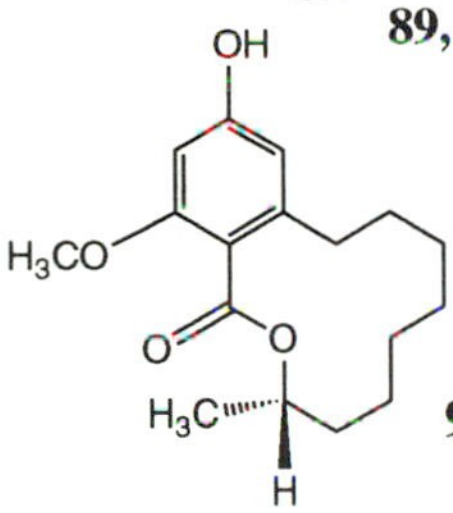

89, (+)-Nortrachelogenin (Wikstromol)

92, Lasiodiplodin

93, Seselidiol

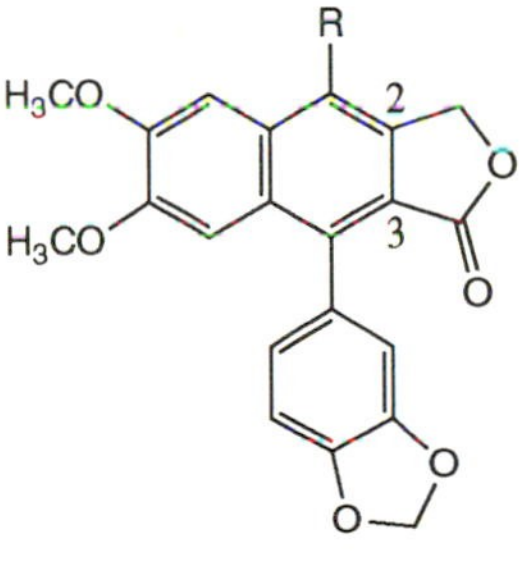

90, Justicidin: $R = OCH_3$
91, Diphyllin: $R = OH$

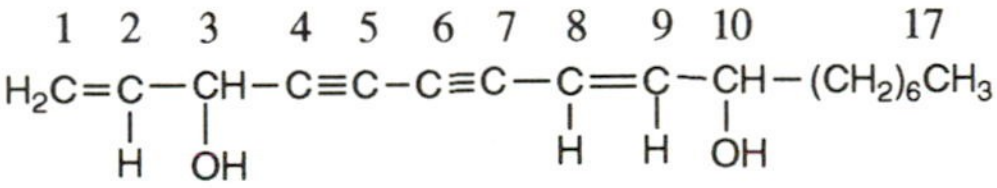

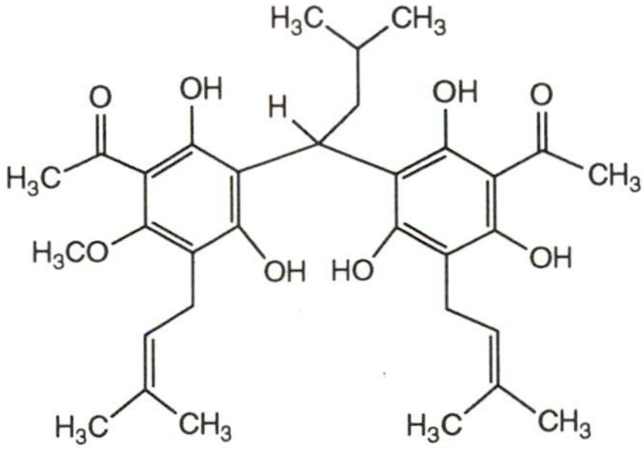

94, Acrovestone

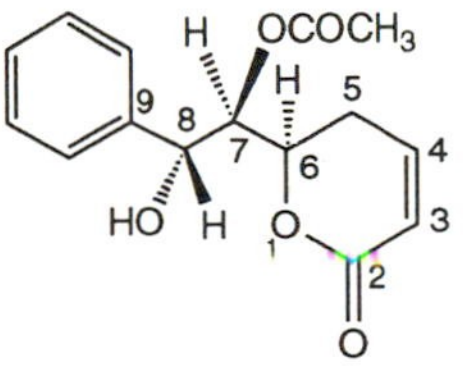

95, Goniodiol 7-monoacetate

Polyacetylenes. A new cytotoxic polyacetylene isolated from *Seseli mairei* (*Chu Yeh Fang Fong*) (Umbelliferae) is seselidiol (**93**) (*110*).

Polyphenols. Acrovestone (**94**), a polyphenol, was isolated from *Acronychia pedunculata* (*Chiang Cheng Hsiang*) (Rutaceae) as a cytotoxic agent (*111*).

Styrylpyrones. *Goniothalamus amuyon* (Annonaceae) afforded a cytotoxic styrylpyrone, goniodiol-7-monoacetate (**95**) (*112*).

Conclusion. In view of the substantial progress which has been made recently in bringing new antineoplastic agents and their analogs from Chinese traditional medicine into clinical use or clinical trials as anticancer agents as mentioned above, the future must be visualized as an optimistic one. New anticancer agents, especially anti-solid tumor agents, will be continuously discovered, based on NCI's new screening strategy. The new leads discussed above could be further developed or modified to yield useful drugs or be subjected to biochemical and pharmacological investigations to increase the understanding of tumor-cell biology. Continuing searches among Chinese medicinal plants and the semi-synthesis of their analogs will undoubtedly lead to further examples of novel plant-derived antineoplastic agents.

Acknowledgments. I would like to thank my collaborators who contributed in many ways to the completion of much of this research work, and who are cited in the accompanying references. This investigation was supported by grants from the National Cancer Institute (CA 17625) and the American Cancer Society (CH 370 and DHP-13E).

Literature Cited

1. Lee, K. H. In *Advances in Chinese Medicinal Materials Research*; Chang, H. M., Ed.; World Scientific Publ. Co.: Singapore, **1985**, 353-367, and literature cited therein.
2. Lien, E. J.; Li, W. Y. In *Anticancer Chinese Drugs*; Oriental Healing Arts Institute: Los Angeles, California, **1985**.
3. Lee, K. H. *Oriental Healing Arts Internat. Bull.* **1986**, *11*, 53-72.
4. Lee, K. H.; Yamagishi, T. *Abstr. Chin. Med.* **1987**, *1*, 606-625.
5. Lee, K. H. *Kaohsiung J. Med. Sci.* **1987**, *3*, 234-250.
6. Xu, B. *Trends Pharmacol. Sci.* **1981**, 271-274.
7. Xiao, P. In *Natural Products as Medicinal Agents*; Beal, J. L.; Reinhard, E., Eds.; Hippokrates Verlag: Stuttgart, W. Germany, **1981**, 351-394.
8. *Antitumor Drugs* in the book of *Yao Ping Zhi*: Shanghai Institute for Medicinal & Pharmaceutical Industries, Ed.; Shanghai Scientific & Technological Publishing Company: Shanghai, China, **1983**, Vol. 5.
9. Neuss, N.; Gorman, M.; Johnson, I. S. In *Methods in Cancer Res.*; Academic Press: New York, **1976**, 633-702, and literature cited therein.
10. Lomax, N. R.; Narayanan, V. L. In *Chemical Structures of Interest to the Division of Cancer Treatment* ; National Cancer Institute, Bethesda, Maryland: **1988**, Vol. 6, 1-28.
11. Barnett, C. J.; Cullinan, G. J.; Gerzon, K.; Hoying, R. C.; Jones, W. E.; Newton, W. M.; Poore, G. A.; Robinson, R. L.; Sweeney, M. J.; Todd, G. C. *J. Med. Chem.* **1978**, *21*, 88-96, and literature cited therein.
12. Cragg, G.; Suffness, M. *Pharmacol. Therap.* **1988**, *37*, 425-461, and literature cited therein.

13. Issell, B. F.; Muggia, F. M.; Carter, S. K., Eds.; *Etoposide (VP-16) - Current Status and New Developments* ; Academic Press: Orlando, Florida, **1984**.
14. Jardine, I. In *Anticancer Agents Based on Natural Product Models*; Cassady, J.; Douros, J., Eds.; Academic Press: New York, **1980**, 319-351, and literature cited therein.
15. Stahelin, H.; von Wartburg, A. In *Progress in Drug Research.*; Jucker, E., Ed.; Birkhauser Verlag: Stuttgart, Germany, **1989**, *33*, 169-266.
16. Wall, M. E.; Wani, M. C. In *Anticancer Agents Based on Natural Product Models*; Cassady, J. M.; Douros, J. D., Eds.; Academic Press: New York, **1980**, 417-436.
17. Xu, B.; Yang, J. L. In *Advances in Chinese Medicinal Materials Research*; Chang, H. M.; Yeung, H. W.; Tso, W. W.; Koo, A., Eds.; World Scientific Publ. Co.: Singapore, **1985**, 377-389.
18 Powell, R. G; Weisleder, D.; Smith, C. R. *J. Pharm. Sci.*, **1972**, *61*, 1227.
19. Kao, Y. S.; Xu, R. S.; Hsieh, Y. Y.; *Acta Chim. Sin.,* **1979**, *5*, 11-21.
20. O'Dwyer, P. J.; King, S. A.; Hoth, D. F.;Suffness, M.; Leyland-Jones, B. *J. Clin. Oncol.*, **1986**, *4*, 1563.
21. Culvenor, C. C. J. *J. Pharm. Sci.* **1968**, *57*, 112.
22. Lo, H. S.; Chou, D. H. *Common Chinese Antitumor Herbal Drugs*; Canton Science & Technology Press: Guan-Zhou, China, **1981**, 68 and 147.
23. Fujita, E.; Murai, F.; *Yakugaku Zasshi* . **1951**, *71*, 1039, and literature cited therein.
24. Tomita, M.; Fujitani, K.; Kishimoto, T.; *Yakugaku Zasshi* . **1962**, *82*, 1148.
25. Okuda, S.; Murakoshi, I.; Kamata, H.; Kashida, Y.; Haginiwa, J.; Tsuda, K. *Chem. Pharm. Bull.* **1965**, *13*, 482-487.
26. Fang, F.D. *Chin J. Intern Med.*, **1982**, *21*, 312-320.
27. Fukushima, S.; Kuroyanagi, M.; Ueno, A.; Akahori, Y.; Saiki, Y.; *Yakugaku Zasshi* . **1970**, *90*, 863.
28. Fujita, E.; Fujita, T.; Shibuya, M. *Tetrahedron Lett.*, **1966,** 3153-3162.
29. Nagao, Y.; Fujita E. *Proc. Symposium WAKAN-YAKU.* **1980**, *13*, 79.
30. Hsiang, Y. H.; Hertzberg, R.; Hecht, S.; Liu, L. F. *J. Biol. Chem.* **1985**, *260*, 14873-14878.
31. Hsiang, Y. H.; Liu, L. F.; Wall, M. E.; Wani, M. C.; Nicholas, A. W.; Manikumar, G.; Kirschenbaum, S.; Silber, R.; Potmesil, M. *Cancer Res.* **1989**, *49*, 4385-4389.
32. Johnson, P. K.; McCabe, F. L.; Faucette, L. F.; Hertzberg, R. P.; Kingsbury, W.D.; Boehm, J. C.; Caranfa, M. J.; Holden, K. G. *Proc. Am. Assoc. Cancer Res.* **1989**, *30*, 623.
33. Wani, M. C.; Nicholas, A. W.; Wall, M. E. *J. Med. Chem.* **1986**, *29*, 2358-2363.
34. Fukuoka, M.; Negoro, S.; Niitani, H.; Taguchi, T. *Proc. Am. Soc. Clin. Oncol.* **1990**, *9*, 874-884.
35. *Washington Insight*, **1989**, *2*, 6 and **1990**, *3*, 6.
36. Chen, G. L.; Yang, L.; Rowe, T. C.; Halligan, B. D.; Tewey, K.; Liu, L. *J. Biol. Chem.* **1984**, *259*, 13560-13566.
37. Ross, W.; Rowe, T.; Glisson, B.; Yalowich, J.; Liu, L. *Cancer Res.* **1984**, *44*, 5857-5866.
38. Rowe, T.; Kuppfer, G.; Ross, W. *Biochem. Pharmacol.* **1985**, *34*, 2483-2487.
39. van Maanen, J. M. S.; De Ruiter, C.; Kootstra, P. R.; De Vries, J.; Pinedo, H. M. *Proc. Am. Assoc. Cancer Res.* **1984**, *25*, 384.

40. van Maanen, J. M. S.; De Ruiter, C.; Kootstra, P. R.; Broersen, J.; De Vries, J.; Laffeur, M. V. M.; Retel, J.; Kriek, E.; Pinedo, H. M. *Proc. Am. Assoc. Cancer Res.* **1986**, *27*, 308.
41. Sakurai, H.; Miki, T.; Imakura, Y.; Shibuya, M.; Lee, K. H. *Molec. Pharmacol.* **1991**, *40*, 965-973.
42. Haim, N.; Roman, J.; Nemec, J.; Sinha, B. K. *Biochem. Biophys. Res. Commun.* **1986**, *135*, 215-220
43. van Maanen, J. M. S.; De Ruiter, C.; De Vries, J.; Kootstra, P. R.; Gobars, G. Pinedo, H. M. *Eur. J. Cancer Clin. Oncol.* **1985**, *21*, 1099-1106.
44. Lee, K. H.; Beers, S. A.; Mori, M.; Wang, Z. Q.; Kuo, Y. H.; Li, L.; Liu, S. Y.; Chang, J. Y.; Han, F. S.; Cheng, Y. C. *J. Med. Chem.* **1990**, *33*, 1364-1368.
45. Wang, Z. Q.; Kuo, Y. H.; Schnur, D.; Bowen, J. P.; Liu, S. Y.; Han, F. S.; Chang, J. Y.; Cheng, Y. C.; Lee, K. H. *J. Med. Chem.* **1990**, *33*, 2660-2666.
46. Chang, J. Y.; Han, F. S.; Liu, S. Y.; Wang, Z. Q.; Lee, K. H.; Cheng, Y. C. *Cancer Res.* **1991**, *51*, 1755-1759.
47. Geran, R. I.; Greenberg, N. H.; MacDonald, M. M.; Schumacher, A. M.; Abbott, J. B. *Cancer Chemother. Rep., Part 3*, **1972**, 1-88.
48. Boyd, M. R. In Cancer: *Principles and Practice of Oncology Updates;* DeVita, V. Y. ; Hellman, S; Rosenberg, S. A., Eds.; J. B. Lipppincott: Philadelphia, **1989**, 1-12.
49. Monks, A.; Scudiero, D.; Skehan, P.; Shoemaker, R.; Paull, K.; Vistica, D.; Hose, C.; Langley, J.; Cronise, P.; Vaigro-Wolff, A.; Gray-Goodrich, M.; Campbell, H.; Mayo, J.; Boyd, M. *J. Natl. Cancer Inst.* **1991**, *83*, 757-766.
50. Lee, K. H., *Program & Abstract - 16th Natl. Med. Chem. Symposium - ACS*, Kalamazoo, Michigan, June 18-22, **1978**, 43-58.
51. Lee, K. H.; Furukawa, H.; Kozuka, M.; Huang, H. C.; Luhan, P. A.; McPhail, A. T. *J. Chem Soc., Chem. Commun.* **1973**, 476-477.
52. Lee, K. H.; Ibuka, T.; Huang, H. C.; Harris, D. L. *J. Pharm. Sci.* **1975**, *64*, 1077-1078.
53. McPhail, A. T.; Onan, K. D.; Lee, K. H.; Ibuka, T.; Kozuka, M.; Shingu, T.; Huang, H. C. *Tetrahedron Lett.* **1974**, *32*, 2739-2741.
54. McPhail, A. T.; Onan, K. D.; Lee, K. H.; Ibuka, T.; Huang, H. C. *Tetrahedron Lett.* **1974**, *36*, 3203-3206.
55. Lee, K. H.; Kimura, T.; Haruna, M.; McPhail, A. T.; Onan, K. D.; Huang, H. C. *Phytochemistry.* **1977**, *16*, 1068-1070.
56. Lee, K. H.; Haruna, M.; Huang, H. C.; Wu, B. S.; Hall, I. H. *J. Pharm. Sci.* **1977**, *66*, 1194-1195.
57. Lee, K. H.; Huang, E. S.; Piantadosi, C.; Pagano, J. S.; Geissman, T. A. *Cancer Res.* **1971**, *31*, 1649-1654.
58. Lee, K. H.; Furukawa, H.; Huang, E. S. *J. Med. Chem.* **1972**, *15*, 609-611.
59. Lee, K. H.; Hall, I. H.; Mar, E. C.; Starnes, C. O.; Elgebaly, S. A.; Waddell, T. G.; Hadgraft, R. I.; Ruffner, C. G.; Weidner, I. *Science* **1977**, *196*, 552-553.
60. Lee, K. H.; Meck, R.; Piantadosi, C.; Huang, E. S. *J. Med. Chem.* **1973**, *16*, 299-301.
61. Lee, K. H.; Ibuka, T.; Sims, D.; Muraoka, O.; Kiyokawa, H.; Hall, I. H.; Kim, H. L. *J. Med. Chem.* **1981**, *24*, 924-927.
62. Page, J. D.; Chaney, S. G.; Hall, I. H.; Lee, K. H.; Holbrook, D. J. *Biochim. Biophys. Acta* **1987**, *926*, 186-194.

63. Williams, W. L.; Chaney, S. G.; Willingham, W.; Considine, R. T.; Hall, I. H.; Lee, K. H. *Biochim. Biophys. Acta* **1983**, *740*, 152-162.
64. Williams, W. L.; Chaney, S. G.; Hall, I. H.; Lee, K. H. *Biochem.* **1984**, *23*, 5637-5644.
65. Kasai, R.; Lee, K. H.; Huan, H. C. *Phytochemistry* **1981**, *20*, 2592-2594.
66. Zhu, D. Y. *Abstr. Chin. Med.* **1987**, *1*, 606.
67. Wu, T. S.; Lin, Y. M.; Haruna, M.; Pan, D. J.; Shingu, T.; Chen, Y. P.; Hsu, H. Y.; Nakano, T.; L; Lee, K. H. *J. Nat. Prod.* **1991**, *54*, 823-829.
68. Pan, D. J.; Li, Z. L.; Hu, C. Q.; Chen, K.; Chang, J. J.; Lee, K. H. *Planta Med.* **1990**, *56*, 383-385.
69. Lee, K. H.; Imakura, Y.; Sumida, Y.; Wu, R. Y.; Hall, I. H.; Huang, H. C. *J. Org. Chem.* **1979**, *44*, 2180-2185.
70. Sasaki, T.; Yoshimura, S.; Tsuyuki, T.; Takahashi, T.; Honda, T.; Nakanishi, T. *Tetrahedron Lett..* **1986**, *27,* 593-596.
71. Kupchan, S. M., Britton, R. W.; Lacadie, J. A.; Ziegler, M. F.; Sigel, C. W. *J. Org. Chem.* **1975**, *40*, 648-654.
72. Cassidy, J. M.; Suffness, M. In *Anticancer Agents Based on Natural Product Models*; Cassady, J. M.; Douros, J., Eds.; Academic Press: New York, **1980**, 201-269.
73. Okano, M.; Lee, K. H. *J. Org. Chem.* **1981**, *46*, 1138-1141.
74. Lee, K. H.; Tani, S.; Imakura, Y. *J. Nat. Prod.* **1987**, *50*, 847.
75. Lee, K. H.; Okano, M.; Hall, I. H.; Brent, D. A.; Soltmann, B. *J. Pharm. Sci..* **1982**, *71*, 338-345.
76. Hall, I. H.; Liou, Y. F.; Okano, M.; Lee, K. H. *J. Pharm. Sci..* **1982**, *71*, 345-348.
77. Liou, Y. F.; Hall, I. H.; Okano, M.; Lee, K. H.; Chaney, S. G. *J. Pharm. Sci.* **1982**, *71*, 430-435.
78. Willingham, W.; Considine, R. T.; Chaney, S. G.; Lee, K. H.; Hall, I. H. *Biochem. Pharmacol..* **1984**, *33*, 330-333.
79. Okano, M.; Fukamiya, N.; Lee, K. H. In *Studies in Natural Products Chemistry*; Atta-ur-Rahman, Ed.; Elsevier Science Publishers B. V.: Amsterdam, **1990**, *7*, 369-404.
80. Nozaki, H.; Suzuki, H.; Lee, K. H.; McPhail, A. T. *J. Chem. Soc., Chem. Commun.* **1982**, 1048-1051.
81. Rice, G. K.; Yokoi, T.; Hayashi, T.; Suzuki, H.; Lee, K. H.; McPhail, A. T. *J. Chem. Soc. Chem. Commun.* **1986**, 1397-1398.
82. Yamagishi, Y.; Zhang, D. C.; Chang, J. J.; McPhail, D. R.; McPhail, A. T.; Lee, K. H. *Phytochemistry* **1988**, *27*, 3213-3216.
83. Lee, K. H.; Lin, Y. M.; Wu, T. S.; Zhang, D. C.; Yamagishi, T.; Hayashi, T.; Hall, I. H.; Chang, J. J.; Wu, K. Y.; Yang, T. H. *Planta Med.* **1988**, *54*, 308-311.
84. Yamagishi, T.; Yan, X. Z.; Wu, R. Y.; McPhail, D. R.; McPhail, A. T.; Lee, K. H. *Chem. Pharm. Bull..* **1988**, *36*, 1615-1617.
85. Yamagishi, T.; Haruna, M.; Yan, X. Z.; Chang, J. J.; Lee, K. H. *J. Nat. Prod.* **1989**, *52*, 1071-1079.
86. Chen, G. F.; Li, Z. L.; Chen, K.; Tang, C. M.; He, X.; Pan, D. J.; Hu, C. Q.; McPhail, D. R.; McPhail, A. T.; Lee, K. H. *J. Chem. Soc. Chem. Commun.* **1990**, 1113-1114.
87. Wu, Y. C.; Lu, S. T.; Wu, S. T.; Lee, K. H. *Heterocycles* **1987**, *26*, 9-12.
88. Wu, Y. C.; Chen, C, H,; Yang, T. H.; Lu, S. T.; McPhail, D. R.; McPhail, A. T.; Lee, K. H. *Phytochemistry* **1989**, *28*, 2191-2195.
89. Wu, Y. C.; Lu, S. T.; Chang, J. J.; Lee, K. H. *Phytochemistry* **1988**, *27*, 1565-1568.

90. Wu, Y. C.; Liou, Y. F.; Lu, S. T.; Chen, C. H.; Chang, J. J.; Lee, K. H. *Planta Med.* **1989**, *55*, 163-165.
91. Wu, T. S.; Liou, M. J.; Lee, C. J.; Jong, T. T.; McPhail, A. T.; McPhail, D. R.; Lee, K. H. *Tetrahedron Lett.* **1989**, *30*, 6649-6652.
92. Kuo, Y. H.; Chen, C. H.; Yang Kuo, L. M.; King, M. L.; Wu, T. S.; Lu, S. T.; Chen, I. S.; McPhail, D. R.; McPhail, A. T.; Lee, K. H. *Heterocycles* **1989**, *29*, 1465-1468.
93. Kuo, Y. H.; Chen, C. H.; Yang Kuo, L. M.; King, M. L.; Wu, T. S.; Haruna, M.; Lee, K. H. *J. Nat. Prod.* **1990**, *53*, 422-428.
94. Tatematsu, H.; Mori, M.; Yang, T. H.; Chang, J. J.; Lee, T. T. Y.; Lee, K. H. *J. Pharm. Sci.* **1990**, *80*, 325-327.
95. Chang, P.; Lee, K. H.; Shingu, T.; Hirayama, T.; Hall, I. H.; Huang, H. C. *J. Nat. Prod.* **1982**, *45*, 206-210.
96. Chang, P.; Lee, K. H. *Phytochemistry* **1984**, *23*, 1733-1736.
97. Chang, P.; Lee, K. H. *J. Nat. Prod.* **1985**, *48*, 948-951.
98. Wu, T. S.; Jong, T. T.; Tien, H. J.; Kuoh, C. S.; Furukawa, H.; Lee, K. H. *Phytochemistry* **1987**, *26*, 1623-1625.
99. Hayashi, T.; Smith, F. T.; Lee, K. H. *J. Med. Chem.* **1987**, *30*, 2005-2008.
100. Wu, T. S.; Tien, H. J.; Yeh, M. Y.; Lee, K. H. *Phytochemistry* **1988**, *27*, 3787-3788.
101. Yan, X. Z.; Kuo, Y. H.; Lee, T. J.; Shih, T. S.; Chen, C. H.; McPhail, D. R.; McPhail, A. T.; Lee, K. H. *Phytochemistry* **1989**, *28*, 1541-1543.
102. Hayashi, N.; Lee, K. H.; Hall, I. H.; McPhail, A. Y.; Huang, H. C. *Phytochemistry* **1982**, *21*, 2371-2373.
103. Lee, K. H.; Tagahara, K.; Suzuki, H.; Wu, R. Y.; Huang, H. C.; Ito, K.; Iida, T.; Lai, J. S. *J. Nat. Prod.* **1981**, *44*, 530-535.
104. Hall, I. H.; Tagahara, K.; Lee, K. H. *J. Pharm. Sci.* **1982**, *71*, 741-744.
105. Liou, Y. F.; Hall, I. H.; Lee, K. H. *J. Pharm. Sci.* **1982**, *71*, 1340-1344.
106. Lee, K. H.; Hayashi, N.; Okano, M.; Nozaki, H.; Motoharu, J. I. *J. Nat. Prod.* **1984**, *47*, 550-551.
107. Lin, Y. M.; Chen, F. C.; Lee, K. H. *Planta Medica.* **1989**, *55*, 166-168.
108. Fukamiya, N.; Lee, K. H. *J. Nat. Prod.* **1986**, *49*, 348-350.
109. Lee, K. H.; Hayashi, N.; Okano, M.; Hall, I. H.; Wu, R. Y.; McPhail, A. T.; *Phytochemistry* **1982**, *21*, 1119-1121.
110. Hu, C. Q.; Chang, J. J.; Lee, K. H. *J. Nat. Prod.* **1990**, *53*, 932.
111. Wu, T. S.; Wang, M. L.; Jong, T. T.; McPhail, A. T.; McPhail, D. R.; Lee, K. H. *J. Nat. Prod.* **1989**, *52*, 1284-1289.
112. Wu, Y. C.; Duh, C. Y.; Chang, F. R.; Chang, G. Y.; Wang, S. K.; Chang, J. J.; McPhail, D. R.; McPhail, A. T.; Lee, K. H. *J. Nat. Prod.* **1991**, *54*, 1077.

RECEIVED March 24, 1993

Chapter 13

Novel Strategies for the Discovery of Plant-Derived Anticancer Agents

Geoffrey A. Cordell[1], Norman R. Farnsworth[1], Christopher W. W. Beecher[1], Djaja Doel Soejarto[1], A. Douglas Kinghorn[1], John M. Pezzuto[1], Monroe E. Wall[2], Mansukh C. Wani[2], Daniel M. Brown[2], Melanie J. O'Neill[3], Jane A. Lewis[3], R. Murray Tait[3], and Timothy J. R. Harris[3]

[1]Program for Collaborative Research in the Pharmaceutical Sciences, Department of Medicinal Chemistry and Pharmacognosy, College of Pharmacy, University of Illinois at Chicago, Chicago, IL 60612
[2]Chemistry and Life Sciences Laboratory, Research Triangle Institute, Research Triangle Park, NC 27709
[3]Biotechnology, Glaxo Group Research Ltd., Greenford Road, Greenford, Middlesex UB6 0HE, United Kingdom

Drug discovery from plants has been a goal of mankind since prehistoric times, and through previous efforts, plants have become a viable source of clinically useful anticancer agents, have afforded leads for synthetic modification and have served as tools for mechanistic studies. In this chapter, strategies for the discovery of anticancer agents from plants will be described in which ethnomedical information is correlated with pertinent published chemical and biological information, as part of an overall plant collection program. Botanically authenticated plants are extracted and the extracts tested in a broad array of nearly 30 human cancer cell, mechanism- and receptor-based assays, through a cooperative research program involving a university, a research institute and a pharmaceutical company. Bioactivity-directed fractionation is carried out at all three sites, with a view to identifying novel compounds which may serve as candidates for preclinical testing.

The discovery of health beneficent agents from natural sources has been an essential quest of mankind since prehistoric times. Because of their relative ease of collection,

0097–6156/93/0534–0191$06.00/0

their potential for sustainable development and the structural and biological diversity of their constituents, terrestrial plants offer a unique and renewable resource for the discovery and development of potential new drugs and biological entities. A particularly germane question, though, is how to find the proverbial needle (active compound) in the haystack (plant kingdom). Current, possibly conservative, estimates indicate that there are about 250,000 species of flowering plants on Earth, and, of these, it is estimated that 155,000 are found in the tropics (*1*). Thus, a rational strategy for drug discovery is required, since it is currently not reasonable to investigate all available plants for even a single biological activity.

It is generally recognized that there are five systematic approaches for the selection of plants that may contain new biological agents from plants: the random, the taxonomic, the chemotaxonomic, the ethnomedical, and the information-managed (*2*). In the random approach, available species are collected from a particular site or sites, without regard to previous knowledge of the uses of a plant. In the taxonomic approach, plants of selected families or genera considered to be of interest are sought from diverse locations. In the chemotaxonomic (phytochemical) approach, a particular compound type, e.g., isoquinoline alkaloids, is judged to be of biological interest, and plants anticipated to produce related alkaloids are collected. In the ethnomedical approach, credence is given to oral or written information on the indigenous medicinal use of the plant, and, based on this information, the plant is collected. In the information-managed approach, ethnomedical, biological and chemical information is collated and prioritized to afford a list of plants for specific collection. A sixth "approach" is serendipity, where during the course of pursuing a particular bioactivity or ethnomedical use one discovers another pertinent bioactivity.

Following collection, the material is evaluated in a range of therapeutically relevant bioassays. Active leads are then prioritized, and those viewed as most active through established criteria are subjected to bioassay-directed fractionation procedures for elucidation of the active principle(s). Depending on the particular circumstances, there are merits to each of these approaches to collection. Theoretically, one could devise discovery programs for any of a broad range of biologically relevant events. Historically, these discovery programs have focussed on the identification of anticancer agents from plants and marine organisms. In this chapter, after briefly summarizing the history of previous drug discovery efforts for new and novel plant anticancer agents, we will describe a new, integrated program of plant selection, collection and evaluation of activity based on a broad biological approach.

There have been several previous efforts to discover new anticancer agents from plants, organized, among others, by the National Cancer Institute and by the Eli Lilly Company in Indianapolis. The latter program utilized folklore information (not necessarily for cancer) with evaluation for CNS, antimicrobial, insecticide and antitumor activity. Four important compounds came out of the 400 plants that were initially analyzed: vincaleukoblastine (vinblastine, VLB), leurocristine (vincristine (VCR), 9-methoxyellipticine and acronycine (*3*).

The National Cancer Institute program was initiated by Dr. Jonathan Hartwell who began the investigation of the Mayapple, *Podophyllum peltatum*, a plant with a long and extensive folkloric use by North American Indians for the treatment of various kinds of cancer and warts. It took many years for the related, semisynthetically derived products etoposide and teniposide, to emerge as promising clinical entities.

Hartwell also assembled the only published compilation of the ethnomedical use of plants against cancer (*4*). Hartwell's list, which is organized by plant family, contains over 3,000 different species of plants and affords considerable detail of the various uses of plants. As one might expect, for much of prehistory there is no documentary evidence of the use of plants as anticancer agents, and the earliest written text is probably the *Ebers Papyrus* which dates to 1550 B.C. Forty plants were recommended in this compilation, including barley, flax, absinth, coriander, figs, onions, garlic, dates, juniper and grapes. Several of these plants are now known to contain compounds toxic to cancer cells (e.g., grapes, juniper and flax) (*5*).

In 1957, the National Cancer Institute, under the leadership of Hartwell, and through the newly created Cancer Chemotherapy National Service Center, initiated the evaluation of plant extracts against a leukemia model (L1210 lymphocytic leukemia in mice), a carcinoma model (adenocarcinoma 755 in mice) and a sarcoma model system (sarcoma 180 in rats). Substantive plant collection for this program was initiated in 1960. In 1966, the Walker 256 carcinosarcoma test system in rats replaced the A755 and S180 assays, but this assay was discovered to be particularly sensitive to phytosterols, saponins and tannins, and was subsequently dropped.

The program burgeoned from being an intramural program in the 1960s to an extramural program with a number of contract sites in the 1970s. Plant materials, provided by the United States Department of Agriculture, were primarily randomly collected, although the Hartwell text was occasionally used as a source of ethnomedical information. The assays employed initially were the P-388 lymphocytic leukemia assay in mice and the KB carcinoma of the nasopharynx in cell culture. These assays were conducted at external contractual sites, and eventually, as delays increased for the testing of fractions obtained in the bioactivity-directed fractionation, it became apparent that an alternative approach was needed. In the early 1980s, this program was terminated by the NCI, who then several years later reintroduced a new intramural program in which the goal is to randomly collect plants from three designated areas of the world and screen them against a broad array of human cancer cell lines (*6-10*).

In the period 1960-1986, the NCI program screened over 35,000 plant species and 108,330 extracts against the murine tumors *in vivo* or *in vitro* or the KB test system; 4,149 (3.83%) were active, representing 1,410 genera and 2,935 species. Of the 2,000 crystalline plant-derived compounds that were tested, 95 were identified for special testing and 11 approved for extensive tumor panel testing (*10,11*). These compounds included bruceantin, maytansine, colubrinol, indicine-*N*-oxide, tripdiolide, taxol, homoharringtonine, ellipticine and bouvardin. A number of other plant-derived compounds, including tylocrebrine, lapachol, camptothecin, acronycine, emetine, thalicarpine and tetrandrine, were examined in this program in various clinical trials.

Thus, it is now well established that plants are very useful sources of clinically relevant anticancer compounds. Indeed the *Catharanthus* alkaloids, vincristine (Oncovin) and vinblastine (Velban), are two of the most important cancer chemotherapeutic agents in current use (*12*), followed by the semisynthetic podophyllotoxin derivative, etoposide. Colchicine derivatives (e.g. demecolcine) are also used to treat cancer in some countries of the world and the pyrrolizidine alkaloid, monocrotaline, is used topically in China to treat skin cancers. A number of other plant-derived antitumor agents have been subjected to recent clinical evaluation, including taxol (*13*), harringtonine (*14*), homoharringtonine (*15-17*), 10-hydroxycamptothecin (*18*), teniposide (*19*), indirubin (*20*) and ellipticine derivatives (*21-23*). Taxol is now a compound of exceptional chemical, biological and clinical interest (*24,25*) and is discussed in detail elsewhere in this volume in the chapter by Kingston.

The previous efforts to discover anticancer agents from plants have led to the emergence of two features: i) that there is a broad range of plant families with active species (*26*), and ii) that there is a broad spectrum of natural product structure types which elicit *in vivo* activity (*27,28*). Indeed, the conclusion has been drawn that representatives from almost every class of terpenoid, alkaloid, lignoid, shikimate-derived or acetate-derived natural product show some form of potential antineoplastic activity (*29,30*). Within a given plant there may be two or even three classes of compounds which display cytotoxic activity.

In terms of future strategies for drug discovery there are some important implications of these conclusions. One is that the extensive prior literature may already permit some rationalization (positively or negatively) with respect to the biological potential of a reputed use. A second important conclusion is that diversity of plant material should not predicate biological potential, i.e., that useful activity might be found anywhere on a botanical or global basis.

In the past, most drug discovery efforts have concentrated on identifying cytotoxic agents from plants, and these have led to some notable discoveries, including camptothecin (*31*) and taxol (*32*). However, individual tumors have developed resistance to chemotherapy or these agents have been too toxic to mammalian systems (e.g. maytansine). The National Cancer Institute has ongoing an intramural program in which randomly collected plants are screened against an extensive battery of cultured cell types (*6-9*). We have therefore chosen a strategy that combines cell and mechanism-based bioassays. The group that is involved in this program is based at the University of Illinois at Chicago, with partners at Research Triangle Institute, and at Glaxo Group Research Ltd (GGR), in England.

New Strategies for Anticancer Drug Discovery

We consider that there are seven key factors in developing a successful program aimed at drug discovery of new anticancer agents from plants based on information management.

i) Rational plant selection and collection, followed by authoritative taxonomic verification.
ii) Availability of diverse, proven cytotoxicity and mechanism-based bioassays.
iii) Unequivocal dedication to bioactivity-directed fractionation.

iv) Effective structure determination of a broad array of novel compounds, including new natural product skeleta, and to synthesize analogues to potentiate activity.
v) Ability to develop new strategies in the botanical, chemical and biological areas.
vi) Critical decision-making capabilities regarding potential candidates for future development, i.e., prioritization of leads.
vii) Knowledge and successful experience in preclinical and clinical development of primary lead compounds.

Here we will describe three novel aspects of the program underway: a) a two-pronged strategy for the selection and collection of the plants to be tested, b) a computerized literature surveillance and evaluation program which is critical for the initial selection of candidate plants for collection, and for the continuous evaluation of the available literature, leading to a more effective decision-making process, and finally, c) a broad array of cell-based and mechanistically oriented bioassays being used for the evaluation of plant extracts.

Plant Selection and Collection

The group has developed two strategies for the collection of plants for the program. One of these coalesces a network of collaborating botanists and collectors in 25 different countries in primarily tropical regions of the world. These collaborators obtain plant samples, usually two or three specimens per species, based on whether the plant is used ethnomedically for the treatment of certain cancer-related, specific diseases, or if the plant is endemic to the country of collection. To assist the botanists in identifying plants for collection, a set of ethnobotanical usages has been developed which reflects disease states bearing some relevance to cancer or a cancer symptom.

The second approach to plant selection takes advantage of the ability of the NAPRALERT database (see below) to conduct relational searches which, based on selected weighting factors involving existing experimental data, can afford a list of plants prioritized for targeted collection. Some aspects of this process will be described subsequently.

The program of plant collection is organized by NRF and/or conducted by experienced local collectors and taxonomists, as well as our own team of collectors, in order to assure success. World-wide field experience and a botanical contact network, coupled with capabilities regarding identification and authentication using indigenous resources, are also essential, particularly since local names of plants may refer to several different species. It is standard practice that complete taxonomic determination and documentation of voucher specimens at established herbaria is required of the collectors.

All plant collectors are informed in advance that should a commercial entity result from the plant materials supplied that they will share in any royalties that might accrue. Thus GGR's existing Material Transfer Agreements make reference to a financial benefit payable to the supplier in the event that GGR is able to develop a commercial product as a consequence of plant extract screening. The magnitude of this payment, which is linked to the net sales of the product, recognizes the relative contribution of the discovery of the active principle in the particular plant to the subsequent development of the commercial product. The consortium understands the impact that unauthorized and/or unrestrained removal of plant materials from their indigenous habitats can have on the environment and economy of a country and that sources in the less developed world may be particularly vulnerable. To that end, the consortium will select suppliers and negotiate plant supply agreements with them on the understanding that a significant portion of the financial benefit accruing to the latter on marketing a commercial product will be returned to the source country or its material benefit.

Literature Surveillance

Another essential aspect of the program concerns the computerized literature surveillance for plants which show activity and for compounds which may be responsible for the observed bioactivity. Experience has shown that such a dereplication process will markedly enhance the ability to isolate novel bioactive compounds, since time and experimental effort expended on the isolation and characterization of known active compounds is minimized. A further phase to our program is the acquisition of plants through a selection and collection program. Coupled with an emphasis on endemic, previously uninvestigated genera with ethnomedical and/or experimental biological activity, it is hoped that this approach will afford a high probability of discovering new skeletal structures possessing biologically interesting activity.

Use of NAPRALERT in the Plant Selection and Dereplication Processes

The crucial element of the plant selection and literature surveillance programs is the NAPRALERT (**NA**tural **PR**oducts **ALERT**) database initiated (by NRF) in 1975. NAPRALERT is a database of natural product information (*33*). It includes records containing the pharmacology, biological activity, ethnomedical uses, taxonomic distribution and chemistry of natural products (*34*). Many aspects of this database, in addition to its subject matter, make it unique, including the fact that complex queries can be structured. The database grows at an average rate of 500 articles each month, culled from the surveillance of some 700 national and international journals; additional scanning of some abstracting services is also performed. The database currently occupies 900 megabytes of disk storage and has records on some 95,000 scientific articles, 85,000 compounds and 40,000 organisms. Information is routinely included on plant extracts and ethnomedical preparations as they appear in the literature. This database is now available on-line through the STN information network.

The primary and secondary scientific literature contains three relevant bodies of information that necessitate evaluation prior to plant selection. There are, for example, many articles that present the use of a plant for specific cancer-related treatments by various indigenous groups. There are also numerous references to plants, their extracts or specific chemical compounds which, when evaluated in experimental systems, demonstrate biological activities potentially associated with cancer chemotherapeutics. Finally, there is a substantial natural products literature regarding the isolation and characterization of compounds from various sources. In certain instances, several database contributors are present in the same article, identifying a particular compound in a particular plant with a particular biological activity, and, as a result offering validity to an ethnomedical or screening observation. In other instances, these three aspects may occur in literature references that are diversely located. NAPRALERT, by recognizing the interrelated existence of these entities and permits substantial retrospective literature analysis on anticancer natural products.

We have developed a search pattern designed to develop a series of datasets of plants and compounds which demonstrate chemotherapeutic activities. These datasets are then correlated against one another in order to determine the plants that have been reported to exhibit a pertinent biological activity in an experimental system where that activity cannot be associated with any chemical known to occur in that plant. Finally, these datasets are correlated with the ethnobotanical data to determine those ethnobotanical observations that can be supported by the experimental literature and those which may suggest that further experimental examination is needed. A scoring system permits prioritization for collection.

Mention should also be made of the data management system which has been established to accumulate and analyze data generated at each of the sites. The system is based at UIC and both acquires and makes available data on-line at the chemistry and biology sites at UIC as well as at RTI and Glaxo. The system is also used for analytical purposes, for preparing reports and for establishing literature databases on the active plants through the NAPRALERT system. All pertinent data related to the collection and identification of the plant, the initial and any subsequent extractions and every biological result, are entered into the database and are available to the whole consortial group. Thus, there is an essentially instantaneous data flow that can occur between the three sites. This has the dual effect of enhanced communication and also enhanced involvement in the program on a day-to-day basis.

Primary and Secondary Bioassay Capabilities

The third crucial aspect to this program concerns the range of the primary and secondary bioassays being conducted. As knowledge has increased concerning the evolution of the cancerous state, and the potential mechanisms of action of agents which could interfere with that process, it has become apparent that a drug discovery process which involves mechanism-based assays, and which selects for specific target processes, provides a complementary strategy to cytotoxicity testing.

One approach then is to consider the search for drugs that could modulate disease-specific processes that are required for progression, for example, metastasis. Since a localized primary tumor (not associated with a vital body organ) can often be definitively resolved by means of surgery and/or radiation therapy, the phenomenon of metastasis is of key importance in leading to the serious consequences associated with cancer. As a result, a number of assays have been constructed to screen for antimetastatic agents (*35,36*). A second important example of ostensible tumor-specificity is the process of angiogenesis. Tumor-derived factors can stimulate the proliferation of endothelial cells and induce angiogenesis (*37*). Consequently, the discovery of specific angiogenesis inhibitors (*38,39*) is a valid endeavor.

The reality of finite resources begs the question: what assay (or battery of assays) will be most decisive for screening plant extracts in a search for new anticancer agents? In a treatise dealing with this question (*40*), it was eventually concluded that no concise answer to this question is available. Beyond the philosophical considerations that are imbedded in the assay systems selected for drug discovery, there are always elements (constraints and opportunities) of personal interest, capability and physical resources. The battery of drug discovery assays that will be briefly mentioned here was selected by our group after extensive analysis. As new information has become available, both from the literature and our personal experience, the nature of this battery of tests has evolved.

A summary of some of the test systems that have been used follows:

- Nine human cancer cell lines (melanoma, sarcoma, astrocytoma, lung, colon, squamous cell carcinoma, hormone-dependent and hormone-independent breast, and hormone-dependent prostate)
- ASK astrocytoma cells [to evaluate antimitotic activity (*11,41*)]
- KB and KB-V1 (drug-resistant) cells
- Reversal of drug resistance using KB-V1 cells
- Topoisomerases I and II
- Inhibition of the synthesis of endocrine hormones
- Inhibition of autophosphorylation by EGF receptor tyrosine kinase activity
- Inhibition of farnesylation of the *ras* protein by farnesyl pyrophosphate transferase
- Agonism and antagonism of androgen and estrogen receptor binding
- Inhibition of human collagenase

Although the assays listed are quite diverse, the use of a combination of cell- and mechanism-based assays was thoroughly contemplated; there are several areas in which the assays are complementary, and other areas wherein we simply did not want to over-restrict our ability to detect potentially important leads. A brief description of some of the more significant elements of this strategy will be presented below.

Some of the mechanism-based *in vitro* assays were designed in direct analogy with the types of molecular responses mediated by known (clinically effective) antitumor agents. Thus, we monitor effects similar to those known to be mediated by vincristine and vinblastine (tubulin depolymerization), taxol (tubulin stabilization), camptothecin (topoisomerase I inhibition), and adriamycin (topoisomerase II inhibition). Similarly, considering recent advances in areas such as intracellular signal transduction, proto-oncogene expression and function, and cell-cell interactions, additional sites for the targeting of drugs become obvious, as do experimental assay systems which are amenable to large numbers of samples. Additionally, antagonism of hormone-dependent tumors is of known clinical value, and our battery of assays reflects this.

An ancillary approach toward the discovery of agents useful in the treatment of cancer involves drug resistance. It is well-known that resistance to multiple drugs often develops in a clinical setting, and essentially all patients who become refractory die of their disease (*42*). It is of interest therefore to compare the cytotoxic potential of substances with multidrug-resistant cells and the parent cell line. Activity of equivalent potency with these cell lines suggests the presence of an agent(s) to which the "resistant" cell line is susceptible, and thus a compound worthy of characterization. A second approach toward therapeutic reversal of drug resistance involves alteration of P-glycoprotein-mediated drug efflux. Most agents that reverse multidrug resistance also inhibit the binding of vinblastine or doxorubicin to the P-glycoprotein (*43*). Thus, attempts are made to enhance vinblastine-mediated cytotoxicity with drug-resistant KB-V1 cells in culture.

Although mechanism-based bioassays offer great potential in drug discovery programs, some possible limitations should also be noted. For example: i) Toxic mechanisms known to be facilitated by an antitumor agent are not limited to malignant cells. Additional factors are involved in providing a demonstrable therapeutic index, and these factors are either unknown or impossible to assess appropriately with uncomplicated *in vitro* test systems. ii) While progress has been made in defining certain mechanisms of antitumor activity, it is unlikely that many known agents function through a single mode of action. Thus, secondary mechanisms may contribute in a synergistic and essential manner toward the observed activity. iii) Additional mechanisms of cancer proliferation which are at present totally unknown probably remain to be uncovered.

Based on considerations such as these, a broad overall scope of our bioassay capability has been retained, including a battery of human tumor cell lines. Although the magnitude of this latter endeavor is more limited in terms of number of cell lines than the recent NCI initiative (*6-9*), the overall philosophy and therapeutic targets are similar. We recognize that cytotoxicity is neither necessary nor sufficient for antitumor activity. However, cytotoxicity is an activity that is consistent with antitumor activity, and interference with any mechanism required for cell survival will mediate a positive response. Thus, in conjunction with the results of other biological assays, these results will aid in establishing correlations, and assist in deciding which materials to subject to fractionation procedures. Based solely on non-specific cytotoxic activity, a plant is typically not given a high priority and therefore is not subjected to fractionation.

We continue to evaluate the primary screen of mechanism-based assays, and a plant placed on hold due to nonmechanistically-defined cytotoxicity could then be re-evaluated. These procedures will help to characterize more fully the biologic potential of the plant materials we select for evaluation, and this is considered a more judicious procedure than ranking a potentially clinically useful material as a "false negative" due to the unavoidable situation of utilizing a highly discriminatory screen.

The use of a battery of human cell lines derived from a variety of human tumor-types also has substantial theoretical and practical value, since the cells can be carried as solid tumors in athymic mice. In this way, more advanced testing of active principles is facilitated. It is anticipated that we will encounter test materials that will demonstrate cell-type specificity, and that this will lead to the isolation and identification of novel, selective cytotoxic agents. Based on structure and activity, such a selective agent would be a candidate for more advanced testing. Irrespective of the outcome, however, the concept of selective cytotoxicity implicitly suggests the presence of a cell-specific receptor that differentiates one tumor-type from another. Such a discovery would be of exceptional interest in terms of developing tumor-specific therapeutic strategies, and a cytotoxic agent specific for one cell-type would greatly aid in biologically identifying the appropriate subcellular target.

Following biological evaluation of the plant extracts in each of the bioassays, some are classified as "active" according to established criteria. In order to rank order plants for further study, a decision network has been developed. Plants, and their associated activity and previous history, are considered on an individual basis, and the collective experience of the personnel in the group involved is crucial. When a plant extract demonstrates a highly significant response in a particular test system (e.g., a particularly intense response, or a particularly interesting dose-response), it is placed into a high priority category for detailed review. Decisions to fractionate, however, are based on quantitative data, standard criteria are utilized, and the designation of the fractionation site is made by the group. Although it is not possible to devise an enduring set of rules, guidelines have been established and are constantly re-evaluated. Since many of the bioassay procedures are novel, intensity of response is continuously discussed by the group as data accumulate. Some of the factors that are taken into account in ranking active plants include:

High Priority

- Active in one test system only. (This type of response suggests the presence of a novel active principle).
- Active in more than one test system in which there is no obvious correlation of the demonstrated activities (e.g., tyrosine kinase inhibition and inhibition of a topoisomerase). (This type of profile suggests the presence of two distinct active principles).
- Active only in test systems wherein mechanism-based activities correlate.
- Active in one cell culture system only and any mechanism-based assay.

Lower Priority

- Active in more than one cell culture system.
- Active in more than one cell culture system, and also active in any mechanism-based assay.

After leads have been rank-ordered, a single bioassay is selected to direct the fractionation to yield the active principle(s). In order to optimize the efficiency of the bioactivity-directed fractionation, this work usually occurs at the site of the bioassay being used. The isolated and structurally defined active principles are screened through the entire battery of assays. These biological results, when taken in conjunction with the chemical structure of active isolates, assist in formulating hypotheses regarding mechanism of action, and contribute toward an informed decision network regarding more advanced antitumor testing.

Finally, it is important to determine the cytotoxic potential of the isolates obtained on the basis of mechanistic assays. Some of the these may not function by mechanisms that are reflected through cytotoxicity (e.g., estrogen antagonism), but others will be identified on the basis of cytotoxicity with a single cell line, or on the basis of mechanisms that are consistent with a cytotoxic response. Cytotoxic potential is initially determined with our primary battery of cell lines, and the evaluation is expanded to include additional human cell lines derived from the same classes of human tumors. This approach provides a strong indication of the type of tumor to be used for an initial evaluation of efficacy using *in vivo* test systems.

Experiences to date

Since the initiation of the program in September, 1990, meetings of the principals of the group and the NCI Coordinator have been held every three-four months at the three experimental sites or at major scientific meetings and have been crucial in establishing the cohesiveness of the group. In addition, they have allowed for the resolution of issues related to efficient functioning, for the establishment of protocols and procedures for the operation of the program, deciding, as a group, which assays to eliminate or modify in some way, developing a prioritized list of plants for the purposes of fractionation, and discussing the proposed royalty reimbursement to the plant suppliers in the event that a commercially viable discovery is made.

To the end of August 1992, 1,010 plant samples, representing 494 plant species in 132 plant families have been collected in 26 countries. In accordance with the distribution plan, 509 samples were sent to RTI and 501 to UIC for extraction and sample preparation. The NAPRALERT database has identified 2,237 plants which have been reported to have an anticancer related activity. Following literature analysis and weighting, 33 of these are being studied for possible collection prior to biological evaluation.

All of the bioassays are fully operational and data are being accumulated in a database management system at UIC, following electronic transfer from RTI and Glaxo, as described above. Over 30 plants are currently under bioactivity-directed fractionation at the three sites based on their biological activity and more than twenty plants have been recollected from their original or a closely related collection site. A number of new, biologically active compounds have been isolated, their structures elucidated and are currently being evaluated in the panel of bioassays.

Based on the initial testing data of the submitted extracts at all three sites, certain of the bioassays which gave especially high "hit" rates were reviewed to either: lower the dose, reduce the "active" cut-off rate or eliminate the assay from the panel.

Some of the issues which have been addressed, or which are currently under discussion, include the transfer of data and technological expertise between the research sites, the degree of selectivity and the level of biological activity needed to warrant a fractionation effort, the need to eliminate tannins from extracts in order to avoid false-positive results in certain of the bioassays, the assays that are of value in the decision-making process, and the question of whether to add new cell lines or other assays.

Conclusions

A new strategy for the discovery of anticancer agents from plants is presently underway in a consortial effort involving groups at the University of Illinois at Chicago, Research Triangle Institute, and Glaxo Group Research. In this program published information is critically evaluated against existing chemical and biological information and plants are prioritized for collection; in addition, plants are collected by a network of botanical collectors around the world according to established protocols. Authenticated plants are extracted and the extracts tested in a broad array of human cancer cell and mechanism-based assays through a cooperative research program involving a university (UIC), a research institute (RTI) and a pharmaceutical company (Glaxo). Over 1,000 plants have been collected and extracts are being biologically evaluated. Plant collection and bioassay data are being stored at a single site. Bioactivity-directed fractionation is being carried out at all three sites on extracts which have demonstrated selective activity. Some novel compounds have been isolated and characterized and may serve as candidates for preclinical testing.

Acknowledgements

This program is supported by the Division of Cancer Treatment, National Cancer Institute, Bethesda, MD under the National Cooperative Natural Products Drug Discovery Program (U01 CA 52956).

Literature Cited

1. Prance, G. T. *Ann. Missouri Bot. Gard.* **1977**, *64*, 659-684.
2. Cordell, G. A. In *Bioactive Natural Products* Atta-ur-Rhaman; Basha, F. Z. (Eds.); Elsevier Science Publishers: Amsterdam, Netherlands, in press.
3. Barrios, V.; Farnsworth, N. R. *Proceedings from a Consultation of the Potentials for the Use of Plants Indicated by Traditional Medicine in Cancer Therapy*; WHO: Geneva, Switzerland, 1979; pp 4-6.
4. Hartwell, J. L. *Lloydia* **1967**, *30*, 363-463; **1968**, *31*, 71-170; **1969**, *32*, 79-107; **1969**, *32*, 153-205; **1969**, *32*, 247-296; **1970**, *33*, 98-194; **1970**, *33*, 288-392; **1971**, *34*, 103-160; **1971**, *34*, 204-255, **1971**, *34*, 310-361 and **1971**, *34*, 386-438.
5. Perdue, R. E., Jr.; Hartwell, J. L. *Morris Arb. Bull.* **1969**, *20*, 35-58.
6. Suffness, M. In *Biologically Active Natural Products*; Hostettmann, K.; Lea, P. J. (Eds.); Clarendon Press: Oxford, U.K., 1987; pp 85-104.
7. Alley, M. C.; Scudiero, D. A.; Monks, A.; Hursey, M. L.; Czerwinski, M. J.; Fine, D. L.; Abbott, B. J.; Mayo, J. G.; Shoemaker, R. H.; Boyd, M. R. *Cancer Res.* **1988**, *48*, 589-601.
8. Shoemaker, R. H.; Monks, A.; Alley, M. C.; Scudiero, D. A.; Fine, D. L.; McLemore, T. L.; Abbott, B. J.; Paull, K. D.; Mayo, J. G.; Boyd, M. R. In *Prediction of Response to Cancer Therapy*; Hall, T.C., Ed.; Alan R. Liss, Inc.: New York, NY, 1988; pp 265-286.

9. Scudiero, D. A.; Shoemaker, R. H.; Paull, K. D.; Monks, A.; Tierney, S.; Nofziger, T. H.; Curren, M. J.; Seniff, D.; Boyd, M. R. *Cancer Res.* **1988**, *48*, 4827-4833.
10. Suffness, M. *Gann Monograph on Cancer Research*, **1989**, *36*, 21-44.
11. Suffness, M.; Douros, J. *J. Nat. Prod.* **1982**, *45*, 1-14.
12. Neuss, N.; Neuss, M. N. In *The Alkaloids*; Vol. 37; Brossi, A.; Suffness, M., Ed.; Academic Press: New York, NY, 1990; pp 229-240.
13. Wiernik, P. H.; Schwartz, E. L.; Strauman, J. J.; Dutcher, J. P.; Lipto, R. B.; Paaietta, E. *Cancer Res.* **1987**, *47*, 2486-2493.
14. Li, Y. H.; Guo, S. F.; Zhou, F. Y.; Xu, S. Z.; Zhang, H. L. *Chung-Hua I Hsueh Tsa Chih* (English ed.) **1983**, *96*, 303-305.
15. Ajani, J. A.; Dimery, I.; Chawaia, P. J.; Pinnameni, K.; Benjamin, R. S.; Legha, S. S.; Krakoff, I. H. *Cancer Res.* **1986**, *70*, 375-379.
16. Legha, S. S.; Keating, M.; Picket, S.; Ajani, J. A.; Ewer, M.; Bodey, G. P.; *Cancer Treat. Rep.* **1984**, *68*, 1085-1091.
17. Neidart, J. A.; Young, D. C.; Kraut, E.; Howenstein, B.; Metz, E. N. *Cancer Res.* **1986**, *46*, 967-969.
18. Hsu, B.; Yang J. L. In *Advances in Chinese Medicinal Materials Research*; Chang, H. M.; Yeung, H. W.; Tso, W. W.; Koo, A. (Eds.); World Scientific Press: Philadelphia, PA, 1984; pp 377-389.
19. Muss, H. B.; Bundy, B. N.; Yazigi, R.; Yordan, E. *Cancer Treat. Rep.* **1987**, *71*, 873-874.
20. Han, J. *J. Ethnopharmacol.* **1988**, *24*, 1-17.
21. Caille, P.; Mondesir, J. M.; Droz, J. P.; Kerbrat P.; Goodman A.; Ducret J. P.; Theodore, C.; Speilmann, M.; Rouesse, J. G.; Amiel, J. L. *Cancer Treat. Rep.* **1985**, *69*, 901-902.
22. Einzig, A. I.; Gralla, R. J.; Leyland-Jones, B. R.; Kelsen, D. P.; Cibas, I.; Lewis, E.; Greenberg, E. *Cancer Invest.* **1985**, *3*, 235-241.
23. Rouesse, J. G.; Le Chevalier, T.; Caille, P.; Mondesir, J. M.; Sancho-Garnier, H.; May-Levin, F.; Speilmann, M.; De Jager, R.; Amiel, J. L. *Cancer Treat. Rep.* **1985**, *69*, 707-708.
24. McGuire, W. P.; Rowinsky, E. K.; Rosenchein, N. B.; Grumbine, F. C.; Ettinger, D. S.; Armstrong, D. K.; Donehower, R. C. *Ann. Intern. Med.* **1989**, *111*, 273-279.
25. Holmes, F. A.; Walters, R. S.; Theriault, R. L.; Forman, A. D.; Newton, L. K.; Raber, M. N.; Buzdar, A. U.; Frye, D. K.; Hortobagyi, G. N. *J. Natl. Cancer Inst.* **1991**, *83*, 1797-1805.
26. Barclay, A. S.; Perdue, R. E., Jr. *Cancer Treat. Rep.* **1976**, *60*, 1081-1113.
27. Cordell, G. A. In *New Natural Products and Plant Drugs with Pharmacological, Biological or Therapeutical Activity*; Wagner, H.; Wolff, P. (Eds.); Springer Verlag: Berlin, Germany, 1977; pp 55-82.
28. Suffness, M. In *Advances in Medicinal Plant Research*; Vlietink, A. J.; Dommisse, R. A. (Eds.); Wissenschaftliche Verlags GmbH: Stuttgart, Germany, 1985; pp 101-133.
29. Cordell, G. A.; Farnsworth, N. R. *Lloydia*, **1977**, *40*, 1-44.

30. Suffness, M.; Douros, J. In *Methods in Cancer Research*; Vol. 16, Part A; DeVita, V. T.; Busch, H. (Eds.); Academic Press: New York, NY, 1979; pp 73-126.
31. Wall, M. E.; Wani, M. C.; Cook, C. E.; Palmer, K. H. *J. Am. Chem. Soc.* **1966**, *88*, 3888-3890.
32. Wani, M. C.; Taylor, H. L.; Wall, M. E.; Coggon, P.; McPhail, A. T. *J. Am. Chem. Soc.* **1971**, *93*, 2325-2327.
33. Loub, W. D.; Farnsworth, N. R.; Soejarto, D. D.; Quinn, M. L. *J. Chem. Inf. Comp. Sci.* **1985**, *25*, 99-103.
34. Farnsworth, N. R.; Loub, W. D.; Soejarto, D. D.; Cordell, G. A.; Quinn, M. L.; Mulholland, K. *Kor. J. Pharmacog.* **1981**, *12*, 98-109.
35. Hendrix, M. J. C.; Seftor, E. A.; Seftor, R. E. B.; Fidler, I. J. *Cancer Lett.* **1987**, *38*, 137-147.
36. Liotta, L. A. *Cancer Res.* **1986**, *46*, 1-7.
37. Folkman, J. *Cancer Res.* **1986**, *46*, 467-473.
38. Crum, R.; Szabo, S.; Folkman, J. *Science* **1985**, *230*, 1375-1378.
39. Folkman, J. *Adv. Cancer Res.* **1985**, *43*, 175-230.
40. Suffness, M.; Pezzuto, J. M. In *Methods in Plant Biochemistry*; Vol. 6; Hostettmann, K., Ed.; Academic Press: London, U.K., 1991; pp 71-133.
41. Swanson, S. M.; Jiang, J.-X.; de Souza, N. J.; Pezzuto, J. M. *J. Nat. Prod.* **1988**, *51*, 929-936.
42. Bellamy, W. T.; Dalton, W.S.; Kailey, J. M.; Gleason, M. C.; McCloskey, T. M.; Dorr, R. T.; Alberts, D. S. *Cancer Res.* **1988**, *48*, 6303-6308.
43. Akiyama, S. -I.; Cornwell, M. M.; Kuwano, M.; Pastan, I.; Gottesman, M. M. *Mol. Pharmacol.* **1988**, *33*, 144-147.

RECEIVED March 15, 1993

Chapter 14

Cancer Chemopreventive Agents

From Plant Materials to Clinical Intervention Trials

John M. Pezzuto

Department of Medicinal Chemistry and Pharmacognosy, Program for Collaborative Research in the Pharmaceutical Sciences, College of Pharmacy, University of Illinois at Chicago, and Specialized Cancer Center, University of Illinois College of Medicine at Chicago, Chicago, IL 60612

Chemopreventive agents are of great interest since they may reduce the incidence of cancer in human populations. Similar to other drug discovery programs, experimental procedures can be devised for the discovery and characterization of cancer chemopreventive agents. Once such agents are found, based on epidemiology, hypothetical efficacy or mechanism of action, they may be entered into clinical intervention trials. These trials may monitor a reduction in the incidence of human cancer or a modulation of putative intermediate endpoints can be determined. Prevention of cancer is a rational way of fighting this horrendous disease, and it is likely that chemoprevention will play an increasing role in this ongoing campaign.

At the current time, cancer claims the lives of approximately seven million people worldwide on an annual basis. In the United States alone, there are approximately one million new cases each year and approximately one-half million succumb to the disease. However, various causes and methods of preventing cancer are now obvious, and this knowledge should be brought to bear by members of an enlightened society. As an example, over 100,000 individuals in the United States die on an annual basis due to the manifestations of lung cancer, and a large percentage of these deaths could undoubtedly be negated by abolishing the smoking of cigarettes. In fact, the National Cancer Institute has devised a campaign in which the goal is to reduce the 1985 cancer mortality rate by 50% by the year 2000 (*1*, *2*). The approach to achieving this goal is obviously comprehensive. In addition to primary prevention strategies (e.g., cessation of cigarette smoking, reduction of exposure to chemical carcinogens), elements such as early diagnosis, dietary modification, and cancer training programs are emphasized.

0097–6156/93/0534–0205$06.00/0

An adjunct approach to reduction in the incidence of cancer is chemoprevention. Cancer chemoprevention is a term which was coined by Dr. M.B. Sporn as part of his classical work dealing with retinoids and cancer prevention. It may be defined in general terms as the prevention of cancer in human populations by ingestion of chemical agents that prevent carcinogenesis. As a scientific discipline, it is important to differentiate cancer chemoprevention from cancer prevention or cancer chemotherapy. As summarized in Table I, agents that serve as cancer chemopreventive agents can be placed into various groups. Of key importance, of course, is the potential of these agents to affect the incidence of cancer in human populations. In general, compounds can be rigorously tested for chemopreventive activity in animal studies, but activity in humans is largely speculative. However, particularly strong epidemiological evidence has been provided suggesting there is an inverse correlation between lung cancer among people who smoke and consume carotene-rich foods (*4*, *5*). One study involved over 250,000 subjects (*6*), and similar conclusions regarding smoking and the carotene consumption have been reached in cohort and case-control studies. These types of observations are not limited to carotenoids, in that other epidemiological evidence suggests an inverse correlation between vitamin C and esophageal and stomach cancers, selenium and various types of cancer, vitamin E and lung cancer, protease inhibitors and breast, colon, prostate and oral cancers, and folic acid and cervical dysplasia.

Table I. Various Groups of Potential Cancer Chemopreventive Agents[a]

Groups	*Examples*
1. Micronutrients	Vitamin A, C, E Selenium, calcium, zinc
2. Intentional food additives	Antioxidants
3. Non-nutritive food molecules	Carotenoids Coumarins Indoles Alkaloids
4. Industrial reagents	Photographic developers Herbicides UV light protectors
5. Pharmaceutical agents	Retinoids Nonsteroidal anti-inflammatory agents Anti-thrombogenic agents Anti-prostaglandins
6. Hormones and anti-hormones	Dehydroepiandrosterone Tamoxifen

[a]Adapted from ref. 3.

Thus, epidemiological leads may be one avenue toward the discovery of cancer chemopreventive agents. However, this approach is of limited scope; laboratory studies are typically more productive. As described previously in the area of cancer chemotherapy (*7*), a broad range of factors needs to be taken into account when devising a drug discovery program. When designing a program in cancer chemoprevention, it is logical to first consider the use of animal models wherein a chemically-induced tumor can be reproducibly generated and wherein blocking agents can be studied under well-defined conditions. A generalized protocol that summarizes the magnitude of a study to assess the chemopreventive potential of a single agent in a single model system is presented in Table II. Given a good lead compound, a study of this type is unquestionably warranted.

Table II. Generalized Protocol for Screening of Chemopreventive Agents as Inhibitors of *N*-Methyl-*N*-Nitrosourea (MNU)-Induced Mammary Carcinogenesis

Group No.	*Animals/Group*[a]	*Carcinogen*	*Agent Dose*
1	20	+	None (basal diet)
2	20	+	None (vehicle)
3	20	+	80% of MTD[b]
4	20	Solvent only	80% of MTD
5	10	None	None

[a]In order to achieve statistic significance, this number assumes 100% control cancer incidence and a decrease in tumor incidence of at least 20% on treatment with the test agent.

[b]MTD, Maximum Tolerated Dose. These determinations required approximately 80 animals and are 6-8 weeks in duration.

For natural product drug discovery, however, it is typically not reasonable to rely on animal models. The standard approach is the use of bioassay-directed fractionation [7, Angerhofer, C.K.; Pezzuto, J.M. In *Biotechnology and the Practice of Pharmacy*; Pezzuto, J.M.; Johnson, M.E.; Manasse, Jr. H.R., Eds.; Chapman and Hall: New York, in press]. As summarized in Table III, a "typical" fractionation procedure using MNU-induced mammary cancer as a model system would require the use of over 6,000 rats and a time period of over four years. Under normal circumstances this would not be acceptable.

Table III. Isolation of Biologically Active Substances From Plant Material Using Inhibition of MNU-Induced Mammary Cancer as a Model System for Bioassay-Directed Fractionation

Sample	*Test Material*	*No. Animals*[a]	*Time Required*
Original extract	2 fractions	270	8 months
Column 1	10 fractions	1,100	8 months
Column 2	10 fractions	1,100	8 months
Column 3	10 fractions	1,100	8 months
Column 4	10 fractions	1,100	8 months
Column 5	10 fractions	1,100	8 months
Isolates	4 compounds	500	8 months
Totals		6,270	56 months

[a]Based on the results of preliminary work and evaluation the test substance at the MTD.

An alternate system for detecting a physiological response that is indicative of cancer chemoprevention involves organ cultures of mammary glands. As summarized in Table IV, there is an excellent correlation between the mammary organ culture test system and inhibition of carcinogen-induced mammary tumors in rodents. Unlike the full-term carcinogenesis inhibition systems, however, evaluations can be performed in the mammary organ culture test system with only a few milligrams of test substance. Moreover, results can be obtained in approximately one month, the expense is relatively low, and the required number of laboratory animals is kept to a minimum. A brief description of the procedure follows.

Female BALB/c mice with immature mammary glands are given β-estradiol and progesterone for 9 consecutive days. On the tenth day (which then becomes day 0), the mice are killed and the second thoracic mammary glands are excised and placed in culture (*9*). The tissue is incubated for 10 days with hormones (insulin, prolactin, hydrocortisone and aldosterone) which promote growth and differentiation of mammary alveoli. Between days 3 and 4, the mammary glands are treated for 24 hr with 7,12-dimethylbenz(a)anthracene (DMBA). On day 10, all hormones except insulin are withdrawn and the cells remain in culture for an additional 14 days. This hormonal deprivation causes regression of the alveolar

Table IV. Effectiveness of Chemopreventive Agents in Mammary Gland Organ Cultures and Comparison with *In Vivo* Rat Mammary Carcinogenesis and Antimutagenesis

Compound	*In Vivo Mammary Carcinogenesis*	*DMBA-Induced Mammary Lesions in Organ Culture*	*Antimutagenic Activity*
Aesculetin	ND[a]	Effective (*8*)	Effective
Ajoene	Ineffective	Ineffective (*8*)	Effective
Brassinin	In Progress	Effective[b]	ND
β-Carotene	Inconclusive	Effective	Effective
Catechin	ND	Effective	Effective
Curcumin	Effective	Effective (*8*)	Effective
Diallyl disulfide	Effective	Ineffective	ND
Eicosapentenoic acid	Effective	Ineffective	ND
Erythoxydiol X	ND	Effective	Effective
Glycyrrhetinic acid	Effective	Effective (*8*)	Effective
Limonene	Effective	Effective (*8*)	ND
Nordihydroguaiaretic acid	Marginally effective	Effective (*8*)	Effective
Purpurin	In Progress	Effective (*8*)	Effective
Silymarin	In Progress	Effective (*8*)	ND
β-Sitosterol	Ineffective	Effective (*8*)	Effective
Taurine	ND	Ineffective	ND
α-Tocopherol acetate	Marginally effective	Marginally effective (*8*)	Effective

[a]ND = Not Determined.

[b]Mehta, R.G.; Constantinou, A.; Moriarty, R.; Pezzuto, J.M.; Moon, R.C. *Proceedings of the 84th Annual Meeting of the American Association for Cancer Research*, Orlando, Florida, May 19-22, 1993.

structures and only mammary ducts remain (*10*). Exposure to DMBA induces nodule-like alveolar lesions (NLAL) which are hormone-independent and therefore do not regress. Experimental evidence indicates that these lesions are preneoplastic (*11*) and morphologically similar to hyperplastic alveolar nodules which develop *in vivo* in mouse mammary glands (*9*). In addition, enzymically-dissociated cells from mammary glands with NLAL produce hyperplastic alveolar outgrowths when transplanted into gland-free mammary fat pads of syngeneic mice. These outgrowths can be maintained in the animals by serial transplantation (*12*) and some become tumorigenic when allowed to progress. At the end of 24 days, the cultured glands are fixed, stained with alum carmine and scored for incidence of mammary lesions (*10*). This system is also used for initiation-promotion studies of mammary carcinogenesis (*11*). In this case, 12-*O*-tetradecanoylphorbol-13-acetate (TPA), a potent tumor promoter, enhances both the incidence and severity of DMBA-induced NLAL when included in the culture medium during days 9-14 of the culture period. Therefore, inhibition of initiation is studied by including the test substance in the medium prior to addition of the carcinogen (between days 1 and 4) and inhibition of the promotional phase is studied by including the test compound in the medium onward from day 9 (*11*). Nonetheless, as summarized in Table V, mammary organ culture is not generally suitable for bioassay-directed fractionation procedures.

Table V. Isolation of Biologically Active Substances from Plant Materials Using Mammary Organ Culture as a Model System for Bioassay-Directed Fractionation

Sample	*Test Material*	*No. Animals*[a]	*Time Required*
Original extract	2 fractions	180	2 months
Column 1	10 fractions	780	2 months
Column 2	10 fractions	780	2 months
Column 3	10 fractions	780	2 months
Column 4	10 fractions	780	2 months
Column 5	10 fractions	780	2 months
Isolates	4 compounds	330	2 months
Totals		4,410	14 months

[a]Based on the use of 15 glands per test substance (5 concentrations).

In general, it is necessary to resort to short-term *in vitro* bioassay procedures. Short-term procedures suitable for the discovery and characterization of cancer chemopreventive agents are not prevalent. However, Cassady *et al.* (*13*) have investigated the effects of chemopreventive agents on carcinogen binding and metabolism with cultured CHO cells, Muto *et al.* (*14*) studied the inhibition of TPA-induced early antigen of Epstein-Barr virus in culture, and Bertram *et al.* (*15*) reported inhibition of C3H/10T½ cell transformation by enhancing gap junction communication.

The experimental approach we have developed is summarized in Table VI. In brief, plant extracts are screened through a battery of *in vitro* systems, and active leads are then evaluated in the mammary organ culture system described above. Active leads in both the mammary organ culture and the *in vitro* test systems are then subjected to bioassay-directed fractionation utilizing the *in vitro* test system as a monitor. Once pure active principles are identified, they are tested in mammary organ culture for efficacy, and substances of sufficient merit are then subjected to more advanced test systems (e.g., full-term testing as summarized in Table II). As recently described for plant antimutagenic compounds (Shamon, L.; Pezzuto, J.M. In *Economic and Medicinal Plant Research*; Vol. 6; Farnsworth, N.R.; Wagner, H., Eds.; Academic Press: London; in press), there is little probability that the random discovery of an agent demonstrating such an activity would lead to the provision of a useful cancer chemopreventive agent. However, correlation of short-term *in vitro* activity with a response that is more physiologically indicative of cancer chemoprevention, *viz.*, inhibition of carcinogen-induced lesions in mammary organ culture, is anticipated to yield more useful compounds. As summarized in Table IV, certain plant-derived antimutagens demonstrate activity in the organ culture system, and we have extended this to include inhibitors of protein kinase C and ornithine decarboxylase (ODC) (Kinghorn, A.D.; Mehta, R.G.; Moon, R.C.; Pezzuto, J.M. *Proceedings of the 84th Annual Meeting of the American Association for Cancer Research*, Orlando, Florida, May 19-22, 1993).

At the present time, approximately 600 "chemopreventive" agents are known, so an important question relates to the selection criteria for the entry of agents into human intervention trials. It is clear that any epidemiological support would need to be taken into account, and experimental evidence for efficacy (e.g., animal studies) and lack of toxicity are also of great importance. As our understanding of chemopreventive agents increases, however, mechanism of action becomes of greater significance. In considering the general stages of chemically-induced carcinogenesis (Figure I), agents can be classified as inhibitors of carcinogen formation (e.g., ascorbic acid, tocopherols, phenols), inhibitors of initiation (e.g., phenols, flavones, aromatic isothiocyanates, diallyldisulfide, ellagic acid, antioxidants, glutathione, S_2O_3), and inhibitors of post-initiation events (e.g., β-carotene, retinoids, tamoxifen, dehydroepiandrosterone, terpenes, protease inhibitors, prostaglandin inhibitors, Ca^{2+}, nerolidal). In addition, certain agents may demonstrate pleiotropic mechanisms of action, and combinations of agents may demonstrate chemopreventive activity in a synergistic manner (*16*).

Table VI. Experimental Approach for the Discovery of Cancer Chemopreventive Agents from Natural Product Source Materials

1. Conduct short-term *in vitro* bioassays
   a. Inhibition of protein kinase C
   b. Inhibition of phorbol ester-induced ODC with cultured mouse 308 cells
   c. Antagonism of estrogen-receptor interactions
   d. Induction of HL-60 cell differentiation
   e. Inhibition of carcinogen-induced mutagenicity
   f. Inhibition of cyclooxygenase activity

2. Evaluate active leads as inhibitors of DMBA-induced lesions in mammary organ culture

3. Active in 1 and 2: Isolate active principles utilizing select short-term *in vitro* tests

4. Test resulting active principles in mammary organ culture

5. Active in 4: Test resulting active principles in full-term tumorigenesis systems

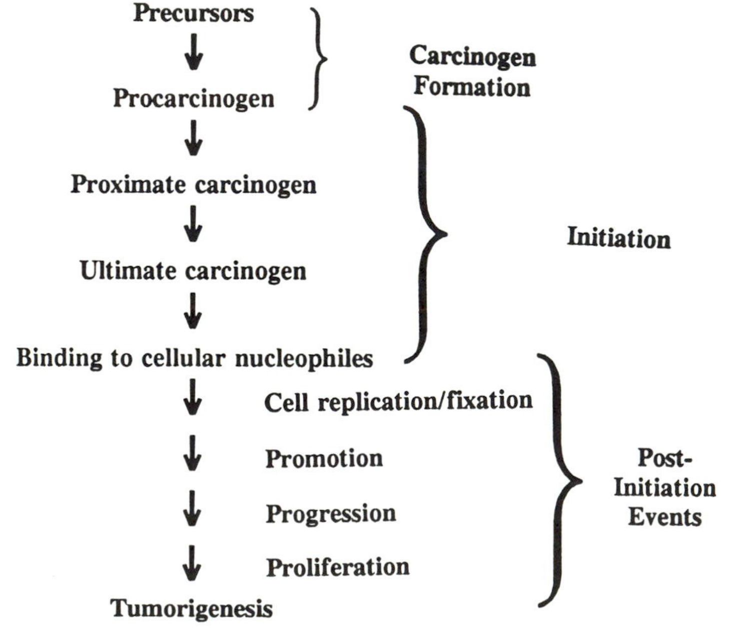

Figure I. Stages of Chemical Carcinogenesis

Once a suitable candidate is identified, it may be entered into a clinical intervention trial. When designing such a trial, it is necessary to carefully consider parameters such as the data accrual period, anticipated decrease in clinical appearance of disease, rate of compliance, etc. (*17*). A controversial trial (*18*) which has recently been initiated will study the potential of tamoxifen to reduce the incidence of breast cancer. This intervention trial will involve approximately 16,000 subjects and the NIH will invest approximately $68 million (*19*).

Obviously, a trial of this magnitude is a major undertaking that cannot be repeated with great frequency. Thus, rather than limiting such studies to clinical endpoints indicative of cancer prevention (e.g., breast tumors), intermediate endpoints have been suggested which can be monitored as indicative of cancer chemopreventive activity. A list of representative intermediate endpoints is presented in Table VII. In addition, as summarized by Boone *et al.* (*21*), a number of chemopreventive agents are presently being evaluated in intervention studies, some of which will monitor intermediate endpoints (Table VIII).

Table VII. Putative Intermediate Endpoints[a]

- Genetic
  - Oncogene activation/suppression
  - Micronuclei
  - Quantitative DNA analysis, DNA ploidy
- Biochemical
  - Ornithine decarboxylase
  - Prostaglandin synthetase
- Cellular
  - Sputum metaplasia/dysplasia
  - Cervical dysplasia
  - Gastric metaplasia
  - Colonic cell proliferation
- Precancerous lesions
  - Colon polyps
  - Bladder papillomas
  - Oral leukoplakia
  - Dysplastic nevi

[a] Adapted from ref. 20.

Table VIII. Examples of Chemoprevention Intervention Studies[a]

Target site organ	*Target/risk group*	*Inhibitory agents*
Cervix	Cervical dysplasia	*trans*-Retinoic acid
Cervix	Cervical dysplasia	Folic acid
Colon	Familial polyposis	Vitamins C and E and fiber
Colon	Familial polyposis	Calcium
Colon	Adenomatous polyps	β-Carotene and vitamins C and E
Colon	Adenomatous polyps	Piroxicam
Lung	Chronic smokers	Folic acid and vitamin B_{12}
Lung	Asbestosis	β-Carotene and retinol
Lung	Cigarette smokers	β-Carotene and retinol
Lung	Smoking males	β-Carotene
Lung	Asbestosis	β-Carotene
Skin	Albinos	β-Carotene
Skin	Basal cell carcinoma	β-Carotene and vitamins C and E
Skin	Basal cell carcinoma	β-Carotene
Skin	Actinic keratoses	Retinol
Skin	Basal cell carcinoma	Retinol and 13-*cis*-retinoic acid
Breast	Adenocarcinoma	*N*-(4-hydroxyphenyl)retinamide
All sites	American physicians	β-Carotene

[a] Adapted from ref. 21.

To conclude, cancer chemoprevention is a field of great promise. It is likely that we exist in a state of homeostasis, and that certain dietary components have contributed to our overall state of well-being including cancer prevention. This has recently been obviated by certain dietary campaigns inaugurated by the American Cancer Society, the National Cancer Institute, and others. For example, it is currently recommended that 5-7 servings of vegetables be consumed on a daily basis, to function as a source of cancer chemopreventive agents. Although it is not reasonable to assume cancer chemopreventive agents will safeguard humans from known carcinogenic risks such as cigarette smoking, it is reasonable to anticipate that they will play an increasing role in cancer prevention strategies.

Acknowledgments

The author is grateful to Dr. R. G. Mehta for information concerning the mammary organ culture test system, Dr. R. C. Moon for information concerning *in vivo* testing procedures, Ms. C. Lewandowski for help in accumulating some of the review data presented in this paper, and Ms. M. Sitt for her assistance in the typing of this manuscript. JMP was the recipient of a Research Career Development Award from the National Cancer Institute and a Research Fellowship from the Alexander von Humboldt Foundation. In our laboratory, experimental work in this area of research is supported in part by the National Cancer Institute through Program Project grant PO1 CA48112.

Literature Cited

1. Greenwald, P; Sondik, E.; Lynch, B. S. *Am. Rev. Public Health* **1986**, *7*, 267-291.
2. Greenwald, P; Cullen, J. W.; McKenna, J. W. *J. Natl. Cancer Inst.* **1987**, *79*, 389-400.
3. Costa, A.; Santoro, G; Assimakopoulos, G. *Acta Oncol.* **1990**, *29*, 657-663.
4. Bjelke, E. *Int. J. Cancer* **1975**, *15*, 561-565.
5. Shekelle, R. B.; Liu, S.; Raynor, Jr., W. J.; Lepper, M.; Makza, C.; Rossof, A. H.; Paul, O.; Shryock, A. M.; Stamiler, J. *Lancet* **1981**, *2*, 1185-1189.
6. Hirayama, T. *Nutr. Cancer* **1979**, *1*, 67-81.
7. Suffness, M.; Pezzuto, J. M. In *Methods in Plant Biochemistry*; Vol. 6; Hostettmann, K., Ed.; Academic Press: London; 1991, pp. 71-133.
8. Mehta, R. G.; Moon, R. C. *Anticancer Res.* **1991**, *11*, 593-596.
9. Lin, F. J.; Banerjee, M. R.; Crump, L. R. *Cancer Res.* **1976**, *36*, 1607-1614.
10. Mehta, R. G.; Steele, V.; Kelloff, G. J.; Moon, R. C. *Anticancer Res.* **1991**, *11*, 587-591.
11. Mehta, R. G.; Moon, R. C. *Cancer Res.* **1986**, *46*, 5832-5835.
12. Telang, N. T.; Banerjee, M. R.; Iyer, A. P.; Kundu, A. B. *Proc. Natl. Acad. Sci. USA* **1979**, *76*, 5886-5890.
13. Cassady, J. M.; Baird, W. M.; Chang, C.-J. *J. Nat. Prod.* **1990**, *53*, 23-41.
14. Muto, Y.; Ninomiya, M.; Fujiki, H. *Jpn. J. Clin. Oncol.* **1990**, *20*, 219-224.
15. Bertram, J. S.; Hossain, M. Z.; Pung, A; Rundhaug, J. E. *Prev. Med.* **1989**, *18*, 562-575.
16. Ip, C.; Ganther, H. E. *Carcinogenesis* **1991**, *12*: 365-367.
17. Love, R. R. *J. Natl. Cancer Inst.* **1990**, *82*, 18-21.
18. Han, X. L.; Liehr, J. G. *Cancer Res.* **1992**, *52*, 1360-1363.
19. Smigel, K. *J. Natl. Cancer Inst.* **1992**, *84*, 669-670.
20. Malone, W. F. *Am. J. Clin. Nutr.* **1991**, *53*: 305S-313S.
21. Boone, C. W.; Kelloff, G. J.; Malone, W. E. *Cancer Res.* **1990**, *50*, 2-9.

RECEIVED March 5, 1993

Anti-infective and Antimicrobial Chemotherapeutic Agents from Plants

Chapter 15

National Cancer Institute Intramural Research on Human Immunodeficiency Virus Inhibitory and Antitumor Plant Natural Products

John H. Cardellina II, Kirk R. Gustafson, John A. Beutler, Tawnya C. McKee, Yali F. Hallock, Richard W. Fuller, and Michael R. Boyd

Laboratory of Drug Discovery Research and Development, Developmental Therapeutics Program, Division of Cancer Treatment, National Cancer Institute, Frederick, MD 21702–1201

The natural products drug discovery program of the United States National Cancer Institute (NCI) was dramatically reorganized and revitalized during the mid-to-late 1980's. One component of the overall strategic approach was the creation of the first-ever intramural, interdisciplinary research unit, the Laboratory of Drug Discovery Research and Development (LDDRD), with a central focus on natural products. LDDRD scientists identify leads from extracts housed in the NCI Natural Products Repository and screened in the high throughput *in vitro* antitumor and AIDS-antiviral bioassays, isolate and identify the antitumor and/or antiviral agents through bioassay-guided fractionation and spectral analyses, and examine in detail the biological activity of the isolated compounds. This report focuses on some selected illustrative examples of LDDRD research on HIV-inhibitory and antitumor plant natural products.

The primary mission of the NCI's intramural Laboratory of Drug Discovery Research and Development (LDDRD) is the discovery and chemical and biological characterization of new antitumor and anti-HIV drug development leads from natural products. The specific discovery strategy is based upon the application of novel screening models to the bioassay-guided isolation and structure elucidation of active constituents from extracts of diverse terrestrial, marine and microbial organisms. The concepts, rationales, historical background and technical details of the primary screening models currently used by LDDRD are reviewed elsewhere (*1-5*). These screening models, which were substantially developed by LDDRD scientists, are now operated on a very large scale by the NCI extramural service screening program for testing of tens of thousands of pure synthetic and naturally-occurring compounds submitted annually to NCI by investigators worldwide.

The main source of extracts for LDDRD research is the NCI Natural Products Repository. This is a key component of a radically restructured and revitalized NCI natural products drug discovery program, launched during the mid-to-late 1980's (*1-7*). Other new components, in addition to the two major new primary screens, include the extensive extramural, contracts-based collections network (described by G. Cragg *et al.*, elsewhere in this volume), a centralized extraction facility, an unprecedented new extramural grants program (National Cooperative Natural Products Drug Discovery Groups), and the first-ever NCI intramural, interdisciplinary research unit (LDDRD) with a central focus upon natural products.

The NCI Natural Products Repository is a unique resource for the broader scientific research community. There is considerable optimism that the continuing development and application of a great variety of novel screening models and research strategies to samples now or eventually to be contained in the repository will lead to important new drug discoveries. Our experiences using the two new NCI primary screening models, which have provided our particular focus to date, lend support for such optimism. Some selected recent examples follow.

Preliminary Evaluation of New Leads

A very substantial number of initial leads issuing from the two primary screens, particularly the anti-HIV screen, necessitated the development of a preliminary evaluation stage prior to full-scale fractionation efforts. Our goals at this stage are the dereplication of known antiviral or antitumor compounds, the prioritization of surviving leads for detailed study, and the gaining of some preliminary insight into the chemical class(es) and chromatographic behavior of the active constituent(s).

In the case of antitumor leads, the dereplication problem encompasses the large number of cytotoxic agents previously reported from natural sources over the past three decades. Careful study of the literature on the genus or family usually provides some indication of whether known cytotoxins might be present. A small-scale (~200 mg crude extract) preliminary fractionation followed by biological testing and ^{1}H NMR analyses of the resulting fractions frequently provides evidence to corroborate the results of the literature search. Aqueous extracts are distributed between *n*-BuOH and H_2O, while organic extracts are subjected to a four-step solvent partitioning scheme modified from a procedure popularized and employed by the Kupchan group some years ago (*8*). An additional key ingredient in this analysis is the application of the computerized *Compare* pattern-recognition algorithms (*2-4,9*), which correlate similar patterns of differential cytotoxicity produced by extracts, fractions and pure compounds tested in the NCI 60 cell line panel. This analysis frequently provides biological support that the cytotoxicity expressed by an extract is likely due to a known class of compounds.

In contrast, many of the initial HIV-inhibitory leads are characterized by the presence of a few known classes of antiviral compounds which are broadly distributed within (sulfolipids, sulfated sterols) and across (anionic polysaccharides, tannins) taxonomic boundaries. Therefore, it has proved necessary to develop simple, efficient protocols to differentiate extracts for which antiviral activity is due solely to these known classes of compounds from those extracts which potentially contained other, uncharacterized HIV-inhibitory compounds. Anti-HIV active organic extracts presently undergo the same solvent-solvent partitioning analyses as the antitumor leads, but the treatment of aqueous extracts is more complex. First, an ethanol precipitation procedure is used to eliminate those extracts in which activity is due exclusively to anionic polysaccharides or other complex, high molecular weight biopolymers. At this stage, the great majority of marine and lichen extracts (85-90%) are eliminated, along with about half the cyanobacterial extracts, but only a small fraction (10-15%) of the plant extracts are set aside. Plant extracts are then analyzed for tannin content; at this stage, about 95% of the remaining plant extracts are eliminated from further consideration. It should also be noted that, in our experience, the HIV-inhibitory activity of more than 50% of the plant organic extracts has been partitioned to polar phases and subsequently traced to tannins. All the active supernatant fractions are then subjected to a "chemical screen" analysis, which preliminarily examines the elution characteristics of the antiviral constituents on several chromatography media in a solid-phase extraction system. The resulting patterns of elution, which provide clues to the general chemical nature of the HIV-inhibitory compounds, have led recently to the identification of sulfated sterols as a commonly occurring antiviral constituent of sponges. The dereplication of polysaccharides (J. A. Beutler *et al.*, *Antiviral Chem. Chemother.*, in press), the chemical screen protocol and tannin dereplication (J. H. Cardellina II *et al.*, *J. Nat. Prod.*, in press), and the identification of sulfated sterols as a recurring class of HIV-inhibitor (*10-12*) have all been described in detail elsewhere.

Plant Antitumor Agents

Our early efforts in this area focused on the investigation of the possible utility of the *Compare* algorithm analyses for the preliminary "biological" dereplication of extracts and fractions. This data analysis methodology had been developed for and shown to be effective in the comparison of standard antitumor agents with the activity profiles of new compounds tested in the NCI *in vitro* screening panel. In our pilot feasibility studies (LDDRD, unpublished data), an extract which subsequently was revealed to be from *Camptotheca acuminata*, was blindly selected for chemical fractionation on the basis of its screening profile which matched closely with that of known topoisomerase I inhibitors (camptothecin and derivatives). Fractionation was guided by the full 60 cell line screen to the purification of camptothecin. The *Compare* algorithms showed high correlations of camptothecin with the crude extract and the most potent fraction at each stage of the separation. Similarly positive feasibility results were obtained with *Trewia* extracts containing maytansinoids, which were tracked on the basis of a screening profile typical of microtuble-interactive antimitotics.

We are currently exploring the use of the *Compare* algorithm to select extracts for study which may likely contain new or known compounds previously unknown to have a mechanism of antitumor action of current interest. For example, using the *Compare* algorithms in this manner to select extracts for fractionation, we have identified centaureidin, **1**, the first flavone known to inhibit tubulin polymerization (J. A. Beutler *et al.*, *BioMed. Chem. Lett.*, in press).

Organic extracts of the plant *Thevetia ahouia* were also analyzed by means of bioassay-guided fractionation. Three cardenolides, neriifolin, 3'-O-methyl evomonoside and 2'-acetylneriifolin, **2**, were isolated and characterized as the major cytotoxins present. Of these, **2** was the most potent and was more than two orders of magnitude more toxic to certain non-small cell lung cancer cell lines than to the less sensitive lines of other subpanels (L. A. Decosterd *et al.*, *Phytother. Res.*, in press).

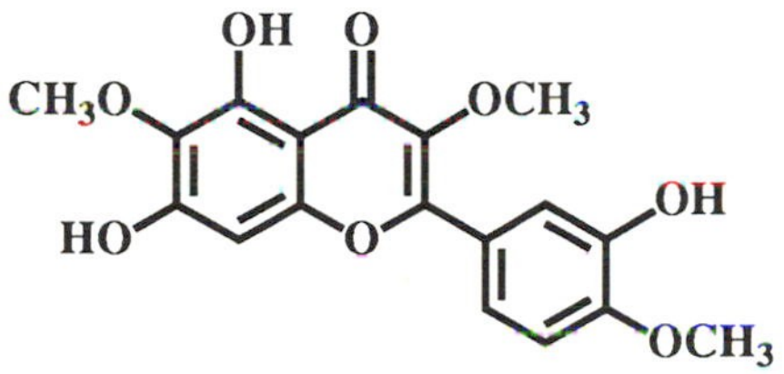

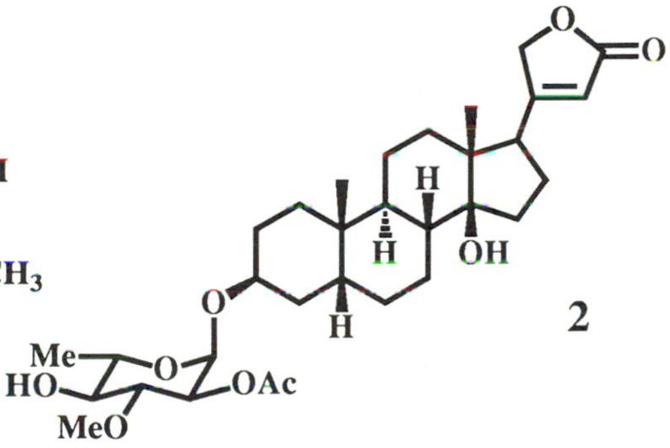

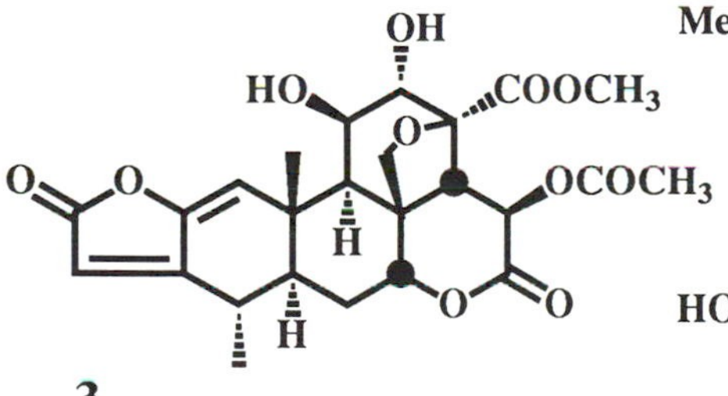

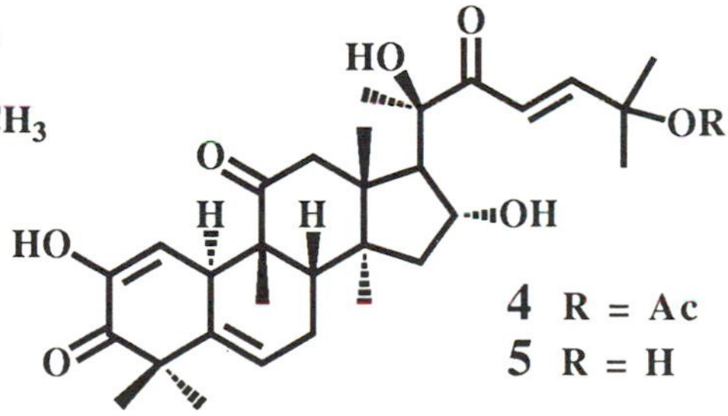

The crude organic and aqueous extracts of the monotypic plant *Cedronia granatensis* were selected for chemical analysis on the basis of potent cytotoxicity with a strong differential response by certain melanoma and colon tumor cell lines. Bioassay-guided fractionation of both extracts provided sergiolide, **3**, and isobrucein B as the responsible compounds. Sergiolide, which has a cyclopentenone ring annulated to the quassinoid framework, was two orders of magnitude more potent than isobrucein B against the more sensitive tumor cell lines (*13*).

We isolated cucurbitacins E and I (**4** and **5**, respectively) from seed extracts of *Iberis amara* based upon a unique pattern of selective cytotoxicity to renal tumor and melanoma cell lines (R. W. Fuller, unpublished). These compounds show nearly a three-order of magnitude differential in potency towards most cell lines in the sensitive subpanels relative to the other subpanels.

HIV-Inhibitors from Plants

To date, our laboratory has isolated and characterized from plants eight different compounds or chemical classes previously unknown to be HIV-inhibitors. A considerable diversity of chemical structures is represented by this group, which includes terpenoids, alkaloids, phenolics and compounds of mixed biogenesis. Because our efforts in this area have been reviewed recently (*14*), this report will focus on results obtained since that review.

First among these isolates was prostratin, **6**, obtained as the major AIDS-antiviral constituent from the small tree *Homalanthus nutans*, the bark of which has been used by Samoan healers (*taulesea*) to treat a variety of ailments (*15*). Initial enthusiasm for prostratin as a possible drug development candidate was tempered by concerns that it, like other phorbol esters, might be a potent irritant and tumor-promoting agent. However, a series of experiments conducted by Blumberg's group at NCI revealed little irritancy or tumor promoting properties by this 12-deoxyphorbol, but actually instead an apparent anti tumor-promoting potential (*16*). Thus, prostratin may have significant promise either as an AIDS-antiviral and/or as an anti tumor-promoter. Continuing research on prostratin is focused upon detailed antiviral biology and mechanism of action studies and on developing analytical methodology for preclinical toxicology and pharmacokinetics studies. In addition, since preliminary analyses of ethnobotanical preparations from *H. nutans* suggested the presence of prostratin (P. Cox *et al.*, unpublished data), careful quantitative analyses of bark preparations from Samoan *taulesea* are being undertaken.

Activity in side fractions obtained during the purification of prostratin from *H. nutans* led us further to isolate HIV-inhibitory atisane diterpenes; a concurrent study of the tropical plant *Chrysobalanus icaco* yielded HIV-inhibitory kaurane diterpenes (*17*). Neither plant contained potent or fully cytoprotective analogs; a survey of a dozen related diterpenes from the NCI pure compound repository provided no analog of increased potency. Consequently, further work in this area has been discontinued.

HIV-inhibitory activity in several genera of trees from the family Guttiferae (Clusiaceae) has been traced to a series of chemically related prenylated benzophenones (*18*). Five new compounds, guttiferones A-E, and two known compounds, xanthochymol and isoxanthochymol, were characterized from species of three genera (*Symphonia*, *Garcinia* and *Clusia*). Guttiferone A, **7**, was the most abundant of these compounds and is distinguished from previously known compounds of this type by a unique pattern of isoprenyl group substitution. The guttiferones were all roughly equipotent, with screening EC_{50} values of 1-10 μg/mL. However, more detailed biological evaluations revealed that, while the HIV-infected host cells remained viable in the presence of the compounds, there were no corresponding decreases in indices of viral replication. This was reminiscent of results with the peltatols (e.g., **8**), which are prenylated catechol dimers of comparable molecular weight (*19*). An important differentiating feature, however, is that the enolized β-diketone of the guttiferones must be essential to their antiviral activity, since isoxanthocaymol, **9**, is devoid of HIV-inhibitory activity. It is intriguing, nevertheless, that both compound classes provided similar modest cytoprotection of host cells, while viral replication appeared undiminished.

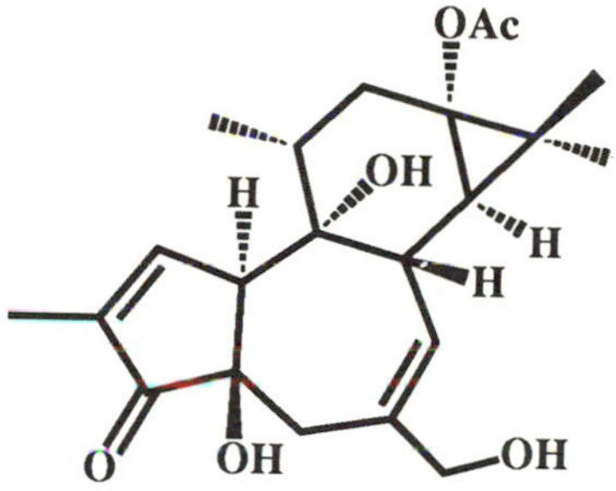

6

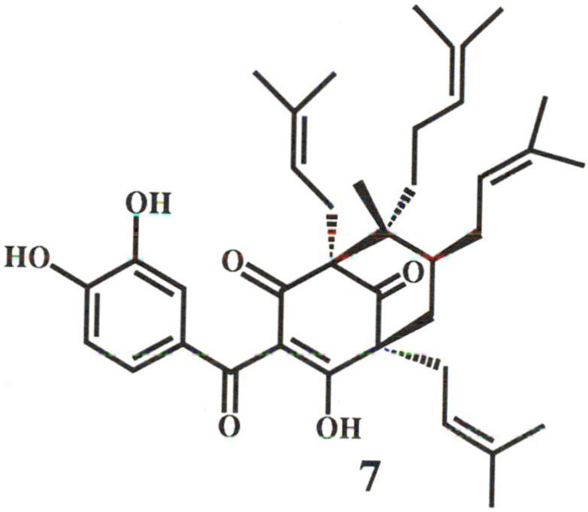

7

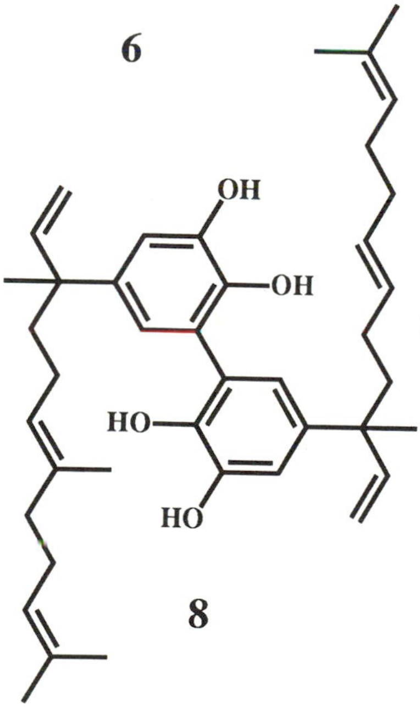

8

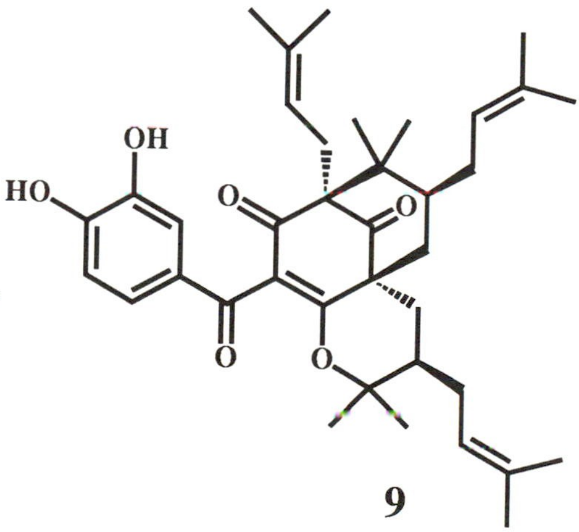

9

A significant discovery of the past year has been the isolation and identification of the calanolides (*20*) as potent inhibitors of HIV *in vitro*. The lead compound of the series is calanolide A, **10**, a new member of a rather large class of prenylated coumarins. Calanolide A exhibited an EC_{50} of 0.1 μM and provided 100% protection against the cytopathic effects of HIV-1 infection over a broad range of concentrations (up to 10 μM). Across the same concentration range, calanolide A completely inhibited all indices of HIV-1 replication. Even more significant was the observation that, while inactive against HIV-2, calanolide A fully protected human lymphoblastoid target cells from the cytopathic effects of AZT-resistant and pyridinone-resistant strains of HIV-1. Evidence presented elsewhere (P. L. Boyer *et al.*, *J. Virol.*, in press; A. Hizi *et al.*, *Antimicrob. Agents Chemother.*, in press) supports the view that calanolide A represents a new pharmacological class of non-nucleoside, HIV-1 specific, reverse transcriptase inhibitors.

Preclinical Development

Once the initial isolation and identification of a new chemical entity representing a new legitimate antitumor or AIDS-antiviral drug development lead has been accomplished, critical additional challenges for successful development of a natural product as a drug must be confronted. Prerequisite to the necessary preclinical toxicology, pharmacokinetics, metabolism and chemical studies to be performed on any new natural product drug development candidate, the supply problem frequently is most formidable. Our recent experiences with three new leads is illustrative.

Michellamine B, **11**, is a unique dimeric naphthalene-tetrahydroisoquinoline alkaloid which we recently reported from a tropical liana tentatively identified as *Ancistrocladus abbreviatus* (*21*). Michellamine B subsequently was selected by NCI as a high-priority candidate for preclinical development. When we set out to isolate additional supplies (100-200 g) of **11** for developmental research we were stunned to find no michellamine B or any other related dimeric alkaloids in collections of *A. abbreviatus* from any locations at or near the original collection site in Cameroon. Careful reexamination of the original voucher specimens and comparison with numerous herbarium specimens indicated that the original michellamine-containing vine from Korup Forest in Cameroon was a previously undescribed species (D. Thomas, unpublished data). Extensive field surveys in Cameroon have suggested a limited distribution of this new species. Fortunately, michellamine B is a major constituent of the leaf extracts; so, for the short term at least, the natural source of michellamine B should be adequate for preclinical research needs. Supplies of **11** are currently generated in our laboratory by means of a scale-up of our original separation protocol. Utilization of a 1.5 liter centrifugal partition chromatograph and preparative-scale (4 x 25 cm column) HPLC is currently providing 1-2 g michellamine B per week. In the meantime, studies have been initiated to determine whether the liana can be brought into cultivation. The absolute configuration of michellamine B has been determined recently (G. Bringmann *et al.*, *Angewandte Chemie*, submitted) via ruthenium-catalyzed oxidative degradations (*22*).

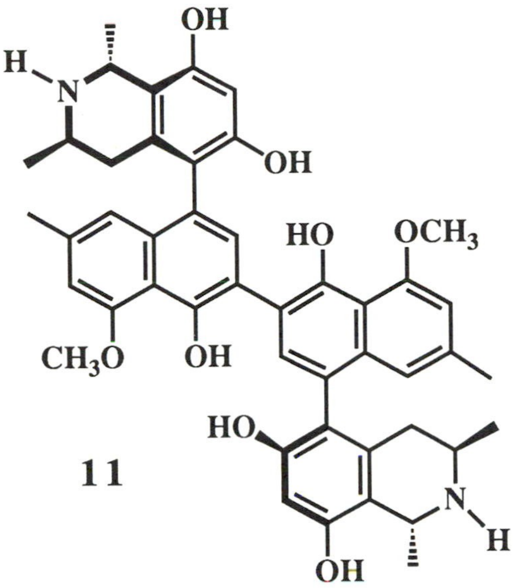

The case of the calanolides is even more challenging and complex. Since the original collections were made from a rather large tree from a well-characterized species (*Calophyllum lanigerum* var. *austrocoriaceum*), we initially felt that supply would not be a substantial problem. To our dismay, the tree which provided the original 1 kg collection of leaves, twigs and fruit had been cut down some time before the NCI field collectors returned for a resupply. Even worse, other trees in the general area, field-identified as the same species, proved to be devoid of calanolides A or B when extracts were analyzed in our laboratory. Subsequently, a number of trees, carefully identified as the species in question, were sampled from several locations, but not a single one contained calanolides A or B (D. Soejarto, unpublished data). As a result, preclinical development of this very important lead compound must await resolution of the supply problem, either through total synthesis or by location of a reliable natural source of the calanolides.

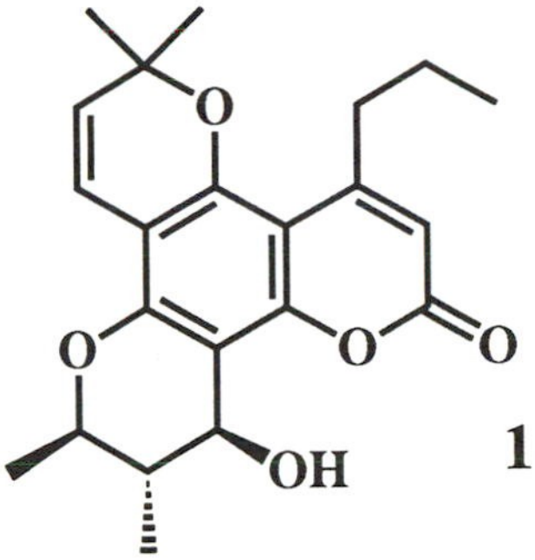

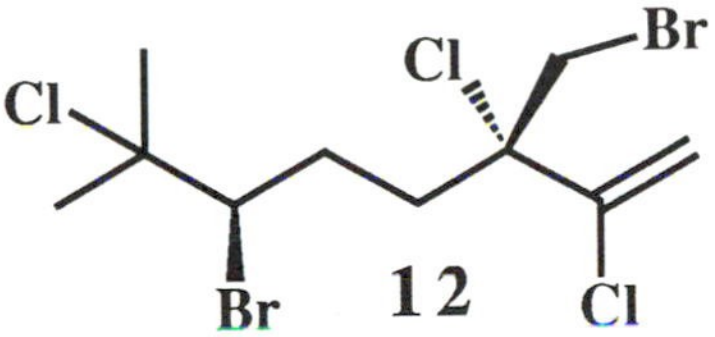

A third example, one representing a novel antitumor agent, is that of halomon, **12**, from the marine plant *Portieria hornemannii* (*23*). This pentahalogenated monoterpene produces a unique pattern of differential cytotoxicity in the NCI 60 cell line human tumor screening panel. In particular, brain tumor cell lines exhibit strong sensitivity to halomon. The original collection site, near Batan Island in the northern Philippines, is not readily accessible throughout the year due to stormy seas. Efforts to secure reasonable supplies of the red alga in other areas of the Philippines have met with success, but **12** has not been found, even in minor amounts, in five different collections. A reexamination of Hawaiian *P. hornemannii* has also failed to provide useful supplies of halomon. A recollection from the original site was successfully made in early 1992 and a 1.1 g sample of **12** was isolated from the extracts. Marine plants, particularly the red algae which produce halogenated monoterpenes, have long been noted for tremendous variation in secondary metabolite content from site to site, and over time at the same site.

Conclusion

These, and other recently reviewed (*14*) examples from LDDRD research, support the view that empirically-based discovery of novel drug development leads from natural products remains a fruitful pursuit. However, from any new lead discovery, the formidable challenges unique to successful drug development from natural products will inevitably follow.

Acknowledgments

We thank G. M. Cragg and K. M. Snader for coordinating the collections, T. McCloud for extractions, L. Pannell and G. Gray for mass spectral analyses, J.-R. Dai, H. Bokesch, J. Lind and J. Collins for technical assistance, A. Monks, D. Scudiero, O. Weislow and D. Clanton for screening data, and J. McMahon, M. Currens, R. Gulakowski, R. Moran, P. Staley, R. Shoemaker and T. Prather for biological and mechanistic data.

Literature Cited

1. Boyd, M.R. In *Accomplishments in Oncology. Vol. 1, No. 1, Cancer Therapy: Where Do We Go From Here?*; Frei, E.J.; Freirelich, E.J., Eds.; Lippincott: Philadelphia, PA, 1986; pp 68-76.
2. Boyd, M.R. In *Cancer: Principles and Practice of Oncology Updates. Vol. 3, No. 10*; DeVita, V.T., Jr.; Hellman, S.; Rosenberg, S.A., Eds.; Lippincott, Philadelphia, PA, 1989; pp 1-12.
3. Boyd, M.R.; Paull, K.D.; Rubinstein, L.R. In *Cytotoxic Anticancer Drugs: Models and Concepts for Drug Discovery and Development*; Valeriote, F.A.; Corbett, T.; Baker, L., Eds.; Kluwer Academic Publishers: Amsterdam, 1992; pp 11-34.
4. Boyd, M.R. In *Current Therapy in Oncology. Section I. Introduction to Cancer Therapy*; Niederhuber, J.E., Ed.; B.C. Decker, Inc.: Philadelphia, PA, 1992; pp 11-22.
5. Boyd, M.R. In *AIDS, Etiology, Diagnosis, Treatment and Prevention*; DeVita, V.T., Jr.; Hellman, S.; Rosenberg, S.A., Eds.; Lippincott: Philadelphia, PA, 1988; pp 305-319.
6. NCI-DCT Approves New Screening Program, Natural Products Concepts. *Cancer Lett.* **1985**, *11*, 4-8.
7. Profile: NCI Natural Products Lab. *Washington Insight* **1991**, *4*, 7.

8. Kupchan, S.M.; Britton, R.W.; Ziegler, M.F.; Sigel, C.W. *J. Org. Chem.* **1973**, *38*, 178-179.
9. Paull, K.D.; Shoemaker, R.H.; Hodes, L.; Monks, A.; Scudiero, D.A.; Rubinstein, L.; Plowman, J.; Boyd, M.R. *J. Natl. Cancer Inst.* **1989**, *81*, 1088-1092.
10. McKee, T.C.; Cardellina, J.H., II; Tischler, M.; Snader, K.M.; Boyd, M.R. *Tetrahedron Lett.* **1993**, *34*, 389-392.
11. Sun, H.H.; Cross, S.S.; Gunasekera, M.; Koehn, F.E. *Tetrahedron*, **1991**, *47*, 1185-1190.
12. Lednicer, D.; Narayanan, V.L. In *Natural Products as Anti-HIV Agents*; Chu, C.K.; Cutler, H., Eds.; Plenum Press, Inc.: New York, NY, 1992; pp 223-238.
13. Tischler, M.; Cardellina, J.H., II; Cragg, G.M.; Boyd, M.R. *J. Nat. Prod.* **1992**, *55*, 667-671.
14. Gustafson, K.R.; Cardellina, J.H., II; Manfredi, K.R.; Beutler, J.A.; McMahon, J.B.; Boyd, M.R. In *Natural Products as Antiviral Agents*; Chu, C.K.; Cutler, H., Eds.; Plenum Press, Inc.: New York, NY, 1992; pp 57-67.
15. Gustafson, K.R.; Cardellina, J.H., II; McMahon, J.B.; Gulakowski, R.J.; Ishitoya, J.; Szallasi, Z.; Lewin, N.E.; Blumberg, P.M.; Weislow, O.S.; Beutler, J.A.; Buckheit, R.W., Jr.; Cragg, G.M.; Cox, P.A.; Bader, J.P; Boyd, M.R. *J. Med. Chem.* **1992**, *35*, 1978-1986.
16. Szallasi, Z.; Blumberg, P.M. *Cancer Res.* **1991**, *51*, 5355-5360.
17. Gustafson, K.R.; Munro, M.H.G.; Blunt, J.W.; Cardellina, J.H., II; McMahon, J.B.; Gulakowski, R.J.; Cragg, G.M.; Boyd, M.R.; Brinen, L.S.; Clardy, J.; Cox, P.A. *Tetrahedron* **1991**, *47*, 4547-4554.
18. Gustafson, K.R.; Blunt, J.W.; Munro, M.H.G.; Fuller, R.W.; McKee, T.C.; Cardellina, J.H., II; McMahon, J.B.; Cragg, G.M.; Boyd, M.R. *Tetrahedron* **1992**, *48*, 10093-10102.
19. Gustafson, K.R.; Cardellina, J.H., II; McMahon, J.B.; Pannell, L.K.; Cragg, G.M.; Boyd, M.R. *J. Org. Chem.* **1992**, *57*, 2809-2811.
20. Kashman, Y.; Gustafson, K.R.; Fuller, R.W.; Cardellina, J.H., II; McMahon, J.B.; Currens, M.J.; Buckheit, R.W., Jr.; Hughes, S.H.; Cragg, G.M.; Boyd, M.R. *J. Med. Chem.* **1992**, *35*, 2735-2742.
21. Manfredi, K.P.; Blunt, J.W.; Cardellina, J.H., II; McMahon, J.B.; Pannell, L.; Cragg, G.M.; Boyd, M.R. *J. Med. Chem.* **1991**, *34*, 3402-3405.
22. Bringmann, G.; Gedder, T.; Rübenacker, M.; Zagst, R. *Phytochemistry*, **1991**, *30*, 2067-2070.
23. Fuller, R.W.; Cardellina, J.H., II; Kato, Y.; Brinen, L.S.; Clardy, J.; Snader, K.M.; Boyd, M.R. *J. Med. Chem.* **1992**, *35*, 3007-3011.

RECEIVED March 26, 1993

Chapter 16

Discovery and Development of Novel Prototype Antibiotics for Opportunistic Infections Related to Acquired Immunodeficiency Syndrome

Alice M. Clark and Charles D. Hufford

Department of Pharmacognosy and Research Institute of Pharmaceutical Sciences, School of Pharmacy, University of Mississippi, University, MS 38677

The major AIDS-related opportunistic fungal and bacterial infections are candidiasis, cryptococcosis and mycobacteriosis. The need for new, more effective and less toxic antibiotics for the treatment of these infections is obvious in light of the significant toxicities and failure rates of the currently available agents. In the past, the discovery of new antibiotics has relied primarily on the isolation of such agents from natural sources. The major advantage of this approach is the likelihood of identifying new prototype drugs with different chemical structures, and hence, possible new mechanisms and less likelihood of similar toxicities and cross-resistance. Although in the past microorganisms have been the primary source of new antibiotics, higher plants are now recognized as important sources of new antimicrobial agents. Recent efforts to discover new prototype antibiotics with potential utility specifically for the treatment of opportunistic disseminated mycoses and mycobacteriosis are discussed. Initial *in vitro* evaluation of higher plant extracts for antifungal and antimycobacterial activity is followed by fractionation and purification of active extracts using a bioassay-directed scheme. Pure compounds with significant *in vitro* activity are evaluated for *in vivo* efficacy in established animal models of disseminated mycosis and mycobacteriosis in order to determine their potential clinical utility.

Acquired immunodeficiency syndrome (AIDS) is an immune disorder resulting from the destruction of cell-mediated immunity by the virus known as Human Immunodeficiency Virus (HIV). Patients with AIDS are particularly susceptible to the development of certain opportunistic infections (OI) that serve as effective indicators of AIDS. Opportunistic infections associated with AIDS can be categorized on the basis of the causative microorganism, i.e., protozoal, viral, bacterial or fungal infections. Prior to the AIDS epidemic, the management of disseminated OI in immunosuppressed patients involved the reduction of immunosuppressive therapy coupled with chemotherapy for the infectious disease. Since the course of immunosuppression in AIDS patients cannot currently be halted or reversed, the only recourse for these individuals is treatment of the OI, even though such infections are a reflection of a more complicated underlying immune disorder. Unfortunately immunosuppressed patients appear to be more resistant to conventional antibiotic therapy than normal individuals (*1*). Among the

0097–6156/93/0534–0228$06.00/0

most important OI are the disseminated mycotic infections cryptococcosis, candidiasis, and histoplasmosis (*2-4*). Disseminated infection with atypical mycobacteria is also now recognized as the major AIDS-related opportunistic bacterial infection (*5-10*).

Cryptococcosis, the most common systemic fungal infection in AIDS patients (*11-13*), affects 7-10% of all patients and usually occurs as meningitis due to *Cryptococcus neoformans* (*14-20*). Cryptococcal meningitis progresses rapidly to death if untreated, and even with treatment, mortality rates remain unacceptably high at ~50%. It has been noted that when cryptococcosis is the case-defining OI, the average life expectancy is only 2 months (*21*).

Another of the major AIDS-related disseminated fungal OI is candidiasis, which is due primarily to *Candida albicans* (*22-25*). Oral candidiasis (thrush) in AIDS patients is an early indicator of imminent progression to full-blown AIDS and occurs in >75% of all AIDS patients at some time. Progression to the more severe esophageal and GI candidiasis is associated with difficulty in swallowing and hence is a cause of dehydration, noncompliance with medication regimens, general malnutrition, etc. (*25*). Although progression to or development of systemic candidiasis is less common, incidence in AIDS patients has been reported as high as 10% (*26,27*).

Currently only four clinically useful antifungal agents are indicated for the treatment of disseminated mycoses and these fall into three structural classes, each with a different molecular target: polyene antibiotics, flucytosine, and synthetic azoles. The polyene antibiotic, amphotericin B (AMB), was the first systemic antifungal antibiotic to be used clinically, and after more than 20 years of use remains the most effective therapy for disseminated mycoses (*3, 4, 24-27, 28-35*). AMB acts by binding sterols in the fungal cell membrane. There are, however, significant drawbacks to its therapy, the most serious being renal damage, which occurs in > 80% of patients and can be permanent in patients receiving large doses of the drug (*36*). The combination of AMB and flucytosine, a synthetic nucleoside that inhibits protein synthesis, is designed to reduce the dosage of AMB (thereby reducing its dose-related toxicity) and to eliminate the development of resistance to flucytosine [an estimated 5% of *C. neoformans* strains exhibit primary resistance to flucytosine and secondary resistance develops rapidly during therapy (*18*)]. However, it has been noted that flucytosine toxicity (leukopenia or thrombocytopenia) may increase when it is used in combination with AMB (*18*), due in part to AMB-induced nephrotoxicity, and hence, lower renal clearance of flucytosine (*28-35*).

The synthetic imidazole antifungals were introduced over 10 years ago as broad spectrum, topically effective agents that act by inhibition of cytochrome P450 sterol demethylase. Ketoconazole, the first orally absorbed antifungal agent for certain systemic mycoses, was later found to be mostly ineffective for disseminated mycoses in immunocompromised patients (*32*). Fluconazole was introduced in the U.S. in 1990 and offers a number of advantages over the older imidazole (*37*). However, its major drawback is its fungistatic mechanism of action, thus requiring immunocompromised hosts to be maintained on daily lifetime therapy. Further, many clinicians doubt the wisdom of initial therapy of acute cryptococcal meningitis with fluconazole alone. A recent study compared AMB/flucytosine combination therapy to fluconazole alone for acute cryptococcal meningitis and concluded that the AMB/flucytosine combination is superior (*38*). However, comparative studies of AMB and fluconazole in AIDS cryptococcal meningitis reported a success rate of only about 50% with either regimen (*39*). Further, a recent investigation into the potential synergistic efficacy (in a murine model) of fluconazole with flucytosine has opened the possibility of the combination of fluconazole with other known antifungal agents in an attempt to improve its success rate, particularly in acute cryptococcal meningitis (*40*). At present, there is still some controversy regarding the most appropriate therapy of cryptococcal infections (*41*).

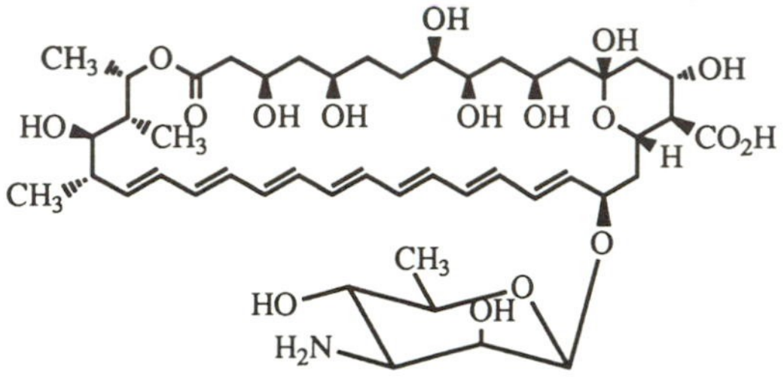

Amphotericin B

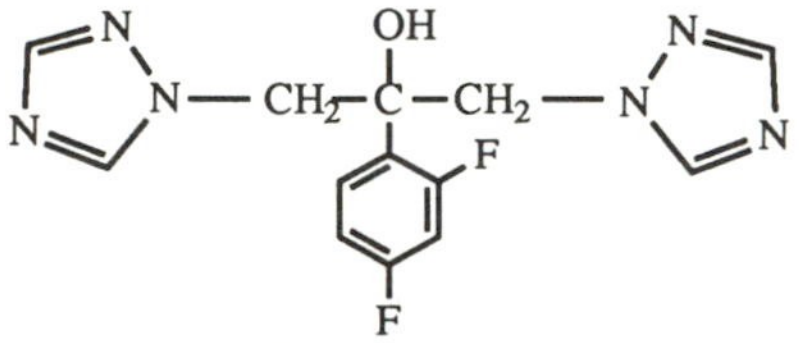

Fluconazole

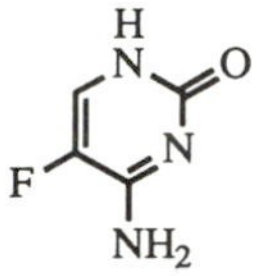

Flucytosine

Both the polyene antibiotics and synthetic azoles are utilized in the therapy of candidiasis, the choices largely dependent upon the extent of the disease. Disseminated candidiasis requires parenteral amphotericin B, sometimes in combination with flucytosine. Flucytosine alone is ineffective for systemic candidiasis. There have been reports of polyene- and azole-resistant strains of *Candida* and it has been noted that candidiasis due to species other than *C. albicans* (with varying susceptibilities to both azoles and polyenes) is an emerging problem (*23, 42-44*).

The treatment of AIDS-related fungal OI is complicated by several factors, including the relative ineffectiveness of available agents, the relatively severe toxicities of such agents, the development of resistance to existing agents, and the clinical status of the patient suffering from an underlying immune disorder. Clearly, the therapy of the AIDS-related fungal OI is inadequate and attention must immediately be focused on the discovery and development of potential new therapies. Each of these systemic antifungal agents suffers from rather significant shortcomings. The problems most often cited are toxicity (particularly regarding amphotericin B and flucytosine), lack of effectiveness (particularly for flucytosine and ketaconazole), and the development of resistance (especially for flucytosine and the azoles). The development of strains resistant to ketaconazole is well documented; however, it is not yet known whether these strains will be cross-resistant with fluconazole or whether resistance development to fluconazole will be a clinically significant problem. In the past, efforts to develop new chemotherapies for fungal systemic infections have focused primarily on the use of existing agents in some manner, i.e.:

1. Combinations of existing drugs to maximize synergistic effects
2. Derivatization of existing drugs to improve the therapeutic ratio
3. Improved delivery of existing drugs to the target site
4. Synthesis of new analogs of existing classes of drugs
5. Discovery of new prototype classes of antibiotics

Some of the most intensive efforts to develop new antifungal agents have centered on the synthesis of analogs of the existing synthetic azole antifungal agents. However, since some of the features related to biological or therapeutic effect are undoubtedly due to skeletal structure, such features tend to remain consistent throughout the class. For example, the azoles presumably all act by the same mechanism, i.e., inhibition of sterol demethylase. While the various azole analogs may exhibit varying degrees of affinity and selectivity for the molecular target, the fundamental mechanism remains the same. Thus, all of the azoles are fungistatic in their action, requiring long-term (lifetime) use. Therefore, it seems unlikely that efforts to modify the structures of existing agents will provide any **new** products that do not share similar toxicity or resistance problems.

The pivotal role of prototype agents in the future of drug discovery in general, and antibiotics in particular, is crucial. Novel antibiotics with novel chemical structures can serve two important functions: (a) as "lead" compounds for structure-activity relationship studies (SAR) and subsequent development of improved agents and (b) as probes for new molecular targets. Historically, most **prototype** bioactive substances have been natural products and it is logical that the search for new prototype antibiotics for AIDS-related OI should also include natural products. The major advantage of this approach over chemical synthesis or modification of existing agents is the likelihood of identifying **novel prototype** drugs with unique and different chemical structures, and hence, less likelihood of similar toxicities, cross-resistance, or even mechanism of action. Over the last several decades the search for antibiotics has been limited primarily to microorganisms, particularly the streptomycetes and some fungi.

In designing a search for novel prototype antibiotics for AIDS-related OI, it seems reasonable to assume that if new agents are to be found that have different structures with different or supplemental activities from the ones in current use or development, then a source other than the more traditional microorganisms must **also** be investigated. In particular, the **higher plants** are a logical choice, chiefly because of their seemingly infinite variety of novel organic molecules, which are often referred to as "secondary metabolites". Antifungal agents are widely distributed among the higher plants (*45-47*), but very few have been evaluated for their activity against human pathogenic fungi and essentially none of these have been evaluated in animal models of disseminated mycoses.

In our program of antifungal drug discovery, we are searching for natural product prototype antibiotics effective against the important AIDS-related opportunistic fungal and bacterial pathogens. Such prototypes, by definition, will consist of novel structures, but is our hope that they also will possess novel mechanism(s) of action. In any event, the identification of a novel structure, with or without a novel mechanism of action, provides a prototype agent from which a new class of antifungal antibiotics may be derived. By the same token, compounds which possess novel mechanism(s) of action, regardless of whether they become clinically useful, may serve as important biochemical tools to probe the critical biochemical pathways of the opportunistic pathogens.

Program Approach

For several years our group has pursued a program designed to detect, isolate, and structurally characterize novel antifungal (and, more recently, antimycobacterial) antibiotics from higher plants. In this program, we utilize a tiered screening system, as represented in Scheme 1. The process begins with sample acquisition. In our program, sources of samples are plants, microorganisms or pure compounds. Efforts have focused primarily on plants because these have not been extensively investigated as sources of novel antifungal antibiotics. Samples are acquired and subjected to evaluation in a primary *in vitro* assay which assesses inhibitory activity against the target pathogens, i.e., *Candida albicans*, *Cryptococcus neoformans*, *Mycobacterium intracellulare*, *Aspergillus fumigatus*, and *Aspergillus flavus*. This assay is rapid, high capacity, sensitive, convenient and cost-effective. The purpose of the primary *in vitro* qualitative assay is simply to identify samples which possess activity. Active plant extracts are then subjected to bioassay-directed fractionation in order to isolate the active constituents. Once the pure active compounds are isolated, then two processes begin simultaneously. One is a secondary assay in which the activity is corroborated and quantified and the spectrum of activity is established. At the same time the structure of the active compounds is determined and structural modifications and total synthesis are carried out. Finally, the most promising candidates move into the last tier of the assay system in which the *in vivo* efficacy and toxicity of the compounds are evaluated. In the following pages is a brief discussion of the individual aspects of each tier of this screening program.

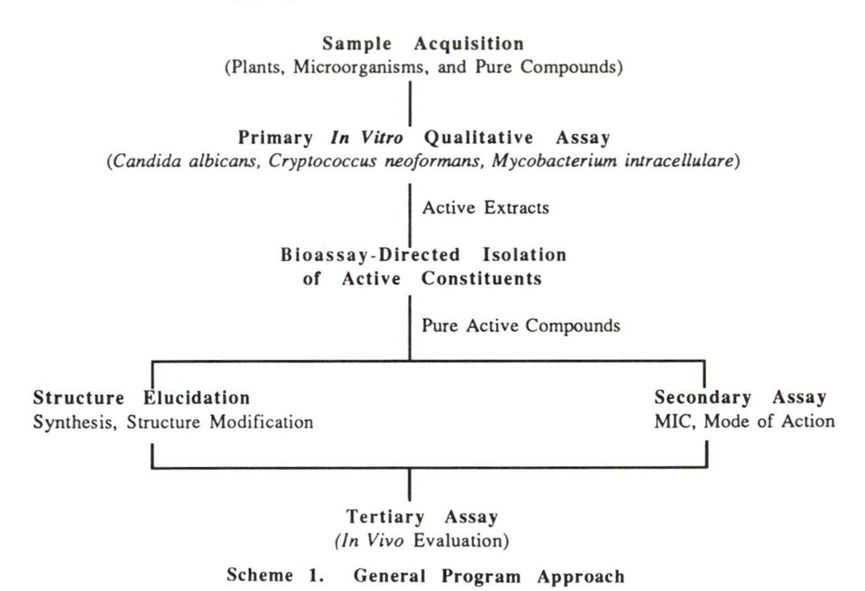

Scheme 1. General Program Approach

Sample Acquisition. The possible sources of novel prototype antifungal agents are natural products or synthetic compounds. Natural sources include, for example, plants, microorganisms, or marine invertebrates. Typically, in the past, microorganisms have been the source of antibiotics. However, in our program to identify prototype agents, higher plants are investigated as a potential source of antibiotic substances. The selection of plants for evaluation may be based on folkloric or traditional medicinal uses for disease states that can be interpreted as infectious diseases. For example, the utilization of a particular plant for the treatment of such ailments as fever, diarrhea, or skin abrasions may be due to the presence of antimicrobial constituents. Additionally, random collection and evaluation may also lead to the identification of active plants. The collected plant part is air-dried, ground, and percolated exhaustively with 95% ethanol. The ethanolic extract is then subjected to antimicrobial evaluation, using the primary agar well-diffusion assay described below.

Primary *In Vitro* Qualitative Assay. The initial evaluation for *in vitro* antimicrobial activity is accomplished by use of an agar-well diffusion assay (*48,49*). It is this assay procedure which initiates the discovery of novel agents. The agar well-diffusion assay is a rapid, sensitive, high capacity assay that simply, but effectively, assesses the presence or absence of desired antibiotic activity against specific target pathogens. New leads are identified when positive inhibitory activity is observed. For the primary assay, test organisms are grown in an enriched liquid culture medium for an appropriate incubation period (e.g., 24 hr at 37° for *C. albicans*). Inoculum is prepared from the fresh culture and used to inoculate the surface of agar petri plates. A small well (typically 11 mm diameter) is then cut into the center of the agar. Solutions or suspensions of test substances (plant extracts, column fractions, or pure compounds) are placed into the wells and the plates are incubated at an appropriate temperature. During the incubation period, the material in the well diffuses into the surrounding agar medium, and after an appropriate period, the plates are observed for the presence or absence of a "zone of inhibition" surrounding the well, i.e., where the antibiotic substance has diffused into the medium and inhibited the growth of the test microorganism. Those samples that exhibit a zone of inhibition are selected for further study. In particular, plant extracts that demonstrate reproducible zones of inhibition are subjected to bioassay-directed fractionation in order to isolate the active constituents.

Isolation of Active Constituents. The isolation of the pure active constituent(s) from a plant extract is accomplished using bioassay-directed fractionation procedures. In this technique, each fractionation step is monitored by bioassay, and only active fractions are pursued for further fractionation. There are several advantages to this approach: (a) there is a higher probability of identifying novel prototype antibiotics, (b) all active constituents are isolated, and (c) it is an efficient and effective means to identify the active constituents. Techniques utilized in the isolation of active constituents include standard chromatographic methodologies, solvent partitionings, and crystallization.

A perpetual problem in the search for new biologically active natural products is the question concerning the identification of known active agents, prior to purification, so that the same agent is not repeatedly isolated, a process often referred to as "dereplication". Dereplication is a major problem in certain areas, particularly in the isolation of antitumor agents from natural sources and in the isolation of antibacterial antibiotics from soil microorganisms. However, there are three important factors that significantly reduce the problem of dereplication in the search for novel antifungals and antimycobacterials from higher plants. First, in the past, most antibiotic substances isolated were active against "normal" pathogenic bacteria. It has been clearly demonstrated that activity against these bacteria does not translate into antimycobacterial activity. Otherwise, known agents would be active against *Mycobacterium avium-intracellulare* and this is not the case. Likewise, relatively few antifungal agents are reported. Second, the source, higher plants, is quite different from the more traditional source, microorganisms. Since it is highly unlikely that the same compounds would be isolated from these two vastly different sources, the problem of dereplication is reduced significantly. In other words, there are simply not very many reports of antifungal/antimycobacterial agents isolated from higher plants; thus it is not likely one will "reisolate" what has not yet been isolated. Finally, the novelty of chemical structure is less important than is the novelty of biological activity. It is in fact possible that a known compound will be isolated, but in the absence of any reported antibiotic activity, such a compound nevertheless represents a **prototype** antibiotic. In consideration of these facts, the problem of "dereplication" for this program is essentially reduced to a careful review of the combined literature on the plant species being investigated and the biological activity in question. Since our goal is to identify prototypic antifungal and antimycobacterial substances, the lack of novelty of chemical structure alone is insufficient cause to discard a lead. Clearly, if a plant species or constituent has been previously reported to possess antifungal or antimycobacterial activity, then it receives a lower priority or is discarded as a lead. Based on the knowledge of the plant species, the HPLC, TLC, and/or bioautographic properties of the extract, and review of the literature, a determination of whether the antimicrobial activity is due to a known substance can be made relatively rapidly. If so, obviously, the extract is not be pursued.

Structure Elucidation. Solvent partitioning of extracts followed by suitable chromatographic techniques monitored by bioassay at every stage allow for the isolation of the active principle(s). Once the active principle(s) has been isolated in pure form physical data are obtained (m.p. or b.p., optical rotation). Structure elucidation is accomplished primarily by analyses of spectroscopic data such as proton and carbon nuclear magnetic resonance (NMR) spectra, infrared (IR) spectra, ultraviolet (UV) spectra, mass spectra and circular dichroism (CD) spectra. High field superconducting broad band NMR (300 MHz ^{1}H and 75 MHz ^{13}C) spectroscopy are used extensively to gain evidence for the skeleton of new compounds via a variety of two-dimensional heteronuclear and homonuclear correlated experiments (e.g., ^{1}H-^{1}H COSY, ^{1}H-^{13}C HETCOR, INADEQUATE, NOESY, or, as needed, HMBC and HMQC). IR and UV spectra supplement the NMR data and provide information regarding the types of functional groups present and the types of chromophores present, respectively. High-resolution mass spectrometry gives the molecular ion, if a parent ion is present, and pertinent fragmentation patterns. If an optically active chromophore is present in the molecule, spectral data from a circular dichroism spectropolarimeter is used to obtain information about stereochemistry. If the compound is a new natural product, confirmation of the assigned structure may require chemical conversions, synthesis, or x-ray diffraction analysis. Once the chemical structure of the active principle has been established then suitable derivatives may also be prepared and tested. Also, all derivatives that are prepared for the structural studies should be routinely tested for antifungal and antimycobacterial activity. This may establish structural features essential for activity. When feasible, proposed structures should be confirmed by total synthesis. However, natural products are generally complex structures with many asymmetric centers and total synthesis is often not readily achievable. Often, however, structural modifications of the natural products or natural product derivatives is possible and, when all of these analogs are evaluated for antimicrobial activity, some preliminary assessment of structure activity relationships can be made.

Secondary Assay. The secondary assay is designed to corroborate the activity observed in the primary assay, to determine the relative potency of the compound by quantification of the activity, to determine the mode of action, and to identify candidates for the next tier of the process, i.e., *in vivo* studies. Generally, quantitation of antibiotic activity refers to a determination of the minimum inhibitory concentration (MIC), i.e., the lowest concentration of the compound that is required to inhibit the growth of the pathogen. The MIC is determined by a two-fold serial broth dilution assay and is expressed in µg/ml. In the two-fold serial broth dilution assay, a serial dilution of the test compound in culture broth provides a series of tubes in which each tube contains half the concentration of the test compound in the preceding tube. Each tube is inoculated with the same volume of inoculum of test organism and incubated under identical conditions. Following appropriate incubation, the minimum concentration of test compound required to inhibit the growth of the test organism is determined (typically, the growth of the test organisms can be easily assessed by visual inspection). The lowest concentration at which there is no growth is recorded as the MIC. In addition to providing the MIC, this assay can be extended to provide information regarding the mode of action, i.e., fungicidal vs. fungistatic or bacteriocidal vs. bacteriostatic, by subculturing from the MIC tubes that show no growth onto drug-free agar medium. For those compounds that are *static* in their mode of action, cultivation onto drug-free medium will result in the growth of the organism once again. Conversely, those that are *cidal* will continue to show no growth. Therefore, the secondary assay establishes both the relative potency (MIC), as well as the mode of action. Information regarding the relative potency and the mode of action, taken together with the availability of the compound, is important in identifying candidates for further biological studies, in particular, *in vivo* efficacy and toxicity.

Tertiary Assay. The assessment of the potential clinical utility of any compound can only be made by evaluation of its activity in an appropriate animal model of the disease state. Additionally, an assessment of general toxicity can be established by monitoring animals during the treatment protocol. Compounds that possess promising *in vitro* activity and are available in sufficient quantity are subjected to evaluation for *in vivo* efficacy and toxicity in appropriate animal models. The efficacy of novel prototype antibiotics is established in murine models of disseminated infection which measure either tissue burden of the organism in target organs or survival time following challenge with a lethal dose of the test pathogen. Typically, mice are infected intravenously with a sufficient dose of the pathogenic organism to induce a lethal disseminated infection. The test drug is then administered intraperitoneally at varying doses and following the infection. In the tissue burden model, mice are sacrificed several days after infection allowing for several doses of the test compound. The target organs are asceptically removed, homogenized and a portion of the homogenates is serially diluted and plated, in triplicate, on appropriate agar medium. These agar plates are then incubated and the number of colonies, each of which arises from a single cell or *colony forming unit* (CFU), are counted and the total CFU/gm of tissue is calculated. The values for drug-treated mice are then compared with vehicle -treated infected control mice using a non-parametric rank sum test ($p < 0.05$ as a test of significance).

In the lethality model, the drugs are administered daily for up to 21 days and animals are monitored daily for lethality. The rate of deaths and percent lethality are compared between vehicle-treated and drug-treated groups using ANOVA or Chi square test.

Antifungal Alkaloids of *Cleistopholis patens*

Over the past eight years, we have used the programmatic approach described above to identify a number of prototype classes of antifungal antibiotics. What follows is a summary of the work accomplished to date on one plant, *Cleistopholis patens* and the identification of the antifungal alkaloids of this plant. *Cleistopholis patens* is a West African medicinal plant of the Annonaceae family. The plant was collected in 1985 on the campus of the University of Ile-Ife in Nigeria. The root bark (3.8 kg) was defatted with *n*-hexane, extracted exhaustively, and the ethanolic extract was partitioned between ethyl acetate and water. The antimicrobial activity was found to reside in the organic layer which, following chromatography over silica gel afforded 5 pooled fractions designated A-E (Scheme 2). Fraction B was antimicrobially active and was subjected to further chromatography over alumina to yield eight fractions, three of which exhibited interesting antifungal activity (B2, B4, B6).

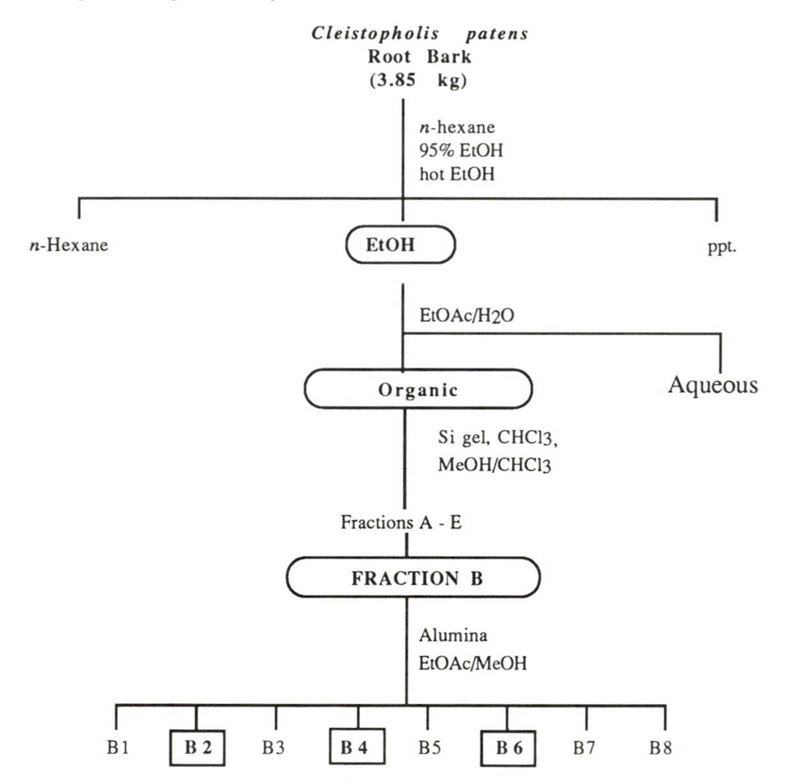

Scheme 2. Fractionation of *Cleistopholis patens*

From fraction B6, the known oxoaporphine alkaloid, liriodenine, was isolated. Liriodenine had been isolated by us earlier as the antifungal constituent in the heartwood of *Liriodendron tulipifera* L. (*49*) and was well known in the literature as a phytochemical constituent of other plants as well; however, its antifungal activity was first reported by Hufford *et al.* (*49*).

From fraction B2 the known azafluoranthene alkaloid, eupolauridine, was isolated (*50*); however, this was the first report of its broad spectrum antifungal activity. It should be noted that eupolauridine was isolated in very small quantities and in order to complete its biological evaluation it was necessary to synthesize additional quantities. The synthesis of eupolauridine had been reported by Bowden *et al.* (*51*) and this procedure was used to prepare sufficient quantities for further biological studies. During the course of the synthesis, the intermediate aza-fluorenone, onychine, was also found to possess significant antifungal activity (*50*). Both eupolauridine and onychine exhibit low acute toxicity in mice and have been demonstrated to show *in vivo* efficacy in murine candidiasis (*52*).

Liriodenine Eupolauridine Onychine

From the third active fraction, B4, the novel copyrine alkaloid, 3-methoxysampangine, was isolated as the antifungal constituent (*53*). Once again, 3-methoxysampangine was isolated in very low yield and it was necessary to prepare additional quantities of this material for further biological evaluation. As part of efforts directed toward the total synthesis of 3-methoxysampangine, the parent unsubstituted alkaloid, sampangine, as well as several other A- and B-ring analogs were prepared. Sampangine had been previously reported as a phytochemical constituent of *Cananga odorata* (*54*), but no biological activity was reported. The general synthetic approach used for the syntheses of substituted sampangines is shown in Scheme 3. In addition, the annelated analog, 4,5-benzosampangine, was prepared utilizing a different synthetic route (Scheme 4).

The *in vitro* antifungal activities of the sampangines are summarized in Table 1. In addition, based on secondary assay results, it appears that sampangine acts by a fungicidal mode of action. The results of time-to-kill studies with sampangine are shown in Fig. 1. Following exposure of 10^6 cells *C. albicans* in 1 ml culture broth to 1 μg of sampangine for 24 hr, no viable cells could be recovered and cultured (Fig 1).

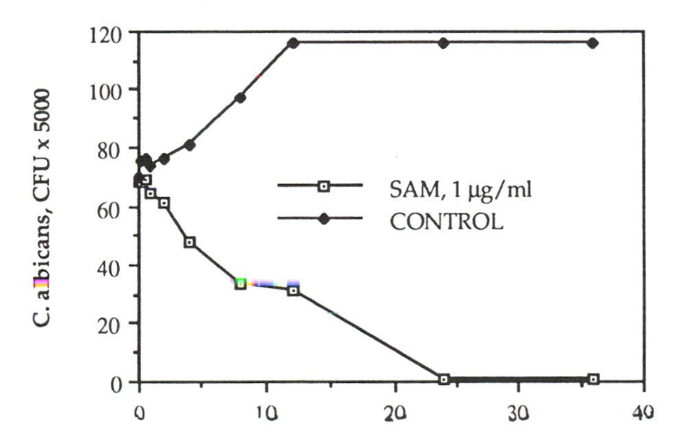

Figure 1. Time-curve lethality study of Sampangine

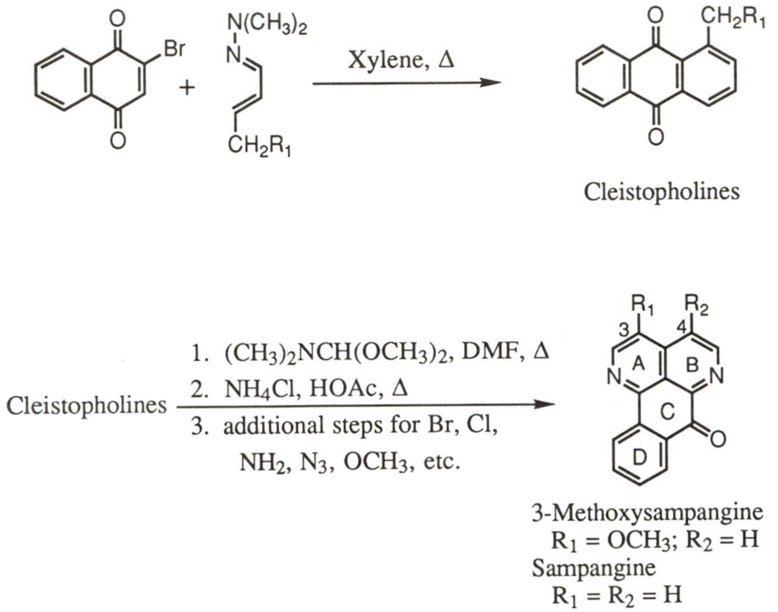

O
Br
O
N(CH3)2
N
+
CH2R1
Xylene, Δ
O CH2R1
O
Cleistopholines

1. $(CH_3)_2NCH(OCH_3)_2$, DMF, Δ
2. NH_4Cl, HOAc, Δ
Cleistopholines
3. additional steps for Br, Cl, NH_2, N_3, OCH_3, etc.

R1 R2
3 4
A B
N N
C
O
D

3-Methoxysampangine
$R_1 = OCH_3$; $R_2 = H$
Sampangine
$R_1 = R_2 = H$

Scheme 3 - General Synthetic Route for Sampangines

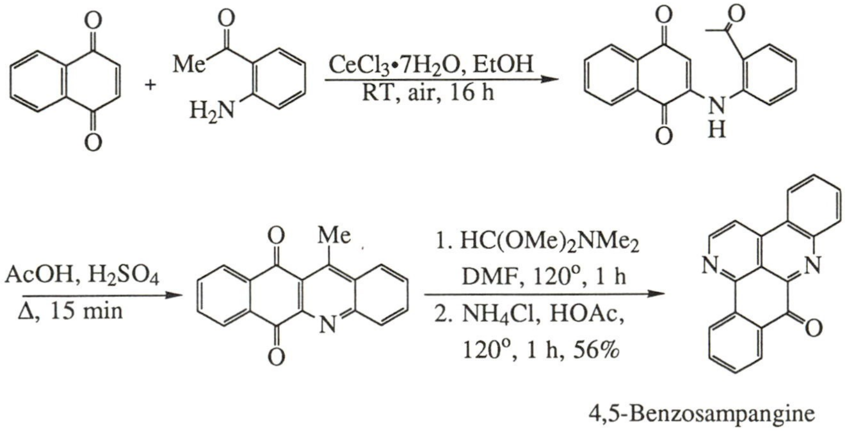

O
O
O
Me
+
H_2N
$CeCl_3{\bullet}7H_2O$, EtOH
RT, air, 16 h
O O
O
N
H

AcOH, H_2SO_4
Δ, 15 min
O Me
N
O
1. $HC(OMe)_2NMe_2$
DMF, 120°, 1 h
2. NH_4Cl, HOAc,
120°, 1 h, 56%
N
N
O

4,5-Benzosampangine

Scheme 4 - Synthesis of 4,5-Benzosampangine

Table 1 - *In Vitro* Antimicrobial Activity* of Sampangines (SAM)

Sample \ Organism	*Candida albicans*	*Cryptococcus neoformans*	*Aspergillus fumigatus*	*Mycobacterium intracellulare*
SAM	1.6	0.8	1.6	0.8
3-MeO-SAM	1.6	0.8	1.6	1.6
3-Me-SAM	0.4	0.4	0.8	0.4
4-Br-SAM	12.5	0.02	50	3.1
4-Cl-SAM	50	0.1	ND§	3.1
4-MeO-SAM	3.1	1.6	ND	3.1
4-NH_2-SAM	100	25	ND	3.1
4-N_3-SAM	25	6.3	ND	ND
BZ-SAM	0.4	<0.20	ND	ND
Amphotericin B	0.4	0.4	0.4	ND

*MIC is measured in micrograms per milliliter.
§ND means not determined.

Summary

The AIDS crisis has served to focus renewed interest in the discovery of agents to treat opportunistic infections in immunocompromised hosts. These infections rarely occur in patients with normally functioning immune systems, but are a leading cause of morbidity and mortality in AIDS patients. Opportunistic infections also occur commonly in patients with any type of immunosuppression, i.e., cancer patients and organ transplant recipients, both of whom receive drugs which suppress the immune system and therefore leave these patients susceptible to life-threatening opportunistic infections. The therapy of opportunistic infections is complicated by several factors, including the relative ineffectiveness of available agents, the relatively severe toxicities of such agents, the development of resistance to existing agents and the clinical status of the patient suffering from an underlying immune disorder. All of these factors contribute to the urgent need for the discovery and development of totally new, prototype antibiotics which may serve as "leads" for the development of new classes of chemotherapeutic agents which do not share the same toxicities and cross-resistance of the existing agents. In the past, natural products have been a rich source of such compounds. Higher plants are being investigated as potential new sources of antibiotics in a program of research that includes the collection of plant materials from locations worldwide, detection of biological activity in crude natural product preparations, the isolation and structural characterization of the active principles from these sources, the synthesis of the bioactive natural products, and the initial evaluation of efficacy and toxicity of active substances in animals so that a preliminary assessment of potential clinical utility can be made.

Acknowledgments: The authors gratefully acknowledge support of this work by the National Institutes of Health, National Institute of Allergy and Infectious Diseases, Division of AIDS, Grant RO1-AI-27094 and Contracts NO1-AI-42549 and NO1-AI-72638.

Literature Cited:

1. AMA Drug Evaluation; 5th ed.; American Medical Association: Chicago, Illinois, **1983**; pp 1779-1788.
2. Jaffe, H.W.; Bergman, D.J.; Selik, R.M. *J. Inf. Dis.* **1983** *148*, 339-345.
3. Leoung, G; Mills, J., Eds., *Opportunistic Infectiouns in Patients with the Acquired Immunodeficiency Syndrome*, Marcel Dekker, New York, **1989**.
4. Vanden Bossche, H; Mackenzie, D.W.R.; Cauwenbergh, G.; Van Custem, J.; Drouhet, E.; Dupont, B., Eds., *Mycoses in AIDS Patients*, **1990**, Plenum Press, NY.
5. Hawkins, C.C.; Gold, J.W.M.; Whimbey, E.; Kiehn, T.E.; Brannon, P.; Cammarata, R.: Brown, A.E.; Armstrong, D. *Ann. Intern. Med.* , **1986**, *105*, 184-188.
6. Green, J.B.; Sidhu, G.S.; Lewin, S.; Levine, J.F.; Masur, H.; Simberkoff, M.S.; Nicholas, P.; Good, R.C.; Zolla-Pazner, S.B.; Pollock, A.A.; Tapper, M.L.; Holzman, R.S. *Ann. Intern. Med.* **1982**, *97*, 539-546.
7. Zakowski, P.; Fligiel, S.; Berlin, O.G.W.; Johnson, B.L. *J. Am. Med. Assoc.*, **1982**, *248*, 2980-2982.
8. Poon, M. -C.; Landay, A.; Prasthofer, E.F.; Stagno, S. *Ann. Intern. Med.*, **1983**, *98*, 287-290.
9. Macher, A.M.; Kovacs, J.A.; Gill, V.; Roberts, G.D.; Ames, J.; Park, C.H.; Straus, S.; Lane, H.C.; Parrillo, J.E.; Fauci, A.S.; Masur, H. *Ann. Intern. Med.*, **1983**, *99*, 782-785.
10. Woods, G.L.; Washington, J.A. *Rev. Inf. Dis.*, **1987**, *9*, 275-294.
11. Dismukes, W.E. *J. Inf. Dis.*, **1988**, *157*, 624-628.
12. Larsen, R.A. *J. Inf. Dis.*, **1990** *162*: 727-730.
13. Sugar, A.M.; Stera, J.J.; Dupont, B. *Rev. Inf. Dis.*, **1990**, *12* (S3): S338-S348.
14. Furio, M.M.; Wordell, C.J. *Clin. Pharm.* **1985**, *4*, 539-554.
15. Whimbey, E.; Gold, J.W.M.; Polsky, B.; Dryjanski, J.; Hawkins, C.; Blevins, A.; Brannon, P.; Kiehn, T.E.; Brown, A.E.; Armstrong, D. *Ann. Intern. Med.* **1986**, *104*, 511-514.
16. Armstrong, D.; Gold, J.W.M.; Dryjanski, J.; Whimbey, E.; Polsky, B.; Hawkins, C.; Brown, A.E.; Bernard, E.; Kiehn, T.E. *Ann. Intern. Med.* **1985**, *103*, 738-743.
17. Zuger, A.; Louie, E.; Holzman, R.S.; Simberkoff, M.S.; Rahal, J.J. *Ann. Intern. Med.* **1986**, *104*, 234-240.
18. Drouhet, E.; Dupont, B. *Rev. Inf. Dis.* **1987**, *9* (Suppl. 1), S4-S14.
19. Jaffe, H.W.; Bergman, D.J.; Selik, R.M. *J. Inf. Dis.* **1983**, *148*: 339-345.
20. Dismukes, W.E.; Cloud, G.; Gallis, H.A.; Kerkering, T.M.; Medoff, G.; Caraven, P.C.; Kaplowitz, L.G.; Fisher, J.F.; Gregs, C.R.; Bowles, C.A.; Shadomy, S.; Stamm, A.M.; Diasio, R.B.; Kaufman, L.; Soon, S.-J.; Blackwelder, W.C. *N. Engl. J. Med.* **1987**, *317*, 334-341.
21. Symposium "Management of Fungal Infections in the Immunocompromised Host" presented following the 30th ICAAC, October **1990**.
22. Bodey, G.P. *Med. Clin. North Amer.* **1988**, *72*, 637-659.
23. Saral, R. *Rev. Inf. Dis.* **1991**, *13*, 487-492.
24. Meyer, R.D.; Holmberg, K. "Fungal Infections in HIV-Infected Patients" in *Diagnosis and Therapy of Systemic Fungal Infections*, Holmberg, K; Meyer, R.; Eds., Raven Press, New York, **1989**, pp 79-100.
25. Greene, S.I. "Treatment of Fungal Infections in the Human Immunodeficiency Virus-Infected Individual" in *Antifungal Drug Therapy: A Complete Guide for the Practitioner*, Jacobs, P.H.; Nall, L.; Eds., Marcel Dekker, Inc., New York, **1990,** pp 237-246.
26. Sobel, J.D. In *Mycoses in AIDS Patents*, Eds. Vanden Bossche, H.; Mackenzie, D.W.R.; Cauwenbergh, G.; Van Custem, J.; Drouhet, E.; Dupont, B. Plenum Press, New York, **1990**, pp 93-100.

27. Threlkeld, M.G.; Dismukes, W.E. In *Opportunistic Infections in Patients with the Acquired Immunodeficiency Syndrome*, Leoung, G; Mills, J.; Ed., Marcel Dekker, Inc.: New York, **1989**, pp 285-314.
28. Rinaldi, M.G. "Histoplasma in AIDS Patients" in *Mycoses in AIDS Patients*; Vanden Bossche, H.; Mackenzie, D.W.R; Cauwenbergh, G.; Van Custom, J.; Drouket, E.; Dupont, B.; Eds, Plenum Press, New York, **1990**,pp 163-169.
29. Graybill, J.R. *J. Inf. Dis.* **1988**, *158*, 623-626.
30. "Discussion: Fungal Infections in the Immunocompromised Host," *Rev. Inf. Dis.* **1991**, *13*: 504-508.
31. Daneshmend, T.K.; Warnock, D.W. *Clin. Pharmacokin.* **1983**, *8*, 17-42.
32. Medoff, G.; Brajtburg, J.; Kobayashi, G.S.; Bolard, J. *Ann. Rev. Pharmacol. Toxicol.* **1983**, 303-330.
33. LeFrock, J.L.; Smith, B.R. *Clin. Pharmacol.*, **1984**, *30*, 162-167.
34. King, K. Ed., *Med. J. Austr.*, **1985**, *143*, 287-290.
35. Young, L.S. *Rev. Inf. Dis.*, **1985**, *7* (suppl. 3): S380-S388.
36. Facts and Comparisons, Kastrup, E.K.; Boyd, J.R. Eds., G.H. Schwach, St. Louis, MO, **1980**, 356.
37. Marriott, M.S.; Richardson, K. In *Recent Trends in The Discovery, Development and Evaluation of Antifungal Agents*, Fromtling, R.A; Ed., J.R. Prous Science Publishers: Barcelona, Spain, **1987**, pp 81-92.
38. Larsen, R.A.; Leal, M.A.E.; Chan, L.S. *Ann. Intern. Med.* **1990**, *113*, 183-187.
39. Dismukes, W.G.; Cloud, G.; Thompson, S.; Sugar, A.; Tuazon, C.; Kaufman, C.; Grieco, M.; Jacobson, J.; Powderly, W.; Robinson, P. *29th* Interscience Conference on Antimicrobial Agents and Chemotherapy, **1989**, Abst. #1065.
40. Allendoerfer, R.; Marquis, A.J.; Rinaldi, M.G.; Graybill, J.R.; *Antimicrob. Agents Chemother.* **1991**, *35*, 726-729.
41. Drutz, D.J.; *Rev. Inf. Dis.* **1987**, *9*, 417-420.
42. Rodriguez-Tudela, J.L.; Laguna, F., Martinez-Suarez, J.V.; Chaves, F; Dronda, F. 32nd Interscience Conference on Antimicrobial Agents and Chemotherapy, **1992**, Abst. #1204.
43. Cairns, M.R. *Electron Microscopy Tech.* **1988**, *8*: 115-131.
44. Rippon, J.W. In *Antifungal Drug Therapy*, Jacobs, P.H.; Nall, L. Eds., Marcel Dekker, New York, **1990** ,pp 247-256.
45. Oliver-Bever, B. *J. Ethnopharmacol.*, **1983**, *9*:1-83.
46. Caceres, A.; Jauregui, E.; Herrera, D.; Logemann, H.*J.Ethnopharm.* **1991**, *33*, 277-283.
47. Dimayuga, R.E.; Garcia, S.G. *J. Ethnopharm.* **1991** *31*, 181-192.
48. Clark, A.M.; El-Feraly, F.S.; Li, W.-S. *J. Pharm. Sci.*, **1981**, *70*, 951-952.
49. Hufford, C.D.; Fundeburk, M.J.; Morgan, J.M.; Robertson, L.W. *J. Pharm. Sci.*, **1975**, *64*, 789-792.
50. Hufford, C.D.;Liu, S.; Clark, A.M.; Oguntimein, B.O. *J. Nat. Prod.* **1987**, *50*, 961-964.
51. Bowden, B.F.; Picker, K.; Ritchie, E.; Taylor, W.C. *Aust. J. Chem.*, **1975**, *28*, 2681-2701.
52. Clark, A.M.; Hufford, C.D. 29th Interscience Conference on Antimicrobial Agents and Chemotherapy, Sept. 17-20, Houston, TX, **1989**, Abstract No. 720.
53. Liu, S.; Oguntimein, B.; Hufford, C.D.; Clark, A.M. *Antimicrob.Agents Chemother.* **1990**, *34*, 529-533.
54. Rao, J.U.M.; Giri, G.S.; Hanamaiah, T.; Rao, K.V. *J. Nat. Prod.*, **1986**, *49*, 346-347.

RECEIVED March 5, 1993

Chapter 17

Artemisia annua

From Weed to Respectable Antimalarial Plant

Daniel L. Klayman[†]

Division of Experimental Therapeutics, Walter Reed Army Institute of Research, Washington, DC 20307–5100

In the early 1970's, a potent and essentially non-toxic antimalarial agent was isolated from *qing hao* (*Artemisia annua*), a plant used in Chinese folk medicine for some 20 centuries. The active compound, artemisinin (qinghaosu), is a sesquiterpene lactone bearing an unusual endoperoxide group that is essential for its activity. Artemisinin is effective against drug-sensitive and -resistant *Plasmodium falciparum* and *P. vivax*. Oil-soluble derivatives of artemisinin, such as artemether and arteether, and water-soluble derivatives, such as sodium artesunate and sodium artelinate, have increased potency and allow fewer recrudescences. The latter two have special application for the treatment of potentially fatal cerebral malaria. Recent animal studies suggest that the artemisinin class of compounds may be effective when administered transdermally.

Malaria is caused by protozoa of the genus *Plasmodium* that enter the circulatory system from the salivary glands of infected female *Anopheles* mosquitos. These insects bite in order to obtain a blood meal during the reproductive phase of their lives. The four species of protozoa that can cause disease in humans are *Plasmodium vivax*, *P. malariae*, *P. ovale*, and *P. falciparum*. Malaria is clinically manifested by recurrent fever, chills, sweating, dizziness, headache, and gastrointestinal disturbances. An advanced stage of the disease caused by *P. falciparum*, called cerebral malaria, can result in the patient lapsing into a coma and, ultimately, to die.

The World Health Organization (WHO) has estimated that annually more than 270 million people acquire malaria, and that 2-3 million die from it, some 1 million of whom are children. The need for new antimalarials is urgent inasmuch as an antimalarial that had widespread usage, chloroquine, is failing in many parts of the world due to the appearance of resistant *P. falciparum* strains. Fortunately, a totally new class of antimalarials from the People's Republic of China, based on a plant-derived constituent, is on the horizon.

[†]Author deceased.

The *Qing Hao* Plant and Artemisinin

For about 20 centuries, Chinese herbalists have known about the medicinal value of the weed, *qing hao*, the source of the new antimalarial agent (*1*). In the *Handbook of Prescriptions for Emergency Treatments* of 340 AD, the plant was prescribed as a treatment for fevers. The feverish patient was advised by the author, Ge Hong, to take a handful of sweet wormwood, soak it in one *sheng* (*ca.* 1 liter) of water, squeeze out the juice, and drink it all (*2*). The sixteenth century herbalist/ pharmacologist Li Shizhen mentions in his *Encyclopedia of Materia Medica* (*3*) the use of *huang hua hao* (meaning yellow flower, identified later as *Artemisia annua*) for children's fever and *qing hao* (*A. apiacea*) for use in malaria. (Most references to *qing hao* today mean *A. annua*.)

In the late 1960's, an effort was made in China to evaluate several traditional herbal remedies for their present day validity. Although *qing hao* was one of them, a hot tea made from the plant disappointingly lacked antimalarial activity. In 1971, a low-temperature extraction with diethyl ether yielded an extract that gave positive antimalarial results in infected mice and monkeys. A biological activity-directed isolation of the specific constituent responsible for the antimalarial properties yielded in 1972 (*2,4*) a compound that was named artemisinin. Alternative names given by the Chinese are *qinghaosu* (meaning extract of the *qing hao* plant), arteannuin, and artemisinine. Although there is an impression that Chinese workers were the first to isolate artemisinin, it appears more likely that Jeremić and co-workers (*5*) from Belgrade, Yugoslavia, actually isolated the material first; however, they reported an incorrect ozonide structure for it.

Artemisinin, after testing in animals, was administered to humans infected with malaria and was found to be an effective blood schizontocide with virtually no toxicity. By the end of 1972, artemisinin and derivatives had already been used in ten regions of China by some 6,000 patients (*2*). Chemically, artemisinin is unlike any previous antimalarial agent, because virtually all of the rest have a nitrogen-containing heterocycle, whereas artemisinin is a sesquiterpene lactone that incorporates an endoperoxide moiety. The latter is a group that is infrequently encountered in natural products. The subject of artemisinin as an antimalarial was reviewed in detail earlier (*6-10*), as well as in briefer surveys (*11-16*). In addition, peroxides with potential application as antimalarials have also been reviewed (*17,18*). A comprehensive updated review is expected to be published in the near future by Klayman and Ziffer (*19*).

Like most naturally-occurring therapeutic agents, artemisinin exists in the plant in very small concentrations. Chinese workers were unable to find artemisinin in about 30 other *Artemisia* species, whereas our group at the Walter Reed Army Institute of Research has studied some 70 species and did not find artemisinin in any of them. A Japanese variety of *A. apiacea* was reported to contain 0.08% w/w of artemisinin (*20*) and recently, it was announced by Luo *et al.* (*21*) that artemisinin was isolated from *A. lancea* together with two hydroperoxide compounds.

Artemisia annua belongs to the large Compositae family. The genus *Artemisia* has over 300 species, many of which have been used as spices, to repel insects, and for their essential oils. *Artemisia annua*, in addition to flourishing throughout much of the People's Republic of China, is found in many places in the temperate zone. In the United States, the weed grows primarily along rivers, although it is not believed to be native to this country. The plant is known here by the common names of annual wormwood, sweet wormwood, and sweet Annie. It is also found in Argentina, and grows wild or is cultivated in many European countries: the former Yugoslavia, Hungary, Bulgaria, Romania, as well as Italy, France (*22*), Spain, Turkey (*23*), and the former Soviet Union (*24*). *Artemisia annua* is cultivated commercially in limited quantities in the U.S. for ornamental purposes and in Hungary for its volatile oil (*25,26*).

The yield of artemisinin from plant material of various localities in the People's Republic of China has been determined (*27-29*). It was observed that plants from the area of Shin Yan produce up to 0.947% w/w, whereas plants from the U. S. (*30*) and India (*31*) have been found to contain considerably less. The growth of *A. annua* in tissue culture has been investigated by Nair *et al.* (*32*), Martinez *et al.* (*33,34*), and Kudakasseril *et al.* (*35*), and others (*36-40*).

Isolation of Artemisinin from the Plant

Although artemisinin was obtained originally in Yugoslavia or the People's Republic of China in the early 1970's, the first published laboratory procedure by Klayman *et al.* (*41*) for its isolation appeared in 1984. Air-dried leaves of *A. annua* were extracted with petroleum ether (bp 30-60°C), which was subsequently removed *in vacuo*. The residue was redissolved in chloroform to which acetonitrile was added to precipitate inert plant components such as sugars and waxes. The concentrated extract was then chromatographed on a column of silica gel. Fractions with a high artemisinin content crystallized readily, with recrystallization being effected with cyclohexane or 50% ethanol. A slightly modified isolation procedure has been patented by El-Feraly and ElSohly (*42*). Acton *et al.* (*43*) employed the Ito multilayer coil separator-extractor to aid in the isolation of artemisinin. A large-scale extraction technique, i.e., starting with 400 g of leaves, has also been described (*44*).

Artemisinin is a poorly water-soluble, crystalline compound whose melting point is 156-157°C, and $[\alpha]_D^{25}$ +66.3° (*c* 1.64, $CHCl_3$) (*4*). Elemental analysis and mass spectrometry led to the empirical formula, $C_{15}H_{22}O_5$ (molecular weight 282), which suggested that it is a sesquiterpene. Its ir spectrum shows absorption bands at 1,745 cm^{-1}, for a δ-lactone (verified by alkali titration), and at 831, 881, and 1,115 cm^{-1}, which were attributed to a peroxide group and verified by titration with triphenylphosphine (*45*). Its ^1H-nmr spectrum shows the presence of one tertiary and two secondary methyl groups and one acetal proton, and its ^{13}C-nmr spectrum includes four methylene carbons, three methine carbons and two quaternary carbons, with one carbonyl carbon (*1,46*). The structure and stereochemistry of artemisinin (**1**) were confirmed by X-ray crystallography (*47*). Sodium borohydride reduced the lactone to a lactol without affecting the peroxide moiety, whereas hydrogenation in the presence of palladium on calcium carbonate removed one of the peroxide oxygens (*48*).

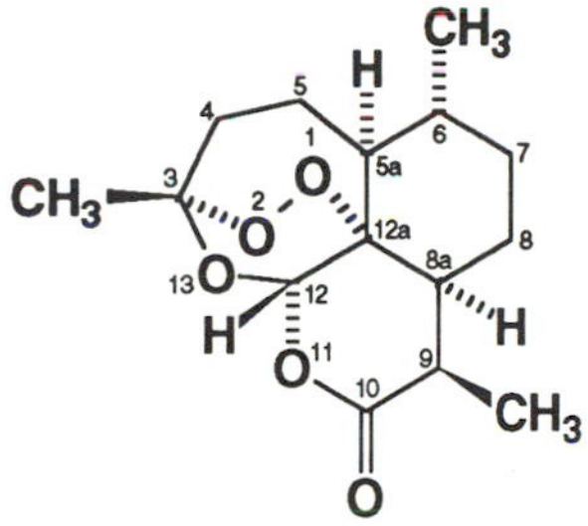

1

The discovery in the People's Republic of China of the utility of artemisinin as an antimalarial has led to intensified interest in the identification of the other constituents of *A. annua*. The components of the essential oil of *A. annua* produced in Hungary (*25,26*), in Oregon (*49*), in the former USSR (*50-52*), Bulgaria (*53-58*), the People's Republic of China (*59*), Mongolia (*60*), and Vietnam (*61*) have been studied. The main volatile constituent of the plant has been identified as artemisia ketone (3,3,6-trimethyl-1,5-heptadien-4-one), occurring to the extent of 35.7% w/w (*49*). In addition, one finds artemisia alcohol, artemisilactone (*62*), 1,8-cineole, α-pinene, β-pinene, myrcene, sabinene, pinocarvone, *trans*-pinocarveol, germacrene D (*49*), borneol acetate, cuminal, β-caryophyllene, stigmasterol, camphene, cadinene, camphor, β-farnesene, isoborneol, *p*-cymol, and β-myrcene (*58,63,64*).

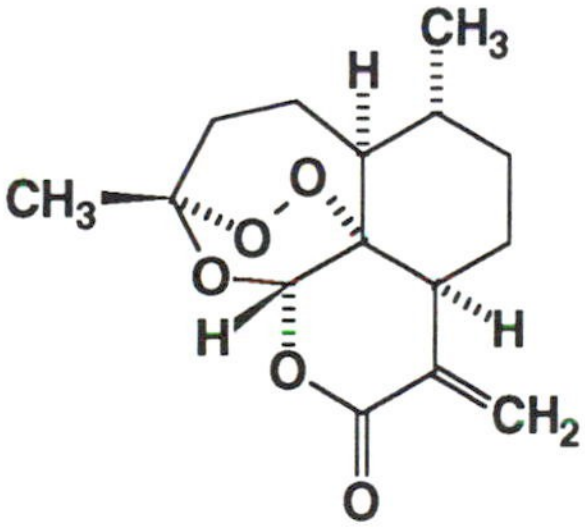

2

Artemisitene (**2**), an endoperoxide closely resembling artemisinin, detected by HPLC-EC, was isolated in low yield from *A. annua* during the flowering season (*65*). It has an *exo*-methylene group in place of the 9-methyl and has been synthesized from artemisinin (*66*) and artemisinic acid (*67*). Its antimalarial activity *in vitro* is lower than that of artemisinin (Milhous W. K.; Walter Reed Army Institute of Research; unpublished results). Dihydroartemisitene, i.e., artemisitene lactol, was found to isomerize on silica gel (*66*). Acton *et al.* (*30*), using HPLC in conjunction with an electrochemical detector, were able to detect and quantify several chemically reducible compounds besides artemisinin. Two of these compounds were later isolated and characterized by Rücker *et al.* (*68*) as α- and β-myrcene hydroperoxides.

The most abundant sesquiterpene in *A. annua* is artemisinic acid (arteannuic acid, qinghao acid) (*69-71*), contained at 8-10 times the abundance of artemisinin (*70,72*). Artemisinic acid, a useful synthon, has been converted into artemisinin (*73,74*), artemisitene (*67*), desoxyartemisinin (deoxyartemisinin) (*75*), deoxoartemisinin (*76*), homodeoxoartemisinin (*77*), and arteannuin B (*78*).

Arteannuin B (qinghaosu-II) (*63,71,79,80*) is probably the second most plentiful sesquiterpene in *A. annua*. In addition, the plant contains the related compound *epi*-deoxyarteannuin B (*81,82*) and numerous other polycyclic compounds: 6,7-dehydroartemisinic acid (*82*), epoxyarteannuic acid (*83,84*), qinghaosu-I (*63*), qinghaosu-III (desoxyartemisinin) (*71*), qinghaosu-IV (4-hydroxydesoxyartemisinin) (*71*), and qinghaosu-V (*71*). The polycyclic sesquiterpenes belong to the amorphane (cadinene, *cis*-decalin) class of terpenes, the subject of a recent review (*85*).

Among the non-terpenes isolated from the plant are two paraffinic alcohols (*71*), coumarins (*63,86*), flavonols (*63,71,87-90*), esters (*66*), phenols (*91*), an alkaloid (aurantiamide acetate) (*92*), and β-glycosidases (*93*).

Thus far, four *de novo* syntheses of artemisinin have been accomplished, each of which requires a final photooxidative step. Schmid and Hofheinz (*94*), the first to synthesize the molecule, did so in 1982 from (-)-isopulegol. In 1986, Xu *et al.* (*95*) reported a synthesis that began with (+)-citronellal. Avery *et al.* (*96*) used 3(*R*)-methyl-6-(phenylsulfoxy)cyclohexanone as a chiral building block, whereas Ravindranathan *et al.* (97) used (+)-car-3-ene. Because of their complexity, the *de novo* syntheses do not appear to be practical or affordable sources of artemisinin.

Antimalarial Properties of Artemisinin

In 1982, Guan *et al.* (*98*), studying artemisinin inhibition of *P. falciparum in vitro*, found an ED_{50} of 1.99 ng/ml, about the same ED_{50} as that of chloroquine. The rapid inhibition by artemisinin of the incorporation of [^{3}H]-isoleucine by *P. falciparum* was reported in 1983 (*99*). In the United States, the initial verification of the antimalarial activity of artemisinin was performed by Milhous (*100*) using the *in vitro* semi-automated micro-dilution method of Desjardins *et al.* (*101*). The IC_{50} (the concentration of artemisinin that inhibited [^{3}H]-hypoxanthine uptake by 50%) was equal for both the Camp (chloroquine-sensitive) and Smith (chloroquine-resistant) strains of *P. falciparum* (*41*). On-going evaluations performed against the Indo-China (W-2) and Sierra Leone (D-6) strains typically yield IC_{50}'s of 0.66 ng/ml and 2.93 ng/ml, respectively (*102*).

As alluded to earlier, the isolation of artemisinin was guided by results obtained from the *P. berghei*-infected mouse test system (*2*). It was noted that parasites in the blood were cleared by comparatively low doses of artemisinin given orally (*4*) and were later observed to be eliminated by 50 mg/kg of artemisinin given by the oral route daily for three days, to yield an overall ED_{50} of 138.8 mg/kg (*4*). Artemisinin, in a single dose, was not found to be active when administered in the typical Rane screen (University of Miami) using *P. berghei*-infected mice; however, when three doses were given 24 hours apart (suspended in peanut oil), artemisinin cured all five mice given 160 mg/kg (three times), and three of five mice at 40 mg/kg (three times) (*102*). In healthy monkeys infected with sporozoites of *P. cynomolgi* and given 200 mg/kg of artemisinin orally, the parasites were cleared in 2-3 days, but the parasitemia recurred in 5/7 of the animals (*4*).

In a 1979 study of 207 human cases of *P. vivax* and *P. falciparum* malaria treated in the People's Republic of China, artemisinin in oil was found to be rapidly acting, effective, and without side effects (*103*). In 1975, all 2,069 patients that were treated with different artemisinin formulations were clinically cured (*4*). Among them were 1,511 *P. vivax* and 588 *P. falciparum* patients that included 143 chloroquine-resistant cases and 141 cases of cerebral malaria. The drug was formulated and administered as a tablet, oil solution, oil suspension, or water suspension. The recrudescence rate within one month averaged 21.4% in the *P. vivax* group, and 10.5% (oil suspension, i.m. injection) and 85% (tablet) in the *P. falciparum* group. Clearance was substantially more rapid than with chloroquine (*4*).

Toxicity of Artemisinin

In an early acute toxicity study of artemisinin, the LD_{50} for mice was 5,105 mg/kg, orally, and 2,800 mg/kg, i.m., as an oil suspension. Acute toxicity of artemisinin in rats, cats, and dogs by various routes indicated that their EEG, ECG, and hepatic functions were unaffected (*4*). The main side effects were to the hemopoietic system; however, these were reversible. It was also shown that artemisinin in mice and rats, when administered in the middle and late stages of gestation, is embryotoxic, i.e., all fetuses were absorbed (*104*). This effect may preclude the administration of artemisinin-related compounds to pregnant women. It should be noted, however, that in China, six pregnant women were treated with artemether without the mothers or progeny suffering ill effects (*105*). In the Ames test, artemisinin was found to be non-mutagenic (*104,106*).

Derivatives of Artemisinin

The occurrence of recrudescences and the poor oil- and water-solubility of artemisinin has stimulated the preparation of some 100 semi-synthetic compounds, mainly in the People's Republic of China and the United States. They all retain the endoperoxide group, which has been determined as being essential for activity (*1*). Many of the derivatives are more potent than the parent compound. The preparation of virtually all modifications depends on reducing the sole amenable functional group of artemisinin, the lactone, to a lactol and altering the hydroxyl group. Overall, the order of activity was: carbonates > esters > ethers > artemisinin (*48,107,108*). In general, alkylation of dihydroartemisinin (**3a**) in the presence of an acidic catalyst gives products with predominantly β-orientation, whereas acylation in alkaline medium preferentially yields α epimers. Among the most promising derivatives obtained are dihydroartemisinin (**3a**), artemether (**3b**), arteether (**3c**), artesunic acid (**3d**), and artelinic acid (**3e**).

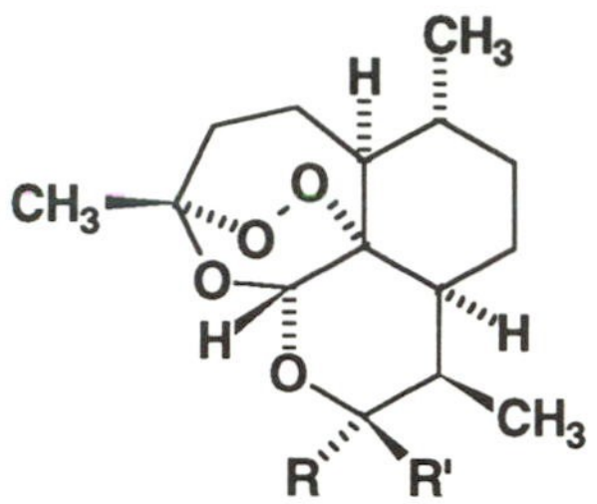

	R	R'
3a	H	OH
	OH	H
3b	H	OCH_3
3c	H	OC_2H_5
3d	$OCOCH_2CH_2COOH$	H
3e	$OCH_2C_6H_4COOH$	H

Artemether (**3b**), the methyl ether of dihydroartemisinin, was synthesized by treating the latter with methanol in the presence of boron trifluoride-etherate (*48,107*). The β anomer (mp 86-88°C) had an SD_{90}=1.02 mg/kg against chloroquine-resistant *P. berghei*, whereas the α anomer (mp 97-100°C) had an SD_{90}=1.16 mg/kg, making each of them considerably more potent than artemisinin (SD_{90}=6.2 mg/kg) (*109*). When artemether was tested as an oil solution in *P. cynomolgi*-infected rhesus monkeys, it was considerably more effective than artemisinin (*109*). In human trials in China reported in 1981 (*110*), artemether had rapid therapeutic action against chloroquine-resistant *P. falciparum* malaria. In 1985, a clinical comparison was made in 61 patients of i.m. artemisinin and artemether, and oral chloroquine (*111*). The artemether in oil solution, administered in a 200 mg dose daily for three days, gave a radical cure rate of 95.7%, whereas artemisinin gave 75%, and chloroquine (1.5 g of base) gave 11.1%. In Thailand, i.m. artemether cured virtually all *P. falciparum* patients from low- and high-resistance areas (*112*). Artemether in oil solution, contained in sealed glass ampoules, is now available commercially in the People's Republic of China.

The synthesis of arteether (**3c**) [β anomer (mp 81-83°C), SM227] was reported by Chinese workers (*48,113*) and again by Brossi *et al.* (*114*). The diastereomeric mixture of α- (an oil) and β-anomers was obtained in a *ca.* 75:25 ratio. The latter derivative was shown to have slightly lower antimalarial activity in chloroquine-resistant *P. berghei* than does artemether. The α-anomer was made stereoselectively and quantitatively from dihydroartemisinin by the addition of ethyl iodide in the presence of freshly prepared Ag_2O at room temperature (*115*). The name arteether without any further designation is assumed to be the β-anomer. Although artemether had been already used effectively in man, arteether is the oil-soluble derivative that is being developed by the WHO (*114*). No important therapeutic or toxicologic differences have been found thus far, between this compound and artemether (Shmuklarsky, M. J.; Klayman, D. L.; Milhous, W. K.; Rossan, R. N.; Ager, A. L., Jr.; Tang, D. B.; Heiffer, M. H.; Canfield, C. J.; Schuster, B. C.; Walter Reed Army Institute of Research; submitted for publication).

Ester derivatives of dihydroartemisinin were prepared by treating the latter with either acid chlorides or acid anhydrides (*108*) in pyridine or with an acid in the presence of dicyclohexylcarbodiimide (*38,48*) or 4-dimethylaminopyridine (DMAP) (*107,116-118*). The most important ester derivative is artesunic acid (**3d**), the half succinic acid ester (hemi-succinate) of dihydroartemisinin, whose sodium salt is readily water soluble. Artesunic acid, mp 138-139°C, is prepared by treating dihydroartemisinin with succinic anhydride in the presence of DMAP (*48*). The ED_{50} of sodium artesunate (804-Na, sodium dihydroartemisinin hemisuccinate) *in vitro* against *P. falciparum* is 0.14 ng/ml (*98*). In a comparison of the efficacy and toxicity of i.v. sodium artesunate *vs.* artemisinin and chloroquine, sodium artesunate was found to be at least five times more suppressive than artemisinin against chloroquine-sensitive and -resistant *P. berghei* in mice (*119*). The compound is well tolerated in test animals, but is nevertheless more toxic than artemisinin. However, it is responsible for far fewer deleterious effects to the heart than chloroquine (*119*). In a comparative study, it was demonstrated by Li *et al.* (*120*) that among artemisinin, quinine, chloroquine, and sodium artesunate, the latter was most rapidly acting in man against chloroquine-resistant *P. falciparum* cerebral malaria, causing patients to regain consciousness typically after 12 hours. Sodium artesunate for injection is now available commercially in the People's Republic of China packaged in two vials: the first contains artesunic acid and the second, 5% sodium bicarbonate solution. The prescribed dosage regimen is 60 mg (or 1.2 mg/kg) for adults to be administered intravenously or intramuscularly, to be repeated 4, 24, and 48 hours after the initial dose. Tablets of artesunic acid are also available in China and are recommended for both the treatment and prophylaxis of *P. falciparum* and *P. vivax*. Sodium artesunate in rats, rabbits, and dogs was found to be transformed into dihydroartemisinin and thus may be considered to be a pro-drug (*121*). Sodium artesunate hydrolyzes at the ester function in alkaline solution after a few hours causing the precipitation of insoluble dihydroartemisinin.

A series of dihydroartemisinin derivatives were made in which the linkage of the water-solubilizing function to dihydroartemisinin was by an ether, rather than an ester, functional group (*104,122*). The most active of this series is artelinic acid (**3e**) (*102*). Artelinic acid was prepared from dihydroartemisinin by reaction with methyl *p*-(hydroxymethyl)benzoate in the presence of boron trifluoride-etherate and the resulting condensation product was then de-esterified with methanolic alkali. Sodium artelinate compares favorably with sodium artesunate both *in vitro* against *P. falciparum* and *in vivo* against *P. berghei*. It has the advantage of being considerably more stable in solution than the former, with virtually no decomposition being detectable after three months (*102*).

Dihydroartemisinin (**3a**), made by the $NaBH_4$ reduction of artemisinin and obtained as a mixture of epimeric lactols (*102*), has been found to be a distinctly more active antimalarial than artemisinin (*48*). Against *P. falciparum*, dihydroartemisinin is reported to be *ca.* 200 times more active than artemisinin *in vitro* and 1.7 times more active against *P. berghei in vivo* (*123*). As indicated above, evidence points to dihydroartemisinin as the main metabolite in the body. The thermal decomposition products of artemisinin (*124*) and dihydroartemisinin have been identified by Lin *et al.* (*125*). Dihydroartemisinin was assayed by means of GC-MS by quantifying one of the pyrolysis products, (2*S*,3*R*,6*R*)-2-(3-oxobutyl)-3-methyl-6-[(*R*)-2-propanal]cyclohexanone (*126*).

Other promising artemisinin derivatives are: the trimethylsilyl ether of dihydroartemisinin (*127*), deoxoartemisinin, i.e., the compound in which the lactone carbonyl of artemisinin is replaced by CH_2 (*128*), and 12-(3'-hydroxy-*n*-propyl)deoxoartemisinin, reported to be five times more potent *in vitro* than artemisinin (*129*) and about 80 times more potent than homodeoxoartemisinin (*77*).

The combined use of artemether, sulfadoxine, pyrimethamine, and primaquine in the treatment of chloroquine-resistant *P. falciparum* malaria has been studied by Naing and co-workers (*130*). Antagonism was shown in *P. berghei*-infected mice when artemisinin was combined with dapsone, sulfadoxine, pyrimethamine, pyrimethamine-sulfadoxine mixture, or cycloguanil, whereas potentiation occurred with primaquine (in a primaquine-resistant strain) and mefloquine (in a mefloquine -resistant strain) (*131*).

Several less typical pharmaceutical forms have been proposed for the administration of artemisinin and its derivatives. Sustained-release tablets of artesunic acid have been investigated and show promise (*132*). Artemisinin in suppositories was safely administered to 100 patients with *P. falciparum* malaria on Hainan Island, People's Republic of China, at a total dosage of 2,800 mg. All patients were cleared of their infection; however, the recrudescence rate was 46% in the patients who were followed up (*133*). This route was effective also for the four cerebral malaria patients in the study. Xuan *et al.* (*134*) showed that artemether and sodium artesunate were well absorbed through the skin to cure *P. berghei*-infected mice. They also demonstrated that sodium artesunate given transdermally over three days cured *P. cynomolgi*-infected rhesus monkeys (*135*). Klayman and co-workers (*136*), in prophylactic and curative experiments with artelinic acid given transdermally against *P. berghei* in mice, were able to prevent the establishment of the parasitemia, as well as to clear the mice of infection without recrudescences. Subsequent studies in mice with artemisinin and seven of its derivatives indicate that all of them are absorbed through the skin. Artemisinin appears to be the least effective (Klayman, D. L.; Ager, A. L., Jr.; Walter Reed Army Institute of Research; unpublished data), whereas several others are more potent than artelinic acid.

Acknowledgments

I thank the following co-workers and collaborators for their excellent cooperation and advice: Ai J. Lin, Nancy Acton Roth, John P. Scovill, Liang-Quan Li, Mark Hoch, Wilbur K. Milhous, Arba L. Ager, Jr., and Moshe Shmuklarsky.

Literature Cited

1. China Cooperative Research Group on Qinghaosu. *J. Trad. Chin. Med.* **1982**, *2*, 3-8.
2. Ximen, L. *China Reconstructs* **1979**, (8), 48.
3. Li, S.-C. *Chinese Medicinal Herbs*; Georgetown Press: San Francisco, California, 1973; p 50.
4. Qinghaosu Antimalaria Coordinating Research Group. *Chin. Med. J.* **1979**, *92*, 811-816.
5. Jeremić, D.; Jokić, A.; Behbud, A.; Stefanović, M. *8th Internat. Sympos. Chem. Nat. Prod.*; New Delhi, India; Feb. 1972; *Abstr.* C-57.
6. Klayman, D. L. *Science (Washington, D.C.)* **1985**, *228*, 1049-1055.
7. Klayman, D. L. *Natural History* **1989**, (10), 18-27.
8. Luo, X.-D.; Shen, C.-C. *Med. Res. Rev.* **1987**, *7*, 29-52.
9. Woerdenbag, H. J.; Lugt, C. B.; Pras, N. *Pharm. Weekblad, Scient. Ed.* **1990**, *12*, 169-181.
10. Zaman, S. S.; Sharma, R. P. *Heterocycles* **1991**, *32*, 1593-1638.
11. Kar, K.; Shankar, G.; Bajpai, R.; Dutta, G. P.; Vishwakarma, R. A. *Indian J. Parasitol.* **1988**, *12*, 209-212.
12. Koch, H. *Pharm. Internat.* **1981**, *2*, 184-185.
13. Fourth Meeting of the Scientific Working Group on the Chemotherapy of Malaria. WHO Report TDR/CHEMAL-SWG(4)/QHS/81.3: Beijing, People's Republic of China; 6 Oct. 1981.
14. Bruce-Chwatt, L. J. *Br. Med. J.* **1982**, *284*, 767-768.
15. Peters, W. *Experientia* **1984**, *40*, 1351-1357.
16. Khomchenovskii, E. I. *Zh. Vses. Khim. O-va. im D. I. Medeleeva* **1986**, *31*, 102-104; *Chem. Abstr.* **1986**, *105*, 90545v.
17. Vennerstrom, J. L.; Acton, N.; Lin, A. J.; Klayman, D. L. *Drug Dev. Deliv.* **1988**, *4*, 45-54.
18. Docampo, R.; Moreno, S. N. J. In *Free Radicals in Biology*; Pryor, W. A., Ed.; Academic Press: New York; Vol. 6; 1984; pp 243-288.
19. Klayman, D. L.; Ziffer, H. *Prog. Chem. Org. Nat. Prod.*, in press.
20. Liersch, R.; Soicke, H.; Stehr, C.; Tullner, H.-U. *Planta Med.* **1986**, *52*, 387-390.
21. Luo, S.; Ning, B.; Hu, W.; Xie, J. *J. Nat. Prod.* **1991**, *54*, 573-575.
22. Desheraud, M. M.; Rochan, R. *Soc. Linn. Lyon Bull. Mens.* **1969**, *38*, 103.
23. Cubukcu, B.; Mericli, A. H.; Ozhatay, N.; Damadyan, B. *Acta Phar. Turc.* **1989**, *31*, 41-42; *Chem. Abstr.* **1989**, *111*, 12390a.
24. Shreter, A. I.; Rybalko, K. S.; Konovalova, O. A.; Derevinskaya, T. I.; Maisuradze, N. I.; Makarova, N. V. *Rastit. Resur.* **1988**, *24*, 66-72; *Chem. Abstr.* **1988**, *108*, 164792j.
25. Galambosi, B. *Herba Hung.* **1979**, *18*, 53-62.
26. Galambosi, B. *Acta Hort.* **1980**, 343.
27. Chen, H. R.; Chen, M.; Zhong, F.; Chen, F. *Chinese Med. Rept.* **1986**, *7*, 393-395.

28. Chen, H.; Chen, M.; Zhong, F.; Chen, F.; Huang, J.; Zhang, M.; Huang, J. *Zhongyao Tongbao* **1986**, *11*, 393-395; *Chem. Abstr.* **1986**, *105*, 149765x.
29. Luo, H.-M.; Chao, P.-P.; Yu, C.-C.; Tai, C.; Liu, C.-W. *Yao Hsueh T'ung Pao* **1980**, *15*, 8-10; *Chem. Abstr.* **1981**, *95*, 68092y.
30. Acton, N.; Klayman, D. L.; Rollman, I. *Planta Med.* **1985**, *51*, 445-446.
31. Singh, A.; Kaul, V. K.; Mahajan, V. P.; Singh, A.; Misra, L. N.; Thakur, R. S.; Husain, A. *Indian J. Pharm. Sci.* **1986**, *48*, 137-138.
32. Nair, M. S. R.; Acton, N.; Klayman, D. L.; Kendrick, K.; Basile, D. V. *J. Nat. Prod.* **1986**, *49*, 504-507.
33. Martinez, I.; Blanca, C. *Diss. Abstr. Int. B* **1989**, *50*, 138; No. DA8907392.
34. Martinez, B. C.; Lam, L.; Staba, E. J. *Proc. VIth Internat. Congr. Pl. Tiss. Cell Cult.* 1986; *Abstr.* p 352.
35. Kudakasseril, G. J.; Lam, L.; Staba, E. J. *Planta Med.* **1987**, *53*, 280-284.
36. Chen, H.-R.; Chen, H.-T.; Ho, S.-S. *Chung Yao T'uong Pao* **1983**, *8*, 5-6.
37. Chen, H.-R.; Chen, H.-T.; Ho, S.-S. *Chung Yao T'uong Pao*, **1984**, *9*, 7-8.
38. Chen, P. *Proc. VIth Internat. Congr. Pl. Tiss. Cell Cult.* 1986; *Abstr.* p 142.
39. Fulzele, D. P.; Sipahimalani, A. T.; Heble, M. R. *Phytother. Res.* **1991**, *5*, 149-153.
40. He, X.; Zeng, M.; Li, G.; Liang, X. *Zhiwu Xuebao* **1983**, *25*, 87-90; *Chem. Abstr.* **1983**, *99*, 50383s.
41. Klayman, D. L.; Lin, A. J.; Acton, N.; Scovill, J. P.; Hoch, J. M.; Milhous, W. K.; Theoharides, A. D.; Dobek, A. S. *J. Nat. Prod.* **1984**, *47*, 715-717.
42. El-Feraly, F. S.; ElSohly, H. S. *U.S. Pat.* 4,952,603; 28 Aug. 1990.
43. Acton, N.; Klayman, D. L.; Rollman, I. J.; Novotny, J. F. *J. Chromatog.* **1986**, *355*, 448-450.
44. ElSohly, H. N.; Croom E. M.; El-Feraly, F. S.; El-Sherei, M. M. *J. Nat. Prod.* **1990**, *53*, 1560-1564.
45. Liu, J.-M.; Ni, M.-Y.; Fan, J.-F.; Tu, Y.-Y.; Wu, Z.-H.; Wu, Y.-L.; Chou, W.-S. *Acta Chim. Sin.* **1979**, *37*, 129-143.
46. Blaskó, G.; Cordell, G. A.; Lankin, D. C. *J. Nat. Prod.* **1988**, *51*, 1273-1276.
47. Qinghaosu Research Group. *Scient. Sin.* **1980**, *23*, 380.
48. China Cooperative Research Group. *J. Trad. Chin. Med.* **1982**, *2*, 9-16.
49. Libbey, L. M.; Sturtz, G. J. *Essent. Oil Res.* **1989**, *1*, 201-202.
50. Dembitskii, A. D.; Krotova, G. I.; Kuchukhidze, N. M.; Yakobashvili, N. Z. *Maslo-Zhir. Prom.-st.* **1983**, 31-34; *Chem. Abstr.* **1983**, *99*, 27798x.
51. Kapelev, I. G. *Byul. Nikitsk. Botan. Sada* **1970**, 40.
52. Kapelev, I. G. *Ekol. Genet. Rast. i Zhivoynykh Tez. Dokl. Vses. Konf.* **1981**, 2.
53. Genov, N.; Georgiev, E. *Nauchni Tr.-Vissh. Inst. Khnit. Vkusova Prom-st., Plovdiv* **1983**, *30*, 141-148; *Chem. Abstr.* **1983**, *101*, 197919q.
54. Georgiev, E.; Genov, N. *Nauchni Tr.-Vissh. Inst. Khnit. Vkusova Prom-st., Plovdiv* **1983**, *30*, 149-160.
55. Toleva, P. D.; Ognyanov, I. V.; Karova, E. A.; Georgiev, V. *Riv. Ital. Essenze, Profumi, Piante Offic., Aromi, Saponi, Cosmet., Aerosol* **1975**, *57*, 620.
56. Tsankova, E.; Ognyanov, I., *Riv. Ital. Essenze, Profumi, Piante Offic., Aromi, Saponi, Cosmet., Aerosol* **1976**, *58*, 502-504.

57. Georgiev, E.; Genov, N., Lazarova, R.; Kuchkova, D.; Ganchev, G. *Int. Congr. Essential Oils*; Tokyo, 1979; pp 233-237; *Chem. Abstr.* **1979**, *92*, 64527j.
58. Georgiev, E.; Genov, N.; Khristova, N. *Rastenievud. Nauki* **1981**, *18*, 95-102; *Chem. Abstr.* **1982**, *97*, 52627g.
59. Liu, Q.; Yang, Z.; Deng, Z.; Sa, G.; Wang, X. *Zhiwu Xuebao* **1988**, *30*, 223-225; *Chem. Abstr.* **1988**, *109*, 134804t.
60. Satar, S. *Pharmazie* **1986**, *41*, 819-820.
61. Nguyen, X. D.; LeClerq, P. A.; Dinh, H. K.; Nuuyen M. T. *Tap Chi Duoc Hoc* **1990**, 11-13; *Chem. Abstr.* **1990**, *115*, 189455j.
62. Zhu, D.; Deng, D.; Zhang, S.; Xu, R. *Huaxue Xuebao* **1984**, *42*, 937-939; *Chem. Abstr.* **1985**, *102*, 21192d.
63. Tu, Y.; Ni, Y.; Zhong, Y.; Li, L.; Cui, S.; Zhang, M.; Wang, X.; Liang, X. *Yaoxue Xuebao* (*Acta Pharm. Sin.*) **1981**, *16*, 366-370; *Chem. Abstr.* **1982**, *97*, 52497q.
64. Tu, Y.; Ni, M.; Chung, Y.; Li, L. *Chung Yao T'ung Pao* **1981**, *6*, 31; *Chem. Abstr.* **1981**, *95*, 175616u.
65. Acton, N.; Klayman, D. L. *Planta Med.* **1985**, *51*, 441-442.
66. El-Feraly, F. S.; Ayalp, A.; Al-Yahya, M. A.; McPhail, D. R.; McPhail, A. T. *J. Nat. Prod.* **1990**, *53*, 920-925.
67. Haynes, R. K.; Vonwiller, S. C. *J. Chem. Soc., Chem. Commun.* **1990**, 451-452.
68. Rücker, G.; Mayer, R.; Manns, D. *J. Nat. Prod.* **1987**, *50*, 287-289.
69. Deng, A.; Zhu, D.; Jao, Y.; Dai, J.; Xu, R. *Kexue Tongbao* **1981**, *26*, 1209-1211; *Chem. Abstr.* **1981**, *96*, 40792y.
70. Roth, R. J.; Acton, N. *Planta Med.* **1987**, *53*, 501-502.
71. Tu, Y.; Ni, Y.; Zhong, Y.; Li, L.; Cui, S.; Zhang, M.; Wang, X.; Ji, Z. *Planta Med.* **1982**, *44*, 143-145.
72. Jung, M.; ElSohly, H. N.; McChesney, J. D. *Planta Med.* **1990**, *56*, 624.
73. Roth, R. J.; Acton, N. *J. Nat. Prod.* **1989**, *52*, 1183-1185.
74. Ye, B.; Wu, Y.-L. *J. Chem. Soc., Chem. Commun.* **1990**, 726-727.
75. Jung, M.; ElSohly, H. N.; Croom, E. M.; McPhail A. T.; McPhail, D. R. *J. Org. Chem.* **1986**, *51*, 5417-5419.
76. Jung, M.; Li, X.; Bustos, D. A.; ElSohly H. N.; McChesney, J. D. *Tetrahedron Lett.* **1989**, *30*, 5973-5976.
77. Bustos, D. A.; Jung, M.; ElSohly, H. N.; McChesney, J. D. *Heterocycles* **1989**, *29*, 2273-2277.
78. Jung, M.; Yoo, Y.; ElSohly, H. N.; McChesney, J. D. *J. Nat. Prod.* **1987**, *50*, 972-973.
79. Jeremić, D.; Jokić, A.; Behbud, A.; Stefanović, M. *Tetrahedron Lett.* **1973**, 3039-3042.
80. Stefanović, M.; Jokić, A.; Behbud, A; Jeremić, D. *8th Internat. Sympos. Chem. Nat. Prod.*; New Delhi, India; Feb. 1972; *Abstr. C-56*.
81. Roth, R. J.; Acton, N. *Planta Med.* **1987**, *53*, 576.
82. El-Feraly, F. S.; Al-Meshal, I. A.; Khalifa, S. I. *J. Nat. Prod.* **1989**, *52*, 196-198.
83. Wu, Z.-H.; Wang, Y.-Y. *Huaxue Xuebao* **1984**, *42*, 596-598; *Chem. Abstr.* **1984**, *101*, 207579.
84. Wu, Z.-H.; Wang, Y.-Y. *Acta Chim. Sin.* **1985**, *43*, 901-903.

85. Bordeloi, M.; Shukla, V. S.; Nath, S. C.; Sharma, R. P. *Phytochemistry* **1989**, *28*, 2007-2037.
86. Aleskerova, A. N. *Izv. Akad. Nauk Az. SSR, Ser. Biol. Nauk* **1985**, (2), 25-26; *Chem. Abstr.* **1985**, *103*, 85125n.
87. Baeva, R. T.; Nabi-Zade, L. I.; Zapesochnaya, G. G.; Karryev, M.O. *Khim. Prir. Soedin.* **1988**, 298-299; *Chem. Abstr.* **1988**, *109*, 70429q.
88. Djermanović, H. M.; Jokić, A.; Mladenović, S.; Stefanović, M. *Phytochemistry* **1975**, *14*, 1873.
89. Liu, K. C.-S.; Yang, S.-L.; Roberts, M. F.; Elford, B. C.; Phillipson, J. D. *Planta Med.* **1989**, *55*, 654-655.
90. Jeremić, D.; Stefanović, M.; Doković, D.; Milosavljević, S. *Glas. Hem. Drus., Beograde* **1979**, *44*, 615-618; *Chem. Abstr.* **1979**, *92*, 211806e.
91. Marco, J. A.; Sanz, J. F.; Bea, J. F; Barbera, O. *Pharmazie* **1990**, *45*, 382-383.
92. Tu, Y.-Y.; Yin, J.-P.; Gi, J.-P.; Li, H. *Zhongcaoyao* **1985**, *16*, 200-201; *Chem. Abstr.* **1985**, *103*, 76120e.
93. Lakhtin, M. V.; Magazova, N. S.; Mosolov, V. V.; Kuaneva, R. N. *Biotekhnol. Biotekh.* **1988**, (3), 40-41; *Chem. Abstr.* **1988**, *109*, 186050p.
94. Schmid, G.; Hofheinz, W. *J. Am. Chem. Soc.* **1983**, *105*, 624-625.
95. Xu, X.-X.; Zhu J.; Huang, D.-Z.; Zhou, W. S. *Tetrahedron*, **1986**, *42*, 819-828.
96. Avery, M. A.; Chong, W. K. M.; Jennings-White, C. *J. Am. Chem. Soc.* **1992**, *114*, 974-979.
97. Ravindranathan, T.; Kumar, M. A.; Menon, R. B.; Hiremath, S. V. *Tetrahedron Lett.* **1990**, *31*, 755-756.
98. Guan, W.-B.; Huang, H.-J.; Zhou, Y.-C.; Gong, J.-Z., *Acta Pharmacol. Sin.* **1982**, *3*, 139-141.
99. Gu, H. M.; Warhurst, D. C.; Peters, W. *Biochem. Pharmacol.* **1983**, *32*, 2463-2466.
100. Milhous, W. K.; Klayman, D. L.; Lambros, C. *Proc. XI Internat. Congr. Trop. Med. Malaria*; Calgary, Canada; Sept. 16-22, 1984.
101. Desjardins, R. E.; Canfield, C. J.; Haynes, D. E.; Chulay, J. D. *Antimicrob. Agents Chemother.* **1979**, *16*, 710-718.
102. Lin, A. J.; Klayman, D. L.; Milhous, W. K. *J. Med. Chem.* **1987**, *30*, 2147-2150.
103. Antimalarial Group of the First and Second Affiliated Hospitals of Kunming. *J. New Med.* **1979**, 49-51.
104. China Cooperative Research Group. *J. Trad. Chin. Med.* **1982**, *2*, 31-38.
105. Wang, T. *J. Trad. Chin. Med.* **1989**, *9*, 28-30.
106. Li, G.; Ji, Y.; Wang, Z.; Yang, L.; Xue, B.; Dai, R.; Yao, B. *J. Trad. Chin. Med.* **1981**, *1*, 113-114.
107. Li, Y.; Yu, P.; Chen, Y.; Li, L.; Gai, Y.; Wang, D.; Zheng, Y. *Yaoxue Xuebao* **1981**, *16*, 429-439; *Chem. Abstr.* **1982**, *97*, 92245n.
108. Li, Y.; Yu, P.-L.; Chen, Y.-X.; Li, L.-Q.; Gai, Y.-Z.; Wang, D.-S.; Zheng, Y.-P. *K'o Hsueh T'ung Pao* **1979**, *24*, 667-669; *Chem. Abstr.* **1979**, *91*, 211376u.
109. China Cooperative Research Group. *J. Trad. Chin. Med.* **1982**, *2*, 17-24.
110. Gu, H.-M.; Liu, M.-Z.; Lu, B.-F.; Xu, J.-Y.; Chen, L.-J.; Wang, M.-Y.; Sun, W.-K.; Xu, B.; Ji, R.-Y. *Chung-kuo Yao Li Hsueh Pao* **1981**, *2*, 138-144; *Chem. Abstr.* **1981**, *95*, 161913b.
111. Wang, T.; Xu, R. *J. Trad. Chin. Med.* **1985**, *5*, 240-242.

112. Bunnag, D.; Viravan, C.; Looareesuwan, S.; Harinasuta *VII Internat. Congr. Parasitol.*; Paris, France; 20 Aug. 1990, *Abstr. S9*.D13.VII.
113. Gu, H.-M.; Lu, B.-F.; Qu, Z.-X. *Acta Pharmacol. Sin.* **1980**, *1*, 48-50.
114. Brossi, A; Venugopalan, B.; Gerpe, L. D.; Yeh, H. J. C.; Flippen-Anderson, J. L.; Buchs, P.; Luo, X.-D.; Milhous, W. K.; Peters, W. *J. Med. Chem.* **1988**, *31*, 645-650.
115. Vishwarkarma, R. A. *J. Nat. Prod.* **1990**, *53*, 216-217.
116. Li, Y.; Yu, P.-L.; Chen, Y.-X.; Ji, R.-Y. *Huaxue Xuebao* **1982**, *40*, 557-561; *Chem. Abstr.* **1983**, *98*, 4420h.
117. Li, Y.; Yu, P.; Chen, I.; Chi, J. *Yao Hseuh T'ung Pao* **1980**, *15*, 38; *Chem. Abstr.* **1982**, *96*, 6883u.
118. Li, Y. *Huaxue Shiji* **1982**, *2*, 88-94; *Chem. Abstr.* **1982**, *97*, 91182c.
119. Yang, Q.; Shi, W.; Li, R.; Gan, J. *J. Trad. Chin. Med.* **1982**, *2*, 99-103.
120. Li, G. Q.; Guo, X.; Jin, R.; Wang, Z.; Jian, H.; Li, Z. *J. Trad. Chin. Med.* **1982**, *2*, 125-130.
121. Li, R.; Zhou, L.-L.; Li, X.; Zhong, J.-J.; Li C.-H.; Liao, Z.-Y. *Acta Pharm. Sin.* **1985**, *20*, 485-490.
122. Lin, A. J.; Klayman, D. L.; Milhous, W. K. *U.S. Pat.* 4,791,135; 13 Dec. 1988.
123. Coordinating Group for Research on the Structure of Qing Hau Sau. *K'o Hsueh T'ung Pao* **1977**, *22*, 142; *Chem. Abstr.* **1977**, *87*, 98788g.
124. Lin, A. J.; Klayman D. L.; Hoch, J. M.; Silverton, J. V.; George, C. F. *J. Org. Chem.* **1985**, *50*, 4504-4508.
125. Lin, A. J.; Theoharides, A. D.; Klayman D. L. *Tetrahedron* **1986**, *42*, 2181-2184.
126. Theoharides, A. D.; Smyth, M. E.; Ashore, R. W.; Halverson, J. M.; Zhou, Z.-M.; Ridder, W. E.; Lin, A. J. *Anal. Chem.* **1988**, *60*, 115-120.
127. Lin, A. J.; Li, L.-Q.; Andersen, S. L.; Klayman, D. L. *J. Med. Chem.* **1992**, *35*, 1639-1642.
128. McChesney, J. D.; Jung, M. *U.S. Pat.* 4,920,147; 24 Apr. 1990.
129. Jung, M.; Yu, D.; Bustos, D.; ElSohly, H. N.; McChesney, J. D. *Bioorg. Med. Chem. Lett.* **1991**, *1*, 741-744.
130. Naing, U. T.; Win, U. H.; Nwe, D. Y. Y.; Myint, U. P. T.; Shwe, U. T. *Trans. Roy. Soc. Trop. Med. Hyg.* **1988**, *82*, 530-531.
131. Chawira, A. N.; Warhurst, D. C.; Robinson, B. L.; Peters W. *Trans. Roy. Soc. Trop. Med. Hyg.* **1987**, *81*, 554-558.
132. Zhao, Y.; Li, A.; Xie, P.; Hou, H.; Gu, W.; Jia, M.; Liu, X.; Lei, T. *J. Trad. Chin. Med.* **1987**, *7*, 287-289.
133. Li, G.-Q.; Guo, X.-B.; Jian, H.-X.; Fu, L.-C.; Shen, L.-C.; Li, R.-S.; Dai, B.-Q.; Li, Z.-L. *J. Trad. Chin. Med.* **1985**, *5*, 159-161.
134. Xuan, W.; Zhao, Y.; Li, A.; Xie, P.; Cai, Y. *J. Trad. Chin. Med.* **1988**, *8*, 282-284.
135. Xuan, W. Y.; Zhao, Y.; Li, A. Y.: Xie, P. S.; Liu X. *Acta Pharm. Sin.* **1990**, *25*, 220-222.
136. Klayman, D. L.; Ager, A. L., Jr.; Fleckenstein, L.; Lin, A. J. *Am. J. Trop. Med. Hyg.* **1991**, *45*, 602-607.

RECEIVED March 5, 1993

Chapter 18

Search for Molluscicidal and Larvicidal Agents from Plants

M. Maillard, A. Marston, and K. Hostettmann[1]

Institut de Pharmacognosie et Phytochimie, Ecole de Pharmacie, Université de Lausanne, B.E.P., CH 1015 Lausanne, Switzerland

One approach to the control of tropical diseases is the elimination of the vectors responsible for their transmission. Schistosomiasis, which affects more than 250 million people worldwide, requires aquatic snails as intermediate hosts. By the use of plant-derived molluscicides, an attempt can be made to reduce the incidence of this illness. The role of saponins and saponin-containing plants in the control of schistosomiasis is outlined. Other tropical diseases such as yellow fever and malaria are transmitted by mosquitoes. The application of larvicidal compounds from tropical plants could be one means of controlling the spread of these diseases. Examples of the isolation and structure determination of molluscicidal and larvicidal compounds from several plants are presented.

Schistosomiasis and Plant Molluscicides

Most of the population of the Third World is affected by tropical diseases and lives in areas where the prevalence of illness is high. For example, schistosomiasis (bilharzia) is a parasitic disease endemic throughout South America, Africa and the Far East. It affects more than 250 million people in over 76 countries (*1*). Water plays an important role in the transmission of schistosomiasis, with the result that communities living near slow-moving water or near the banks of dams or rivers are the most affected by this illness. Paradoxically, development programs involving dam constructions or establishment of irrigation schemes provide ideal breeding sites for transmission and lead to an increase in the incidence of the disease. Thus, rural and agricultural populations are principally involved.

There are different kinds of schistosomiasis, but in all cases the reproductive cycle involves a stage implicating aquatic snails, in which the parasite multiplies into thousands of cercariae. The cercariae, after leaving the

[1]Corresponding author

0097–6156/93/0534–0256$06.00/0

snails, can penetrate the skin of humans who come into contact with the water source in question. Once through the skin, they change gradually into the mature trematodes (or flukes), known as schistosomes. The schistosomes (the three main species of which are *Schistosoma haematobium*, *S. mansoni* and *S. japonica*) mate and lay eggs, which are carried away with faeces or urine. Eggs from infected individuals enter water and produce miracidia, which locate snails of the appropriate species and the cycle begins again (Figure 1).

Chemotherapy with orally administered anti-schistosomal drugs remains the best method to cure people infected by the parasites. However, this approach is rather expensive and most Third World countries are not able to support the necessary outlay. Under these conditions, molluscicides represent crucial factors in integrated programs to control schistosomiasis. At the present time, Bayluscide (2,5'-dichloro-4'-nitrosalicylanilide or niclosamide) is the only molluscicide recommended by the World Health Organization (W.H.O.). However, once again, synthetic compounds are expensive and thus they pose a problem in schistosomiasis control programs when the deciding factor is the cost. Synthetic molluscicides, like Bayluscide, in addition, may result in problems of toxicity to non-target organisms and deleterious long-term effects in the environment. The possible development of resistance in schistosomiasis-transmitting snails is another important factor. In contrast, the use of plants with molluscicidal properties is a simple, inexpensive and appropriate technology for local control of the snail vector (*2*).

Plant molluscicides will probably be most useful in areas where transmission is predominantly focal, rather than widespread through large regions. It is believed that they could play an important role in programs, involving community participation, at the village level. In fact, in many areas, a combination of snail control and chemotherapy is now considered to be the quickest and the most efficient way to reduce the prevalence and intensity of infection and to control transmission (*1*).

Since the discovery of highly potent saponins in the berries of *Phytolacca dodecandra* L'HERIT. (Phytolaccaceae) (*3-6*), naturally occurring molluscicides have been receiving considerable attention (*2*, *7*) and the number of reports on the use of plant molluscicides has increased considerably (*8*).

Criteria for Efficient Plant Molluscicides. More than 1000 plant species have been tested for molluscicidal activity (*9*). However, studies on long-term toxicity against non-target organisms (including man) and observations on the mutagenicity/carcinogenicity of these plant molluscicides are rare (*10*).

In addition, only a small proportion of the numerous plants showing activity in laboratory trials are liable to represent a good alternative in the field control of the vector of schistosomiasis since there are several prerequisites for a viable candidate plant molluscicide (*2*, *11*):

a) The molluscicidal activity should be high. The crude extract from which the compounds are obtained should have an activity at concentrations lower than 100 mgL^{-1}. The activity of one of the strongest synthetic molluscicides is around 1 mgL^{-1}; e.g., for Bayluscide, the LC_{100} value against the schistosomiasis-transmitting snail *Biomphalaria glabrata* is 0.3 mgL^{-1} after an exposure time of 24 h. Similar values are obtained for *Bulinus* snails. Consequently, to be effective competitors of synthetic molluscicides, plant-derived molluscicides must have LC_{100}/24 h values of this order of magnitude. It is advantageous if the molluscicide also kills snail eggs.

b) The plant in question should grow abundantly in the endemic area. Either the plant should be of high natural abundance, or alternatively, easily cultivated. In addition, regenerating plant parts (fruits, leaves or flowers) should be used and, if possible, not the roots, since this leads to destruction of the plant.

c) Extraction of the active constituents by water is an advantage. The cost of organic solvents and accompanying extraction apparatus could be prohibitive for schistosomiasis-control programs in Third World countries.

d) Application procedures should be simple and safe for the operator. In addition, formulations and storage must be straightforward.

e) The plant extract or molluscicide should possess low toxicity to non-target organisms (including humans). The discovery of compounds more specifically toxic to snails would be a great advantage. Furthermore, isolation of the active compounds from those molluscicidal plants which are potentially applicable to field trials is important for toxicological and environmental studies.

f) Costs should be low.

Regarding these different criteria of selection, it is obvious that the probability of finding a unique molluscicidal plant which can be used in all endemic areas is very small. Judicious solutions have to be established from case to case, according to the native flora and the local situation. Thus, for example, among the different promising natural molluscicides, Endod (*Phytolacca dodecandra*; Phytolaccaceae) has been intensively studied. The fruits of this plant contain triterpenoid saponins with high molluscicidal activities (*4*, *12*, *13*) and promising field trials to control schistosomiasis have been undertaken with Endod in Ethiopia (*14*). However, *Phytolacca dodecandra* has a limited distribution (in Ethiopia and surrounding areas) and has to be cultivated if required for use in other countries. Therefore, additional plants have to be taken into account and investigated.

Other field trials have been performed with *Ambrosia maritima* (Asteraceae) in Egypt (*15*), *Acacia nilotica* (Leguminosae) in Sudan (*16*) and *Anacardium occidentale* (Anacardiaceae) in Mozambique (*17*). Once again, however, these different plants, which exhibit promising molluscicidal activities, show restricted geographical distribution patterns. Some of them are not necessarily abundant in the areas where schistosomiasis is endemic. In addition, there are complications arising from the allergenicity of constituents, particularly in the case of *A. occidentale*. This emphasizes the need to search for other plant molluscicides which will meet the prerequisites defined above.

Saponin-Containing Plants used for the Local Control of Schistosomiasis. Until a few years ago, relatively little was known about the natural products responsible for the molluscicidal activity of plants. However, during the intervening period, a number of active compounds have been isolated and characterized. As mentioned above, isolation and identification of these constituents are essential for the study of their toxicities and stabilities under field conditions, for dosage purposes, for structure-activity investigations and for effects on snail metabolism or physiology. Once the compounds have been identified, their content in different plant parts can also be determined.

An update on plant molluscicide research has been published recently by Hostettmann and Marston (*8*), who reviewed the plant constituents possessing snail-killing properties. Among the classes of natural products with recognized molluscicidal activity, triterpenoid saponins appear to be at the forefront.

In general, active saponins possess an aglycone of the oleanane type with a sugar chain attached at position C-3 (monodesmosidic saponins). Bidesmosidic saponins carrying an additional glycosidic chain linked through an ester bond at position 28 are devoid of molluscicidal activity (*18*). In this context, the extraction procedure plays an important role, as extraction with water provides predominantly monodesmosidic saponins, and extraction with methanol provides larger quantities of bidesmosidic saponins (*12*). During the water-extraction procedure, hydrolysis of the more labile ester-linked glycosides occurs, leaving the ether-linked glycosides intact. The molluscicidal activities of saponins also vary with the nature of the sugar chains, the sequence of the sugars, the interglycosidic linkages and the substitution patterns of the aglycone. Indeed, diglycosides are more active than monoglycosides, which are in turn more potent than triglycosides. In addition, the presence of substituents on the oleanane skeleton decreases the toxicity of saponins to the snails.

Among different African plants with molluscicidal activity (*19-21*), *Swartzia madagascariensis* DESV. (Leguminosae) was selected for one of our investigations. This tree is very common in many regions of Africa. Since 1939, there have been reports that fruits of this plant have been used effectively in controlling the populations of schistosomiasis-transmitting snails in natural ponds (*22, 23*).

Large quantities (up to 40 kg) of the inedible fruits can be obtained from each tree, and 100 mgL^{-1} of dry seed pods in water kill over 90% of *Bulinus globosus* snails exposed for 24 hours to the molluscicidal extract. As many of the criteria for plant molluscicides are fulfilled by *S. madagascariensis*, it was selected for limited trials in the field (*24*).

Phytochemical investigation showed that molluscicidal activity was linked to the saponin content of this plant. Five compounds, saponins **1** - **5** (Figure 2), were chromatographically isolated from the molluscicidal aqueous extract of the dried ground pods. These compounds were shown to be glucuronides of oleanolic acid and of gypsogenin (*25*).

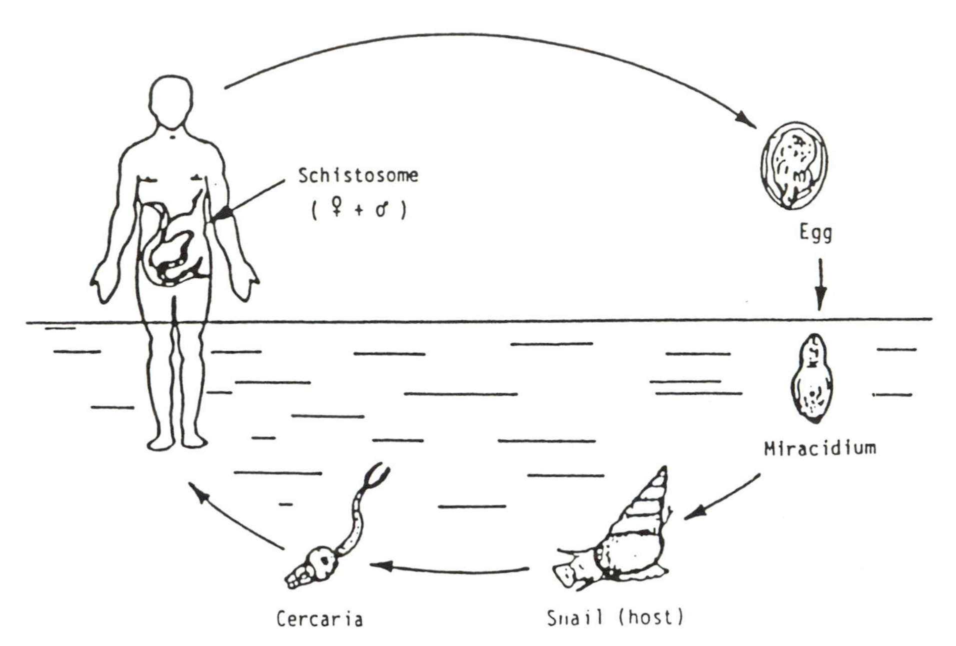

Figure 1. Life cycle of *Schistosoma* species (Reproduced with permission from ref. 2. Copyright 1985 Pergamon Press Ltd.).

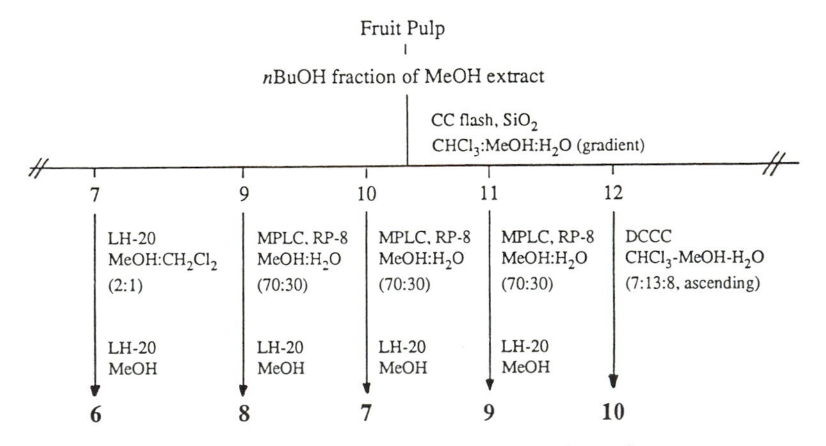

Figure 2. Isolation of molluscicidal saponins from *Tetrapleura tetraptera*.

The results of biological testing demonstrated that saponin **1** presented the highest molluscicidal activity of the isolated compounds against schistosomiasis-transmitting snails. This saponin showed a toxicity of LC_{100} = 3 mgL^{-1} to *Bulinus globosus* and *Biomphalaria glabrata* after an exposure of 24 h. This is within the range of the molluscicidal activity of Endod (*P. dodecandra)*.

Two field trials were performed in order to test the molluscicidal activity of *S. madagascariensis* pod extracts in a natural habitat known to harbour *B. globosus*. These field trials were carried out at Ifakara, Tanzania, in collaboration with the Swiss Tropical Institute (Basel) and its Field Laboratory in southeastern Tanzania. The trials were undertaken at the end of the dry season, when the water level in the ponds was very low. Pods of *S. madagascariensis* were easily obtained in large quantities, dried in the sun and then efficiently extracted by water, a simple procedure that does not demand any sophisticated apparatus or specially trained people. The application of extract was easy, and more than one ton of fruits was used to treat the water of two ponds (*24*)

The results of both trials showed that an initial molluscicide concentration of not less than 100 mgL^{-1} was reached. Complete haemolysis of human erythrocytes (a biological activity semi-quantitatively linked to the concentration of saponins) by treated water was observed during the first 12 hours after the application, and all the exposed, encaged snails died within the same period. The analysis of water samples by TLC parallelled these findings. The densities of aquatic snails dropped to zero one week after a single application of *S. madagascariensis* extract. The snails were observed only at low densities and never reached the initial density during the short-term and long-term follow-up period of five months.

In conclusion, these studies confirmed the molluscicidal properties of *S. madagascariensis* pods. However, consideration must also be given to the toxicity of *S. madagascariensis* pod extracts to non-target organisms such as fish or even man. Focal and seasonal mollusciciding schedules are likely to be the rule and thus will minimize these problems. Nevertheless, investigations on the toxicity and/or mutagenicity of this plant molluscicide are currently being undertaken in conjunction with the W.H.O.

In the course of our work on molluscicidal plants, another legume attracted our attention. *Tetrapleura tetraptera* TAUB. (Leguminosae-Mimosoideae), known locally as "Aridan" in the Yoruba language, is a large tree growing throughout the rain forest belt of West Africa. The plant has many traditional uses, mainly in the management of convulsions, leprosy, inflammation, and rheumatic pains (*26*). Molluscicidal activity of this plant has been reported by several authors (*27-29*). These investigators postulated that the molluscicidal activity was linked to triterpenoid saponins and coumarinic compounds.

Because of the strong molluscicidal properties exhibited by the fruits of *T. tetraptera* in laboratory experiments, field trials have been carried out, and this tree is now considered to be a promising plant for the local control of schistosomiasis (*11*).

Previous investigations of the molluscicidal extract of the fruits of *T. tetraptera* resulted in the isolation and identification of a novel oleanolic acid glycoside, aridanin, **6** (*30*). More recently, Maillard *et al.* undertook a reinvestigaton of the plant (*31, 32*). Four saponins (**7** - **10**), together with aridanin (**6**) were isolated (Figure 2 and 3). With the exception of the inactive saponin **8**, these compounds, responsible for the molluscicidal activity of the fruits, are *N*-acetylglycosides, with either oleanolic acid (**6**, **9**, **10**) or echinocystic acid (**7**) as aglycone. Glycosides **9** and **10** are among the most powerful natural molluscicides and have similar potencies to those isolated from *P. dodecandra* (*13*) and *S. madagascariensis* (*25*). In field experiments, lethal concentrations of *T. tetraptera* varied from 15 to 100 mgL^{-1}, depending on the mode of extraction or the formulation used. The potency of this plant in the local control of schistosomiasis has already been evaluated several times. A lengthy study, over a two-year period, was carried out in the area of Ile-Ife (Nigeria). The water supply of four villages was treated every three months with aqueous extracts of Aridan. By this method, the density of snails was reduced by a factor of 30 during the first weeks after application, and the transmission sites were kept free from cercariae for a minimum of one month after application of molluscicide (*33*).

As mentioned in the introduction, schistosomiasis is also endemic in southeastern Asia. As a part of a research program on Indonesian plants, the tree *Sapindus rarak* DC. (Sapindaceae) has been investigated. This species satisfied some of the criteria defined for promising molluscicides, and preliminary screening for toxicity against the snail *B. glabrata* showed that the methanolic extracts of the pericarp of *S. rarak* exhibited molluscicidal activity. A separation scheme involving normal- and reversed-phase low- and medium-pressure liquid chromatography, followed by semipreparative HPLC, was used to isolate a series of four saponins, among them three *O*-acetylglycosides (Figure 4). Structure elucidation of these four compounds was carried out principally with the aid of ^{1}H-NMR. Complete assignments were achieved with double-quantum filtered phase-sensitive COSY; NOE difference spectroscopy provided unambiguous proofs for the positions of acetyl groups and interglycosidic linkages (Hamburger, M.; Dorsaz, A.-C.; Slacanin, I.; Hostettmann, K.; Dyatmiko, W.; Sutarjadi. *Phytochem. Anal.*, in press.)

The four saponins exhibited similar molluscicidal activities, with LC_{100}/24 h values of 6.25 (**11**, **12**, **14**; Figure 5) and 12.5 mgL^{-1} (**13**, Figure 5). Although they are somewhat less active than the most potent saponins characterized so far, field trials with *S. rarak* are planned in Indonesia, where issues such as feasibility, efficacy and biodegradation of saponins and other constituents will be addressed.

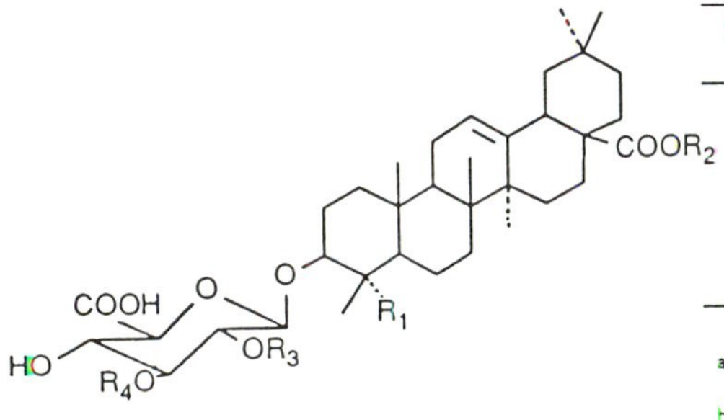

Saponin[a]	R_1	R_2	R_3	R_4	Molluscicidal activity[b]
1	CH_3	H	H	Rha	3
2	CHO	H	H	Rha	25
3	CH_3	H	Glc	Rha	25
4	CHO	H	Glc	Rha	> 50
5	CH_3	Glc	Glc	Rha	no activity

[a] Glc = β-D-Glucopyranosyl, Rha = α-L-Rhamnopyranosyl

[b] Molluscicidal activity against *Biomphalaria glabrata* [mgL^{-1}].

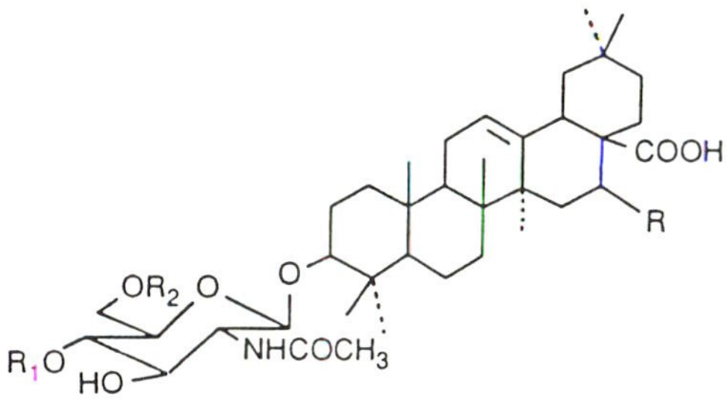

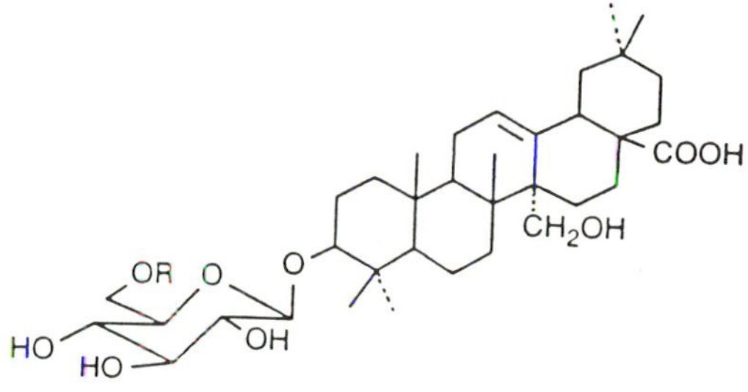

6 $R = R_1 = R_2 = H$, Aridanin

7 $R = \alpha\text{–}OH, R_1 = R_2 = H$

9 $R = R_1 = H, R_2 = Gal$

10 $R = H, R_1 = Glc, R_2 = H$

8 R = Glc

Figure 3. Structures of compounds isolated from *Swartzia madagascariensis* (1-5) and *Tetrapleura tetraptera* (6-10).

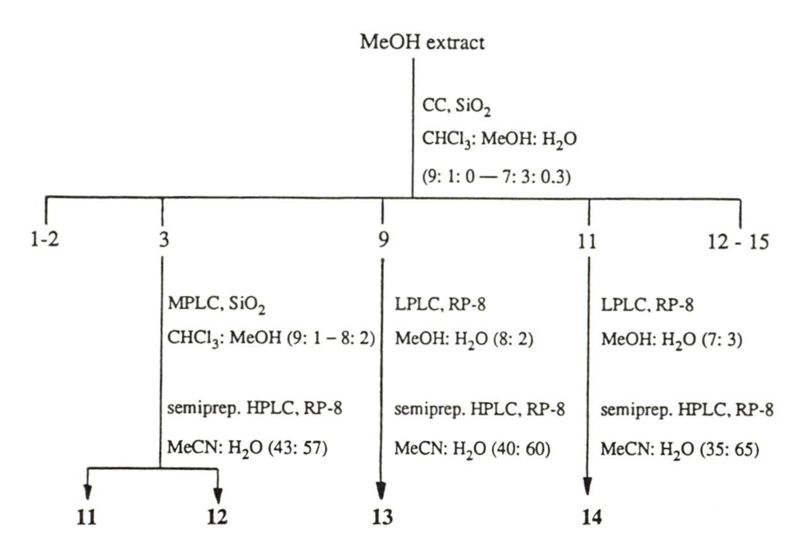

Figure 4. Isolation of saponins from *Sapindus rarak* fruit pericarp.

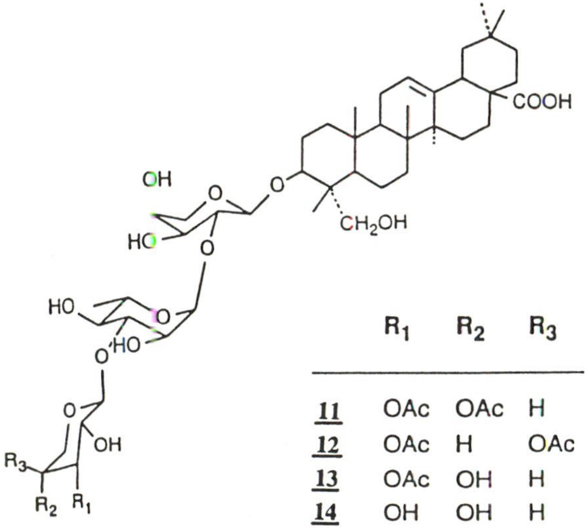

	R_1	R_2	R_3
11	OAc	OAc	H
12	OAc	H	OAc
13	OAc	OH	H
14	OH	OH	H

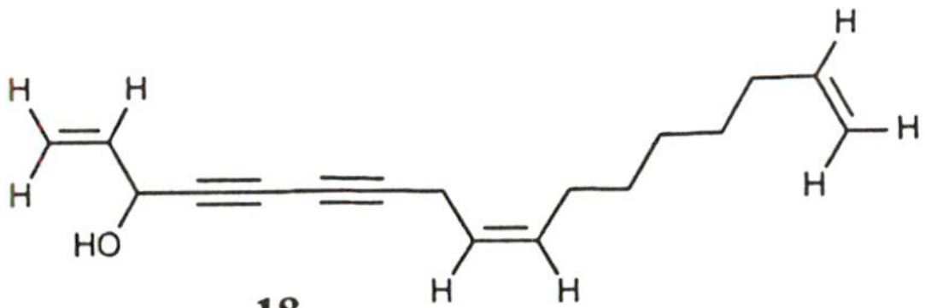

18

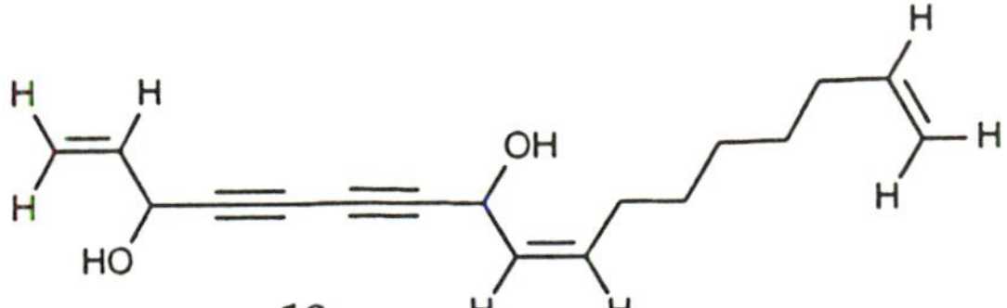

19

Figure 5. Structures of compounds isolated from *Sapindus rarak* (11-14) and *Artemisia borealis* (18, 19).

Analytical Aspects of Plant Molluscicides

As shown in the first part of this paper, some of the most promising molluscicidal plants for the local control of schistosomiasis contain saponins (*P. dodecandra*, *S. madagascariensis*, *T. tetraptera*, *S. rarak*, etc.). However, care must be taken to investigate the mammalian toxicity and the impact on the environment of these natural products before successful snail eradication schemes can be considered. Thus, for any plant which proven itself to be a useful candidate molluscicide, a reliable analytical method is required to measure the content of active principles at a given stage in the exploitation of the vegetable material.

An HPLC method has been developed for the analysis of saponins in *P. dodecandra* (*34*). This method is also suitable for the quantitative determination of saponins in other molluscicidal plants.

An efficient analytical method is necessary to investigate plant material from different strains and geographical locations. Quantitative analysis is also important for the monitoring of extraction procedures. These differ according to the solvent system, temperature and time; and it is important to maximise the yield of saponins for the most effective applications to infected sites. Finally, information on the content of active saponins in treated water is essential for biodegradation and toxicological studies.

Separations on reversed-phase columns have previously been carried out with detection at 206 nm owing to the poor UV absorption of saponins at higher wavelengths (*34*). Consequently, there are limitations concerning the solvents and gradients that can be used. For example, an HPLC chart of an aqueous *P. dodecandra* extract with an acetonitrile-water gradient demonstrates the problem of baseline drift at 206 nm (Figure 6).

The bidesmosidic saponins (**15b**, **16b** and **17b**) elute between 10 and 20 min but the monodesmosidic saponins (**15**, **16**, and **17**; R_2=H in Figure 6.) elute later and are much more difficult to quantify. When a gradient solvent system is used, refractive index detection is not practicable, thus an alternative is to introduce a chromophore in the saponins which facilitates UV detection at 254 nm. As the monodesmosidic saponins from *P. dodecandra*, as well as those from *S. madagascariensis* or *T. tetrapleura* (which are responsible for the molluscicidal activity) possess a free carboxyl group at the C-28 position, derivatization can be carried out at this function. Encouraging results have been obtained by derivatization of the saponins with 4-bromophenacyl bromide in presence of a crown ether (Figure 7). This method has previously been employed for the analysis of fatty acids and prostaglandins. Details of the procedure for saponins have been published (*35*).

Analysis of a derivatized extract of *P. dodecandra* was achieved by HPLC, without any baseline drift, on a reversed-phase octadecylsilyl column using a gradient of acetonitrile in water and UV detection at 254 nm (Figure 8).

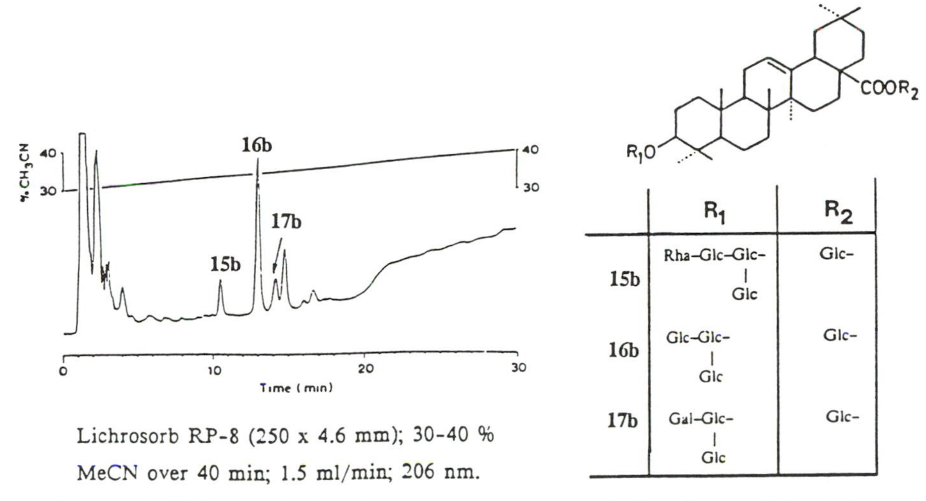

	R_1	R_2
15b	Rha–Glc–Glc– (Glc)	Glc–
16b	Glc–Glc– (Glc)	Glc–
17b	Gal–Glc– (Glc)	Glc–

Figure 6. HPLC analysis of an aqueous extract of *Phytolacca dodecandra* berries (Adapted from ref. 35).

Figure 7. Derivatization of saponins with 4-bromophenacyl bromide.

For quantification of the saponins, two methods can be considered: (a) use of derivatives of previously isolated saponins as external standards; (b) use of naphthalene as internal standard. Comparison of the results showed virtually identical values for the two methods. Consequently, method (a) was employed throughout. Mixtures of selected derivatized saponins (**15a**, **16a** and **17a**) were injected at different concentrations onto the HPLC column, and the surface area under each peak was plotted against concentration in order to obtain calibration curves. Calculation of the percentages of saponins in each extract is therefore a relatively straightforward matter, after derivatizing the extract and performing a preliminary purification step (simple filtration over RP-18) before HPLC analysis.

The bidesmosidic saponins lack a free carboxyl functionality and therefore cannot be derivatized by the method employed for the monodesmosidic saponins. Although this class of saponins was inactive in the molluscicidal assay (*18*), it is important to quantify these glycosides since they may be transformed under certain conditions into monodesmosides. For example, enzymes present in crude plant extracts may cleave the ester linkage (*12*). Furthermore, during large-scale applications in the field, hydrolysis may occur in the treated water source. HPLC analyses of bidesmosides were performed at 206 nm, without any chemical derivatization. By constructing calibration graphs with selected pure bidesmosidic glycosides, the percentages of each of them in the extracts could be ascertained.

These two methods (with and without derivatization of the extracts) were used to quantify saponins in the molluscicidal plant, *P. dodecandra*. A comparison of different aqueous extraction procedures for the berries of this plant showed that measurable amounts of monodesmosidic saponins were obtained only at ambient temperatures. Extractions with hot water produced only bidesmosidic saponins, presumably owing to inactivation of the enzymes responsible for cleaving the glycosidic chain in position C-28 of the triterpene moiety. This important observation is relevant to the problem of schistosomiasis as obviously only cold water extracts (which contain monodesmosidic active saponins) will have any application as plant molluscicides. Thus, pounding the berries with cold water, the most practicable method of obtaining a vegetable molluscicide in endemic regions, conveniently provides the greatest concentration of saponins for application to sites of infestation by transmitter snails (*35*).

For toxicological and other investigations, it is therefore easy to reproduce field extraction conditions of molluscicidal plants and then rapidly, by means of HPLC, obtain standard extracts for their respective active saponins, before submission to required tests.

The HPLC procedure involving derivatization with 4-bromophenacyl bromide can be extended to any saponins, provided either the aglycone or the sugar moiety contains a free carboxyl group. For instance, the saponins from the fruits of *S. madagascariensis* can easily be derivatized by the method described (*36*).

An alternative method was developed for the estimation of the saponin content of *S. rarak* pericarp extracts. Since acidic hydrolysis of the methanolic extract afforded only hederagenin as triterpene aglycone, the quantitative determination of this compound with the aid of an external standard was chosen as a means for a simple and rapid analysis. Under isocratic conditions, the hederagenin peak is well separated from the minor constituents, which have longer retention times. As the major saponins of *S. rarak* are all glycosides of hederagenin and have similar molecular weights (882-966 amu), determination of the total saponin content of the pericarps by this method is sufficiently accurate for analytical purposes (Hamburger, M.; Dorsaz, A.-C.; Slacanin, I.; Hostettmann, K.; Dyatmiko, W.; Sutarjadi. *Phytochem. Anal.*, in press.).

Search for New Larvicidal Compounds

A wide range of tropical diseases are transmitted by mosquitoes of several genera. Malaria, for instance, still remains endemic in more than 100 countries and half of the world population is at risk (*1*). The search for new methods to destroy the vector of this parasitic disease, which affects 200 million people every year and may kill more than 1 million (*38*), is thus of major importance.

Yellow fever is an arbovirus (*ar*thropod-*bo*rne *virus*) disease common in West Africa and South America, whose fatality rate varies from 10 to 80%. Although carried by different mosquitoes, it is mainly *Aedes aegypti* which is responsible for its transmission. This same mosquito is also responsible for the transmission of dengue hemorrhagic fever (endemic to southeast Asia, the Pacific area, Africa and the Americas) (*39*).

As is the case for malaria, the current strategy postulated by the W.H.O. for the control of these tropical diseases is to destroy their vectors. For this reason, we have included a simple bench-top assay in our routine screening of crude plant extracts to detect larvicidal activity against *A. aegypti*.

Larvicidal Assay and Preliminary Results. *Aedes aegypti* eggs are very easy to handle and can be kept in a controlled atmosphere (26-28°, 70-80% rel. humidity) for up to six months. Larvae hatch readily when put into tap water and incubated for 24 h. The assay consists of exposing approximately 20 second instar larvae to various geometrical dilutions of the extracts, previously solubilised in DMSO. Mortality is evaluated with the naked eye after 30 min and 24 h (Figure 9) (*40*).

Screening of plant extracts for larvicidal activity by this procedure has recently been introduced in our laboratory and some interesting results have been recorded. Since the introduction of this test in our panel of assays for bioactivity, 386 extracts from plants of different origin have been screened. Of these, 46 extracts (12%) showed toxicity against *A. aegypti* larvae (Krause, F.; Hamburger, M.O.; Sordat, B.; Hostettmann, K. *J. Ethnopharmacol.*, submitted). Activity-guided fractionation of the most promising plants is currently in progress. For example, the assay detected larvicidal toxicity of a

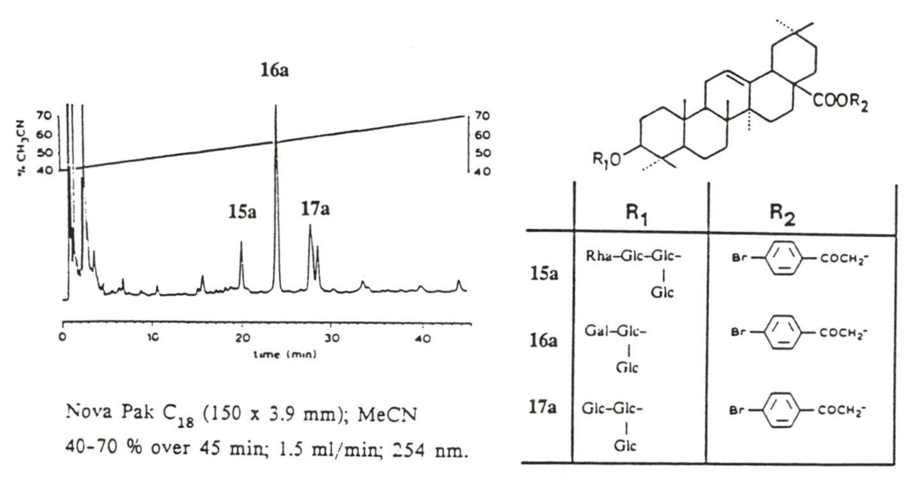

Figure 8. HPLC analysis of a derivatized aqueous extract of *Phytolacca dodecandra* berries.

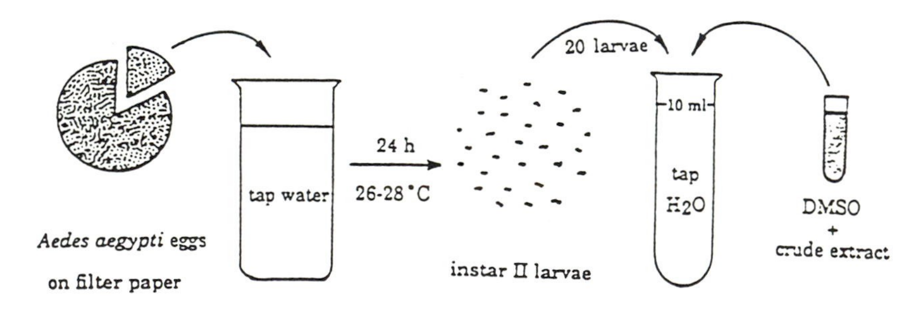

Figure 9. Larvicidal assay using *Aedes aegypti* larvae.

dichloromethane extract of *Artemisia borealis* PALLAS whole plant (Asteraceae). This extract (LD_{100} = 80 mgL^{-1}) killed all the larvae within 24 h. Activity-guided fractionation of the extract yielded two polyacetylene derivatives, **18** and **19** (Figure 5) (*41*). When testing these compounds, 100% mortality was observed at 100 mgL^{-1} . Although active, these polyacetylenes contribute only in part to the larvicidal activity of the extract which seems to contain additional, more potent larvicidal compounds. Furthermore, due to their low stability, these polyacetylenes have no potential for practical use in controlling the transmission of yellow fever. Fractionation and phytochemical investigation of other tropical plants is presently underway.

Conclusions

The screening of extracts of vegetable material has shown that many plants have molluscicidal and/or larvicidal activity. If these activities are high enough and if the compounds responsible for the toxicity are stable and do not pose too great a risk to non-target organisms, plant-derived natural products can be of immense potential for the control of schistosomiasis and other vector-borne diseases. A simple test with the carrier of yellow fever, *Aedes aegypti*, is effective for the discovery of new larvicidal components, while aquatic snails are employed to find plant extracts and compounds with molluscicidal activity.

Despite the large number of new plant molluscicides that have been documented, very few actually satisfy the criteria for effective large-scale application (*7*). Many simply do not have sufficient activity and cannot compete with synthetic molluscicides. Consequently, there is a need to find more potent and more selective natural molluscicides. Extracts of plants for application to infected sites should originate from regenerating parts, such as fruits and leaves, in order not to destroy the whole plant. Of the natural products with the most potential in the fight against schistosomiasis, the triterpene glycosides appear to be in the forefront at the moment, especially as some plant parts can contain as much as 30% saponin. Saponin-containing plants which have already been tested in the field include *Phytolacca dodecandra, Swartzia madagascariensis* and *Tetrapleura tetraptera*, while *Sapindus rarak* will soon undergo trials in Indonesia. However, the search for natural highly active molluscicides remains an interesting challenge, given that more than 20,000 compounds had to be screened in order to discover the synthetic molluscicide, Bayluscide (*37*). Considering that the activities of the recently isolated saponins approach those of Bayluscide, it would seem that the investigation of plant-derived compounds is certainly a worthwhile endeavour.

Acknowledgements

Financial support from the UNDP/WORLD BANK/WHO Special Program for Research and Training in Tropical Diseases and the Swiss National Science Foundation is gratefully acknowledged. Thanks are also due to Dr. K.E. Mott (W.H.O) for his encouragement of this work.

Literature Cited

(*1*) *Tropical Diseases: Progress in International Research 1987-1988, Ninth Program Report*; UNDP/World Bank/WHO Special Program for Research and Training in Tropical Diseases (TDR): Geneva, Switzerland, 1989.

(*2*) Marston, A.; Hostettmann, K. *Phytochemistry* **1985**, *24*, 639-652.

(*3*) Lemma, A. *Ethiopian Med. J.* **1965**, *3*, 187-190.

(*4*) Parkhurst, R.M.; Thomas, D.W.; Skinner, W.A.; Cary, L.W. *Phytochemistry* **1973**, *12*, 1437-1442.

(*5*) Parkhurst, R.M.; Thomas, D.W.; Skinner, W.A.; Cary, L.W. *Indian J. Chem.* **1973**, *11*, 1192-1195.

(*6*) Parkhurst, R.M.; Thomas, D.W.; Skinner, W.A.; Cary, L.W. *Can. J. Chem.* **1974**, *52*, 702-706.

(*7*) Mott, K. In *Plant Molluscicides*; Mott, K., Ed.; Wiley: Chichester, U.K., 1987; preface.

(*8*) Hostettmann, K.; Marston, A. In *Plant Molluscicides*; Mott, K., Ed.; Wiley: Chichester, U.K., 1987; pp 301-322.

(*9*) Kloos, H.; McCullough, F.S. *Planta Med.* **1982**, *46*, 195-209.

(*10*) McCullough, F.S.; Mott, K. *WHO/Schisto*/83.72. 1983.

(*11*) Hostettmann, K. In *Economic and Medicinal Plant Research*; Wagner, H.; Hikino, H., Farnsworth, N.R., Eds.; Academic Press: London, 1989; vol. 3, pp 73-103.

(*12*) Domon, B.; Hostettmann, K. *Helv. Chim. Acta* **1984**, *67*, 1310-1315.

(*13*) Dorsaz, A.-C.; Hostettmann, K. *Helv. Chim. Acta*, **1986**, *69*, 2038-2047.

(*14*) Goll, P.H.; Lemma, A.; Duncan, J.; Mazengia, B. *Tropenmed. Parasit.* **1983**, *34*, 177-183.

(*15*) El Sawy, M.F.; Duncan, J.; Marshall, T.F.; Shehata, M.A.R.; Brown, N. *Tropenmed. Parasit.* **1984**, *35*, 100-104.

(*16*) Hussein Ayoub, S.M. *Planta Med.* **1982**, *46*, 181-183.

(*17*) Webbe, G.; Lambert, J.D.H. *Nature* **1983**, *302*, 754.

(*18*) Hostettmann, K.; Kizu, T.; Tomimori, T. *Planta Med.* **1982**, *44*, 34-35.

(*19*) Haerdi F. *Acta trop.* **1964**, *Suppl. 8*, 241-244.

(*20*) Farnsworth, N.R.; Henderson, T.O.; Soejarto, D.D. In *Plant Molluscicides*; Mott, K.E., Ed., Wiley: Chichester, U.K., 1987; pp 131-204.

(*21*) Kloos, H.; McCullough, F.S. In *Plant Molluscicides*; Mott, K.E., Ed., Wiley: Chichester, U.K., 1987; pp 45-108.

(*22*) Mozley, A. *Trans. Roy. Soc. Edinburgh* **1939**, *59*, 687-730.

(*23*) Mozley, A. *Molluscicides*. Lewis: London, 1952; pp 1-310.

(*24*) Suter, R.; Tanner, M.; Borel, C.; Hostettmann, K.; Freyvogel, T.A. *Acta trop.* **1986**, *43*, 69-83.

(*25*) Borel, C.; Hostettmann, K. *Helv. Chim. Acta* **1987**, *70,* 570-576.

(*26*) Ojewole, J.A.O.; Adesina, S.K. *Planta Med.* **1983**, *49*, 99-102.

(*27*) Adesina, S.K.; Adewunmi, C.O.; Marquis, V.O. *J. African Med.* **1980**, *3*, 7-15.

(*28*) Adewunmi, C.O.; Sofowora, E.A. *Planta Med.* **1980**, *39*, 57-65.
(*29*) Adewunmi, C.O.; Marquis, V.O. *J. Parasitol.* **1981**, *67*, 713-716.
(*30*) Adesina, S.K.; Reisch, J. *Phytochemistry* **1985**, *24*, 3003-3006.
(*31*) Maillard, M.P.; Adewunmi, C.O.; Hostettmann, K. *Helv. Chim. Acta* **1989**, *72*, 668-674.
(*32*) Maillard, M.P.; Adewunmi, C.O.; Hostettmann, K. *Phytochemistry* **1992**, *31*, 1321-1324.
(*33*) Adewunmi, C.O. *J. Anim. Prod. Res.***1984**, *4*, 73-84.
(*34*) Domon, B.; Dorsaz, A.-C.; Hostettmann, K. *J. Chromatogr.* **1984**, *315*, 441-446.
(*35*) Slacanin, I.; Marston, A.; Hostettmann, K. *J. Chromatogr.* **1988**, *448*, 265-274.
(*36*) Borel, C. *Etude des saponines molluscicides de deux espèces du genre Swartzia: Leur rôle pour le contrôle de la schistosomiase*, Ph.D. Thesis, University of Lausanne, 1987; pp 150-161.
(*37*) Andrews, P.; Thyssen, J.; Lorke, D. *Pharmacol. Therap.* **1983**, *19*, 245-295.
(*38*) Hamburger, M.; Marston, A.; Hostettmann, K. In *Advances in Drug Research*, Testa, B., Ed.; Academic Press: London, U.K., 1991; vol. 20, pp 167-215.
(*39*) Manson-Bahr, P.E.C.; Apted, F.I.C. *Manson's Tropical Diseases 18th ed.*, Baillière Tindall: Eastbourne, U.K., 1983; pp 270-274.
(*40*) Hamburger, M.; Hostettmann, K. *Phytochemistry* **1991**, *30*, 3864-3874.
(*41*) Wang, Y.; Toyota, M.; Krause, F.; Hamburger, M.; Hostettmann, K. *Phytochemistry* **1990**, *29*, 3101-3106.

RECEIVED May 21, 1993

Promising Plant-Derived Natural Products with Multiple Biological Activities

Chapter 19

Algal Secondary Metabolites and Their Pharmaceutical Potential

Gabriele M. König and Anthony D. Wright

Department of Pharmacy, Swiss Federal Institute of Technology (ETH) Zurich, CH–8092, Zurich, Switzerland

Algae are a rich and varied source of pharmacologically active natural products. In this paper the biological properties of natural products derived from algae of the divisions Chlorophyta, Rhodophyta, Phaeophyta and Cyanophyta are discussed. The compounds described belong to a wide range of structural classes, e.g., alkaloids, aromatic compounds, macrolides, peptides and terpenes, all of which exhibit some biological activity.

This paper provides an overview of the literature relating to algae of the divisions Chlorophyta, Rhodophyta, Phaeophyta and Cyanophyta that have afforded biologically active natural products. Within each division, information is subdivided according to the structural classes in which the active compounds belong. It is considered beyond the scope of this paper to mention every algal metabolite shown to possess some biological activity. Thus, examples were chosen in such a way as to demonstrate the range of chemical structures and biological activities encountered in algae, with an emphasis on the more recent literature. Prior to this compilation, many interesting and detailed reports have been published relating either to algal metabolites (*1-10*) or to marine natural products in general (*11, 12*), but no recent report has attempted to provide an overall view of bioactive secondary metabolites from algae.

Biologically Active Compounds from Green Algae (Chlorophyta)

The division Chlorophyta comprises eight to nine thousand species, occurring in salt and fresh water habitats (*13*). Phytochemical and biological studies carried out on green algae to date have placed emphasis on members of the families Caulerpaceae and Udoteaceae. Within these two families the genera investigated are *Caulerpa* from the family Caulerpaceae and *Chlorodesmis*, *Halimeda*, *Avrainvillea*, *Penicillus*, *Tydemania* and *Udotea* from the family Udoteaceae. Some metabolites are also known from the families Dasycladaceae, Cladophoraceae, Ulvaceae, Codiaceae and Charophyceae. The isolated compounds are predominantly terpenoid or aromatic. Reported biological activities for these compounds include antimicrobial, antiinflammatory, antiviral, antimutagenic, insecticidal and HMG-CoA inhibition properties.

Sesquiterpenes. Acyclic and monocyclic sesquiterpenes, often with a 1,4-diacetoxybuta-1,3-diene (*bis*-enol acetate) moiety, are frequently encountered in green

0097–6156/93/0534–0276$06.00/0

algae of the genus *Caulerpa*. The majority of these exhibit *in vitro* antimicrobial activity and cytotoxic effects (*5*). One such compound, caulerpenyne (**1**), first obtained from *C. prolifera* and later identified in many other *Caulerpa* species, showed activity against KB cells (*14*) and in a bioassay employing fertilized sea urchin eggs (*5, 15*). Caulerpenyne (**1**) was also isolated from *C. vanbosseae* during an extensive algal screening program, and was shown to inhibit prenyl transferase and 5-lipoxygenase, and to have notable activity towards strains of drug-resistant *Staphylococcus aureus* (*16*).

Algae of the families Udoteaceae and Codiaceae contain closely related sesquiterpenoid metabolites for which a similar range of activities has been demonstrated (*17-19*). The first halogenated sesquiterpenes from Chlorophyta were obtained from *Neomeris annulata* (Dasycladaceae). Compounds **2** and **3** are decalin-based sesquiterpene bromo alcohols, while **4** (neomeranol) was the first example of a sesquiterpene, also a bromo alcohol, having a neomerane skeleton. All three compounds showed moderate toxicity toward brine shrimp (**2**, IC_{50} 9 µg/ml; **3**, IC_{50} 8 µg/ml; **4**, IC_{50} 16 µg/ml) but were inactive toward KB cells (*18*).

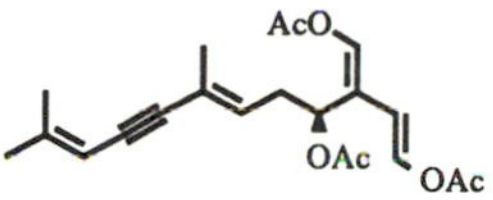

1 Caulerpenyne 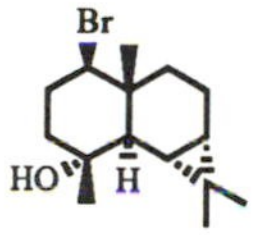 **2** 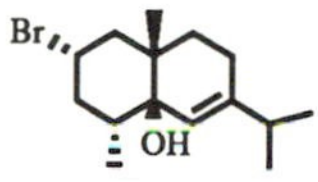 **3** 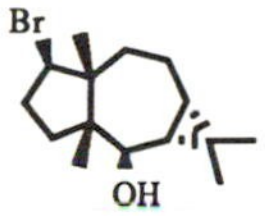 **4** Neomeranol

Diterpenes. Linear diterpenoids with enol-acetate functionalities or related aldehyde groups, similar to compounds isolated from *Caulerpa* species, also occur in many algae of the family Udoteaceae. These metabolites possess biological activities analogous to their sesquiterpenoid counterparts (e.g., **1**) (*5, 19*) such as antimicrobial and cytotoxic properties. Several cyclic diterpenes from green algae have been reported. Two of the more interesting of these are udoteatrial (**5**), isolated from *Udotea flabellum* (*20*), and halimedatrial (**6**), a bicyclic trialdehyde, identified as a constituent of many *Halimeda* species (*21*). Udoteatrial has been shown to be antimicrobial while halimedatrial demonstrated strong toxicity in a sea urchin egg assay (ED_{100}, 1 µg/ml) (*22*). An example of the pharmacological potential of such diterpenoid aldehydes is the *in vitro* antiviral activity of halitunal (**7**), a metabolite isolated from *H. tuna*. A bioassay guided approach to fractionation, using the *in vitro* activity against the murine coronavirus strain A59 on mouse liver cells, led to the isolation of the major antiviral metabolite, halitunal. This compound possesses a unique cyclopentadieno[*c*]pyran ring system, the biosynthesis of which is hypothesised to involve the diterpenoid tetraacetate, **8**, a major metabolite of *H. tuna* (*23*).

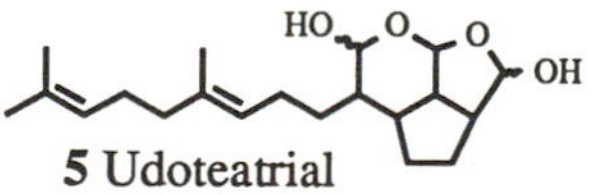

5 Udoteatrial 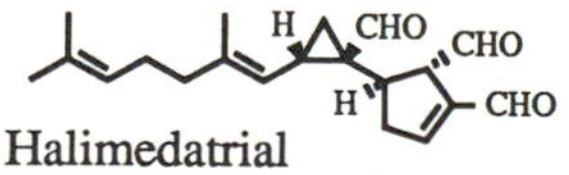 **6** Halimedatrial

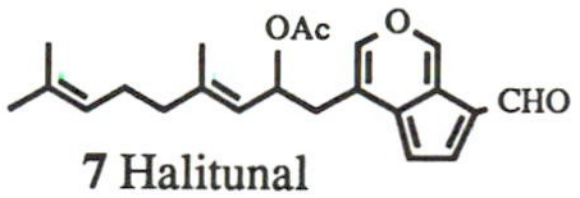

7 Halitunal 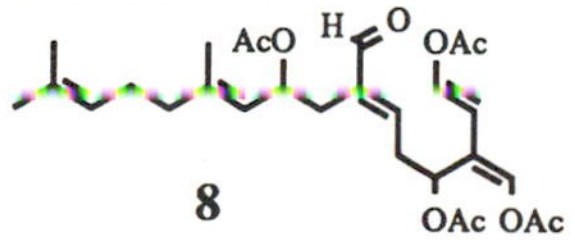 **8**

Aromatic compounds. In contrast to the oxygenated terpenes produced by the majority of green algae, the genus *Avrainvillea* (Udoteaceae) is distinguished by the occurrence of brominated aromatic phenols. Avrainvilleol (**9**), a diphenylmethane derivative, was first isolated from *A. longicaulis* and shown to have antibacterial activity (*24*). Investigation of *A. nigricans* from Puerto Rican waters (*25*) yielded besides avrainvilleol (**9**), 5'-hydroxyisoavrainvilleol (**10**) and 3-bromo-4,5-dihydroxybenzyl alcohol (**11**). Compounds **9, 10** and **11** all show antibacterial activity, with **11** having the most potent and also the broadest spectrum of activity, as well as showing cytotoxicity towards KB cells (ED_{50}, 8.9 μg/ml). Rawsonol (**12**), consisting of two units related to avrainvilleol (**9**), was obtained from *A. rawsoni*. It has been shown to inhibit the activity of human 3-hydroxy-3-methylglutaryl coenzyme A (HMG-CoA) reductase, an enzyme involved in cholesterol biosynthesis, with an IC_{50} of 5 μM, a weak but new type of activity for this structural class of compound (*26*). An extract of *Cymopolia barbata* (Dasycladaceae) was reported to exhibit antibiotic, antifungal, cytotoxic (*27*) and antimutagenic (*28*) activity. Phytochemical investigations yielded a group of prenylated bromohydroquinones, the cymopols. Cymopol (**13**) and some of its derivatives were reported in 1976 (*29*). From the same algal species the tricyclic cymopols, cymobarbatol (**14**) and 4-isocymobarbatol (**15**), were isolated using a modified Ames test (*30*), for antimutagenic activity, as a guide during isolation procedures. Cymopol (**13**) proved to be toxic towards *Salmonella typhimurium* and was thus not further investigated. Compounds **14** and **15** were non-toxic and exhibited potent antimutagenic activities (*28*). Compounds structurally related to the cymopols have been isolated from tunicates and reported to be cancer-preventive agents (*31*).

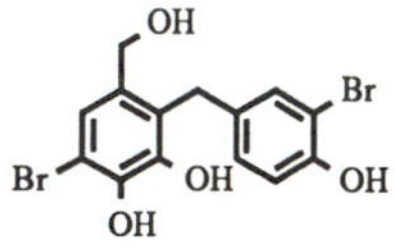

9 Avrainvilleol

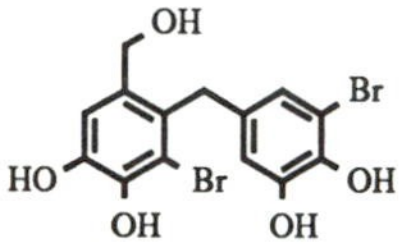

10

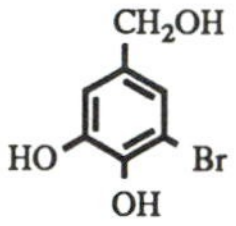

11

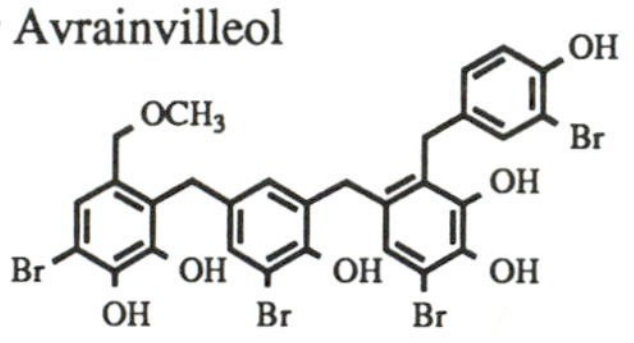

12 Rawsonol

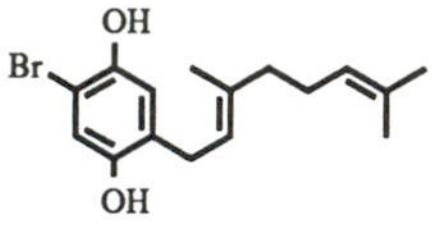

13 Cymopol

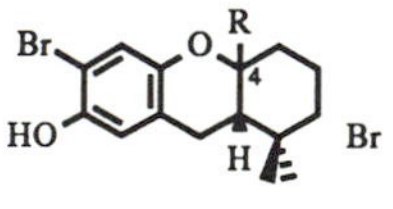

14 R=β-CH_3 Cymobarbatol

15 R=α-CH_3 4-Isocymobarbatol

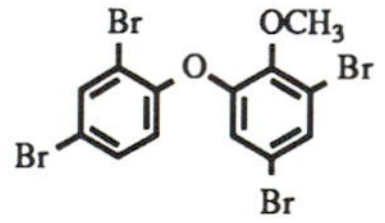

16

A study of marine algae from Okinawan waters resulted in the isolation of a polybrominated diphenylether (**16**) from the green alga *Cladophora fascicularis* (Cladophoraceae). Compound **16** is closely related to metabolites isolated from the sponge *Dysidea herbacea*, one of which has been shown to have a strong coronary vasodilative activity (*32*), and from the acorn worm *Ptychodera flava laysanica*, which exhibited strong antibacterial activities (*33*). The *Cladophora* metabolite was demonstrated to have antibacterial activity and exhibited potent anti-inflammatory activity using the inflammatory Habu venom, the poison of a deadly snake native to Okinawa, in mice (*34*).

Miscellaneous. Species of the genus *Ulva* (Ulvaceae) have been reported as sources of anthelmintic, antimicrobial and cardio-inhibitory agents (*4, 35, 37*). The ethanolic extract of *U. fasciata*, from the Indian Ocean, showed *in vitro* and *in vivo* antiviral activity against the Semeliki forest virus in mice. Bioassay-guided fractionation resulted in the isolation of a sphingosine derivative (**17**), which showed antiviral activity *in vitro* and *in vivo* (*36*).

One of the few fresh water algae investigated is the characean alga, *Chara globularis*, noted for its insecticidal, herbicidal and pronounced antibiotic activity. Phytochemical investigations of this alga yielded two biologically active compounds, charatoxin (**18**) (*38-40*), and 4-azoniaspiro[3,3]heptane-2,6-diol (**19**, charamin) (*41*). Charatoxin (**18**) is structurally similar to nereistoxin, an insecticide obtained from an annelid worm which is suggested to block acetylcholine receptors. It has been demonstrated that the effects of charatoxin and nereistoxin on the sartorius neuromuscular junction of the frog are similar (*42*).

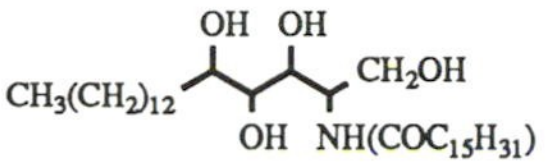

17 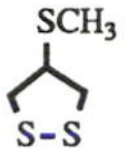 **18** Charatoxin  **19** Charamin

Biologically Active Compounds from Red Algae (Rhodophyta)

Red algae have been extensively studied, with the main focus being on plants of the orders Gigartinales, Ceramiales, Cryptonemiales and Bonnemaisoniales. Various species are reported to be used in folkmedicine (*4,35,43,44*). Phytochemical investigations of red algae have resulted in the identification of a predominance of halogenated metabolites, which are mainly of terpenoid, acetogenic or aromatic origin (*12*). Marine red algae have also been recognized as a source of structurally varied icosanoic acid derivatives (*45-50*). Biological activities discovered for red algal metabolites include antimicrobial, cytotoxic, anti-inflammatory and antiviral effects, as well as inhibition of canine Na^+/K^+ ATPase, human lymphocyte 5-lipoxygenase and human neutrophil degranulation (*48*).

Monoterpenes. Halogenated monoterpenes, e.g., **20** (*51*), occur in algae of the families Ceramiaceae, Rhizophyllidaceae and Plocamiaceae (*6*). Biological activities of such polyhalogenated monoterpenes range from antibacterial, antifungal, molluscicidal (*51, 52*) and cytotoxic (*53*), to insecticidal and acaricidal (*54*) effects. Several acyclic halogenated monoterpenes, derived from *Portieria hornemannii* (Rhizophyllidaceae) (*55*) were tested for their cytotoxic, antimicrobial and antimalarial activities. One of these compounds, **21**, exhibited marginal antimalarial activity (IC_{50} 4 µg/ml) without showing cytotoxicity to cultured cell lines (ED_{50} values > 20 µg/ml) (*56*). A further metabolite, **22**, showed activity against human breast cancer cells (ZR-751, ED_{50}, 1 µg/ml) (G. M. König *et al.*, unpublished data). *P. hornemannii* (cf. *Desmia hornemannii*) was also found to contain cytotoxic cyclohexanones, e.g., **23**. The semi-synthetic acetyl derivative of **23** has antiviral activity (*11*). Tetrachloromertensene, **24**, a secondary metabolite of *Plocamium hamatum* (*57*), was able to inhibit the chemiluminescence response induced by treatment of human granulocytes with phorbol ester, and was also able to induce an independent chemiluminescence response (*56*). Compound **24** has been identified as the substance responsible for the allelopathic effects between *P. hamatum* and the octocoral, *Sinularia cruciata* (*58*).

Sesquiterpenes. Some sesquiterpenes derived from algae of the genus *Laurencia* (Rhodomelaceae) exhibit moderate cytotoxic properties, e.g., the chamigrene, *iso*-obtusol acetate (**25**) (HeLa cells, ED_{50}, 4.5 µg/ml) (*53*), aplysistatin (**26**) (KB cells, ED_{50}, 2.4 µg/ml) (59, 60), palisadin A (**27**) (KB cells ED_{50}, 6.9 µg/ml) (*56*) and palisadin B (**28**) (KB cells ED_{50}, 1.7 µg/ml) (*56*). One of these compounds, palisadin A (**27**), derived from *Laurencia implicata* (*61*), is notable for its cytotoxicity profile.

The ED_{50} values observed were approximately ten times lower in the multidrug-resistant KB-V cell line compared to the parent KB line. This result indicates a possible involvement of **27** in the phenomenon of multi-drug resistance. Aplysistatin (**26**) and 5-hydroxyaplysistatin, although cytotoxic, are also active towards the malaria parasite *Plasmodium falciparum* (IC_{50} 0.4-0.9 μg/ml) (G. M. König *et al.*, unpublished data). Some cuparenes, e.g. cyclolaurenol (**29**) *(62)* and related compounds, isolated from sea hares and from *Laurencia* species (*63*), showed antimicrobial activity.

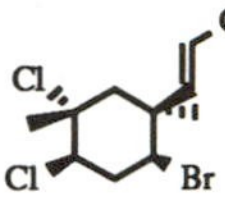

20 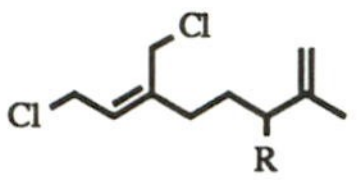 **21** R=Cl **22** R=OH **23** 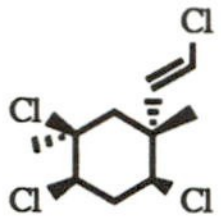 **24** Tetrachloromertensene

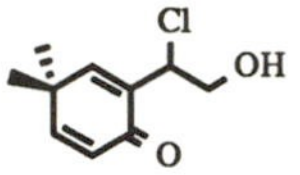

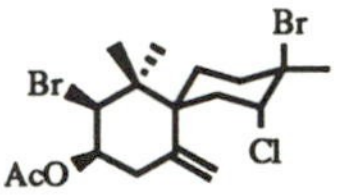

25 *Iso*-obtusol acetate 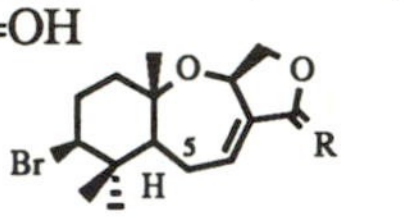 **26** R=O, Aplysistatin **27** R=H_2, Palisadin A **28** Palisadin B

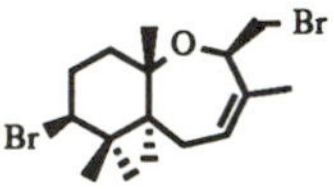

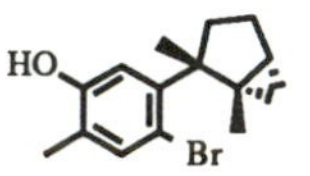

29 Cyclolaurenol 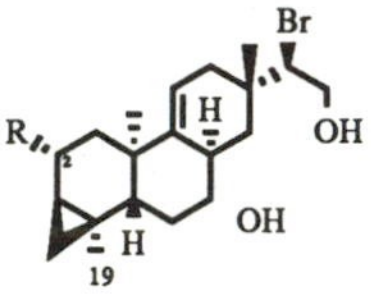 **30** R=OAc **31** R=H, Deoxyparguerol **32** Sphaerococcenol A

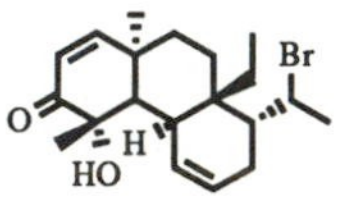

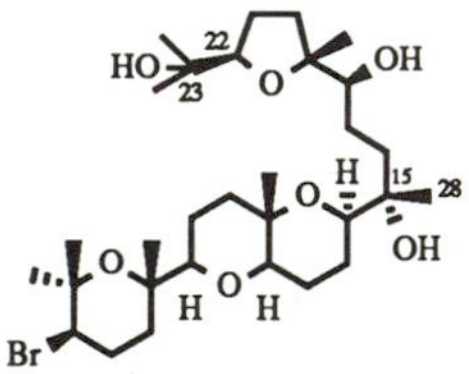

33 Thyrsiferol 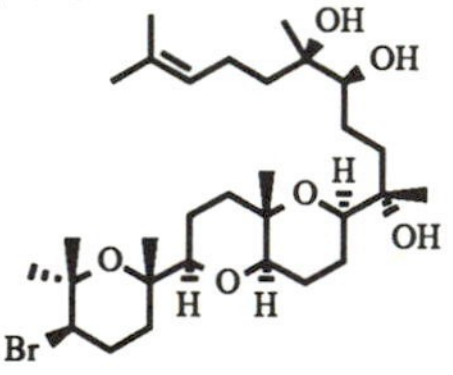 **34** Magireol A

Diterpenes. The alga *Laurencia obtusa* yielded parguerane derivatives (*64*), related to cytotoxic metabolites previously obtained from sea hares (*65*). Of these compounds deoxyparguerol (**30**) (*65*) was shown to be the most potent cytotoxic agent (P388, ED_{50}, 0.38 μg/ml), while its 2-deacetoxyl derivative (**31**) proved to be inactive (*64*). The antibacterial bromo-diterpene, sphaerococcenol A (**32**), was obtained from *Sphaerococcus coronopifolius* (Sphaerococcaceae) using a bioautographic assay system to guide isolation procedures (*66*).

Triterpenes. Squalene-derived polyethers from algae of the genus *Laurencia* proved to be potent biologically active compounds. *Laurencia obtusa* and *L. venusta* yielded thyrsiferol (**33**) (P388 ED_{50} 0.01 μg/ml), thyrsiferyl 23-acetate (P388, ED_{50}, 0.0003 μg/ml), 15(28)-anhydrothyrsiferyl diacetate (P388, ED_{50}, 0.05 μg/ml), 15-anhydrothyrsiferyl diacetate (P388, ED_{50}, 0.1 μg/ml), magireol A (**34**), magireol B and magireol C (both P388, ED_{50}, 0.03 μg/ml) (*67, 68*). Thyrsiferol (**33**), thyrsiferyl 23-acetate and the structurally related metabolite venustatriol, a stereoisomer of thyrsiferol, have also been reported to possess a strong antiviral activity (*69*).

Aromatic compounds. Bromophenols are typical metabolites derived from algae of the family Rhodomelaceae *(70-73, 75)*. Most of these metabolites are reported to possess antimicrobial properties. *Symphyocladia latiuscula* yielded *bis*-(2,3,6-tribromo-4,5-dihydroxybenzyl)ether (**35**) upon extraction with hot acetone, an antimicrobial compound which has been described as an artefact derived from 2,3,6-tribromo-5-hydroxybenzyl-1',4-disulfate (**36**) during extraction procedures (*72, 73*). A further antimicrobial sulphur-containing phenol, **37**, was obtained from *Grateloupia filicina* (Halymeniaceae). Compound **37** is the first sulfone derivative reported from the algae (*74*). The brominated indole (**38**), derived from *Laurencia brongniartii*, has been shown to possess cytotoxic and antimicrobial properties (*76*). A further indole-based compound isolated from *Prionitis lanceolata* is 3-(hydroxyacetyl)indole (**39**). Prior to its isolation as a natural product this compound had been obtained synthetically and was shown to be a potent stimulant of the central nervous system (*77*). The bromophenols vidalol A (**40**) and B (**41**), isolated from *Vidalia obtusaloba* (Rhodomelaceae) were shown to have anti-inflammatory and antibiotic activity (*78*). Both compounds demonstrated 96 % inhibition of phospholipase A_2 from bee venom at a concentration of 1.6 µg/ml, and were also shown to demonstrate significant *in vivo* anti-inflammatory activity using a phorbol-ester mouse ear assay (*78*). Vidalol A (**40**) and B (**41**) are *ortho*-catechols and are thus sensitive to oxidation resulting in formation of the corresponding *ortho*-quinones. It has therefore been suggested that the quinones are the biologically active analogues.

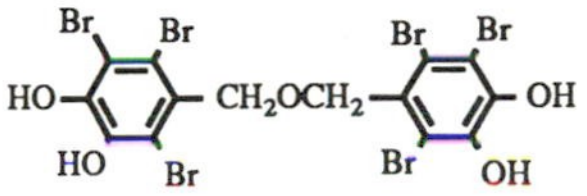

35  **36** 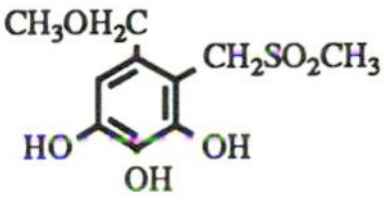 **37** 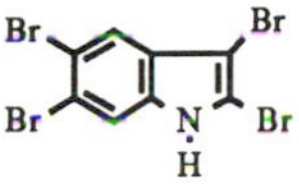 **38**

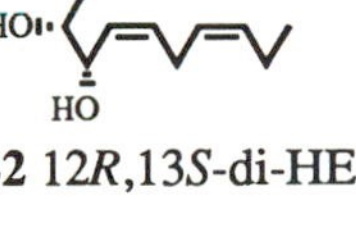

39 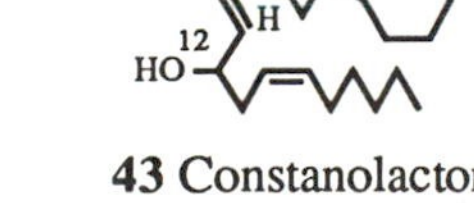 **40** Vidalol A 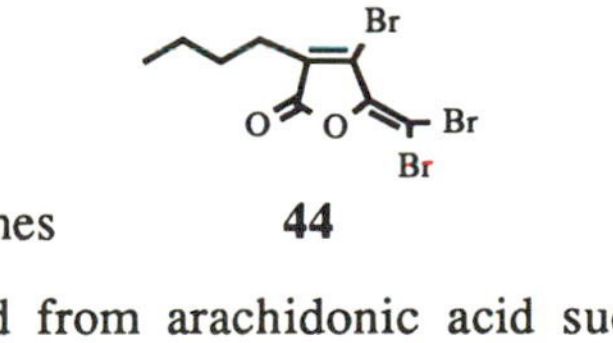 **41** Vidalol B

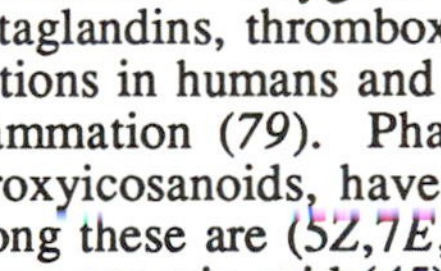

42 12*R*,13*S*-di-HEPE 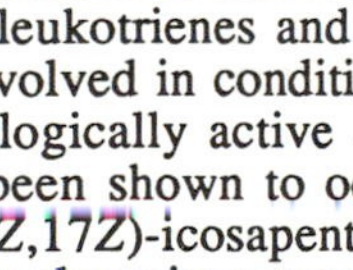 **43** Constanolactones 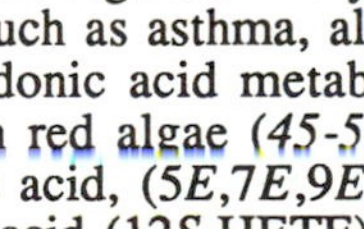 **44**

Icosanoids. Oxygenated metabolites derived from arachidonic acid such as prostaglandins, thromboxanes, leukotrienes and lipoxins regulate many metabolic functions in humans and are involved in conditionss such as asthma, allergies and inflammation (*79*). Pharmacologically active arachidonic acid metabolites, e.g. hydroxyicosanoids, have also been shown to occur in red algae (*45-50, 80, 81*). Among these are (5*Z*,7*E*,9*E*,14*Z*,17*Z*)-icosapentaenoic acid, (5*E*,7*E*,9*E*,14*Z*,17*Z*)-icosapentaenoic acid (*45*), 12*S*-hydroxyicosatetraenoic acid (12*S*-HETE) (*49*), 12*S*-hydroxyicosapentaenoic acid (12*S*-HEPE) (*47*), 12*R*,13*S*-dihydroxyicosatetraenoic acid (12*R*,13*S*-di-HETE) (*48, 81*), 12*R*,13*S*-dihydroxyicosapentaenoic acid (12*R*,13*S* -di-HEPE) (**42**) (*48, 50, 83*), 11-hydroxy-16-oxo-(5Z,8Z,12*E*,14*E*,17*E*)-icosapentaenoic acid (*46*), 6-*trans*-leukotriene B4 ethyl ether (*82*), and hepoxilin B_3 (*82*). In the red alga, *Constantinea simplex*, the cyclopropyl- and lactone-containing icosanoids, constanolactones A and B (**43**), two compounds epimeric at C 9, were found (*49*). From *Gracilariopsis lemaneiformis* galactolipids containing icosanoid moieties were described (*50*). From the same algal species, 12-lipoxygenase activity, involved in the biosynthesis of these arachidonic acid metabolites, was demonstrated (*82*). Further experiments suggested the existence of a hydroperoxidase isomerase enzyme in *G. lemaneiformis* (*83*).

Icosanoids are potent pharmacologically active compounds. A mixture of 12*R*,13*S*-dihydroxyicosatetraenoic acid and 12*R*,13*S*-dihydroxypentaenoic acid (**42**), isolated from *Farlowia mollis* (Cryptonemiales) was shown to inhibit human lymphocyte 5-lipoxygenase, dog kidney Na^+/K^+ ATPase, and to modulate fMLP-induced superoxide anion generation in human neutrophils (*48*). 12*S*-HEPE is a strong inhibitor of platelet aggregation and a mediator of inflammation (*47*). Recent studies with synthetic icosapentaenoic acids showed that these compounds selectively modulate the human platelet thromboxane A_2/prostaglandin H_2 receptor (*84*). Some of these unsaturated arachidonic acid metabolites also show antimicrobial activity (*46, 48*). Icosanoids have also been obtained from marine sponges (*85, 86*) and from starfish (*87*). Prostaglandins have been identified as the hypotensive principles of *Gracilaria lichenoides* (Gracilariaceae) (*88*).

Miscellaneous. *Furcellaria lumbricalis* has a contracting effect on the guinea pig ileum, which was traced to the presence of histamine in the ethanolic extract of the alga (*89, 90*). The fimbrolides isolated from *Delisea pulchra* (cf. *D. fimbriata*) (Bonnemaisoniaceae), e.g., **44** (*91*), and related compounds (*92-94*) are potent *in vitro* antimicrobial agents. Although their *in vitro* activity is comparable to that of standard antibiotics, e.g., ampicillin, and they are only moderately toxic to mice, they unfortunately failed to exhibit any activity during *in vivo* antimicrobial tests (*95*).

Biologically Active Compounds from Brown Algae (Phaeophyta)

The Dictyotales and Fucales are the most thoroughly phytochemically investigated orders of the brown algae. In contrast to metabolites derived from red algae natural products elaborated by brown algae are in general not halogenated. Within the order Dictyotales the family Dictyotaceae has yielded diterpenoids of many different structural types, e.g., extended sesquiterpenes (hydroazulenoids), dolabellanes, dolastanes and xenicanes (*96, 97*), while in the family Cystoseiraceae (Fucales) natural products of mixed biosynthesis, consisting of a terpenoid and an aromatic moiety, are common (*98-100*). Algae from the order Fucales also contain polyhydroxylated phenols consisting of phloroglucinol moieties (*101-105*). Biological activities reported for brown algal metabolites are predominantly antimicrobial and cytotoxic.

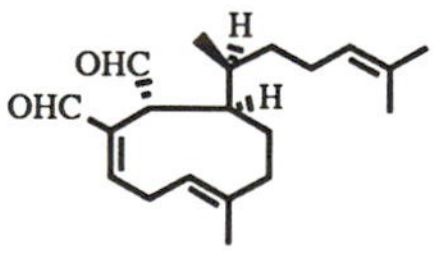

45 Dictyodial 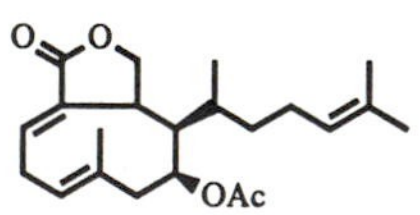 **46** 4-Acetoxydictyolactone 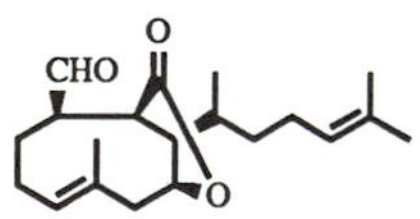 **47** Dictyotalide A

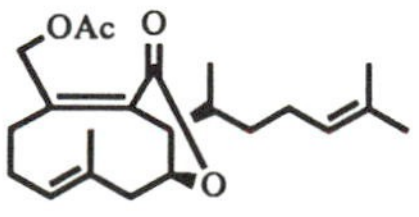

48 Dictyotalide B 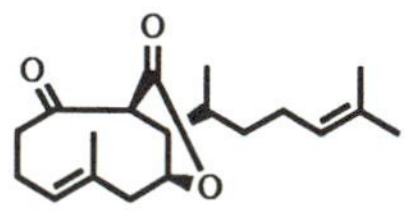 **49** Nordictyotalide 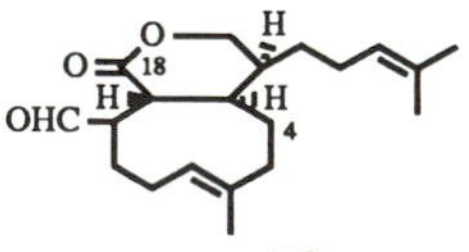 **50**

Diterpenes. Xenicane diterpenes were first observed in the soft coral genus *Xenia*; subsequently, related compounds were obtained from algae of the family Dictyotaceae, e.g., dictyodial (**45**) a substance with antibacterial and antifungal activities (*106*). The unusual xenicane and norxenicane lactones 4-acetoxydictyolactone (**46**), dictyotalide A

(**47**), dictyotalide B (**48**) and nordictyotalide (**49**), were isolated from *Dictyota dichotoma* and found to exhibit cytotoxicity towards B16 melanoma cells with ED_{50} values between 0.6 and 2.6 µg/ml (*107*). Rare xeniolides, with the carbonyl of an ester-function at C 18 (**50**), were isolated from *D. divaricata* (*97*). Compound **50** and its 4-hydroxyl derivative were tested in a battery of human cancer cells and found to exhibit cytotoxicity with some selectivity (G. M. König *et al.*, unpublished results). The crenulides, which are suggested to be biosynthetically related to the xenicanes, consist of a cyclooctane ring fused to a cyclopropane ring. The antimicrobial compound crenuladial (**51**) was obtained from *Dilophus ligulatus* (*108*) and is similar to the crenulacetals isolated from *Dictyota dichotoma* (*109*).

Among the biologically active diterpenes based on the hydroazulenoid ring system is dictyol H (**52**), which showed weak cytotoxicity (KB, ED_{50}, 22 µg/ml) and antibacterial properties (*110*), while dictyol C (**53**), isolated from *D. divaricata* (*97*), inhibited cyclooxygenase from sheep seminal vesicle microsomes with an IC_{50} value of 18.1 µg/ml, and phorbol ester-induced chemiluminescence in human granulocytes, suggesting an anti-inflammatory activity by this metabolite (*56*). The related compounds dictyotriol A and dityotriol B were obtained from the *in vivo* anti-inflammatory extract of *D. indica* from the Yellow Sea (*111*). Diterpenes possessing a dolabellane skeleton are also frequently encountered constituents in marine algae of the family Dictyotaceae. From an unidentified *Dictyota* species from the Italian coast the cytotoxic compound **54** was isolated (*112, 113*).

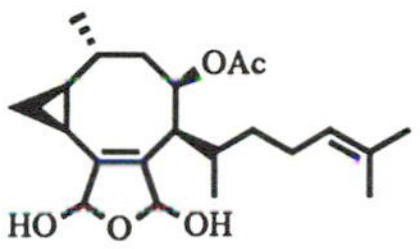

51 Crenuladial 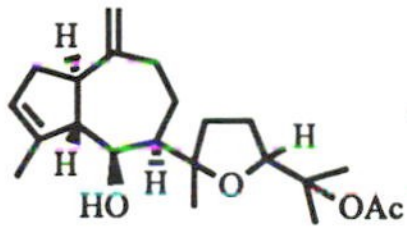 **52** Dictyol H 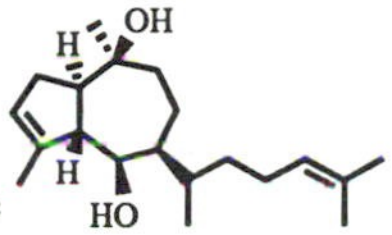 **53** Dictyol C 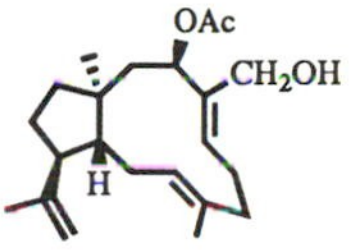 **54**

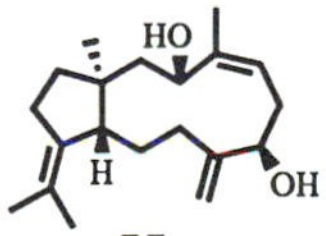

55 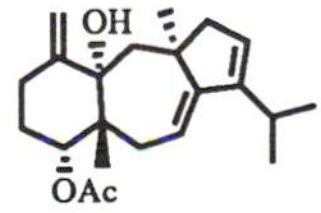 **56** 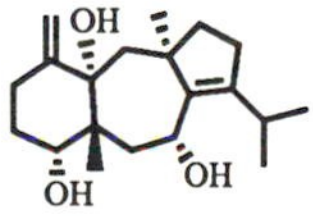 **57**

Compound **55**, a dolabellane isolated from *D. pardalis* (114), exhibited specific antimalarial activity with an IC_{50} value of 0.15 µg/ml towards chloroquine-resistant strains of *Plasmodium falciparum* (c.f., KB cells, IC_{50}, >20 µg/ml) (G. M. König *et al.*, unpublished data). In an investigation of the antifungal and molluscicidal activities of diterpenoids from brown algae of the family Dictyotaceae, six dolabellane derivatives were found to be moderately active against the fungus *Cladosporium cucumerinum*. Two dolabellanes were found to exhibit weak molluscicidal activity towards *Biomphalaria glabrata*, a fresh-water snail involved in the transmission of schistosomiasis (*115*). Further diterpenes having a dolabellane skeleton from *Dictyota dichotoma* were reported to possess antimicrobial activity towards gram-positive and gram-negative bacteria (*116*).

Dolastane diterpenes are tricyclic compounds, which may be biosynthetically derived from the dolabellanes. Dolastanes with interesting pharmacological properties were obtained from *D. divaricata* and *D. linearis*. Compound **56** has been shown to possess reversible histamine antagonism on guinea pig ileum preparations, while the related metabolite **57** inhibited cell division in fertilized sea urchin eggs (*117*). Yet another structural type of diterpene was isolated from *D. dichotoma* and *Pachydictyon coriaceum*. These algae afforded bicyclic diterpenoids with a decalin skeleton, e.g. dictyotin A (**58**). The cytotoxicity (ED_{50}) of these metabolites towards murine B16 melanoma cells ranged from 3-19 µg/ml (*118*). Further cytotoxic diterpenoids from brown seaweeds are based on the spatane skeleton. Spatol (**59**) isolated from *Spatoglossum schmittii*, the first reported spatane-type compound, was shown to have cytotoxic properties against T242 melanoma and 224C astrocytoma neoplastic cell lines (*119*). Related spatane derivatives without the epoxide functionalities of **59** were found to be much less active (*120*).

Meroditerpenoids. *Cystoseira mediterranea* is the source of several metabolites of mixed biosynthesis (*121-127*). Mediterraneol A (**60**), consisting of a *p*-hydroquinone and an unprecedented bicyclo[4.2.1]nonane based diterpenoid moiety, was the first such metabolite reported (*121*). The second compound of this type, mediterraneol B, is the C 7 epimer of **60**. The cyclised analogues, mediterraneols, C and D (**61**), are also epimeric at C 7. These mediterraneols were shown to inhibit the mobility of the sperm and the cell division of the fertilized sea urchin eggs with ED_{50} values of about 2 μg/ml, while mediterraneols A and B have also been shown to exhibit marginal *in vivo* activity in the P388 leukemia test (T/C = 128 % at 32 mg/kg) (*122*).

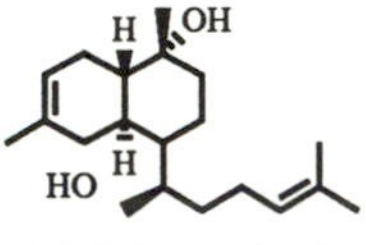

58 Dictyotin A 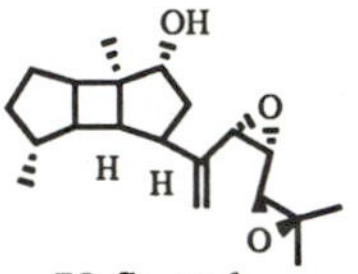 **59** Spatol 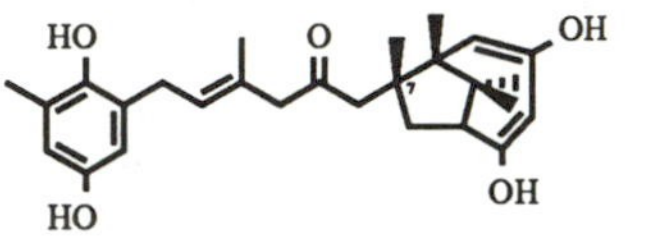 **60** Mediterraneol A

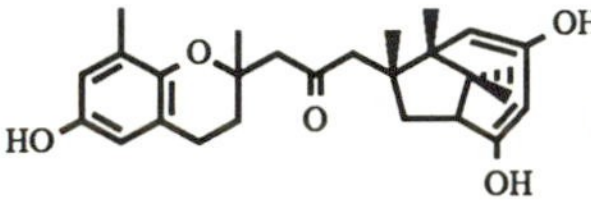

61 Mediterraneol D 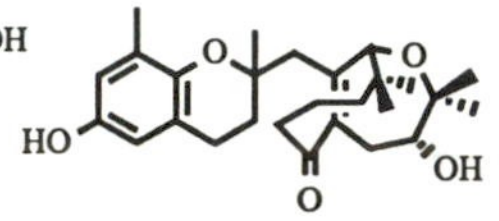 **62** Cystoseirol B 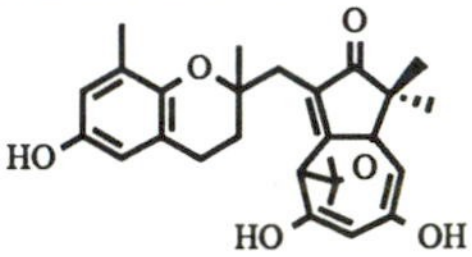 **63** Mediterraneol E

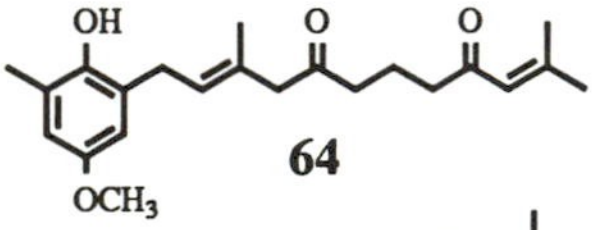

64 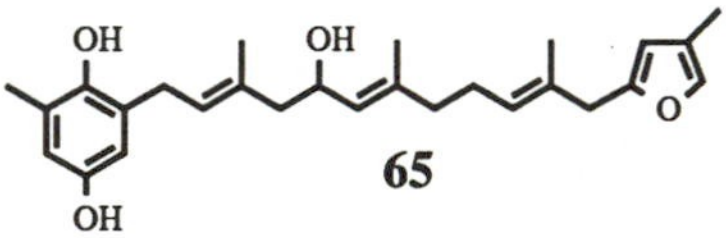 **65**

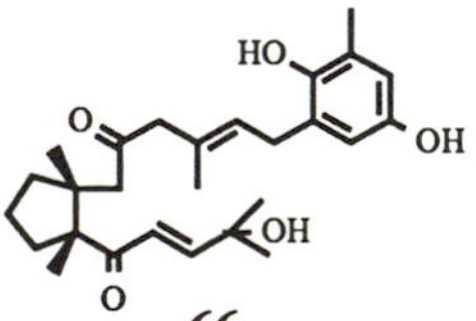

66 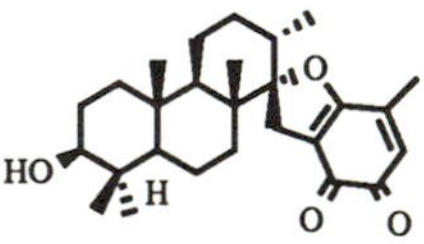 **67** Stypoldione

Cystoseirols (*123, 124*), which contain a novel oxabicyclo[5.4.1]dodecane ring, were isolated from various collections of algae from the family Cystoseiraceae from the French coast, e.g., *C. mediterranea*, *C. tamariscifolia* and *C. stricta*. Cystoseirol B (**62**) showed activity in a crown-gall potato disc assay (*125, 126*). The same test was applied to the activity-guided isolation of mediterraneol E (**63**) and mediterraneone (**64**), a norsesquiterpene (*125, 127*). Mediterraneol E is also claimed to have some inhibitory action on the enzyme elastase (*127*). The crude lipophilic extract of *C. spinosa* var. *squarrosa* exhibited antibacterial properties. This activity was associated with the presence of tetraprenyltoluquinol derivatives in the algal crude extract. Of the compounds isolated 5-hydroxycystofuranoquinol (**65**) was found to be the most active (*128*). Bifurcarenone (**66**), a further metabolite of mixed biosynthesis, consisting of a hydroquinone and a diterpenoid moiety, was isolated from *Bifurcaria galapagensis* (Cystoseiraceae) (*129*). It is an inhibitor of mitotic cell division and exhibited an IC_{50} of 4 μg/ml in a sea urchin egg assay (*129*). Recently its structure was revised based on synthesis (*130*), the absolute stereochemistry being determined by means of ORD analysis (*131*).

Stypoldione (**67**), an *ortho*-quinone, derived by oxidation of the naturally occurring stypotriol from *Stypopodium zonale* (Dictyotaceae) (*132*), exhibits a wide range of biological activities including growth inhibition of cultured tumour cells and P388 lymphocytic leukemia in mice (*133*). *In vitro* tests showed the inhibition of polymerisation of bovine brain tubulin into microtubules by this compound (*134*). Experiments which investigated the bonding of stypoldione to the sulfhydryl groups of proteins and amino acids suggest that the activities of this compound may be caused by its covalent binding to thiol groups of biologically important cellular molecules (*133*). In a fertilized sea urchin egg assay, the effect of stypoldione differed depending on the concentration used. At low concentrations, cell division was inhibited without affecting spindle organisation or chromosome movement, while higher concentrations were found to influence spindle formation, amino acid uptake and protein synthesis (*135, 136*). Stypoldione has also been found in the sea hare, *Aplysia dactylomela*, which grazes on *S. zonale* (*137*). Structurally similar meroditerpenes have also been shown to occur in sponges (*138*).

Aromatic compounds. Eckols, isolated from *Ecklonia kurome*, represent a new type of phlorotannin with a dibenzo-1,4-dioxin skeleton (*103-105*). Besides eckol (**68**) and 2-phloroeckol (**69**), *E. kurome* afforded several dimeric eckols (*103-105*). These compounds have been described as antiplasmin inhibitors, with eckol (**68**) being the most potent and specific inhibitor of α_2-macroglobulin and α_2-plasmin inhibitor. Structural requirements for the inhibitory activity of eckols are the 1,4-dioxane ring and the presence of at least three free hydroxyl groups in rings A and B: the C ring appears not to be essential for activity (*104*). Eckol and related compounds are considered promising lead structures for antithrombotic agents. Some of the dimeric compounds, while also being antiplasmin inhibitors, additionally inhibited plasmin activity itself, thus having less advantageous properties concerning fibrinolysis (*105*).

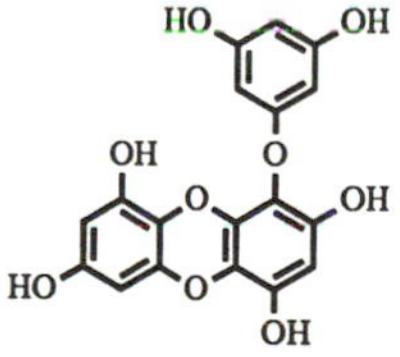

68 Eckol

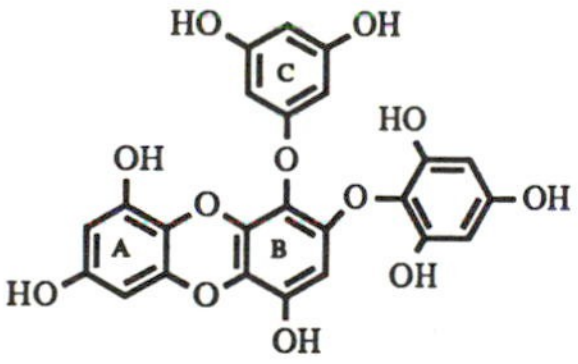

69 2-Phloroeckol

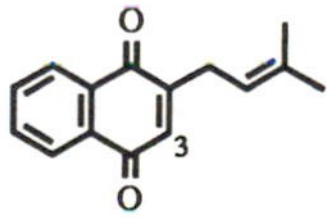

70 Deoxylapachol

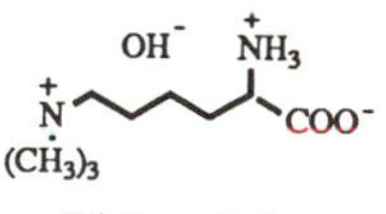

71 Laminine

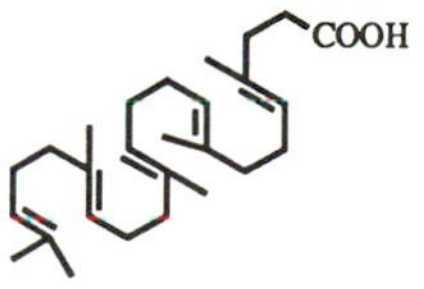

72 Turbinaric acid

The brown algal flora of New Zealand contains many species which exhibit cytotoxicity toward P388 leukemia cells (*139*). One of them, the endemic alga, *Landsburgia quercifolia* (Cystoseiraceae), yielded four aromatic compounds, one of which, deoxylapachol (**70**) proved to be responsible for the cytotoxic activity (P388, ED_{50}, 0.6 μg/ml) of the crude algal extract. Lapachol, the 3-hydroxy derivative of **70**, has *in vivo* antitumour activity (*139*).

Miscellaneous. Laminine (**71**) is a basic amino acid widely distributed in algae of the family Laminariaceae. These algae are known for their traditional use as hypotensive agents (*140*). *In vivo* experiments on rats have shown that laminine at high doses has a ganglion-blocking effect, which results in hypotension. The concentration of laminine in the algae, however, appears to be too low to explain their hypotensive action (*140*). Another investigation of Japanese *Laminaria* species, known to induce hypotension, detected histamine as the possible active principle (*141*). A further Japanese alga, *Turbinaria ornata* (Sargassaceae), showed moderate cytotoxicity against mouse melanoma cells (*142*). The active compound was identified as a secosqualene carboxylic acid, turbinaric acid (**72**), which was weakly cytotoxic towards human colon caricinoma cells (ED_{50}, 12.5 μg/ml). Prior to its first isolation from a natural source, metabolite **72** had been described as a synthetic compound which inhibited the enzyme squalene-2,3-epoxide cyclase.

Biologically Active Compounds from Blue-green Algae (Cyanophyta)

Blue green algae are a rich source of potent biologically active compounds. Structural classes elaborated by algae belonging to this group include alkaloids, acetogenins, macrolides, peptides, and terpenes among others.

Cyclic peptides. Cyclic peptides are responsible for the hepatotoxic effects of many blue-green algae (*143*). Microcystins, cyclic heptapeptides, have been found in *Microcystis* (*144*), *Anabaena* (*145*), *Oscillatoria* and *Nostoc* (*146*) species, and have recently been shown to inhibit protein phosphatase activity, especially 1 and 2A, in a similar manner to okadaic acid (*147*). A terrestrial *Anabaena* species from Hawaii yielded puwainaphycin C, a cardioactive cyclic peptide with positive inotropic effects (*148*). Calcium antagonistic activity was shown for scytonemin A, a cyclic peptide from a *Scytonema* species. This compound was also found to exhibit weak antimicrobial, antiprotozoal and cytotoxic activity (*149*). A cytotoxic and antibacterial undecapeptide, hormothamnin A, has recently been described from the Caribbean cyanobacterium, *Hormothamnion enteromorphoides* (*150, 151*).

Macrolides. The scytophycins are potent cytotoxic and fungicidal macrolides found in algae of the genera *Scytonema*, *Tolypothrix* (Scytonemataceae) (*152*) and *Cylindrospermum* (Nostocaceae) (*153*). Related compounds, the swinholides, have been isolated from sponges, possibly being derived from symbiotic micro-organisms (*154*). Tolytoxin (**73**), the major cytotoxic and antifungal constituent of *Scytonema mirabile* (*155*), exhibits ED_{50} values against human tumour cells ranging from 0.52 to 8 nM. It also exhibited slightly more potent activity than that of nystatin against a spectrum of fungi (*153*). Oscillariolide, a highly oxygenated brominated macrolide from the blue-green alga, *Oscillatoria* sp., was shown to inhibit the cell division of fertilized starfish eggs (*156*).

Miscellaneous. *Scytonema mirabile* also produces a complex mixture of other cytotoxins including the tantazoles, e.g., tantazole A (**74**), with murine solid tumour-selective cytotoxicity (*157*), mirabilene isonitriles exhibiting cytotoxicity in the range 1-10 μg/ml and weak antimicrobial activity (*158*). Mirabimides from the same algal species were found to be moderately cytotoxic (*159*).

Blue-green algae of the family Stigonemataceae contain fungicidal indole alkaloids of the hapalindole type (*160, 161*). Recent additions to this structural class were the ambiguine isonitriles, e.g., ambiguine E isonitrile (**75**), which were obtained from *Fischerella ambigua, Hapalosiphon hibernicus* and *Westiellopsis prolifica*, and proven to have fungicidal properties (*162*). The alkaloid anatoxin-a(s) (**76**), a further neurotoxin isolated from *Anabaena flos-aquae*, was found to be a potent anticholinesterase agent (*143, 163*). The CH_2Cl_2/2-propanol extract of the blue-green alga, *Tolypothrix nodosa*, was able to reverse the multi-drug resistance of a vinblastine-resistant human ovarian adenocarcinoma cell line. Subsequent investigations yielded the multi-drug resistance-reversing agent, tolyporphin (**77**), as dark-purple crystals. Doses of 1 μg/ml of tolyporphin potentiate the cytotoxicity of adriamycin and vinblastine in the aforementioned cell line (*164*). Blue-green algae belonging to the genera *Cylindrospermum* (*165*) and *Nostoc* (*166*) (Nostocaceae) yielded the first naturally occurring [m.n]paracyclophanes. Cylindrocyclophane A (*165*) and nostocyclophanes A-D (*166*) show moderate cytotoxic.

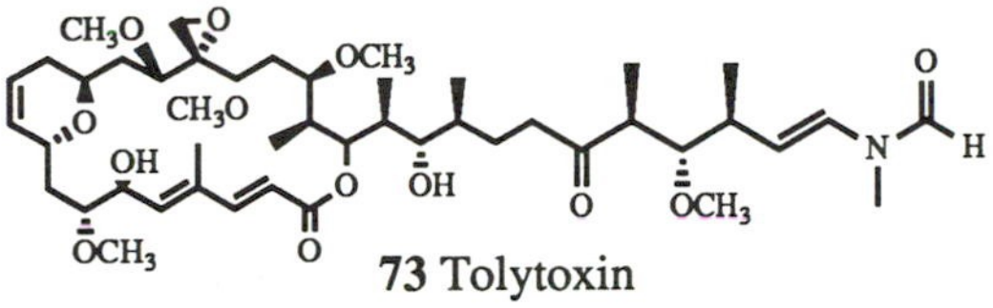

73 Tolytoxin

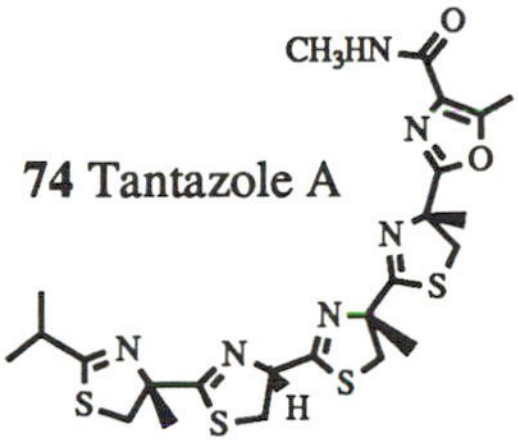

74 Tantazole A

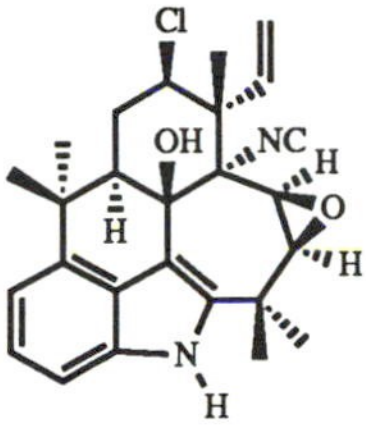

75 Ambiguine E

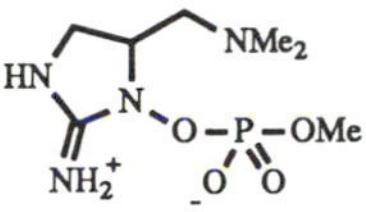

76 Anatoxin-a(s)

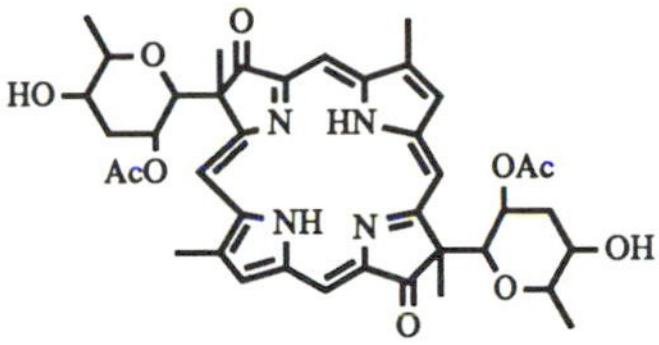

77 Tolyporphin

Conclusions and Future Directions

Relatively few algal secondary metabolites have undergone a detailed pharmacological evaluation. In most cases the only testing performed has been against microorganisms or for the evaluation of cytotoxicity against cancer cell lines. Improved bioassay techniques now allow the screening of crude materials and the subsequent activity-guided fractionation for the active principles. Testing of pure natural products may also be performed in many specific *in vitro* assays even with small amounts of sample (< 5 mg). Thus, algal metabolites already described in the literature, but as yet uninvestigated pharmacologically, warrant screening for their biological activities. Also, phytochemical investigations of algae have tended so far to concentrate on only a few genera; future work should therefore aim to explore the chemistry and associated biological activities of lesser known taxa. The possibility of culturing algal species, in particular blue-green alga, and thus securing a reliable source of material for investigation, may be very useful in the future for many algal researchers since such algae have to date yielded the most potent biologically active compounds. Polysaccharides also constitute a further type of algal metabolite with potential applications as antithrombic (*167-169*), antiviral (*170, 171*) and antitumor (*172, 173*) agents, and as such a group well worth investigating.

Acknowledgments

We would like to thank Drs. J. M. Pezzuto and C. K. Angerhofer, Program for Collaborative Research in the Pharmaceutical Sciences, University of Illinois at Chicago, Chicago, Illinois, for performing many of the biological assays associated with our research and Drs. C. Goez and R. de Nys, Department of Pharmacy, Swiss Federal Institute of Technology, Zürich, Switzerland, for providing useful suggestions relating to the preparation of this chapter.

Literature Cited

1. Bhakuni, D. S.; Silva, M. *Bot. Mar.* **1974**, *17*, 40-51.

2. Baker, J. T. *Hydrobiologia* **1984**, *116/117*, 29-40.
3. Fenical, W. In *Proceedings of the Joint China-U. S. Phycology Symposium*; 16 20 Nov. 1981, Tseng, C. K., Ed.; Science Press: Beijing, China, **1983**, pp 497-521.
4. Hoppe, H. A.; Levring, T., Eds. *Marine Algae in Pharmaceutical Science*; Walter de Gruyter: Berlin, New York, **1979**, Vol. 1; **1982**, Vol. 2.
5. Paul, V. J.; Fenical, W. In *Bioorganic Marine Chemistry*; Scheuer, P.J., Ed.; Springer Verlag: Berlin, Heidelberg, New York, London, Paris, Tokyo, **1987**, Vol. 1; pp 1-29.
6. Naylor, S.; Hanke, F. J.; Manes, L. V.; Crews, P. In *Progress in the Chemistry of Organic Natural Products*; Herz, W.; Grisebach.; H., Kirby, G. W., Eds.; Springer Verlag: Wien, New York, **1983**, Vol. 44; pp 189-241.
7. Moore, R. E. In *Marine Natural Products Chemistry, Chemical and Biological Perspectives*; Scheuer, P. J., Ed.; Academic Press: New York, **1978**, Vol. 1; pp 43-124.
8. Martin, J. D.; Darias, J. In *Marine Natural Products, Chemistry Chemical and Biological Perspectives*; Scheuer, P. J., Ed.; Academic Press: New York, **1978**, Vol. 1; pp 125-173.
9. Fattorusso, E.; Piattelli, M. In *Marine Natural Products Chemistry, Chemical and Biological Perspectives*; Scheuer, P. J., Ed.; Academic Press: New York, **1980**, Vol. 3; pp 95-140.
10. Erickson, K. L. In *Marine Natural Products Chemistry, Chemical and Biological Perspectives*; Scheuer, P. J., Ed.; Academic Press: New York, **1983**, Vol. 5; pp 131-257.
11. Munro, M. H. G.; Luibrand, R. T.; Blunt, J. W. In *Bioorganic Marine Chemistry*; Scheuer, P.J., Ed.; Springer Verlag: Berlin, Heidelberg, New York, London, Paris, Tokyo, **1987**, Vol. 1; pp 93-176.
12. Faulkner, D. J. *Nat. Prod. Rep.* **1991**, *8*, 97-147, and references therein.
13. Clayton, M. N.; King, R. J., Eds. *Biology of Marine Plants*; Longman Cheshire: Melbourne, Australia, **1990**.
14. Hodgson, L. M. *Bot. Mar.* **1984**, *27*, 387-390.
15. Fusetani, N. In *Bioorganic Marine Chemistry*; Scheuer, P.J., Ed.; Springer Verlag: Berlin, Heidelberg, New York, London, Paris, Tokyo, **1987**, Vol.1; pp 61-92.
16. Schwartz, R. E.; Hirsch, C. F.; Sesin, D. F.; Flor, J. E.; Chartrain, M.; Fromtling, R. E.; Harris, G. H.; Salvatore, M. J.; Liesch, J. M.; Yudin, K. *J. Ind. Microbiol.* **1990**, *5*, 113-124.
17. Sun, H. H.; Fenical, W. *Tetrahedron Lett.* **1979**, 685-688.
18. Barnekow, D. E.; Cardellina II, J. H.; Zektzer, A. S.; Martin, G. E. *J. Am. Chem. Soc.* **1989**, *111*, 3511-3517.
19. Paul, V. J.; Fenical, W. *Tetrahedron* **1984**, *40*, 2913-2918.
20. Nakatsu, T.; Ravi, B.N.; Faulkner, D. J. *J. Org. Chem.* **1981**, *46,* 2435-2438.
21. Paul, V. J.; Fenical, W. *Tetrahedron* **1984**, *40*, 3053-3062.
22. Paul, V. J.; Fenical, W. *Science* **1983**, *221*, 747-749.
23. Koehn, F. E.; Gunasekera, S. P.; Niel, D. N.; Cross, S. S. *Tetrahedron Lett.* **1991**, *32*, 169-172.
24. Sun, H. H.; Paul, V. J.; Fenical, W. *Phytochemistry* **1983**, *22*, 743-745.
25. Colon, M.; Guevara, P.; Gerwick, W. H.; Ballantine, D. *J. Nat. Prod.* **1987**, *50*, 368-374.
26. Carte, B. K.; Troupe, N.; Chan, J. A.; Westley, J. W.; Faulkner, D. J. *Phytochemistry* **1989**, *28*, 2917-2919.
27. McConnell, O. J.; Hughes, P. A.; Targett, N. M. *Phytochemistry* **1982**, *21*, 2139-2141.
28. Wall, M. E.; Wani, M. C.; Manikumar, G.; Taylor, H.; Hughes, T.J.; Gaetano, K.; Gerwick, W. H.; McPhail, A. T.; McPhail, D. R. *J. Nat. Prod.* **1989**, *52*, 1092-1099.

29. Högberg, H.; Thomson, R. H.; King, T. J. *J. Chem. Soc. Perkin Trans. I* **1976**, 1696-1701.
30. Wall, M. E.; Wani, M. C.; Hughes, T. J.; Taylor, H. *J. Nat. Prod.* **1988**, *51*, 866-873.
31. Fenical, W. In *Food-Drugs from the Sea Proceedings*; Webber, H. H.; Ruggieri, G. D., Eds.; Marine Technology Society: Washington D.C., **1976**, pp 388-394.
32. Endo, M.; Nakagawa, Y.; Hamamoto, J. ; Ishihama, H. *Pure and Appl. Chem.* **1986**, *58*, 387-394.
33. Iguchi, M.; Nishiyama, A.; Etoh, H.; Okamoto, K.; Yamamura, S.; Kato, Y. *Chem. Pharm. Bull.* **1986**, *34*, 4910-4915.
34. Kuniyoshi, M.; Yamada, K.; Higa, T. *Experientia* **1985**, *41*, 523-524.
35. Blunden, G.; Gordon, S. M. *Pharm. Internat.* **1986**, *7*, 287-290.
36. Garg, H. S.; Sharma, M.; Bhakuni, D. S.; Pramanik, B. N.; Bose, A. K. *Tetrahedron Lett.* **1992**, *33*, 1641-1644.
37. Yamada, K.; Shizuri, Y.; Ishida, Y.; Shibata, S. *J. Pharm. Sci.* **1983**, *72*, 945-946.
38. Anthoni, U.; Cristophersen, C.; Øgård Madsen, J.; Wium-Andersen, S.; Jacobsen, N. *Phytochemistry* **1980**, *19*, 1228-1229.
39. Anthoni, U.; Christophersen, C.; Jacobsen, N.; Svendsen, A. *Tetrahedron* **1982**, *38*, 2425-2427.
40. Jacobsen, N.; Pedersen, L. K. *Pestic. Sci.* **1983**, *14*, 90-97.
41. Anthoni, U.; Nielsen, P. H.; Smith-Hansen, L, Wium-Andersen, S.; Christophersen, C. *J. Org. Chem.* **1987**, *52*, 694-695.
42. Nielsen, L. E.; Pedersen, L. K. *Experientia* **1984**, *40*, 186-188.
43. Chengkui, Z.; Junfu, Z. *Hydrobiologia* **1984**, *116/117*, 152-154.
44. Abe, S.; Kaneda, T. In *Proceedings of the Seventh International Seaweed Symposium*; Sapporo, Japan, Aug. 8-12, **1971**; University of Tokyo Press: Tokyo, Japan, **1972**, pp 562-565.
45. Lopez, A.; Gerwick, W. H. *Lipids* **1987**, *22*, 190-194.
46. Lopez, A.; Gerwick, W. H. *Tetrahedron Lett.* **1988**, *29*, 1505-1506.
47. Bernart, M.; Gerwick, W. H. *Tetrahedron Lett.* **1988**, *29*, 2015-2018.
48. Solem, M. L.; Jiang, Z. D.; Gerwick, W. H. *Lipids* **1989**, *24*, 256-260.
49. Nagle, D. G.; Gerwick, W.H. *Tetrahedron Lett.* **1990**, *21*, 2995-2998.
50. Jiang, Z. D.; Gerwick, W. H. *Phytochemistry* **1990**, *29*, 1433-1440.
51. König, G. M.; Wright, A. D.; Sticher, O. *J. Nat. Prod.* **1990**, *53*, 1615-1618.
52. McConnell, O. J.; Fenical, W. *J. Org. Chem.* **1978**, *43*, 4238-4241.
53. Gonzalez, A. G.; Daria, V.; Estevez, E. *Planta Med.* **1982**, *44*, 44-46.
54. San-Martin, A.; Negrete, R.; Rovirosa, J. *Phytochemistry* **1991**, *30*, 2165-2169.
55. Wright, A. D.; König, G. M.; Sticher, O. *Tetrahedron* **1991**, *47*, 5717-5724.
56. König, G. M.; Wright, A. D.; Sticher, O.; Jurcic, K.; Offermann, F.; Redl, K.; Wagner, H.; Angerhofer, C. D.; Pezzuto, J. M. International Research Congress on Natural Products, Chicago, Illinois, July 21-26, **1991**, P-129.
57. Coll, J. C.; Skelton, B. W.; White, A. H.; Wright, A. D. *Aust. J. Chem.* **1988**, *41*, 1743-1753.
58. de Nys, R.; Coll, J. C.; Price, I. R. *Marine Biology* **1991**, *108*, 315-320.
59. Pettit, G. R.; Herald, C. L.; Allen, M. S.; Von Dreele, R. B.; Vanell, L. D.; Kao, J. P. Y.; Blake, W. *J. Am. Chem. Soc.* **1977**, *99*, 262-263.
60. Capon, R.; Ghisalberti, E. L.; Jefferies, P. R.; Skelton, B. W.; White, A. H. *Tetrahedron* **1981**, *37*, 1613-1621.
61. Wright, A. D.; König, G. M.; Sticher, O. *J. Nat. Prod.* **1991**, *54*, 1025-1033.
62. Ichiba, T.; Higa, T. *J. Org. Chem.* **1986**, *51*, 3364-3366.
63. Caccamese, S.; Rinehart Jr., K. L. In *Drugs and Food from the Sea, Myth or Reality ?*; Kaul, P. N.; Sindermann, C. J., Eds.; The University of Oklahoma Press: Norman, Oklahoma, **1978**; pp 187-197.

64. Suzuki, T.; Takeda, S.; Hayama, N.; Tanaka, I.; Komiyama, K. *Chem. Lett. Jpn.* **1989**, 969-970.
65. Schmitz, F. J.; Michaud, D. P.; Schmidt, P. G. *J. Am. Chem. Soc.* **1982**, *104*, 6415-6423.
66. Caccamese, S.; Cascio, O.; Compagnini, A. *J. Chromatogr.* **1989**, *478*, 255-258.
67. Suzuki, T.; Suzuki, M.; Furusaki, A.; Matsumoto, T.; Kato, A.; Imanaka, Y.; Kurosawa, E. *Tetrahedron Lett.* **1985**, *26*, 1329-1332.
68. Suzuki, T.; Takeda, S.; Suzuki, M.; Kurosawa, E.; Kato, A.; Imanaka, Y. *Chem. Lett. Jpn.* **1987**, 361-364.
69. Sakemi, S.; Higa, T.; Jefford, C. W.; Bernardinelli, G. *Tetrahedron Lett.* **1986**, *27*, 4287-4290.
70. Kurata, K.; Amiya, T. *Bull. Chem. Soc. Jpn.* **1980**, *53*, 2020-2022.
71. Suzuki, M.; Kowata, N.; Kurosawa, E. *Bull. Chem. Soc. Jpn.* **1980**, *53*, 2099-2100.
72. Kurata, K.; Amiya, T. *Phytochemistry* **1980**, *19*, 141-142.
73. Kurata, K.; Amiya, T. *Chem. Lett. Jpn.* **1980**, 279-280.
74. Nozaki, H.; Minohara, K.; Miyazaki, I.; Kondo, H.; Shirane, F.; Nakayama, M. *Agric. Biol. Chem.* **1988**, *52*, 3229-3230.
75. Valdebenito, H.; Bittner, M.; Sammes, G.; Silva, M.; Watson, W. H. *Phytochemistry* **1982**, *21*, 1456-1457.
76. Carter, G. T.; Rinehart, Jr., K. L.; Li, L. H.; Kuentzel, S. L.; Connor, J. L. *Tetrahedron Lett.* **1978**, 4479-4482.
77. Bernart, M.; Gerwick, W. H. *Phytochemistry* **1990**, *29*, 3697-3698.
78. Wiemer, D. F.; Idler, D. D.; Fenical, W. *Experientia* **1991**, *47*, 851-853.
79. Parker, C. W. *Annu. Rev. Immunol.* **1987**, *5*, 65-84.
80. Moghaddam, M. F.; Gerwick, W. H.; Ballantine, D. L. *Prostaglandins* **1989**, *37*, 303-308.
81. Lumin, S; Falck, J. R. *Tetrahedron Lett.* **1990**, *31*, 2971-2974.
82. Moghaddam, M. F.; Gerwick, W. H. *Phytochemistry* **1990**, *29*, 2457-2459.
83. Gerwick, W. H.; Moghaddam, M. F.; Hamberg, M. *Arch. Biochem. Biophys.* **1991**, *290*, 436-444.
84. Parent, C. A.; Lagarde, M.; Venton, D. L.; Le Breton, G. C. *J. Biol. Chem.* **1992**, *267*, 6541-6547.
85. Guerriero, A.; D'Ambrosio, M.; Pietra, F.; Ribes, O.; Duhet, D. *J. Nat. Prod.* **1990**, *53*, 57-61.
86. Carballeira, N. M.; Shalabi, F.; Negron, V. *Lipids* **1989**, *24*, 229-232.
87. Bruno, I.; D'Auria, M. V.; Iorizzi, M.; Minale, L.; Riccio, R. *Experientia* **1992**, *48*, 114-115.
88. Gregson, R. P.; Marwood, J. F.; Quinn, R. J. *Tetrahedron Lett.* **1979**, 4505-4506.
89. Andersson, L.; Bohlin, L. *Acta Pharm. Suec.* **1984**, *21*, 373-376.
90. Barwell, C. *Bot. Mar.* **1979**, *22,* 399-401.
91. Kazlauskas, R.; Murphy, P. T.; Quinn, R. J.; Wells, R. J. *Tetrahedron Lett.* **1977**, 37-40.
92. Rose, A. G.; Pettus, J. A.; Sims, J. J. *Tetrahedron Lett.* **1977**, 1847-1850.
93. Ohta, K. *Agric. Biol. Chem.* **1977**, *41*, 2105-2106.
94. Pettus, J. A.; Wing, R. M.; Sims, J. J. *Tetrahedron Lett.* **1977**, 41-44.
95. Reichelt, J. L.; Borowitzka, M. A. *Hydrobiologia* **1984**, *116/117*, 158-167.
96. König, G. M.; Wright, A. D.; Sticher, O. *Phytochemistry* **1991**, *30,* 3679-3682.
97. König, G. M.; Wright, A. D.; Sticher, O. *Tetrahedron* **1991**, *47*, 1399-1410.
98. Piattelli, M. *New J. Chem.* **1990**, *14*, 777-782.
99. Amico, V.; Neri, P.; Oriente, G.; Piattelli, M. *Phytochemistry* **1989**, *28*, 215-219.

100. Amico, V.; Piattelli, M.; Cunsolo, F.; Neri, P.; Ruberto, G. *J. Nat. Prod.* **1989**, *52*, 962-969.
101. Glombitza, K. W.; Rauwald, H. W.; Eckhard, G. *Planta Med.* **1977**, *32*, 33-45.
102. Glombitza, K. W. In *Marine Algae in Pharmaceutical Science*; Hoppe, H. A.; Levring, T., Eds.; Walter de Gruyter: Berlin, New York, **1979**, Vol. 1; pp 303 -342.
103. Fukuyama, Y.; Miura, I.; Kinzyo, Z.; Mori, H.; Kido, M.; Nakayama, Y.; Takahashi, M.; Ochi, M. *Chem. Lett. Jpn.* **1985**, 739-742.
104. Fukuyama, Y.; Kodama, M.; Miura, I.; Kinzyo, Z.; Kido, M.; Mori, H.; Nakayama, Y.; Takahashi, M. *Chem. Pharm. Bull.* **1989**, *37*, 349-353.
105. Fukuyama, Y.; Kodama, M.; Miura, I.; Kinzyo, Z.; Mori, H.; Nakayama, Y.; Takahashi, M. *Chem. Pharm. Bull.* **1989**, *37*, 2438-2440.
106. Finer, J.; Clardy, J.; Fenical, W.; Minale, L.; Riccio, R.; Battaile, J.; Kirkup, M.; Moore, R. E. *J. Org. Chem.* **1979**, *44*, 2044-2047.
107. Ishitsuka, M. O.; Kusumi, T.; Kakisawa, H. *J. Org. Chem.* **1988**, 53, 5010-5013.
108. Tringali, C.; Oriente, G.; Piattelli, M.; Geraci, C.; Nicolosi, G.; Breitmaier, E. *Can. J. Chem.* **1988**, *66*, 2799-2802.
109. Kusumi, T.; Muanza-Nkongolo, D.; Goya, M.; Ishitsuka, M.; Iwashita, T.; Kakisawa, H. *J. Org. Chem.* **1986**, *51*, 384-387.
110. Alvarado, A. B.; Gerwick, W. H. *J. Nat. Prod.* **1985**, *48*, 132-134.
111. Niang, L. L.; Hung, X. *Hydrobiologia* **1984**, *116/117*, 168-170.
112. Tringali, C.; Piattelli, M.; Nicolosi, G. *Tetrahedron* **1984**, *40*, 799-803.
113. Tringali, C.; Oriente, G.; Piattelli, M.; Nicolosi, G. *J. Nat. Prod.* **1984**, *47*, 615-619.
114. Wright, A. D.; König, G. M.; Sticher, O.; Lubini, P.; Hofmann, P.; Dobler, M. *Helv. Chim. Acta* **1991**, *74*, 1801-1807.
115. Tringali, C.; Piattelli, M.; Nicolosi, G.; Hostettmann, K. *Planta Med.* **1986**, *52*, 404-406.
116. Amico, V.; Oriente, G.; Piattelli, M.; Tringali, C.; Fattorusso, E.; Magno, S.; Mayol, L. *Tetrahedron* **1980**, *36*, 1409-1414.
117. Crews, P.; Klein, T. E.; Hogue, E. R.; Myers, B. L. *J. Org. Chem.* **1982**, *47*, 811-815.
118. Ishitsuka, M. O.; Kusumi, T.; Ichikawa, A.; Kakisawa, H. *Phytochemistry* **1990**, *29*, 2605-2610.
119. Gerwick, W. H.; Fenical, W.; Van Engen, D.; Clardy, J. *J. Am. Chem. Soc.* **1980**, *102*, 7991-7993.
120. Gerwick, W. H.; Fenical, W. *J. Org. Chem.* **1983**, *48*, 3325-3329.
121. Francisco, C.; Banaigs, B.; Valls, R.; Codomier, L. *Tetrahedron Lett.* **1985**, *26*, 2629-2632.
122. Francisco, C.; Banaigs, B.; Teste, J.; Cave, A. *J. Org. Chem.* **1986**, *51*, 1115-1120.
123. Francisco, C.; Banaigs, B.; Codomier, L.; Cave, A. *Tetrahedron Lett.* **1985**, *26*, 4919-4922.
124. Francisco, C.; Banaigs, B.; Rakba, M.; Teste, J.; Cave, A. *J. Org. Chem.* **1986**, *51*, 2707-2711.
125. Fadli, M.; Aracil, J. M.; Jeanty, G.; Banaigs, B.; Francisco, C. *J. Nat. Prod.* **1991**, *54*, 261-264.
126. Galsky, A. G.; Wilsey, J. P.; Powell, R. G. *Plant Physiol.* **1980**, *65*, 184-185.
127. Fadli, M.; Aracil, J. M.; Jeanty, G.; Banaigs, B.; Francisco, C.; Moreau, S. *Tetrahedron Lett.* **1991**, *32*, 2477-2480.
128. Amico, V.; Cunsolo, F.; Neri, P.; Piattelli, M.; Ruberto, G. *Phytochemistry* **1988**, *27*, 1327-1331.

129. Sun, H. H.; Ferrara, N. M.; McConnell, O. J.; Fenical, W. *Tetrahedron Lett.* **1980**, *21*, 3123-3126.
130. Mori, K.; Uno, T. *Tetrahedron* **1989**, *45*, 1945-1958.
131. Mori, K.; Uno, T.; Kido, M. *Tetrahedron* **1990**, *46*, 4193-4204.
132. Gerwick, W. H.; Fenical, W. *J. Org. Chem.* **1981**, *46*, 22-27.
133. O'Brien, E. T.; Asai, D. J.; Groweiss, A.; Lipshutz, B.H.; Fenical, W.; Jacobs, R. S.; Wilson, L. *J. Med. Chem.* **1986**, *29*, 1851-1855.
134. O'Brien, E. T.; Jacobs, R. S.; Wilson, L. *Mol. Pharmacol.* **1983**, *24*, 493-499.
135. White, S. J.; Jacobs, R. S. *Molec. Pharmacol.* **1983**, *24*, 500-508.
136. O'Brien, E. T.; Asai, D. J.; Jacobs, R. S.; Wilson, L. *Molec. Pharmacol.* **1989**, *35*, 635-642.
137. Gerwick, W. H.; Whatley, G. *J. Chem. Ecol.* **1989**, *15*, 677-683.
138. Salva, J.; Faulkner, D. J. *J. Org. Chem.* **1990**, *55*, 1941-1943.
139. Perry, N. B.; Blunt, J. W.; Munro, M. H. G. *J. Nat. Prod.* **1991**, *54*, 978-985.
140. Girard, J. P.; Marion, C.; Liutkus, M.; Boucard, M.; Rechencq, E.; Vidal, J. P.; Rossi, J. C. *Planta Med.* **1988**, *54*, 193-196.
141. Funayama, S.; Hikino, H. *Planta Med.* **1981**, *41*, 29-33.
142. Asari, F.; Kusumi, T.; Kakisawa, H. *J. Nat. Prod.* **1989**, *52*, 1167-1169.
143. Carmichael, W. W.; Mahmood, N. A.; Hyde, E. G. In *Marine Toxins*; Hall, S.; Strichartz, G., Eds.; ACS Symposium Series 418; American Chemical Society: Washington, DC, **1990**; pp 87-106.
144. Namikoshi, M.; Rinehart, K. L.; Sakai, R.; Stotts, R.R.; Dahlem, A. M.; Beasley, V. R.; Carmichael, W. W.; Evans, W. R. *J. Org. Chem.* **1992**, *57*, 866-872.
145. Harada, K.; Ogawa, K.; Kimura, Y.; Murata, H.; Suzuki, M.; Thorn, P. M.; Evans, W. R.; Carmichael, W. W. *Chem. Res. Toxicol.* **1991**, *4*, 535-540.
146. Namikoshi, M.; Rinehart, Jr., K. L.; Sakai, R.; Sivonen, K.; Carmichael, W.W. *J. Org. Chem.* **1990**, *55*, 6135-6139.
147. MacKintosh, C.; Beattie, K. A.; Klumpp, S.; Cohen, P.; Codd, G. A. *FEBS* **1990**, *264*, 187-192.
148. Moore, R. E.; Bornemann, V.; Niemczura, W. P.; Gregson, J. M.; Chen, J.; Norton, T. R.; Patterson, G. M. L.; Helms, G. L. *J. Am. Chem. Soc.* **1989**, *111*, 6128-6132.
149. Helms, G. L.; Moore, R. E.; Niemczura, W. P.; Patterson, G. M. L.; Tomer, K. B.; Gross, M. L. *J. Org. Chem.* **1988**, *53*, 1298-1307.
150. Gerwick, W. H.; Mrozek, C.; Moghaddam, M. F.; Agarwal, S. K. *Experientia* **1989**, *45*, 115-121.
151. Gerwick, W. H.; Jiang, Z. D.; Agarwal, S. K.; Farmer, B. T. *Tetrahedron* **1992**, *48*, 2313-2324.
152. Ishibashi, M.; Moore, R. E.; Patterson, G. M. L.; Xu, C.; Clardy, J. *J. Org. Chem.* **1986**, *51*, 5300-5306.
153. Jung, J. H.; Moore, R. E.; Patterson, G. M. L. *Phytochemistry* **1991**, *30*, 3615-3616.
154. Tsukamoto, S.; Ishibashi, M.; Sasaki, T.; Kobayashi, J. *J. Chem. Soc. Perkin Trans. I* **1991**, 3185-3188.
155. Carmeli, S.; Moore, R. E.; Patterson, G. M. L. *J. Nat. Prod.* **1990**, *53*, 1533-1542.
156. Murakami, M.; Matsuda, H.; Makabe, K.; Yamaguchi, K. *Tetrahedron Lett.* **1991**, *32*, 2391-2394.
157. Carmeli, S.; Moore, R. E.; Patterson, G. M. L. *J. Am. Chem. Soc.* **1990**, *112*, 8195-8197.
158. Carmeli, S.; Moore, R. E.; Patterson, G. M. L.; Mori, Y.; Suzuki, M. *J. Org. Chem.* **1990**, *55*, 4431-4438.
159. Carmeli, S.; Moore, R. E.; Patterson, G. M. L. *Tetrahedron* **1991**, *47*, 2087-2096.

160. Moore, R. E.; Cheuk, C.; Yang, X. G.; Patterson, G. M. L.; Bonjouklian, R.; Smitka, T. A.; Mynderse, J. S.; Foster, R. S.; Jones, N. D.; Swartzendruber, J. K.; Deeter, J. B. *J. Org. Chem.* **1987**, *52*, 1036-1043.
161. Moore, R. E.; Cheuk, C.; Patterson, G. M. L. *J. Am. Chem. Soc.* **1984**, *106*, 6456-6457.
162. Smitka, T. A.; Bonjouklian, R.; Doolin, L.; Jones, N. D.; Deeter, J. B.; Yoshida, W. Y.; Prinsep, M. R.; Moore, R. E.; Patterson, G. M. L. *J. Org. Chem.* **1992**, *57*, 857-861.
163. Matsunaga, S.; Moore, R. E.; Niemczura, W. P.; Carmichael, W. W. *J. Am. Chem. Soc.* **1989**, *111*, 8021-8023.
164. Prinsep, M. R.; Caplan, F. R.; Moore, R. E.; Patterson, G. M. L.; Smith, C. D. *J. Am. Chem. Soc.* **1992**, *114*, 385-387.
165. Moore, B. S.; Chen, J.; Patterson, G. M. L.; Moore, R. E.; Brinen, L. S.; Kato, Y.; Clardy, J. *J. Am. Chem. Soc.* **1990**, *112*, 4061-4063.
166. Chen, J.; Moore, R. E.; Patterson, G. M. L. *J. Org. Chem.* **1991**, *56*, 4360-4364.
167. Grauffel, V.; Kloareg, B.; Mabeau, S.; Durand, P.; Jozefonvicz, J. *Biomaterials* **1989**, *10*, 363-368.
168. Maeda, M.; Uehara, T.; Harada, N.; Sekiguchi, M.; Hiraoka, A. *Phytochemistry* **1991**, *30*, 3611-3614.
169. Nishino, T.; Kiyohara, H.; Yamada, H.; Nagumo, T. *Phytochemistry* **1991**, *30*, 535-539.
170. Nakashima, H.; Kido, Y.; Kobayashi, N.; Motoki, Y.; Neushul, M.; Yamamota, N. *Antimicrob. Agents Chemother.* **1987**, *31*, 1524-1528.
171. Neushul, M. *Hydrobiologia* **1990**, *204/205*, 99-104.
172. Noda, H.; Amano, H.; Arashima, K.; Nisizawa, D. *Hydrobiologia* **1990**, *204/205*, 577-584.
173. Hori, K.; Ikegami, S.; Miyazawa, K.; Ito, K. *Phytochemistry* **1988**, *27*, 2063-2067.

RECEIVED March 5, 1993

Chapter 20

Activation of the Isotypes of Protein Kinase C and Other Kinases by Phorbol Esters of Different Biological Activities

Fred J. Evans[1] and Nahed M. Hassan[2]

[1]Department of Pharmacognosy, School of Pharmacy, University of London, 29–39 Brunswick Square, London WC1N 1AX, United Kingdom
[2]Department of Pharmaceutical Sciences, National Research Centre, Dokki, Cairo, Egypt

The Ca^{2+}- and phospholipid-dependent protein kinase C (PKC) has been recognised as the major receptor of the toxic plant compounds known as the phorbol esters. PKC exists as a number of related isotypes which are variously distributed in tissues. The abilities of a number of phorbol esters of different biological activities to stimulate these isotypes were investigated. The tumour-promoting agent, tetradecanoylphorbolacetate (TPA), activated the purified PKC isotypes α, β^1, γ in a Ca^{2+}-dependent manner, but was also capable of the activation of PKC-δ and ε in a Ca^{2+}-independent manner. The non-promotor, sapintoxin-A (SAPA), failed to activate PKC-δ, whilst 12-deoxyphorbol-phenylacetate-20-acetate (DOPPA) only stimulated PKC-β^1 in a Ca^{2+}-dependent manner. A tissue screen of the responsiveness to TPA using rat brain PKC as separated by hydroxyapatite chromatography produced seven peaks of kinase activity, of which immunological analysis by Western blotting confirmed that peaks 2 to 4 (in order of elution) consisted of PKC-α, β^1, γ, δ and ε. Peaks 1, 5, 6 and 7 failed to respond to these antibodies and may represent new phorbol-kinase receptors. Similarly, a screen of human mononuclear cells, human neutrophils and mouse macrophages using resiniferatoxin (Rx) produced a late peak of activity. This peak, known as Rx-kinase, was immunologically distinct from PKC and was shown to be Ca^{2+} inhibited.

The natural plant toxins in the phorbol ester class, are of considerable importance as tools or probes to investigate one arm of the signal transduction pathway known as the phosphatidylinositol cycle (*1*). The observation that tetradecanoylphorbolacetate (TPA) (*2*-*4*) can stimulate the calcium- and phospholipid-dependent protein kinase C (PKC) and at nM concentrations mimic the actions of the supposed endogenous activator of this kinase, diacylglycerol, has led to the extensive use of TPA to study the physiological roles of PKC.

0097–6156/93/0534–0294$06.00/0

PKC was originally isolated from rat brain (*5*) as a pro-enzyme and further work demonstrated that it was activated in the presence of calcium ions, phospholipid and diolein. Subsequently, PKC was purified to homogeneity from bovine brain (*6*) and found to be a monomeric protein of 79,000 Da. PKC is now known to exist in the form of a number of closely related isotypes, and these isotypes may be variously distributed in mammalian tissues with at present ill-defined physiological functions. The isotypes α, β (β_1 and β_2) and γ are defined with their co-factor requirements for Ca^{2+} and phospholipid (*7,8*), but the less Ca^{2+}-dependent forms δ, ε and ξ have been described also (*9*). Recently, an isotype termed PKC-η has been isolated from lung tissues (*10*). These kinases are thought to be involved in the expression of a range of important membrane receptors (*11*) and a number of protein substrated have been investigated (for example, see *12*). Further studies involved with cellular events including proliferation and differentiation have additionally concentrated on translocation (to the membrane and or the nucleus) and down-regulation. Thus, the cloning and sequencing of cDNA encoding PKC has established the enzyme as a molecular diverse family (*13*) and intense efforts are currently under way to establish physiological function in mammalian tissues.

TPA has a range of biological responses including, for example, induction of inflammation (*14*), mitogenesis of T-lymphocytes (*15*), differentiation of HL-60 cells (*16*), platelet aggregation (*17*), expression of Epstein Barr early antigen (*18*), stimulation of prostaglandin production (*19*) and induction of ornithine decarboxylase (*20*), but is best known as a tumour-promoting constituent isolated from *Croton tiglium* seed oil (*21*). Certain compounds from the extended phorbol ester family have more restricted biological effects, and, for example, are non-promoting agents (*22*), and pharmacological/structure-activity analysis suggests receptor heterogeneity (*23*). This hypothesis has been supported by binding studies which indicate from Scatchard plots at least three classes of binding site (*24*). The occurrence of PKC as a family of enzymes could explain this receptor heterogeneity and in this chapter the abilities of selected phorbol esters with full and/or partial biological responses have been reviewed for their abilities to stimulate the purified isotypes of PKC α, β_1, γ, δ and ε. The isotypes ξ and η were not available for these studies. In an effort to further investigate the biochemical effects of phorbol esters, tissue screens for responsiveness and activation assays combined with immunological analysis were also conducted for the isotypes described above.

The Phorbol Esters

In the studies outlined, TPA was used as a "full spectrum" phorbol ester probe because of its plethora of biological effects (see above). From the range of compounds which have been isolated from natural plant sources or semisynthesised (for a review see *25*), four other phorbol-related compounds were chosen (Figure 1). Sapintoxin-A (SAPA) was isolated from the ripe seeds of *Sapium indicum* (*26*) and is a non-promoting agent in Berenblum tests (*27*), but is promoting in the presence of calcium ionophore (*28*) and has several biological effects in common with TPA (*29*). The daphnane ortho-ester Thy-

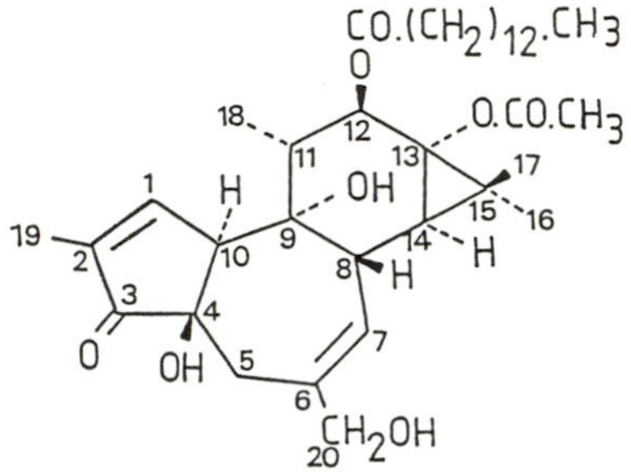

Tetradecanoylphorbolacetate (TPA)

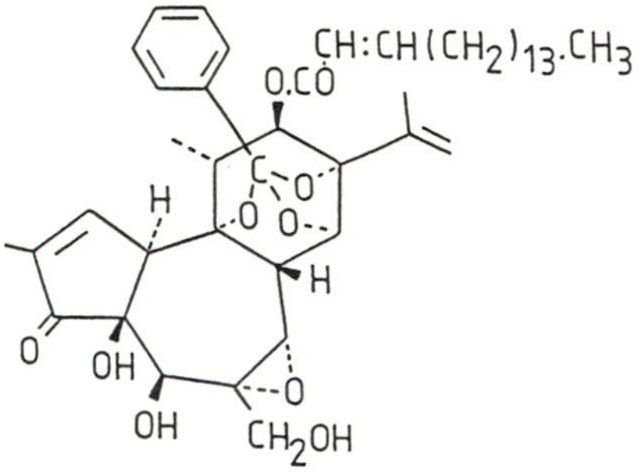

Thymeleatoxin (Tx)

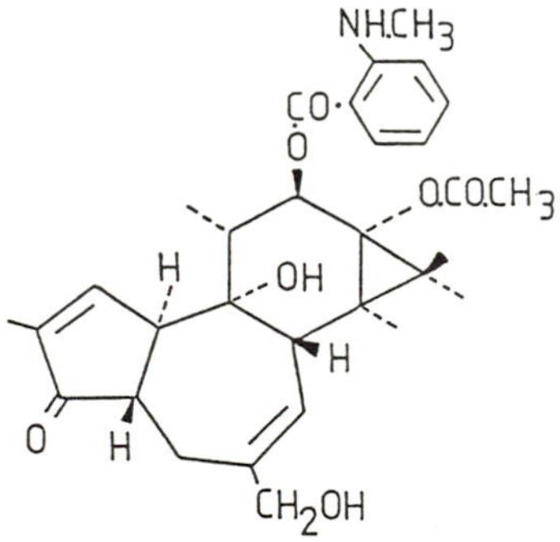

Sapintoxin-A (SAPA)

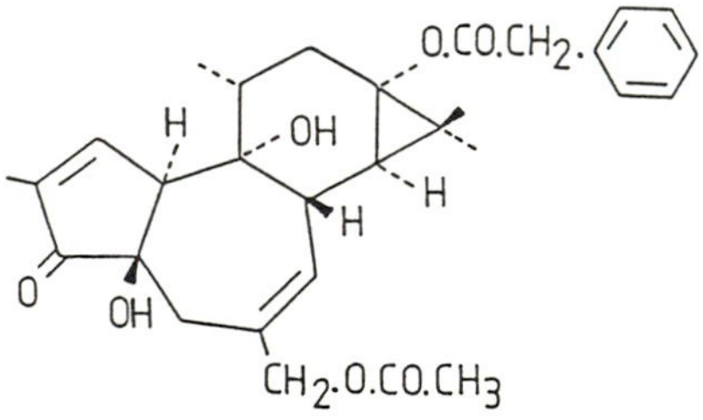

12-Deoxyphorbolphenylacetate-20-acetate (DOPPA)

Figure 1. Phorbol esters used in the biochemical experiments described in Figures 2, 3 and 4. Tetradecanoylphorbolacetate (TPA); Thymeleatoxin (Tx); Sapintoxin-A (SAPA); and 12-Deoxyphorbolphenylacetate-20-acetate (DOPPA).

meleatoxin (Tx) was obtained from the rare desert flower, *Thymelaea hirsuta* (*28*), and is a second-stage tumour-promotor (*29*), while 12-deoxyphorbol-phenylacetate-20-acetate (DOPPA) was isolated from Nigerian *Euphorbia poisonnii* and *E. unispina* (*29*). DOPPA is restricted in its biological effects: it is non-promoting on mouse skin up to a dose of 100 nM per animal over a 16 week period (*22*) but irritand to skin. From the same source the daphnane, resiniferatoxin (Rx) was isolated (*29*). This compound is non-promoting, but is the most pro-inflammatory substance known with an effective dose 50% (ED_{50}) on mammalian skin of 0.0001 μg per animal (*29*). These compounds also have various other selected biological effects in comparison to TPA (see *23* for a review).

Phorbol ester activation of purified PKC-Isotypes

Whether or not phorbol-related compounds of different biological properties have selective abilities to stimulate or otherwise affect the various isotypes of PKC is an important question to be answered if phorbol esters are to be used to study PKC physiological function. Even more consequential is when these enzymes are to be used as targets for novel drug development programmes.

Purified protein kinase C isotypes α, β_1 and γ were prepared from bovine brain as described elsewhere (*30*). PKC-δ was isolated from COS-1 cells transfected with plasmid bovine DNA vectors for PKC and PKC- ε was expressed and purified as previously outlined (*31*). The activation assays as reported in Figure 2 involve the phosphorylation of histone IIIs using ^{32}P-ATP in the micellar assay as modified from Hannan and others (*32*) but in the specific case of the δ-isotype a δ-pseudosubstrate site peptide was used. The results are presented as percent maximal activation produced by TPA vs. concentration of probe ng ml^{-1} up to a maximum of 1000 ng ml^{-1}, both in the presence and absence of added Ca^{2+} (100 mM).

TPA (Figure 2a) was found to activate the PKC forms α, β_1, γ in a Ca^{2+}-dependent manner, confirming the classical Ca^{2+} co-factor requirements for these proteins. It was also capable of the activation of the PKC isotypes δ and ε both in the presence and absence of Ca^{2+}. This action is clearly distinct from PKC-α, β_1 and γ, and reflects the co-factor differences for activation of the latter forms by this probe, a fact that must be kept in mind when interpreting pharmacological data based upon TPA responsiveness. The non-promoting, but highly biologically active, derivative SAPA (Figure 2b) was capable of fully stimulating PKC-α, β_1 and γ, but the requirements for Ca^{2+} as co-factor were greater than for TPA in that the Ca^{2+} curve was well displaced to the left. This probe was also capable of partial activation of the ε-isotype but in a non-Ca^{2+}-dependent manner. It was significant that SAPA was not capable of activation of PKC-δ, up to a concentration of 1000 $ngml^{-1}$. The second-stage promoting agent, Tx, also demonstrated significant differences to TPA in its ability to stimulate PKC (Figure 2c), in that the stimulation of PKC-α, β_1 and was Ca^{2+}-dependent to the extent that in the absence of Ca^{2+} little significant activation was recorded. This compound failed to activate the isotypes δ and ε. Thus, the three related probes when used in combination provide a possible means for the further investigation of the functions of the isotypes in intact cells. The 12-deoxyphorbol derivative, DOPPA, provided the most significant select-

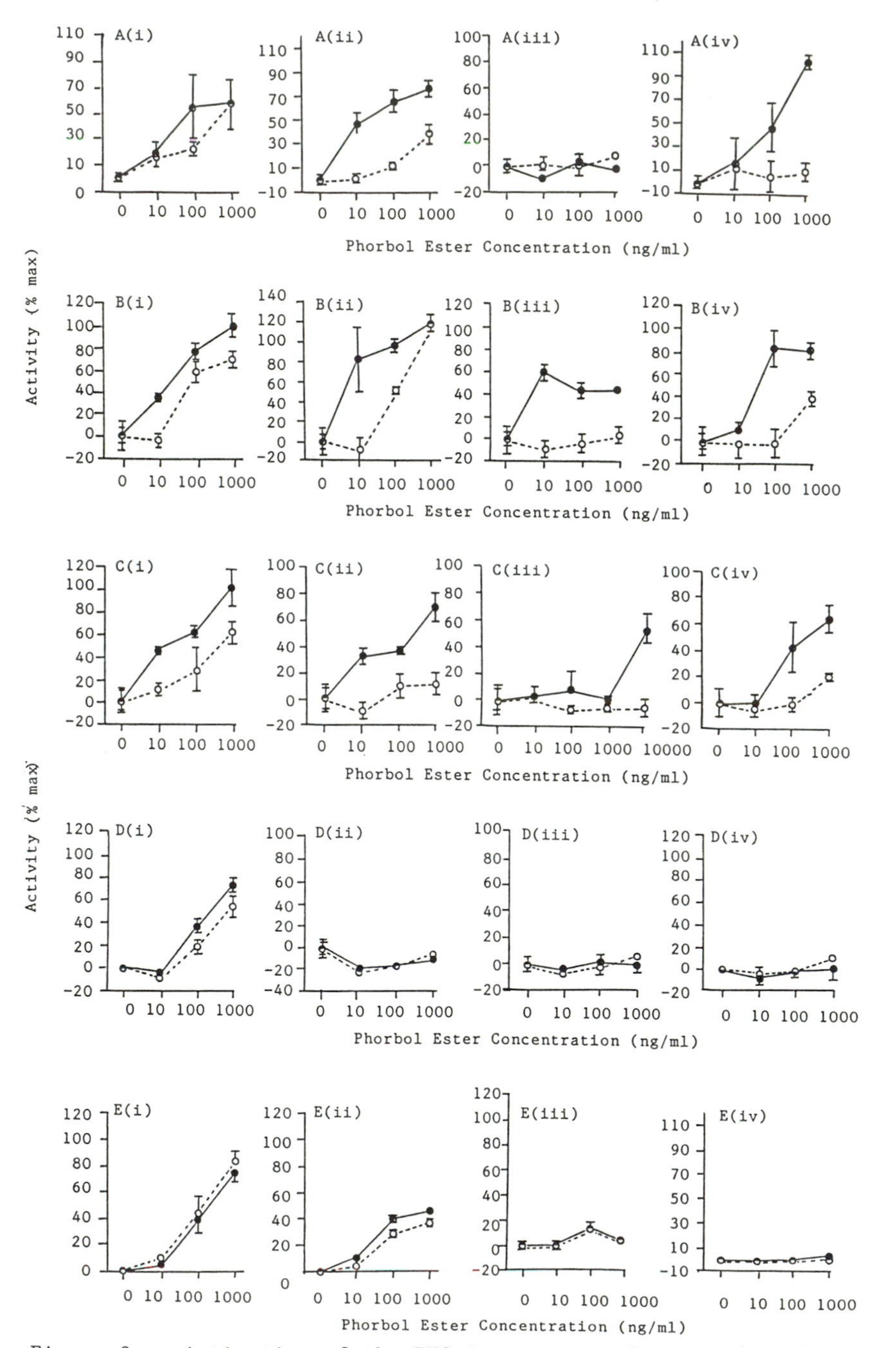

Figure 2. Activation of the PKC isotypes α, β_1, γ, δ and ε by phorbol derivatives [(●) in the presence of 100 mM Ca^{2+} and (○)] in the absence of Ca^{2+}. A, PKC α; B, PKC-β_1; C, PKC-γ; D, PKC-δ; E, PKC-ε. (i), TPA; (ii) SAPA; (iii) DOPPA; (iv) Tx. The results are expressed as percent maximal activation produced by TPA vs. concentration of phorbol esters (ng ml^{-1}).

ivity concerning PKC stimulation (Figure 2d) in that this compound was only capable of β_1-PKC activation in the presence of 100 nM Ca^{2+}. Such biochemical specificity is of interest for the investigation of this isotype despite the fact that in these biochemical experiments the effects of translocation and down regulation remain unknown and possibly important in studies involving intact cells.

Estimation of Tissue Responsiveness to Selective PKC-probes

The experiments summarised above (*33*) involve purified isotypes, but the diverse biological actions of phorbol esters could involve other as yet unknown receptors or could be due to factors such as differences in lipid solubility or metabolism. Cellulose stripping experiments on mouse skin have previously indicated that the compounds used are capable of penetrating skin, and time-course studies over 24 hours indicate that metabolism is negligible (*22*). In an effort to determine the kinase responsiveness to these compounds (Figure 1), a screen of various tissues was conducted. For these experiments, rat brains, human mononuclear cells, human neutrophils and mouse lung macrophages were used for the characterisation of phorbol ester-responsive activity.

The purification of PKC involves a number of chromatographic steps to achieve homogeneity. Such techniques include DE-52, phenyl-Sepharose S-200 and threonine-Sepharose chromatography followed by hydroxyapatite chromatography to separate the isotypes. Prolonged purification of PKC as currently described would lead to the loss of other receptor proteins which may be phorbol ester stimulated. We have developed a system of rapid kinase separation which involves applying the cellular supernatant directly to hydroxyapatite connected to fast liquid protein chromatgraphy (FPLC) and eluting in a gradient of 20-500 mM phosphate buffer. Fractions (1 ml) were collected and used in activation assays as described above. These same fractions were further subjected in parallel experiments to immunological analysis by means of Western blotting.

Figure 3 shows the kinase-stimulated activity profile produced by FPLC separation of the cytosolic fraction from about 8 gm of fresh rat brain. This profile is characteristic and reproducible when TPA is used in the micellar assays at 100 ng ml^{-1} and shows activation throughout the gradient in the presence and absence of 100 mM Ca^{2+}. Seven areas where peaking of kinase activity was observed were designated 1 to 7 in order of elution. A prominent central area of kinase activity, termed peaks 2 to 4 from this profile, is consistent with the isotypes γ, β_1 and α respectively, by comparison with published profiles using hydroxyapatite-FPLC (*34*,*35*). However, a well resolved peak 1, eluting before the central region, and the doublet of peaks 6 and 7 eluting after the central region, were also observed. Additionally, a poorly resolved area, peak 5, was detected. The composition of all seven areas of kinase activity was investigated by immunological analysis. Part of each of the fractions was subjected to SDS-PAGE using a 7.5% polyacrylamied gel on the Protean II system. Samples were transferred onto nitrocellulose by the Western technique and probed with specific PKC-isotype antibodies. Anti-sera to PKC α, β_1, γ, δ and ε were isolated from rabbit blood using synthetic peptides corresponding to the ν_5 region of each iso-

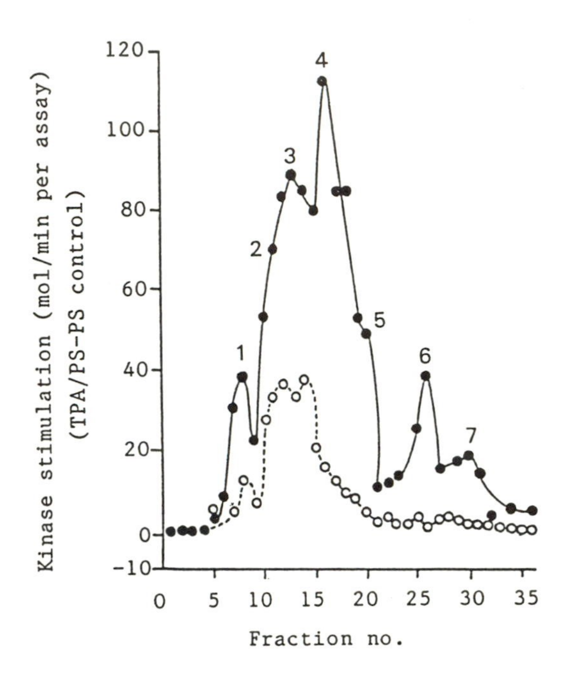

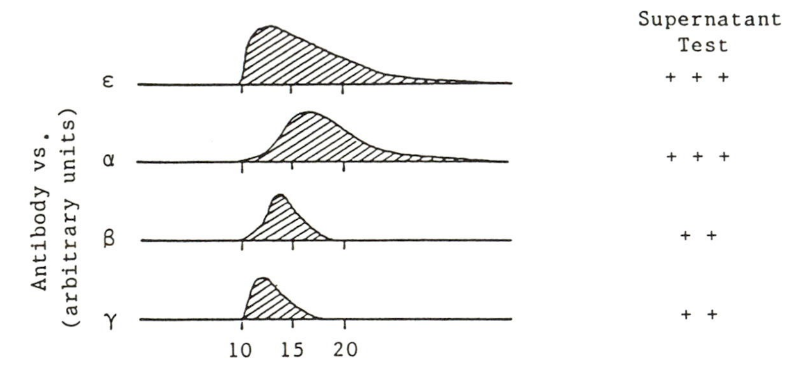

Figure 3. Profile of the phorbol ester stimulated kinase activity separated on a hydroxyapatite column by means of FPLC and eluted in a phosphate gradient of 20-500 mM. Fractions were assayed for kinase activity using TPA as the stimulation agent in the presence (●) and absence (○) of 100 mM Ca^{2+}. Fractions were also subjected to Western blotting using anti-PKC α, $β_1$, γ and ε antisera.

type (*36*,*37*). The peptides used corresponded to the following residues: α,PQFVHPILQSAV; β_1,SEFLKPVKS; γ,PDARSPIJPTPVPVM; ε,NQEEFK-GFSYFGEDLMP: and δ,MNRRGSIKQAKI. The analysis (Figure 3) indicated that the central region of the rat brain profile (peaks 2-4) consisted of the PKC isotypes α, β_1, γ, δ and ε. Immunoreaction with specific anti-sera for all isotypes available could be competed with by prior incubation with their corresponding epitopes. Thus, the central region consisted of at least these isotypes overlapping within peaks 2 to 4. The other peaks of activity were immunologically distinct from PKC isotypes, with Ca^{2+} dependent kinase activities, and possibly represent new phorbol ester receptors. The position of ξ-PKC remains unknown from this study but previous reports using hydroxyapatite chromatography claim that ξ co-elutes with PKC-β (*38*) and would thus also occur in the central region of our profile.

Using a rapid separation without pre-purification followed by hydroxyapatite chromatography with a phosphate gradient designed to concentrate PKC within a central region provided the possible evidence, other than pharmacological or binding analysis, for the presence of other phorbol ester target proteins. This technique was also used to investigate the kinase responsiveness of other tissues. Human mononuclear cells (HMNC) separated from whole blood collected via venupuncture were also screened in this way using TPA (100 ng ml^{-1}) and Rx (100 ng ml^{-1}) separately in the micellar assays for kinase-activity profile of the fractions produced. Figure 4a shows the kinase-activity profile produced from HMNC when the fractions were stimulated with TPA in the presence and absence of 100 mM Ca^{2+}. Three peaks of kinase activity labelled 1 to 3 correspond in order of elution to PKC-β_1, β_2 and α, and this was confirmed by immunological analysis as before. However, a late peak of activity, designated 4, was eluted at high phosphate concentration and this peak was immunologically distinct from the α and β isotypes of PKC. The profile generated when Rx was used in place of TPA is shown in Figure 4b. From our previous results (*33*) it is known that this pro-inflammatory daphnane ester fails to stimulate the purified isotypes of PKC-α, β_1, γ, δ, ε, and η at a concentration of 100 ng ml^{-1} as used here, and the tissue screen confirms this result in that no central region of PKC-activity was recorded. Rx did, however, stimulate the kinase activity peak 4 to a greater extent than TPA. This kinase appears to be independent of Ca^{2+} as a co-factor for activation and subsequent results (not shown) have demonstrated that the activity is in fact Ca^{2+} inhibited. Peak 4 was designated Rx-kinase (*39*), as being distinct both in Ca^{2+} dependence and immunologically from, PKC isotypes currently described. Rx-kinase activity was also demonstrated in a kinase-activity screen of starch-elicited mouse macrophages (*40*) and of human neutrophils, and the results of the immunological analysis of these profiles are given in Table 1. Our recent results concerning purification of Rx-kinase (not shown) suggest that the protein has a molecular weight of about 92kDa, considerably higher than that of PKC.

Conclusions

The cloning and sequencing of cDNA encoding PKC has now established PKC as a family of isotypes, and the varying distribution of these

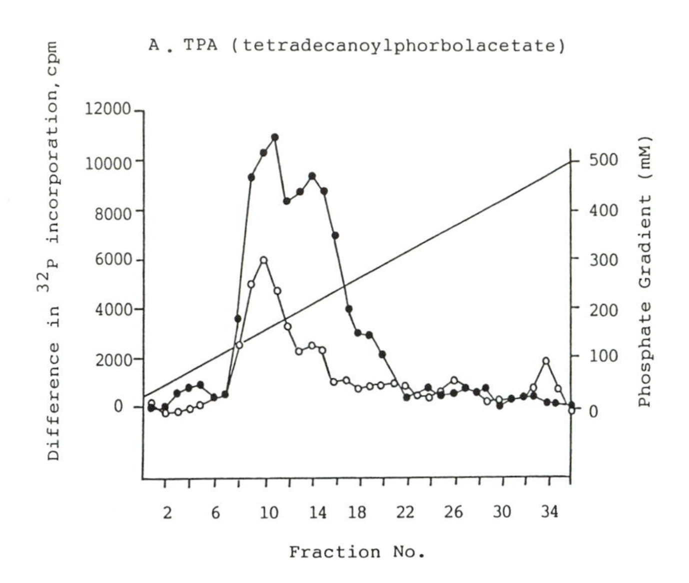

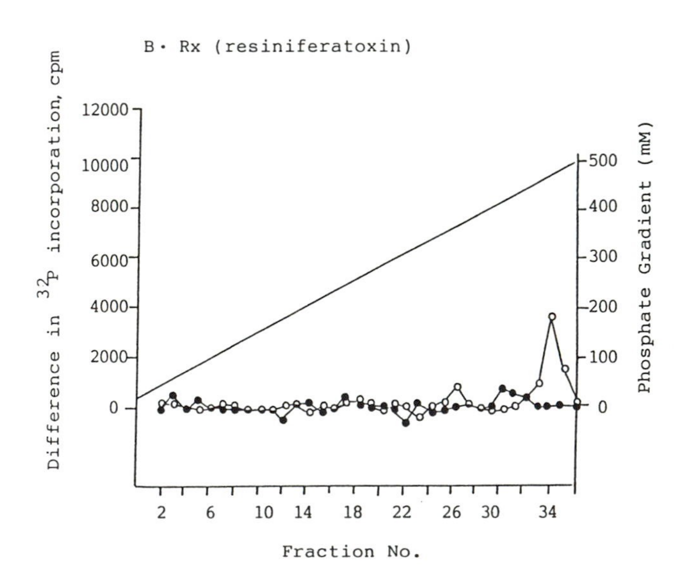

Figure 4. Profile of the phorbol ester-stimulated kinase activity separated on a hydroxyapatite column by means of FPLC and eluted in a phosphate gradient (200-500 mM). Fractions were assayed for kinase activity using (A) TPA and (B) Rx in the presence (●) and absence (○) of 100 mM Ca^{2+}.

Table 1. Immunological Results

Cell Type	Kinase Activity
Human β-lymphocytes	Peak 1, PKC-β_1, PKC-α, peaks 6,7
Human T-lymphocytes	Peak 1, PKC-β_1, PKC-α
Human neutrophils	Peak 1, PKC-β_1, PKC-α, peaks 6,7 Rx-kinase
Mouse elicited macrophages	Peak 1, PKC-β_1, PKC-α, peaks 6,7 Rx-kinase

Separation of kinase activities from tissues as indicated using hydroxyapatite in a phosphate gradient (20-500 mM) combined with immunological analysis for PKC isotypes. Peaks numbered as for Figure 2 ± Rx-kinase activity.

forms in different tissues is a possible explanation for the diverse biological effects of the family of diterpenes known as the phorbol esters. These compounds may, according to their structure, have selective pharmacological activities, and it is significant that they also have various abilities to activate the purified isotypes of PKC. In this respect it is encouraging that at least two phorbol derivatives fail to stimulate certain PKC isotypes. For example, SAPA fails to stimulate PKC-δ, whilst Tx fails to stimulate either PKC-δ or PKC-ε. The 12-deoxyphorbol derivative, DOPPA, appears to be specific for the activation of PKC-β and should prove a valuable probe in cellular function studies of that isotype. The fact that these probes are amphiphatic in nature and capable of penetrating the cell membrane provides an alternative strategy in cell biology for the investigation of PKC, in that isotype insertion techniques require the permeabilisation by means of saponin of the membrane and the physiological consequences of this approach to the problem are unknown. When using phorbol esters in intact cells, the problem also remains of the possibility, as indicated from binding studies, that the compounds may have receptors other than the forms of PKC described here. Tissue screens of the kinase-stimulated activity from rat brain provide some evidence that other kinase-phorbol ester receptors may occur, because several kinase peaks were identified which were immunologically distinct from PKC and in HMNC, neutrophils and mouse macrophages a new Ca^{2+}-inhibited kinase activity, Rx-kinase, was shown to be specifically activated by the pro-inflammatory diterpene Rx. The fact that Rx-kinase is now known to activate the NADPH-oxidase complex and induce superoxide anion production (*40*) suggests that Rx-kinase may be involved at an early stage in the inflammatory response.

Acknowledgements

We are grateful to Dr. P.J. Parker of the Imperial Cancer Research Fund Laboratories for supplies of the purified PKC-isotypes described here, and for preparation and purification of PKC-antibodies. We are also grateful to Drs. A.T. Evans, W.J. Ryves, P. Gordge and P. Sharma for many of the experimental details. F.J.E. acknowledges project grants from MRC, SERC and the Leukemia Research Fund to support this work and N.H.H. ia grateful to The Royal Society for a travel and maintenance grant.

Literature Cited

1. Aitken, A.; *Bot. J. Linn. Soc.* 1987, *94*, 247-263.
2. Castagna, M.; Takai; Kaibuchi, K.; Sano, K.; Kikkawa, V.; Nishizuka, J.; *J. Biol. Chem.* 1982, *257*, 7848-7851.
3. Kikkawa, V.; Takai, Y.; TAnaka, Y.; Miyake, R.; Nishizuka, Y. *J. Biol. Chem.* 1983, *258*, 11442-11445.
4. Niedel, J.E.; Kuhn, L.J.; Vandenbank, G.R. *Proc. Natl. Acad. Sci., U.S.A.* 1983, *80*, 36-40.
5. Kikkawa, V.; Takai, Y.; Minakuchi, R .; Ohohora, S.; Nishizuka, Y. *J. Biol. Chem.* 1982, *257*, 13341-13348.
6. Parker, P.J.; Stabel, S.; Waterfield, M.D. *EMBRO J.* 1984, *3*, 953-959.
7. Ono, Y.; Kurokawa, T.; Fujii, T.L.; Kawahara, K.; Igarashi, K.; Kikkawa, V.; Ogita, K.; Nishizuka, Y. *FEBS Lett.* 1986, *20*, 347-352.
8. Parker, P.J.; Coussens, L.; Totty, N.; Rhee, L.; Young, S.; Chen, E.; Stabel, S.; Waterfield, M.D.; Ullrich, A. *Science* 1986, *233*, 853-859.
9. Ono, Y.; Fujii, T.; Ogita, K.; Kikkawa, V.; Igarishi, K.; Nishizuka, Y. *J. Biol. Chem.* 1988, *263*, 6927-6932.
10. Osada, S.; Mizumo, K.; Saido, T.C.; Akita, Y.; Suzuki, Y.; Kuroki, T.; Ohno, S. *J. Biol. Chem.* 1990, *265*, 22330-22434.
11. Drummond, A.H.; Hughes, P.J. *Phytotherapy Res.* 1987, *1*, 1-16.
12. Brooks, S.F.; Gordge, P.C.; Toker, A.; Evans, A.T.; Evans, F.J.; Aitken, A. *Eur. J. Biochem.* 1990, *188*, 431-437.
13. Nishizuka, Y. *Nature* 1988, *334*, 661-665.
14. Hecker, E.; Schmidt, R. *Prog. Chem. Org. Nat. Prod.* 1974, *31*, 377-467.
15. Touraine, J.L.: Hadden, J.W.; Touraine, F.; Estensen, R.D.; Good, R.A. *J. Exp. Med.* 1977, *145*, 460-465.
16. Abraham, J.; Rovera, G. *Mol. Cell Biochem.* 1980, *31*, 165-175.
17. Zucker, M.B.; Troll, W.; Belman, S. *J. Cell. Biol.* 1974, *60*, 325-336.
18. Davies, A.H.: Grand, R.J.A.; Evans, F.J.; Rickinson, A.B. *J. Virol.* 1991, *65*, 6838-6844.
19. Ohuchi, K.; Levine, L. *Prostagland. Med.* 1978, *1*, 421-431.
20. O'Brien, T.G.; Simsiman, R.C.; Boutwell, R.K. *Cancer Res.* 1975, *35*, 1662-1670.
21. Hecker, E. *Bot. J. Linn. Soc.* 1987, *94*, 197-219.
22. Brooks, G.; Evans, A.T.; Aitken, A.; Evans, F.J. *Carcinogenesis* 1989, *10*, 283-288.
23. Evans, F.J.; Edwards, M.C. *Bot. J. Linn. Soc.* 1987, *94*, 231-246.
24. Driedger, P.E.; Blumberg, P.M. *Proc. Natl. Acad. Sci. U.S.A.* 1980, *77*, 567-571.
25. Evans, F.J.; Taylor, S.E. *Prog. Chem. Org. Nat. Prod.* 1983, *44*, 1-99.
26. Taylor, S.E.; Gafur, M.A.; Choudhury, A.K.; Evans, F.J. *Experientia* 1981, *37*, 681-682.
27. Brooks, G.: Evans, A.T.; Aitken, A.; Evans, F.J. *Cancer Lett.* 1987, *38*, 165-170.

28. Rizk, A.R.; Hammouda, F.M., Ismail, S.E.; El-Missiry, M.M.; Evans, F.J. *Experientia* 1984, *40*, 808-809.
29. Evans, F.J.; Kinghorn, A.D.; Schmidt, R.J. *Acta Pharmacol. Toxicol.* 1975, *37*, 250-256.
30. Marias, R.M.; Parker, P.J. *Eur. J. Biochem.* 1989, *182*, 129-137.
31. Schaap, D.; Parker, P.J. *J. Biol. Chem.* 1990, *265*, 7301-7307.
32. Hannan, Y.A.; Loomis, C.R.; Bell, R.M. *J. Biol. Chem.* 1985, *260*, 10039-10043.
33. Ryves, W.J.; Evans, A.T.; Olivier, A.R.; Parker, P.J.; Evans, F.J. *FEBS Lett.* 1991, *288*, 5-9.
34. Kikkawa, V.; Uno, Y.; Ogita, K.; Fujii, T.; Asoaka, Y.; Sekiguishi, K.; Kosaka, Y.; Igarashi, K.; Nishizuka, Y. *FEBS Lett.* 1987, *217*, 227-231.
35. Pelosin, J.M.; Vilgrain, I.; Chambaz, E.M. *Biochem. Biophys. Res. Commun.* 1987, *147*, 382-391.
36. Schaap, D.; Parker, P.J.; Bristol, A.; Kriz, R.; Knopf, J. *FEBS Lett.* 1989, *243*, 351-357.
37. Kosaka, Y.; Ogita, K.; Ashe, K.; Nomura, H.; Kikkawa, V.; Nishizuka, Y. *Biochem. Res. Commun.* 1988, *151*, 973-981.
38. Stasia, M.J.; Strulovici, B.; Daniel-Issankani, S.; Pelosin, J. M.; Dianoux, A.C.; Chamrez, E.M.; Vignais, P.V. *FEBS Lett.* 1990, *274*, 61-64.
39. Ryves, W.J.; Garland, L.G.; Evans, A.T.; Evans, F.J. *FEBS Lett.* 1989, *245*, 159-163.
40. Evans, A.T.; Sharma. P.; Ryves, W.J.; Evans, F.J. *FEBS Lett.* 1990, *267*, 253-256.

RECEIVED March 30, 1993

Chapter 21

Bioactive Organosulfur Compounds of Garlic and Garlic Products

Role in Reducing Blood Lipids

Larry D. Lawson

Murdock Healthcare, 10 Mountain Springs Parkway, Springville, UT 84663

Garlic (*Allium sativum* L.) and some of its derived products have well-established biological activities as antibiotics and as agents in reducing the risk factors of cardiovascular disease, and have received growing support as potential anticancer agents. However, because garlic's organosulfur compound composition varies greatly with the method of processing and because analysis of most of these compounds has only been recently achieved, there has been considerable variation in the experimental conditions, and subsequent results, in the biological studies with garlic and garlic products. This chapter discusses the importance of garlic's organosulfur compounds and the many changes that occur in their chemistry when garlic is crushed, cooked, ingested, metabolized, or commercially processed. The large variation in composition of commercial garlic products necessitates standardization of these products, especially for use in biological studies. Rapid pre-hepatic physiological changes occur to as yet undetected compounds in the blood, thus making *in vitro* and organ perfusion studies of limited value. The clinical studies on the cardiovascular effects of garlic preparations are summarized, and evidence is presented that low doses of allicin (allyl 2-propenethiosulfinate) and the vinyl-dithiin oils are significant blood lipid-reducing agents.

Garlic (*Allium sativum* L.) has been considered a valuable healing agent by people of many different cultures for thousands of years. Even today it is commonly used by much of the world, especially in eastern Europe and Asia, for its medicinal benefits. In the modern country of Germany, sales of garlic preparations rank with those of the leading prescription drugs. Garlic is also one of the most researched medicinal plants. Since 1960 (through December 1992) there have been approximately 940 research papers published (1200 in the past 100 years) on the chemistry (260) and biological effects (680) of garlic and its related compounds and preparations. About 75% of

0097–6156/93/0534–0306$07.25/0

these investigations have been published since 1980. The biological studies since 1960 have focused chiefly on the cardiovascular (234), antimicrobial (126), and anticancer (143) effects of garlic, and, to a lesser extent, on its hypoglycemic (13), heavy-metal poisoning antidote (13), and liver-protective (8) effects. Interest in the therapeutic potential of garlic has produced three recent international symposia on garlic held in Lüneburg, Germany in 1989, Washington, D.C. in 1990, and Berlin in 1991.

The most extensive review of the biological and chemical studies on garlic up through 1987 was published in German by Koch and Hahn (*1*). Less extensive reviews on the biological effects of garlic through 1987 have been published by Fenwick and Hanley (*2*), Hanley and Fenwick (*3*), Sprecher (*4*), Reuter (*5*), Abdullah *et al.* (*6*), and Petkov (*7*). Koch has recently reviewed the hormonal actions (*8*) and metabolism (*9*) of garlic. Recent reviews on the organosulfur chemistry of garlic and garlic products have been published by Block (*10*) and Sticher (*11*). Other reviews on the chemistry of garlic have been authored by Fenwick and Hanley (*12*), Block (*13, 14*), Lancaster and Boland (*15*), Whitaker (*16, 17*), Lutomski (*18*), and Raghavan *et al.* (*19*). Reviews on the effects of garlic on cardiovascular disease (*20-25*) and cancer (*26-28*) up to 1990 have also been published.

Although garlic has been traditionally eaten as the cooked or chopped clove, it is currently frequently consumed in the form of encapsulated commercial products, particularly when used for health reasons, because of convenience, consistency of dose, and avoidance of garlic odor. In fact, a large percentage of the clinical studies on garlic have employed a processed garlic product. Thus, any discussion of the composition and biological effects of garlic must also include the various commercial products of garlic. Unfortunately, the several forms of processed garlic are frequently called "garlic" in both the lay and scientific literature, even though their composition often differs greatly from whole or crushed garlic cloves, qualitatively and quantitatively, depending on the processing method utilized and the source (*29*).

It is the purpose of this paper to describe the current knowledge of the chemistry, content, and metabolism of the organosulfur compounds from garlic and garlic preparations (which knowledge has improved dramatically in the past four years) and to relate this knowledge to the different types of biological activities, especially the blood lipid-lowering effects, that have been reported for garlic and its preparations.

Chemistry and Analysis: Organosulfur Compounds of Garlic and Processed Garlic

Reasons for the Interest in Garlic's Organosulfur Compounds. The vast majority of the chemical and biological studies on garlic have focused on its organosulfur compounds because *in vitro* and animal studies have long shown that removal of thiosulfinates from crushed garlic by solvent (ether) extraction, or inhibition of thiosulfinate formation, also removes the antibacterial (*30, 31*), antiatherosclerotic (*32*), blood lipid-lowering (*33, 34*), and, more recently, the whole-blood platelet aggregation inhibiting (*35*) effects of garlic. Furthermore, the blood pressure-lowering effect of crushed garlic, which is not removable by solvent extraction (*36*), also appears to be due to another class of organosulfur compounds in garlic, the γ-glutamyl-*S*-alkylcysteines, as indicated by the inhibition of the blood pressure-regulating hormone, angiotensin-converting enzyme (*37, 38*). Therefore, this review on the bioactive compounds of garlic will deal almost strictly with its organosulfur compounds.

Selenium, an element of the same group as sulfur and which can replace the sulfur of organosulfur compounds, has been occasionally cited as an important medicinal component of garlic. However, the selenium content of garlic cloves is extremely low, 0.01-0.2 μg/g fresh wt. (*39, 40*), compared to the Recommended Dietary Allowance (*41*) of 70 μg/day (equivalent to 1 kg garlic/day). Nevertheless, the selenium content of garlic has been increased to 50 μg/g by cultivation in selenium-enriched soil for the purpose of improving the antitumor effects of garlic (*40*). Germanium has likewise been cited in the lay literature as an important element in garlic, but we have found garlic cloves to contain only 0.004 μg/g of germanium.

Organosulfur Compounds of Whole Garlic Cloves. All of the organosulfur compounds of intact garlic cloves contain the amino acid cysteine (except for trace amounts of methionine), and include approximately equal amounts of the *S*-alkylcysteine sulfoxides (cys-S(O)-R) and γ-glutamyl-*S*-alkylcysteines (γ-glu-cys-S-R), the alkyl groups being strictly allyl, methyl, and *trans*-1-propenyl (*42*) (see Table I).

Table I. Principal Organosulfur Compounds in Whole and Crushed Garlic Cloves

Compound	Whole Garlic	Crushed Garlic
	(mg/g)	
<u>*S*-(+)-Alkyl-L-cysteine sulfoxides</u>[a]		
Allylcysteine sulfoxide (alliin)	7-14	nd[b]
Methylcysteine sulfoxide	0.5-2	nd
trans-1-Propenylcysteine sulfoxide (isoalliin)	0.1-2	nd
<u>γ-L-Glutamyl-*S*-alkyl-L-cysteines</u>[c]		
γ-Glutamyl-*S*-*trans*-1-propenylcysteine	3-9	3-9
γ-Glutamyl-*S*-allylcysteine	2-6	2-6
γ-Glutamyl-*S*-methylcysteine	0.1-0.4	0.1-0.4
<u>Alkyl alkanethiosulfinates</u>		
Allyl 2-propenethiosulfinate (allicin)	nd	2.5-4.5
Allyl methyl thiosulfinates (2 isomers)	nd	0.3-1.2
Allyl *trans*-1-propenyl thiosulfinates (2 isomers)	nd	0.05-1.0
Methyl *trans*-1-propenyl thiosulfinates (2 isomers)	nd	0.02-0.2
Methyl methanethiosulfinate	nd	0.05-0.15

[a] Determined by calculation from the thiosulfinates (*48*).
[b] Not detectable.
[c] Range for 8 varieties of garlic less than 2 months after harvesting.

Although it has been known for several decades that garlic contains *S*-allyl and *S*-methyl derivatives of cysteine sulfoxide (*43, 44*) and γ-glutamylcysteine (*45, 46*), the presence of the *trans*-1-propenyl homologues was not known until 1990 (*47*). *S-trans*-1-Propenylcysteine compounds were previously thought to be unique to onions and other non-garlic alliums (*12, 15*). The presence of *S-trans*-1-propenylcysteine sulfoxide (isoalliin) was first indicated by the discovery of *trans*-1-propenyl thiosulfinates in garlic homogenates (*42, 47, 48*), followed by its isolation and structure elucidation (*49*). γ-Glutamyl-*S-trans*-1-propenylcysteine, the most abundant γ-glutamylcysteine and the second most abundant sulfur compound in garlic, was first found in chive seeds (*Allium schoenoprasum* L.) by Virtanen (*50*). It was first discovered and isolated from garlic and its structure determined chemically by Lawson *et al.* (*42, 47*). The structure has subsequently been confirmed with spectral data by Mütsch-Eckner *et al.* (*51*).

The γ-glutamylcysteines, which serve as reserve compounds for the formation of cysteine sulfoxides (*50*), are gradually hydrolyzed during wintering and sprouting by increased levels of γ-glutamyl transpeptidase (*15, 52, 53*) and then oxidized to the cysteine sulfoxides [in onions, oxidation occurs first, resulting in accumulation of γ-glutamyl-*S-trans*-1-propenylcysteine sulfoxide, followed by transpeptidase hydrolysis to isoalliin (*54*)]. The same process occurs when garlic bulbs are stored at room temperature and is faster when stored cool (*42, 48*). The result is that store-purchased garlic cloves contain less γ-glutamylcysteines and more cysteine sulfoxides than fresh-picked garlic (*42, 48*).

The total sulfur content of garlic is about 1.0% of its dry weight or 0.35% of its fresh weight (*55, 56*). The cysteine sulfoxides (8-18 mg/g fresh wt.) and γ-glutamylcysteines (5-15 mg/g) account for about 82% of the total sulfur, with most of the remainder being present in soluble protein (3%), sulfate (5%), and insoluble compounds (6%) (Lawson, L.D.; unpublished data). Small amounts of *S*-2-carboxypropylglutathione (0.09 mg/g) (*57, 58*), γ-glutamyl-*S*-allylmercaptocysteine (0.01 mg/g) (*59*), γ-glutamylmethionine (*50*), methionine (0.023 mg/g) (Lawson, L.D.; unpublished data), and scordinins (0.03 mg/g) (*60*) are also found. Undetected (<0.02 mg/g) in garlic cloves are free cysteine (*61*), *S*-allylcysteine [(*61*) and work in this laboratory with twelve varieties of garlic, only three of which contained detectable amounts, about 0.026 mg/g], *S*-1-propenylcysteine, cystine, γ-glutamylcysteine, glutathione, and oxidized glutathione (Lawson, L.D.; Wang, Z.J.; unpublished data). However, when sliced garlic cloves are oven dried at 50-60°C or dried by lyophilization for the production of garlic powders, *S*-allylcysteine (0.7-1.1 mg/g dry wt.) and *S*-1-propenylcysteine (0.1-0.3 mg/g dry wt.) increase considerably as hydrolysis products of the respective γ-glutamyl-*S*-alkylcysteines (Lawson, L.D.; unpublished data).

Formation of the Thiosulfinates. When garlic cloves are crushed, chewed, or cut (or dehydrated, pulverized garlic is exposed to water), the vacuolar enzyme, alliinase or alliin lyase (EC 4.4.1.4), rapidly lyses the cytosolic cysteine sulfoxides (*62, 63*) to form sulfenic acids (R-SOH) (*64*), which immediately condense to form the alkyl alkanethiosulfinates (R_1-SS(O)-R_2), the compounds which constitute the odor of fresh-cut garlic. The formation of allicin (allyl 2-propenethiosulfinate or diallyl thiosulfinate), the dominant thiosulfinate formed, is complete in 0.2-0.5 min at room temperature, while formation of the methyl thiosulfinates takes 1.5-5 min (*65*). All possible combinations of 2-propene-, l-propene-, and methanesulfenic acids result in thiosulfinates (*48*), except for di-l-propenethiosulfinate, which is unstable and, in onion homogenates, rapidly transforms to other compounds (*66, 67*). Alliinase has a known 448 amino acid sequence and is a major component of garlic total protein (*68*). It is active at pH 4.5-9 (*65, 69*), but is immediately and irreversibly inactivated at pH 3 or below (*65, 70*); however, allicin is unaffected by low (1-3) pH values (*65*). Therefore, garlic that is not chewed, such as garlic powder tablets, must be made stomach acid-resistant (enteric-coated) to prevent inactivation of alliinase by stomach acid, which typically has a pH value of 1-3. Alliinase and alliin are very stable when dry and garlic powders stored up to 5 years show little change in allicin yielding potential (*71, 72*).

Allicin (allyl-SS(O)-allyl) represents 70-80 wt.% of the total thiosulfinates formed, followed by allyl-SS(O)-methyl (6-16%), methyl-SS(O)-allyl (3-9%), *trans*-1-propenyl-SS(O)-allyl(1-7%), allyl-SS(O)-*trans*-1-propenyl(0.2-4%), *trans*-1-propenyl-SS(O)-methyl plus methyl-SS(O)-*trans*-1-propenyl (0.1-2.5%), and methyl-SS(O)-methyl (2%) thiosulfinates (*48*). The total allyl thiosulfinates represent about 95-98% of the total thiosulfinates. The methyl thiosulfinates represent 10-30% of the total thiosulfinates and appear to be less abundant from garlic grown in cooler climates, such as New York State, than from garlic grown at its typical commercial sources of California and Mexico (*73, 74*). Interestingly, the methyl thiosulfinates and allyl thiosulfinates released from *Allium ursinum* L. (wild garlic or ramson) are equally abundant (*73, 75, 76*). The *trans*-1-propenyl thiosulfinates are formed upon lysis of *trans*-1-propenyl cysteine sulfoxide (isoalliin), which is nearly absent in cloves of fresh-picked garlic (*48*). As previously mentioned, the isoalliin and alliin content increase upon storage, especially when stored cool, such that the *trans*-l-propenyl thiosulfinate content of the homogenate of typical store-purchased garlic is 7-18% of the total thiosulfinates (*48*). Interestingly, the allicin content does not increase upon storage of garlic because the alliin and isoalliin formed during storage are used to form the allyl *trans*-l-propenyl thiosulfinates (*48*), which are produced more rapidly than allicin (*65*).

Stability of the Thiosulfinates. The thiosulfinates released from crushed garlic are rather reactive molecules and undergo a number of transformations depending on the temperature, pH, and solvent conditions. We have found that the allyl-*S*-thiosulfinates (allicin, allyl-SS(O)-methyl and allyl-SS(O)-*trans*-l-propenyl) are the least stable of garlic's eight thiosulfinates because they form thioacrolein (CH_2=CH-CH=S) (*13, 48, 77, 78*) and allylmercaptan (allyl-SH) (*77*) as intermediates for further reactions. On the other hand, the allyl-S(O) thiosulfinates (methyl-SS(O)-allyl and *trans*-1-propenyl-SS(O)-allyl), which cannot form thioacrolein by elimination, are fairly stable. For example, in ether, allyl-SS(O)-methyl has the same half-life as allicin (3 hours), whereas methyl-SS(O)-allyl shows no measurable loss after 6 days. The thiosulfinates are most stable in polar solvents. The half-life of allicin (0.1-0.4 mg/ml) at room temperature is 10 days in 1 mM citric acid (pH 3), 4 days in water (also *72*), 48 hours in methanol or chloroform, 30 hours in dichloromethane, 24 hours in ethanol or acetonitrile, 3 hours in ether, 2 hours in hexane, and 16 hours in the absence of solvent (Lawson, L.D.; unpublished data). Allicin is an oil which is partially soluble in water (about 1-2%) and hexane, but is very soluble in most other organic solvents (*79*).

Transformations of the Thiosulfinates. The principal transformation products found after incubation of the thiosulfinates in water are diallyl trisulfide, diallyl disulfide, and allyl methyl trisulfide (*80, 81*) (see Figure 1). The rate of formation of sulfides is greatly accelerated by heat, such as during steam distillation (*82, 83*). Kinetic studies with heated garlic homogenate (*81, 82*) as well as comparisons of freshly steam-distilled garlic oil with commercial steam-distilled garlic oil indicate that the trisulfides are formed first, followed by slow transformation to the somewhat more stable disulfides as well as to the tetrasulfides (*82, 84, 85*). Incubation of allicin or allyl methanethiosulfinate in low-polarity solvents (chloroform, ether, hexane, vegetable oil), or without solvent, produces mainly 1,3-vinyl-dithiin (2-vinyl-4*H*-1,3-dithiin) (51%), 1,2-vinyl-dithiin (3-vinyl-4*H*-1,2-dithiin) (19%) and lesser amounts of ajoene (*E,Z*-4,5,9-trithiadodeca-1,6,11-triene 9-oxide) (12%) and sulfides (18%) (*29, 80, 86, 87*).

Incubation of thiosulfinates in alcohols gives variable results. Incubation of allicin, as garlic homogenate, in ethanol produces mainly diallyl trisulfide (73%) and some diallyl disulfide (8%) and ajoene (8%), whereas ajoene (55%) and vinyl-dithiins (34%) and some sulfides (0-11%) are produced when pure allicin is used (*80*). We have confirmed these results in ethanol and have also found that incubation of garlic homogenate in methanol produces principally diallyl trisulfide (83%) and diallyl disulfide (10%), but only small amounts of ajoene (1%) and vinyl-dithiins (1%), with about 5% unidentified compounds. Incubation of pure allicin in methanol gave only 8% diallyl trisulfide and about equal amounts (20-25%) of ajoene and the vinyl-dithiins, with about 25% unidentified compounds (Lawson, L.D.; unpublished data).

Alkaline Hydrolysis of the Thiosulfinates. Thiosulfinates also undergo immediate alkaline hydrolysis to form disulfides (*78, 88, 89*) and sulfur dioxide (*90*). We have found that alkaline hydrolysis (sodium hydroxide at pH 11-12 in 50% methanol) of one mole of allicin gives 0.5 moles of diallyl disulfide, whereas one mole of allyl-SS(O)-methyl thiosulfinate gives 0.25 moles each of diallyl disulfide and allyl methyl disulfide, while one mole of methyl-SS(O)-allyl thiosulfinate gives 0.25 moles each of dimethyl disulfide and allyl methyl disulfide. No trisulfides or monosulfides were found (Lawson, L.D.; unpublished data).

Effects of Cooking on Garlic's Organosulfur Compounds. Since garlic is more often eaten cooked rather than raw, the effects of cooking on its organosulfur compounds is an important concern and a common question. We have examined several methods of cooking. We found that boiling garlic cloves for 20 min caused a 12% loss of the γ-glutamylcysteines to *S*-allylcysteine and *S*-1-propenyl cysteine, and a similar loss of alliin without production of *S*-allylcysteine. [Yu *et al.* (*91*) have recently reported that frying garlic, particularly microwave-treated garlic, at 180°C converts alliin to allyl alcohol and cysteine.] When garlic is chopped into tiny pieces or is crushed, most of the cysteine sulfoxides are transformed into thiosulfinates. Heating the crushed garlic at boiling temperature for 20 min in a closed container caused complete conversion of the thiosulfinates to mainly diallyl trisulfide and lesser

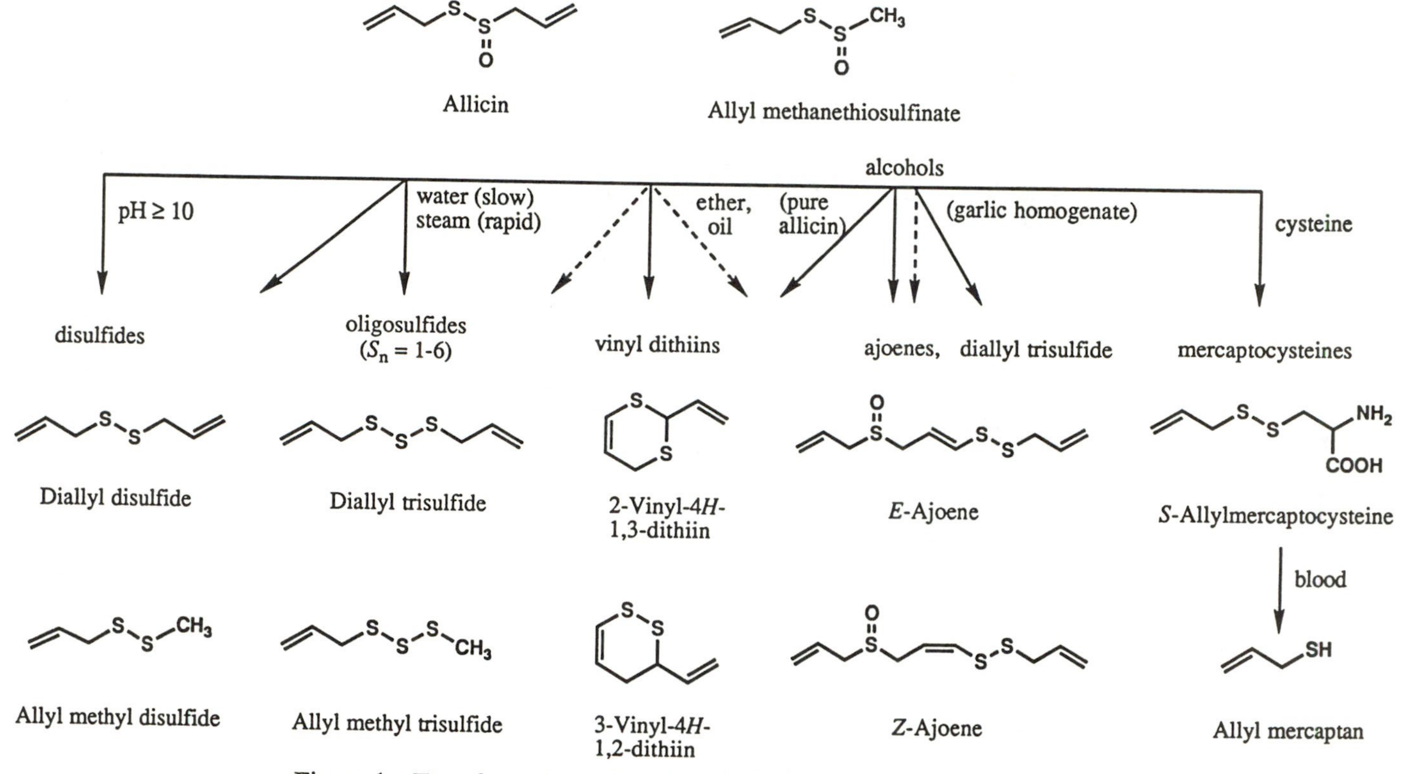

Figure 1. Transformations of the Principal Thiosulfinates of Crushed Garlic.

amounts of diallyl disulfide and allyl methyl trisulfide. Cooking in an open container for 10 or 20 min retained 21 or 7% of the thiosulfinates, respectively, but 97% of the sulfides had evaporated. Cooking in whole milk in an open container caused even more rapid loss of allicin and the other thiosulfinates (only 0.5% remained after 10 minutes), but 70% of the sulfides formed were retained, even after 40 min, presumably by the milk fat. Stir-frying chopped and smashed garlic cloves in hot (near the smoke point) soybean oil for 1 min in a Chinese wok retained about 16% of the sulfides (the main ones being present in equal amounts) in the oil, but no allicin remained (Lawson, L.D.; Wang, Z.J.; unpublished data).

Organosulfur Compounds of Commercially Processed Garlic. Commercial garlic products are sold as food flavorings (garlic powder, garlic salt) or as pills for therapeutic purposes. Those sold as pills are usually a garlic powder tablet or a garlic oil capsule. An exception is garlic aged in aqueous alcohol, which is sold as a brown liquid or as a tablet. Most of the processing methods have been reviewed (*1, 11, 19, 92*).

Garlic Powders. Since garlic powders are simply dehydrated, pulverized garlic cloves, their composition and alliinase activity can be identical to those of fresh garlic, if processed carefully (Table II). If the cloves could be dried whole, this would be true; however, in practice, the cloves must be chopped somewhat for efficient drying. Chopping causes immediate release of alliinase-generated thiosulfinates (which are lost to evaporation upon drying) from damaged cells. Therefore, it is important to dry garlic in pieces as large as possible (*11, 93*). Most garlic powders are prepared by drying chopped garlic at 50-60°C. Some garlic powders are prepared by spray-drying homogenized garlic, resulting in a complete loss of alliin and allicin. Garlic powders have also been prepared by freeze-drying garlic to avoid possible heat damage (*94*), but we have found that oven drying at 60°C does not affect the yield of the main thiosulfinates (allicin and the allyl methyl thiosulfinates) (*65*). Garlic powders usually contain 20-50% less γ-glutamyl-*S*-*trans*-1-propenylcysteine than garlic cloves because of losses sustained during the heating of this somewhat unstable compound (*42*). Most garlic powder tablets contain about two-thirds excipients, similar to the two-thirds water content of fresh cloves, and therefore, represent about an equivalent weight of garlic. As mentioned earlier, garlic powder products must be enteric coated to prevent the rapid and irreversible inhibition of alliinase (allicin formation) by stomach acid (*65*).

Oils of Processed Garlic. There are three types of garlic oils: the oil of steam-distilled crushed garlic, the oil of crushed garlic macerated in a vegetable oil, and the oil produced by ether extraction of the thiosulfinates of crushed garlic followed by solvent removal. None of these oils are present in garlic cloves. Garlic cloves contain only tiny amounts (0.2%) of natural oil - mainly triglycerides, phospholipids and glycolipids, none of which contain sulfur (*12*). Steam-distilled garlic oil is frequently referred to as the "essential oil of garlic," but this is erroneous because the volatile oil of steam-distilled garlic consists exclusively of the sulfides generated from the heated thiosulfinate precursors, which are not formed until after garlic is crushed.

Table II. Total Known Organosulfur Compounds Released from Commercial Garlic Products

	Cysteines[a]		Thiosulfinates		Thiosulfinate transformation compounds[d]			Total
	γ-glutamyl[b] cysteines	S-alkyl[b] cysteines	allicin	others[c]	dialkyl sulfides	vinyl dithiins	ajoene	
Product type (n brands)	(mg/g of product)[e]							
Garlic cloves								
Fresh-picked	14.5±3.1	nd[f]	3.6±0.4	1.1±0.2	nd	nd	nd	19.2
Store-purchased	9.4±2.9	nd	3.7±0.9	1.7±0.4	nd	nd	nd	14.8
Powder tablets								
Typical (n=4)	12.1±1.8	0.5±0.2	2.9±0.6	0.8±0.5	0.05±0.03	nd	nd	16.4
Range (n=9)	2-20		0.2-3.6	0.03-1.3	0.01-0.13	nd	nd	
Steam-distilled oils								
Typical (n=5)	nd	nd	nd	nd	4.4±0.5	nd	nd	4.4
Range (n=9)	nd	nd	nd	nd	0.2-7.3	nd		
Oil-macerates								
Typical (n=5)	nd	nd	nd	nd	0.16±0.05	0.61±0.1	0.10±0.04	0.9
Range (n=9)	nd	nd	nd	nd	0.06-0.22	0.07-0.7	0.02-0.12	
Aged in aqueous alcohol[g] (n=1)	0.5	0.6	nd	nd	nd	nd	nd	1.1

[a] Not included are the *S*-alkylcysteine sulfoxides (alliin, etc.), because they are rapidly converted to thiosulfinates upon homogenization in water.
[b] Includes the *S*-allyl, *S*-1-propenyl, and *S*-methyl derivatives.
[c] Includes allyl methyl, methyl allyl, 1-propenyl allyl, allyl 1-propenyl, 1-propenyl methyl, methyl 1-propenyl, and dimethyl thiosulfinates.
[d] Includes diallyl, allyl methyl and dimethyl mono- to heptasulfides; 2-vinyl-4*H*-1,3-dithiin and 3-vinyl-*H*-1,2-dithiin; (*E*)- and (*Z*)-ajoene.
[e] Mean ± std. dev. for (n) number of brands. Typical values represent the highest 4-5 values.
[f] Not detected (nd). Limit of detection is 0.02 mg/g for the cysteines, 0.005 mg/g for the thiosulfinates, and 0.001 mg/g for the transformation compounds.
[g] Only one brand of this Japanese product could be found. It is sold in liquid form and tablet form. Three lots of each form were found to give similar results. The coefficient of variation for the six lots was 15% for the *S*-alkylcysteines and 52% for the γ-glutamyl-*S*-alkylcysteines.

Steam-distilled garlic oil is by far the most common garlic oil found in the U.S. It consists of the diallyl (57%), allyl methyl (37%), and dimethyl (6%) mono- to hexasulfides. A typical commercial steam-distilled garlic oil contains 26% diallyl disulfide, 19% diallyl trisulfide, 15% allyl methyl trisulfide, 13% allyl methyl disulfide, 8% diallyl tetrasulfide, 6% allyl methyl tetrasulfide, 3% dimethyl trisulfide, 3% monosulfides, 4% pentasulfides, and 1% hexasulfides (*29*); however, freshly distilled garlic oil contains predominantly diallyl trisulfide (*82, 85*). The 1-propenyl sulfides are not usually found in commercial oils since garlic products are usually produced with recently harvested garlic. Commercial steam-distilled garlic oils are diluted (>99%) with vegetable oils to obtain a practical sized pill and to stabilize the sulfides, which gradually form higher sulfide polymers (e.g., trisulfide conversion to tetrasulfide) in the absence of solvent.

Steam-distilled garlic oil has been frequently reported to contain significant amounts of propyl sulfides, particularly allyl propyl disulfide (*1, 12, 16, 82, 95*); however, this is an error originating from the 1892 investigations of Semmler (*84*). Recent investigations have demonstrated the absence of *S*-propylcysteine sulfoxide (*61*) in garlic cloves, γ-glutamyl-*S*-propylcysteine (*42*) and propyl thiosulfinates (*48, 73*) in garlic homogenates, and propyl sulfides (*29*) in garlic oils.

Oil-macerated garlic products are rare in the U.S. but are very common in Europe (*1, 29, 80*). They are produced by homogenizing chopped garlic in a vegetable oil (soybean oil is typical), followed by separation from the oil-insoluble components. The thiosulfinates generated during chopping are very soluble in the oil (much more so than in water) and are subsequently transformed to vinyl-dithiins (70%), sulfides (18%), and ajoenes (12%) (*29, 80, 86, 87*). The oil-macerated garlic oils are the only commercial garlic products which contain the vinyl-dithiins and ajoenes. Unfortunately, the commercial oil-macerates contain only about 20% of the total organosulfur compounds of the steam-distilled garlic oils or of the thiosulfinates of crushed garlic (Table II). The reason for this is not known, but it may be due to insufficient thiosulfinate release prior to oil-maceration, or to volatilization loss.

A third type of garlic oil, nearly identical in qualitative composition to the oil-macerates, is the oil of ether-extracted garlic homogenate. Ether extracts all of the thiosulfinates; however, in ether, as well as in the absence of solvent, the oily thiosulfinates are rapidly converted to the vinyl-dithiins, sulfides, and ajoenes, as previously mentioned. The advantage of this type of oil, compared to the oil-macerate products, is that it is a highly concentrated preparation of the same compounds and can be further diluted with vegetable oil or glycerin to a desired concentration. Though probably not commercially available, this is the type of garlic oil that Bordia has used in his many clinical studies on the lowering of blood lipids (*95-100*). We have found his oil to contain nine times as much of the vinyl-dithiins (5.7 mg/g) and allyl sulfides (1.4 mg/g) and four times as much of the ajoenes (0.4 mg/g) as a typical commercial oil-macerate (Lawson, L.D.; unpublished data).

Aged Alcoholic Garlic Extract. A final type of commercial garlic product which is widely sold is garlic aged in aqueous alcohol, a product of Japan. This product is produced by storing sliced garlic in 15-20% ethanol for 20 months, followed by filtration and concentration (*101,102*). The final liquid product contains 10% w/v (12% v/v) ethanol, 0.05% total sulfur, and 0.29% total nitrogen (garlic cloves typically contain 0.35% (1.0% dry wt.) total sulfur (*55, 56*) and 1.1% (3.1% dry wt.) total nitrogen. It does not contain cysteine sulfoxides or any alliin-derived transformation products (thiosulfinates, allyl sulfides, vinyl-dithiins, or ajoenes), since these volatile products are lost in the long incubation period (*29*). It does, however, contain small amounts of γ-glutamyl-*S*-allylcysteine (0.28 mg/g), γ-glutamyl-*S*-1-propenylcysteine (0.17 mg/g), γ-glutamyl-*S*-methylcysteine (0.02 mg/g), and their hydrolysis products, *S*-allylcysteine (0.30 mg/g), *S*-1-propenylcysteine (0.15 mg/g), *S*-methylcysteine (0.11 mg/g), as well as *S*-allylmercaptocysteine (0.04 mg/g) and cystine (0.01 mg/g), but no (<0.02 mg/g) cysteine or glutathione. The content of the organosulfur compounds in the tableted product was very similar to those in the liquid product (Lawson, L.D.; unpublished data).

Comparative Analysis of Commercial Garlic Products. Several brands of each type of garlic product were purchased and analyzed for their organosulfur compounds. The results, shown in Table II, reveal that there is a broad range of quantitative composition for all types of garlic products. For example, among the garlic powder tablets, there was an 18-fold range in allicin release, and a 10-fold range in γ-glutamylcysteines content. For the steam-distilled oils, there was a 35-fold range in the content of sulfides and for the oil-macerates there was a 10-fold range in the content of vinyl-dithiins. Furthermore, there was considerable variation in amounts of total organosulfur compounds between the each type of garlic product. Hence, it is essential that any biological studies employing a garlic product state exactly what type of product it is and the content or yield of at least one of its major organosulfur compounds.

Reactions with Cysteine. It has been known since its discovery in 1944 that allicin reacts rapidly with the sulfhydryl group of the amino acid, cysteine, to form two moles of *S*-allylmercaptocysteine (allyl-SS-cysteine) plus water (*90*). It was also shown that allicin inhibits a large number of enzymes *in vitro* which contain cysteine at their active sites, but very few that do not (*103*). In fact, many of the explanations given for the biological effects of garlic focus on the ability of allicin to react with sulfhydryl enzymes as well as the sulfhydryl group of acetyl-CoASH, the building block of cholesterol and triglyceride synthesis (*104-106*).

Cysteine reacts not only with allicin, but also with all of the allicin transformation products containing a dithioallyl (allyl-SS-) group. It has been known for some time that diallyl disulfide (*106*) and allyl propyl disulfide (*107*) react with cysteine to form *S*-allylmercaptocysteine. A recent preliminary report found that ajoene reacts very rapidly with cysteine, although no reaction product was reported (*108*). Under artificial physiological conditions (37°C, pH 7, 0.045M sodium phosphate buffer) and at 0.5 mM concentration, we have found that ajoene and diallyl trisulfide react very rapidly with cysteine to form *S*-allylmercaptocysteine, while diallyl disulfide reacts much slower. One mole of *E,Z*-ajoene was found to react with 2 moles of cysteine (half-life <2 min in the presence of 100% excess cysteine) to form one mole of *S*-allylmercaptocysteine [elutes 1.5 min after γ-glutamyl-*S*-*trans*-1-propenylcysteine by C_{18}HPLC (*42*)] and a new compound which eluted one min after *S*-allylmercaptocysteine and which gave a UV-spectrum nearly identical to that of the very characteristic UV-spectrum of *E,Z*-ajoene (*29, 80*), indicating that it contains the

$CH_2=CHCH_2$-S(O)-$CH_2CH=CH$-S- moiety of ajoene. We have named the new compound *ajocysteine* (*E,Z*-2-amino-9-oxo-4,5,9-trithiadodeca-6,11-dienoic acid) and propose its structure to be $CH_2=CHCH_2$-S(O)-$CH_2CH=CH$-SS-$CH_2CH(NH_2)COOH$. Ajocysteine is stable, except in the presence of excess cysteine where it has a half-life of 30 min at a cysteine/ajoene ratio of 4. One mole of diallyl trisulfide (20% methanol added to the buffer to solubilize) in the presence of 100% excess cysteine was found to react with one mole of cysteine (half-life < 2 min) to form one mole of *S*-allylmercaptocysteine, 0.6 mole of allyl mercaptan, and an unidentified compound; however, in the presence of 8-fold excess cysteine, two moles of *S*-allylmercaptocysteine were formed per mole of diallyl trisulfide. Diallyl disulfide was also found to react with cysteine to give one mole of *S*-allylmercaptocysteine and one mole of allyl mercaptan. Although the reaction for diallyl disulfide was slow (half-life of 45 min in presence of 100% excess cysteine), it was increased considerably to a half-life of 15 min by increasing the cysteine/diallyl disulfide ratio to 4 or by increasing the pH to 8, indicating the difficulty of breaking the CHS-SCH bond. We have found the half-life of the reaction of cysteine with allicin to be less than one minute and have confirmed the formation of two moles of *S*-allylmercaptocysteine (*90*) and have also confirmed that no reaction takes place for diallyl monosulfide (*106*) and have found no reaction for 1,3-vinyl-dithiin and no significant reaction for 1,2-vinyl-dithiin. With 16-fold excess cysteine, 1,2-vinyl-dithiin gave a half-life of about 150 min, but with 4-fold excess cysteine, its half-life (about 8 h) was no different than in the absence of cysteine (Lawson, L.D.; Wang, Z.J.; unpublished data). The physiological importance of these reactions with cysteine will be discussed next.

Physiological Fate of Garlic-Derived Organosulfur Compounds

Even though a large number of biological studies have been conducted on garlic, very little progress has been made on identifying new organosulfur compounds in the blood or tissues after consuming garlic or garlic products (see review *9*). Neither the γ-glutamylcysteines nor their metabolites have been reported in blood, but *N*-acetyl-*S*-allylcysteine and *N*-acetyl-*S*-(2-carboxypropyl)cysteine have been found in human urine after garlic consumption (*109*). Even though the γ-glutamylcysteines are dipeptides, they are probably absorbed intact, as is the γ-glutamyl tripeptide, glutathione (*110*). We have found them to be completely stable to artificial gastric fluid at 37°C for at least four hours. Animal studies with alliin indicate that it is well absorbed and rapidly metabolized to diallyl disulfide, which also rapidly metabolized (*111*). Diallyl disulfide has been shown in mice (*112*) to accumulate in the liver as sulfate, although liver perfusion studies have shown that it may first be converted to allyl mercaptan (*113*). Animal studies have also shown that the vinyl-dithiins are well-absorbed and accumulate without further transformation in several tissues, although the 1,2-vinyl-dithiin is cleared more slowly from the blood and accumulates less in the liver than the 1,3-vinyl-dithiin (*114*).

Allicin has never been found in the blood of people who have consumed garlic, indicating that it is rapidly converted to other compounds (*35, 115*). Allyl mercaptan has been shown to be the dominant sulfur compound in human breath after consuming garlic cloves (*116, 117*), which is a good indication of what at least some of the allicin may be transformed to in the body. In an attempt to find allicin or its metabolites in the body, we have fed volunteers up to 25 g of finely-chopped raw garlic in a single dose and were unable to find detectable serum or urine levels of allicin, ajoene, vinyldithiins, or diallyl sulfides, using methods that could have detected 0.1% retention of these compounds (*35*), nor were we able to detect a garlic odor in the blood or urine. Furthermore, after adding limited amounts of allicin directly to blood plasma (0.4 mM) and then immediately extracting the plasma with chloroform, we were unable to recover any allicin, although most of it was recoverable if the plasma was first deproteinated (*35*), indicating that allicin rapidly reacts with protein.

The previously mentioned reaction of allicin, ajoene, diallyl trisulfide, and diallyl disulfide with cysteine to form *S*-allylmercaptocysteine (see *Reactions with Cysteine*) is probably very important toward identifying their *in vivo* metabolites, and hence the biologically active compounds, after ingestion of garlic or garlic products. The reaction rates are rapid enough to form quantitative amounts of *S*-allylmercaptocysteine prior to intestinal absorption due to the presence of protein-bound cysteine as well as free cysteine released during protein digestion. Nevertheless, *S*-allylmercaptocysteine is probably only a very temporary metabolite. We have recently found that incubation of 2 mM *S*-allylmercaptocysteine with fresh whole blood (human) resulted in a rapid (half-life of 3 min at 37°C) and quantitative transformation to allyl mercaptan. Blood cells were responsible for this transformation, since serum gave almost no reaction. The allyl mercaptan required acetonitrile extraction and was stable in the blood for at least four hours, and caused the blood to have a strong garlic odor. A compound similar to *S*-allylmercaptocysteine, allithiamine (allyl-SS-thiamine), the reaction product of allicin with thiamine, has also been reported to rapidly decrease in blood (*118*).

Even if the intestinal reactions with cysteine were to be circumvented, allicin, ajoene, diallyl trisulfide, diallyl disulfide, and 1,2-vinyl-dithiin do react with components in the blood. We have incubated these compounds (0.5 mM) in fresh whole blood maintained at 37°C and found allicin (half-life <1 min), ajoene (half-life 1 min), and diallyl trisulfide (half-life 4 min) to rapidly decrease, while 1,2-vinyl-dithiin (half-life 15 min) and diallyl disulfide (half-life 60 min) decreased more slowly, and no decline was found in two hours for 1,3-vinyl-dithiin, diallyl monosulfide, alliin, or the γ-glutamylcysteines. Allyl mercaptan was detected as a nearly quantitative reaction product (1.6 moles/mole allicin and 0.8 moles/mole for the others) in the blood. It formed immediately with allicin and ajoene, but a 30-45 min time delay was found in its formation after disappearance of diallyl trisulfide and diallyl disulfide (Lawson, L.D.; Wang, Z.J.; unpublished data).

Thus, it appears that allyl mercaptan, or a further metabolite of it, is probably the vehicle of the systemic (non-topical, non-enteral) biological effects of allicin, ajoene, diallyl trisulfide and diallyl disulfide. Determination of allyl mercaptan in blood or tissues after consumption of garlic or garlic products will be necessary to assess the validity of these results *in vivo*; however, this will prove difficult unless a thiol-derivatizing reagent is used, since allyl mercaptan is very volatile and cannot be concentrated. Consumption of allicin from 10 g of garlic could result in formation of 37 mg of allyl mercaptan in the blood at a maximum concentration of only 0.1 mM. The metabolic fate of these compounds *in vivo* is probably not fully represented in these *in vitro* experiments since their respective garlic products do not produce the same biological effects, as will be shortly discussed.

Considerations in Studying the Biological Effects of Garlic

Even though the large number of publications on the chemistry and biological effects of garlic may give the impression that garlic is a thoroughly investigated medicinal plant, the truth lies far short of this expectation. A major problem in garlic research is the fact that the biological studies have proceeded at a far greater pace than the chemical/analytical studies, with the frequent use of variously processed, but chemically undefined, garlic products (and often undescribed processing methods). This is a particularly important problem with garlic, since the thiosulfinates released from crushed garlic can be rapidly converted to a variety of other organosulfur compounds, depending upon treatment conditions and because of the large variation in content of these compounds among the various types and sources of commercial garlic products (Table II). Furthermore, accurate and convenient methods for quantifying allicin (*48, 73, 76, 119-122*), the other thiosulfinates (*48, 73*), alliin and other cysteine sulfoxides (*61*), the γ-glutamylcysteines (*42*), and ajoene (*29, 76, 80*) did not become available until the past two or three years.

There is also the problem of natural variation of the alliin content (or allicin-releasing potential or allicin yield when cloves are crushed) due to the different varieties of garlic (*48, 71*) as well as to the soil-sulfate content (*71, 123*). The variation in allicin yield from bulb to bulb within a field (7%) and from farm to farm (25%) is relatively small (*48*), whereas the variation among *Allium sativum* L. varieties grown around the world (*48, 71, 120*) is about five-fold (0.13-0.66%). About 80% of these 40 varieties fall within the 2.5-fold range of a 0.21-0.54% allicin yield. Soil and climate conditions may have more of an effect on variation than variety. For example, we have recently determined the allicin yield of 34 varieties of both of the major subspecies of garlic [*Allium sativum* L. *sativum* or softneck garlic (11 varieties) and *Allium sativum* L. *ophioscorodon* or hardneck or topset garlic (23 varieties)] grown in close proximity on the same small parcel of land by Grace Reynolds in Troy, NY. There were no significant differences in the levels of allicin or the other thiosulfinates between the two subspecies and only a 1.6-fold range of variation among all 34 varieties, although the softneck varieties contained 1.94 times as much γ-glutamyl-*S*-allylcysteine as the hardneck varieties ($P<0.001$), but the same amount of γ-glutamyl-*S*-*trans*-1-propenylcysteine. This is considerably less variation than is found among commercial garlic products, such as allicin yield of garlic-powder tablets (18-fold) or the sulfides of steam-distilled garlic oils (35-fold) (*29* and Table II). The magnitude of these variations necessitates using garlic or garlic products of defined yield of allicin or other marker compounds (such as allicin for garlic powder products, diallyl sulfides for distilled oils and vinyl-dithiins for the oil-macerates) when biological studies are conducted. Unfortunately, of the 680 biological studies conducted on garlic and processed garlic in the past 32 years, only about 20 of them (mostly the recent clinical trials on blood lipid-lowering effects) used a product of known content of a suspected active or marker compound.

The proposal that garlic consumption can have significant cardiovascular, antibiotic, and perhaps anticancer effects is almost unquestionable because of the wealth of scientific literature supporting these effects, which include both animal and human studies. Far more questionable is the identity of the specific compounds from garlic or garlic products that are responsible for their biological effects and their mechanisms of action *in vivo*. There have been no clinical studies and almost no animal studies conducted with reasonable doses of purified garlic compounds. However, some studies have been conducted using extracts or fractions of variously treated garlic which have provided good clues to the identities of the active compounds.

There have been many *in vitro* as well as several organ perfusion studies using pure garlic-related organosulfur compounds (*35, 113, 124-127*); however, except for intestinal and topical antibiotic effects, these studies are probably meaningless for evaluating the *in vivo* effects of allicin, ajoene, the trisulfides, and perhaps the disulfides. As previously mentioned (see *Reactions with Cysteine*), these compounds probably form *S*-allylmercaptocysteine in the intestinal lumen or epithelial cells followed by formation of allyl mercaptan in the blood. Therefore, until the actual compounds found in the blood or tissues after consumption of allicin, ajoene, or the tri- and disulfides are verified, *in vitro* and organ perfusion studies designed to study the biological activities of these compounds will be of limited value.

Nevertheless, it may be safely said that both the *in vivo* and appropriate *in vitro* biological studies indicate that probably all of the garlic-related organosulfur compounds have some degree of biological activity, although some are clearly more active than others. Allicin has been well-established *in vitro* as the antibiotic principle of crushed garlic (*30, 31*) and is effective topically and enterally. *In vivo*, orally administered allicin has been shown in animals to have hypoglycemic effects equal to the common drug, tolbutamide, at the same high dose of 250 mg/kg (*128*). The allyl methyl thiosulfinates have recently been shown *in vitro* to have similar, but somewhat reduced, antibiotic activity compared to allicin (*31*). The 1-propenyl thiosulfinates have been shown *in vitro* and *in vivo* to have anti-asthmatic activity (*67*) and the γ-glutamylcysteines may have a role in reducing blood pressure (*36-38*). The thiosulfinates also appear to be effective in preventing spontaneous mammary cancer in mice, since feeding dietary levels of fresh garlic caused this effect, but feeding alliinase-inactivated garlic did not (*129*). Adenosine has been proposed as an important antithrombotic compound from garlic (*1*), but the adenosine content of freshly crushed garlic is small (0.02 mg/g) (*48*) and adenosine has no anti-thrombotic activity in whole blood because it is rapidly assimilated by erythrocytes (*35*). The antithrombotic activity of ajoene has been indicated from post-absorption *in vivo* studies (*130*); however, the concentration of ajoene in current garlic products is too small to have physiological effects (*35*). While it is not possible in this short chapter to discuss the anticancer studies, crushed garlic, steam-distilled garlic oils, and large doses of diallyl disulfide, diallyl monosulfide, ajoene, alliin, and *S*-allylcysteine have all been shown to reduce tumor growth in animals (*26-28*). Certainly, more studies with reasonably low doses are needed. The remainder of this chapter will now briefly discuss the general results of the cardiovascular effects of garlic and garlic products and the probable active compounds.

Cardiovascular and Lipid-Lowering Effects of Garlic

Consumption of moderate amounts of garlic cloves, garlic powder, and certain garlic preparations has consistently produced significant favorable effects on blood lipids (total cholesterol, triglycerides, LDL-cholesterol, and HDL-cholesterol), blood pressure, fibrinolysis, and blood circulation in both animals and man (see reviews *1, 2, 20-25*). The lowering of elevated blood lipids in animals fed garlic homogenates or ether-extracted garlic oils has been demonstrated in almost 40 studies from 1933 to 1985. Recent double-blind, placebo-controlled clinical studies have also shown that garlic consumption significantly increases capillary circulation (*131*) and inhibits spontaneous platelet aggregation (*132*). Very few of these studies used purified compounds. One study used pure allicin (*133*) and found a 27% decrease in serum cholesterol and serum triglycerides after 8 weeks of feeding normal rats a very high level of allicin (100 mg/kg/day) (equivalent to 1400 g of garlic cloves/70 kg person/day). Unfortunately, a lower dose of allicin was not tested. Pure alliin and methylcysteine sulfoxide were fed to cholesterol-fed rats with a resultant 20-45% decrease in plasma cholesterol; however, the only doses studied were extremely high (200-400 mg/kg/day) (*134, 135*). Feeding pure diallyl disulfide at 100 mg/kg/day to sucrose-fed rats also resulted in a 28% decrease in serum cholesterol (*136*). All of these studies with pure compounds were conducted at such high levels that no realistic biological activity can be discerned from them in relation to garlic preparations.

Animal studies with fractions of garlic homogenate and steam-distilled garlic have shown that while the aqueous homogenate decreased blood or arterial lipids, neither boiled aqueous extract (*33*) (which boiling would have caused the thiosulfinates, the only known volatile compounds present, to rapidly evaporate) nor the steam-distillation residue (the non-volatiles) (*33, 34*) nor the oil of steam-distilled garlic (the sulfides) (*137, 138*) showed any effect. Therefore, these animal studies suggest that the thiosulfinates may be the agents responsible for the lipid-lowering effects of garlic homogenate. However, other animal studies which have used ether-extracted oil of garlic (*136, 137, 139*) indicate that in addition to allicin, some of the transformation products of allicin (other than the sulfides) also have considerable lipid-lowering ability. As mentioned earlier (see *Oils of Processed Garlic*), the ether-extracted oil of garlic rapidly transforms allicin to vinyl-dithiins and small amounts of ajoene and trisulfides. Therefore, the animal studies have indicated that allicin and the vinyl-dithiins and/or ajoene, but neither the sulfides nor the non-volatile compounds of garlic, have significant lipid-lowering effects when taken orally. In a recent report, blood pressure in dogs has been significantly reduced for several hours when given a small dose of garlic powder (as low as 2.5 mg/kg bw) introduced intragastrically as gelatin capsules (*140*). Since dissolution of the capsules in the stomach would have inactivated alliinase (*65*), the blood pressure-lowering effect cannot be due to allicin, nor is it due to proteins, since only the external dialysate (Martín, N., Concepcion University, Chile; personal communication, 1992) of the garlic homogenate was shown to have activity (*141*). The γ-glutamylcysteines are the only compounds in garlic known as yet from *in vitro* studies that may lower blood pressure, as indicated by their ability to inhibit angiotensin-converting enzyme (*37, 38*).

Clinical Studies on Reduction of Serum Lipids. Since 1975 there have been 32 human studies on the lipid-lowering effects of garlic and garlic products (*98, 99, 142-171*). The vast majority of these studies, particularly those using pills rather than raw garlic, were randomized, double-blind, placebo-controlled studies which lasted 4-16 weeks and used hyperlipidemic patients. Almost all of the studies showed significant decreases in serum cholesterol and serum triglyceride. Those studies which did not show significant decreases had test-material deficiencies, which will be shortly discussed, that give important clues for identifying the active compounds. All of the studies employed assays for serum cholesterol and most for serum triglycerides. Only about one-third of the studies measured lipoproteins (*147-149, 155, 156, 158-161*), but significant favorable changes in LDL-cholesterol (11-26% decrease) and HDL-cholesterol (0-14% increase) were consistently found in those studies showing a decrease in total cholesterol. A few of the studies measured blood pressure changes (*149-154, 156, 162*), which was found to decrease by 5-12% in all cases. A recent study employing a single-dose of acid-protected, standardized garlic powder tablets (900 mg powder) accompanying a high fat meal showed a significant decrease in increased postprandial serum triglycerides, an effect which increased upon daily consumption of the garlic preparation (*161*). A summary of the results of these studies is given in Table III.

Consumption of both raw garlic (which releases allicin immediately upon chewing) and stomach acid-resistant garlic powder tablets yielding high, standardized levels of allicin consistently and significantly reduced serum lipid values (*142-161*). Most of the clinical studies using garlic powder tablets have used a single brand (*147-157, 161*) which does not claim to be enteric coated; however, we have found that these tablets do release 90% and 65% of their allicin potential into artificial intestinal fluid after being present in artificial gastric juice for 30 and 60 min, respectively, according to standard procedures (*172, 173*)

Four studies using garlic powder tablets showed no effects on serum lipids (*162-165*). Two of these studies (*162, 163*) used commercial products known to have low allicin yields (*28, 174*), while the other two studies (*164, 165*) used garlic powder prepared by spray-drying homogenized garlic, a method which releases only and all of the thiosulfinates from garlic into the atmosphere. Therefore, these four negative studies, along with the previously mentioned animal studies, provide good evidence that allicin and the other thiosulfinates released from garlic are solely responsible for the lipid-lowering effects of raw garlic. One single-dose study with 10 volunteers has been reported (*171*) in which both 50g of crushed garlic and 50g of boiled garlic (hence, without allicin production) prevented a four-hour rise in serum cholesterol levels induced by a 100 g fat meal, indicating that at very high levels, garlic may contain cholesterol-lowering compounds other than allicin. More positive proof for the role of the thiosulfinates could be obtained in clinical studies by comparing stomach acid-resistant pills to non-resistant pills made from the same garlic powder or, perhaps even better, to acid-resistant pills made from the same powder after releasing and removing all of the thiosulfinates from the powder by wetting it and then drying it. Again, the thiosulfinates themselves are probably not acting directly on the processes involved in the lowering of serum lipids, but rather, allicin-derived allyl mercaptan or further metabolites of allyl mercaptan.

Although the thiosulfinates appear to be solely responsible for the lipid-lowering effects of garlic cloves or garlic powder products, some other compounds of processed garlic also appear to be very effective, as evidenced by the studies using ether-extracted garlic oil (*98, 99*) and oil-macerates (*166, 167* and Table III). Both of these oils contain predominantly vinyl-dithiins and considerably lesser amounts of ajoene and the trisulfides and were effective at low doses.

Table III. Summary of Human Studies on the Effects of Garlic and Garlic Products on Elevated Serum Lipids

Garlic product (report years)	N[a]	Dose[b]	Duration (weeks)	Effects: Serum cholesterol	Effects: Serum triglyceride	References
Garlic cloves (1975-1979)	5	3-10 g	3-8	↓10-29%	↓24-34%	*(142-146)*
Garlic powder tablets -allicin standardized, stomach acid-resistant (1984-1991)	15	0.6-0.9 g	4-16	↓6-21%	↓8-27%	*(147-161)*
-not standardized (1984-1988)	4	0.3-1.3g	4-12	NC[c]	NC	*(162-165)*
Ether-extracted oil (1981, 1982)	1[d]	15 mg	26	↓14%	↓36%	*(98)*
	1[e]	15 mg	40	↓18%[f]	↓32%[f]	*(98)*
	1	15 mg	3	↓7%	↓36%	*(99)*
Oil-macerate (1986, 1991)	1	2 mg[g]	12	↓15%	NC	*(166)*
	5	?[h]	4-12	HS[i]	NC	*(167)*
Steam-distilled oil (1981, 1985)	1	4 mg	12	NC	NC	*(168)*
	1	100 mg	20	↑20%	NC	*(169)*
Aged in aqueous alcohol (1987)	1[d]	4 ml	26	NC[f]	NC[f]	*(170)*
	1[e]	4 ml	26	↓11%[f]	↓15%[f]	*(170)*

[a] Number of studies.
[b] Daily dose: cloves (fresh wt.), tablets (garlic powder wt.), oils (undiluted wt.).
[c] No change.
[d] Patients with normal serum lipid values
[e] Patients with high serum lipid values.
[f] Lipid values increased in the first two months before declining.
[g] Determined by our own analysis.
[h] A preliminary report which did not include many details.
[i] Described as highly significant.

Steam-distilled garlic oil (allyl sulfides) appears to have no effect on lowering blood lipids, as indicated in two clinical trials (*168, 169*). Although these studies did not indicate the composition or type of garlic oil used, we have obtained the commercial brand name from the author and have analyzed two recent lots of this oil and found it to contain 1.9 mg/g of diallyl and allyl methyl oligosulfides, but no vinyl-dithiins, which demonstrates that it is a steam-distilled garlic oil product of moderate quality (see Table II).

The only clinical study using aged alcoholic garlic extract (*170*) found no lipid-lowering effects for persons with normal blood lipid levels, but did show significant effects for persons with high blood lipid levels; however, significant effects were not found until taking the product for 5-6 months. In fact, in the first two months of consuming the aged garlic product, cholesterol and triglyceride levels increased significantly in all patients. This initial rise in blood lipids was also seen when the ether-extracted garlic oil was given to hyperlipidemics, but not when given to normolipidemics (*98*). It has been proposed (*98, 170*) that the initial two-month rise in blood lipids is due to their being cleared from the arteries and other body tissues; however, no initial rise has ever been reported in the studies using raw garlic or garlic powder tablets (*142-161*), even though several of the studies determined lipid levels monthly. The effective compounds in the aged alcoholic garlic extract are unknown, since it contains only very small amounts of organosulfur compounds (see Table II) and since the non-volatile compounds of fresh garlic appear to have no effect (*164, 165*).

What is remarkable about the lipid-lowering effects of garlic and its effective products are the low doses required. When compared to commonly prescribed drugs (Table IV), garlic and garlic powder, as whole-plant products, compare favorably to

Table IV. Agents Effective for Reducing Elevated Human Serum Cholesterol Levels

Agent	Dose to give 15-20% reduction (mg/70 kg person/day)
Clofibrate	2,000
Neomycin	500
Probucol	1,000
Cholestyramine resin	16,000
Soluble fiber (pectins)	15,000
Raw garlic (dry weight) (5 trials)[a]	1,200 (1 clove)
Garlic powder (14 trials)[a]	900 (1 clove)
Allicin (released from garlic powder)[b]	5
Garlic oils containing vinyl-dithiins and ajoene (3 trials)[a]	15

[a] References given in Table III.

[b] Based on the standardized allicin levels used in the 14 garlic powder tablet trials and on the assumption that allicin and the other thiosulfinates are solely responsible for the cholesterol-lowering effects.

clofibrate, neomycin, and probucol and are much more effective than cholestyramine resin and soluble fiber. However, the amounts of allicin (5 mg/person/day, assuming it to be the sole active agent of garlic and garlic powder) and the vinyl-dithiin/ajoene-containing garlic oil (15 mg/person/day) necessary to lower serum cholesterol levels are two-orders of magnitude lower than for clofibrate, neomycin, and probucol. These low effective doses indicate that inhibition must be at the enzymatic level rather than at the substrate level and, therefore, cannot be due to acetyl-CoASH binding, as has been previously proposed (*104-106*). Furthermore, the commonly recommended therapeutic dose of one small garlic clove (2-3 g fresh weight) per day (*92*) has been substantiated by these clinical trials for blood lipid-lowering effects.

Mortality/Morbidity Study with a Vinyl-dithiin Garlic Oil. A mortality/morbidity study on patients with heart disease using a garlic product has recently been conducted which gives a direct relationship between blood lipid-lowering effects and longevity. Bordia has conducted a three-year randomized, placebo-controlled study in India with 432 patients who had experienced prior heart attacks. The patients were given either 0.1 mg/kg/day (about 6 mg/person) of ether-extracted garlic oil (for content see *Oils of Processed Garlic*) in gelatin capsules or they were given placebo capsules. After three years of consuming this high vinyl-dithiin garlic oil, there was a 35% reduction in new heart attacks and a 45% reduction in total deaths compared to those consuming the placebo (*175, 176*). Significant reductions in the serum lipids were also noted. The results of this study are very noteworthy, especially considering the low dose used, and provide excellent evidence that consumption of certain garlic-derived organosulfur compounds can significantly decrease heart disease. It is hoped that more studies of this type will be conducted in the future and that such studies will also include a standardized, allicin-yielding garlic product.

Literature Cited

1. Koch, H. P.; Hahn, G. *Knoblauch*; Urban and Schwarzenberg: München, Baltimore, 1988.
2. Fenwick, G. R.; Hanley, A. B. *CRC Crit. Rev. Food Sci. Nutr.* **1985**, *23*, 1-73.
3. Hanley, A. B.; Fenwick, G. R. *J. Plant Foods* **1985**, *6*, 211-238.
4. Sprecher, E. *Pharm. Ztg.* **1986**, *131*, 3161-3168.
5. Reuter, H. D. *Z. Phytother.* **1986**, *7*, 99-106.
6. Abdullah, T. H.; Kandil, O.; Elkadi, A.; Carter, J. *J. Natl. Med. Assoc.* **1988**, *80*, 439-445.
7. Petkov, V. *J. Ethnopharmacol.* **1986**, *15*, 121-132.
8. Koch, H. P. *Z. Phytother.* **1992**, *13*, 177-188.
9. Koch, H. P. *Z. Phytother.* **1992**, *13*, 83-90.
10. Block, E. *Angew. Chem. Intl. Ed. Engl.*, **1992**, *31*, 1135-1178; *Angew. Chem.*, **1992**, *104*, 1158-1203.
11. Sticher, O. *Dtsch. Apoth. Ztg.* **1991**, *131*, 403-413.
12. Fenwick, G. R.; Hanley, A. B. *CRC Crit. Rev. Food Sci. Nutr.* **1985**, *22*, 273-377.
13. Block, E. *Sci. Am.* **1985**, *252*, 114-119.

14. Block, E. In *Folk Medicine: The Art and the Science*; Steiner, R. P., Ed.; American Chemical Society: Washington, DC, 1986, pp 125-137.
15. Lancaster, J. E.; Boland, M. J. *In Onions and Allied Crops,* Vol. 3 ; Brewster, J. L.; Rabinowitch, H. D., Eds., CRC Press: Boca Raton, Florida, 1990,; pp. 33-72.
16. Whitaker, J. R. *Adv. Food Res.* **1986**, *22*, 73-133.
17. Whitaker, J. R. In *Food Enzymology*, Vol. 1; Fox, P. F., Ed., Elsevier: London, 1991; pp. 479-497.
18. Lutomski, J. *Components and Biological Properties of Some Allium Species*; Institute of Medicinal Plants: Poznan, 61-707 Poland, 1987.
19. Raghavan, B.; Abraham, K.O.; Shankaranarayana, M. L. *J. Sci. Ind. Res.* **1983**, *42*, 401-409.
20. Kleijnen, J.; Knipschild, P.; Ter Riet, G. *Br. J. Clin. Pharmacol.* **1989**, *28*, 535-544.
21. Lau, B. H. S. *Curr. Top. Nutr. Dis.* **1989**, *22*, 295-325.
22. Turner, M. *J. Roy. Soc. Health* **1990**, *110*, 90-93.
23. Greenwood, T. W., Ed.; *Br. J. Clin. Prac.* **1990**, *44* (suppl. 69), 1-39.
24. Kendler, B. S. *Prev. Med.* **1987**, *16*, 670-685.
25. Ernst, E. *Pharmatherapeutica* **1987**, *5*, 83-89.
26. Lau, B. H. S.; Tadi, P. P.; Tosk, J. M. *Nutr. Res.* **1990**, *10*, 937-948.
27. Sumiyoshi, H.; Wargovich, M.J. *Asia Pacific J. Pharmacol.* **1989**, *4*, 133-140.
28. Dausch, J. G.; Nixon, D. W. *Prev. Med.* **1990**, *19*, 346-361.
29. Lawson, L. D.; Wang. Z. J.; Hughes, B. G. *Planta Med.* **1991**, *57*, 363-370.
30. Cavallito, C. J.; Bailey, J. H.;Buck, J. S. *J. Am. Chem. Soc.* **1945**, *67*, 1032-1033.
31. Hughes, B. G; Lawson, L. D. *Phytother. Res.* **1991**, *5*, 154-158.
32. Silber, W. *Klin. Wochenschr.* **1933**, *12*, 509.
33. Billau, H. Dissertation; 1961; University of Giessen, Germany.
34. Kamanna, V. S.; Chandrasekhara, N. *Ind. J. Med. Res.* **1984**, *79*, 580-583.
35. Lawson, L. D.; Ransom, D. K.; Hughes, B. G. *Thromb. Res.* **1992**, *65*, 141-156.
36. de Torrescacasna, E. U. *Rev. Espan. Fisiol.* **1946**, *2*, 6-31; *Chem. Abstr.* **1947**, *41*, 2172.
37. Sendl, A.; Elbl, G.; Steinke, B.; Redl, K.; Breu, W.; Wagner, H. *Planta Med.* **1992**, *58*, 1-7.
38. Elbl, G. Dissertation; 1991; University of Munich, Germany.
39. Koch, H. P.; Jäger, W.; Brauner, A.; Roth, S. *Dtsch. Apoth. Ztg.* **1988**, *19*, 993-995.
40. Ip, C.; Lisk, D. J.; Stoewsand, G. S. *Nutr. and Cancer* **1992**, *17*, 279-286.
41. *Recommended Dietary Allowances, 10th Ed.*, National Research Council, Subcommittee on the Tenth Edition of the RDAs; National Academy Press: Washington, D. C., 1989.
42. Lawson, L. D.; Wang, Z. J.; Hughes, B. G. *J. Nat. Prod.* **1991**, *54*, 436-444.
43. Stoll, A.; Seebeck, E. *Experientia* **1947**, *3*, 114-115.
44. Fujiwara, M.; Yoshimura, M.; Tsuno, S.; Murakami, F. *J. Biochem.* **1958**, *45*, 141-149.
45. Virtanen, A. I.; Mattila, I. *Suom. Kemistil. B* **1961**, *34*, 44.

46. Suzuki, T.; Sugii, M.; Kakimoto, T. *Chem. Pharm. Bull.* **1961**, *9*, 77-78.
47. Lawson, L. D.; Hughes, B. G. *Planta Med.* **1990**, *56*, 589.
48. Lawson, L. D.; Wood, S. G.; Hughes, B. G.; *Planta Med.* **1991**, *57*, 263-270.
49. Mütsch-Eckner, M. Dissertation; 1991; ETH Zürich, Switzerland.
50. Virtanen, A. I. *Phytochemistry* **1965**, *4*, 207-228.
51. Mütsch-Eckner, M.; Meier, B.; Wright, A. D.; Sticher, O. *Phytochemistry* **1992**, *31*, 2389-2391.
52. Lancaster, J. E.; Shaw, M. L. *Phytochemistry* **1991**, *30*, 2857-2859.
53. Ceci, L. N.; Curzio, O. A.; Pomilio, A. B. *Phytochemistry* **1992**, *31*, 441-444.
54. Lancaster, J. E.; Shaw, M. *Phytochemistry* **1989**, *28*, 455-460.
55. Pentz, R.; Guo, Z.; Kress, G.; Müller, D.; Siegers, C. P. *Planta Med.* **1990**, *56*, 691.
56. Alfonso, N.; Lopez, E. *Z. Lebensm. Unters.* **1960**, *111*, 410-413; *Chem. Abstr.* **1960**, *54*, 13487.
57. Suzuki, T.; Sugii, M.; Kakimoto, T. *Chem. Pharm. Bull.* **1961**, *9*, 77-78.
58. Tsuboi, S.; Kishimoto, S.; Ohmori, S. *J. Agric. Food Chem.* **1989**, *37*, 611-615.
59. Sugii, M.; Suzuki, T.; Nagasawa, S.; Kawashima, K. *Chem. Pharm. Bull.* **1964**, *12*, 1114-1115.
60. Kominato, K. *Chem. Pharm. Bull.* **1969**, *17*, 2193-2200.
61. Ziegler, S. J.; Sticher, O. *Planta Med.* **1989**, *55*, 372-378.
62. Lancaster, J. E.; Collin, H. A. *Plant Sci. Lett.* **1981**, *22*, 169-176.
63. Stoll, A.; Seebeck, E. *Helv. Chim. Acta* **1949**, *32*, 197-205.
64. Block, E.; O'Connor, J. *J. Am. Chem. Soc.* **1974**, *96*, 3929-3944.
65. Lawson, L. D.; Hughes, B. G. *Planta Med.* **1992**, *58*, 345-350.
66. Bayer, T.; Wagner, H.; Block, E.; Grisoni, S.; Zhao, S.H.; Neszmelyi, A. *J. Am. Chem. Soc.* **1989**, *111*, 3085-3096.
67. Bayer, T.; Breu, W.; Seligmann, O.; Wray, Y.; Wagner, H. *Phytochemistry* **1989**, *28*, 2373-2377.
68. Van Damme, E. J. M.; Smeets, K.; Torrekens, S.; Van Leuven, F.; Peujmans, W. J. *Eur. J. Biochem.* **1992**, *209*, 751-757.
69. Stoll, A.; Seebeck, E. *Adv. Enzymol.* **1951**, *11*, 377-400.
70. Blania, G.; Spangenberg, B. *Planta Med.* **1991**, *57*, 371-375.
71. Pfaff, K. *Dtsch. Apoth. Ztg.* **1991**, *131* (suppl. 24), 12-15.
72. Sreenivasamurthy, V.; Sreekantiah, K. R.; Johar, D. S. *J. Sci. Ind. Res.* **1961**, *20C*, 292-295.
73. Block, E. *J. Agric. Food Chem.* **1992**, *40*, 2418-2430.
74. Saghir, A. R.; Mann, L. K.; Yamaguchi, M. *Plant Physiol.* **1965**, *40*, 681-685.
75. Sendl, A; Wagner, H. *Planta Med.* **1991**, *57*, 361-362.
76. Wagner, H.; Sendl, A. *Dtsch. Apoth. Ztg.* **1990**, *130*, 1809-1815.
77. Block, E.; Ahmad, S.; Catalfamo, J. L.; Jain, M. K.; Apitz-Castro, R. *J. Am. Chem. Soc.* **1986**, *108*, 7045-7055.
78. Block, E.; Ahmad, S.; Jain, M. K.; Crecely, R. G.; Apitz-Castro, R.; Cruz, M. R. *J. Am. Chem. Soc.* **1984**, *106*, 8295-8296.
79. Cavallito, C. J.; Bailey, J. H. *J. Am. Chem. Soc.* **1944**, *66*, 1950-1951.

80. Iberl, B.; Winkler, G.; Knobloch, K. *Planta Med.* **1990**, *56*, 202-211.
81. Yu, T.-H.; Wu, C.-M. *J. Food Sci.* **1989**, *54*, 977-981.
82. Miething, H. *Phytother. Res.* **1988**, *3*, 149-151.
83. Yu, T.-H.; Wu, C.-M. ; Liou, Y.-C. *J. Agric. Food Chem.* **1989**, *37*, 725-730.
84. Semmler, F.W. *Arch. Pharm.* **1892**, *230*, 434-443.
85. Jirovetz, L.; Jäger, W.; Koch, H. P.; Remberg, G. *Z. Lebensm. Unters.* **1992**, *194*, 363-365.
86. Brodnitz, M. H.; Pascale, J. V.; Van Derslice, L. *J. Agric. Food Chem.* **1971**, *19*, 273-275.
87. Voigt, M.; Wolf, E. *Dtsch. Apoth. Ztg.* **1986**, *126*, 591-593.
88. Kice, J. L.; Rogers, T. E. *J. Am. Chem. Soc.* **1974**, *96*, 8009-8015.
89. Müller, B. *Dtsch. Apoth. Ztg.* **1989**, *129*, 2500-2504.
90. Cavallito, C. J.; Buck, J. S.; Suter, M. *J. Am. Chem. Soc.* **1944**, *66*, 1952-1954.
91. Yu, T.-H.; Wu, C.-M.; Ho, C.-T. *J. Agric. Food Chem.* (in press).
92. Koch, H. P. *Dtsch. Apoth. Ztg.* **1988**, *128*, 408-412.
93. Aye, R. D. *Cardiol. Prac.* **1989**, *7* (suppl. 10), 7-8.
94. Pruthi, J. S.; Singh, L. J.; Kalbag, S. S.; Lal, G. *Food Sci.* **1959**, *8*, 444-448.
95. Bordia, A.; Bansal, H. C.; Arora, S. K., Singh, S. V. *Atherosclerosis* **1975**, *21*, 15-19.
96. Bordia, A.; Bansal, H. C. *Lancet* **1973**, *2*, 1491-1492.
97. Bordia, A.; Joshi, H. K.; Sanadhya, Y. K.; Bhu, N. *Atherosclerosis* **1977**, *28*, 155-159.
98. Bordia, A. *Am. J. Clin. Nutr.* **1981**, *34*, 2100-2103.
99. Bordia, A.; Sharma, K. D.; Parmar, Y. K.; Verma, S. K. *Ind. Heart J.* **1982**, *34*, 86-88.
100. Bordia, A. *Cardiol. Prac.* **1989**, *7* (suppl. 10), 14.
101. Nakagawa, S.; Kasuga, S.; Matsuura, H. *Phytother. Res.* **1989**, *3*, 50-53.
102. Hirao, Y.; Sumioka, I.; Nakagami, S.; Yamamoto, M.; Hatono, S.; Yoshida, S.; Fuwa, T.; Nakagawa, S. *Phytother. Res.* **1987**, *1*, 161-164.
103. Wills, E. D. *Biochem. J.* **1956**, *63*, 514-519.
104. Bailey, J. H.; Cavallito, C. J. *J. Bacteriol.* **1948**, *55*, 175-182.
105. Augusti, K. T.; Mathew, P. T. *Experientia* **1974**, *30*, 468-470.
106. Papageorgiou, C.; Carbet, J. P.; Menezes-Brandao, F.; Pecegueiro, M.; Beneza, C. *Arch. Dermatol. Res.* **1983**, *275*, 229-234.
107. Augusti, K. T. *Ind. J. Exptl. Biol.* **1978**, *15*, 1223-1224.
108. Winkler, G.; Iberl, B.; Knobloch, K. *Planta Med.* **1992, 58** (suppl. 1), A665.
109. Jandke, J.; Spiteller, G. *J. Chromatog.* **1987**, *421*, 1-8.
110. Hagen, T. M.; Wierzbicka, G. T.; Sillau, A. H.; Bowman, B. B.; Jones, D. P. *Am. J. Physiol.* **1990**, *259*, G524-G529.
111. Pentz, R.; Guo, Z.; Siegers, C.-P. *Med. Welt* **1991**, *42* (suppl. 7a), 46-47.
112. Pushpendran, C. K.; Devasagayam, T. P. A.; Chintalwar, G. J.; Banerji, A.; Eapen, J. *Experientia* **1980**, *36*, 1000-1001.
113. Egen-Schwind, C.; Eckard, R.; Kemper, F. H. *Planta Med.* **1992**, *58*, 301-305.

114. Egen-Schwind, C.; Eckard, R.; Jekat, F. W.; Winterhoff, H. *Planta Med.* **1992**, *58*, 8-13.
115. Reuter, H. D. *Z. Phytother.* **1991**, *12*, 83-91.
116. Laakso, I.; Seppänen-Laakso, T.; Hiltunen, R.; Müller, B.; Jansen, H.; Knobloch, K. *Planta Med.* **1989**, *55*, 257-261.
117. Minami, T.; Boku, T.; Inada, K.; Morita, M.; Okazaki, Y. *J. Food Sci.* **1989**, *54*, 763-765.
118. Fujiwara, M.; Watanabe, H.; Matsui, K. *J. Biochem.* **1954,** *41*,29-39.
119. Jansen, H.; Müller, B.; Knobloch, K. *Planta Med.* **1987**, *53*, 559-562.
120. Iberl, B.; Winkler, G.; Müller, B.; Knobloch, K. *Planta Med.* **1990**, *56*, 320-326.
121. Müller, B. *Planta Med.* **1990**, *56*, 589-590.
122. Müller, B.Dissertation;1991; University of Erlangen-Nürnberg, Germany.
123. Freeman, G. G.; Mossadeghi, N. *J. Sci. Food Agric.* **1971**, *22*, 330-334.
124. Orellana, A.; Kawada, M. E.; Morales, M. N.; Vargas, L.; Bronfman, M. *Toxicol. Lett.* **1992**, *60*, 11-17.
125. Sendl, A.; Elbl, G.; Steinke, B.; Redl, K.; Breu, W.; Wagner, H. *Planta Med.* **1992,** *58*, 1-7.
126. Gebhardt, R. *Arzneim. Forsch./Drug Res.* **1991**, *41*, 800-804.
127. Sendl, A.; Schliak, M.; Löser, R.; Stanislaus, F.; Wagner, H. *Atherosclerosis* **1992**, *94*, 79-85.
128. Mathew, P. T.; Augusti, K. T. *Ind. J. Biochem. Biophys.* **1973**, *10*, 209-212.
129. Kröning, F. *Acta Unio Intern. Contra Cancrum* **1964**, *20*, 855-856.
130. Apitz-Castro, R.; Ledezema, E.; Escalante, J.; Jorquera, A.; Pinăte, F. M.; Moreno-Rea, J.; Carillo, G.; Leal, O.; Jain, M. K. *Arzneim. Forsch./Drug Res.* **1988**, *38* (II), 901-904.
131. Jung, E. M.; Jung, F.; Mrowietz, C.; Kiesewetter, H.; Pindur, G.; Wenzel, E. *Arzneim. Forsch./Drug Res.* **1991**, *41*, 626-630.
132. Jung, F.; Kiesewetter, H.; Pindur, G.; Jung, E. M.; Mrowietz, C.; Wenzel, E. *Med. Welt* **1991**, *42* (suppl. 7a), 18-19.
133. Augusti, K.T.; Mathew, P.T. *Experientia* **1974**, *30*, 468-470.
134. Fujiwara, M.; Itokawa, Y.; Uchino, H.; Inoue, Y. *Experientia* **1972**, *28*, 254-255.
135. Itokawa, Y.; Inoue, K.; Sasagawa, S.; Fujiwara, M. *J. Nutr.* **1973**, *103*, 88-92.
136. Adamu, I.; Joseph, P.K.; Augusti, K.T. *Experientia* **1982**, *38*, 899-901.
137. Silber, W. *Klin. Wochenschr.* **1933**, *12*, 509.
138. Fields, M.; Lewis, C. G.; Lure, M. D. *J. Am. Coll. Nutr.* **1992**, *11*, 334-339.
139. Bordia, S.; Arora, S. K.; Kothari, L. K.; Jain, K. C.; Rathore, B. S.; Rathore, A. S.; Dube, M. K.; Bhu, N. *Atherosclerosis* **1975**, *22*, 103-109.
140. Pantoja, C. V.; Chiang, Ch.; Norris, B. C.; Concha, J. B. *J. Ethnopharmacol.* **1991**, *31*, 325-331.
141. Martín, N.; Bardisa, L.; Pantoja, C.; Román, R.; Vargas, M. *J. Ethnopharmacol.* **1992**, *37*, 145-149.
142. Kerekes, M. F.; Feszt, T. *Artery* **1975**, *1*, 325-326.
143. Augusti, K. T. *Ind. J. Exptl. Biol.* **1977**, *15*, 489-490.
144. Jain, R. C. *Am. J. Clin. Nutr.* **1977**, *30*, 1380-1381.

145. Sucur, M. *Diabetol. Croat.* **1980**, *9*, 323-338.
146. Bhushan, S.; Sharma, S. P.; Singh, S. P.; Agrawal, A.; Indrayan, A.; Seth, P. *Ind. J. Physiol. Pharmacol.* **1979**, *23*, 211-214.
147. Ernst, E.; Weihmayr, T.; Matrai, A. *Br. Med. J.* **1985**, *291*, 139.
148. Ernst, E.; Weihmayr, T.; Matrai, A. *Aerztl. Praxis* **1986**, *38*, 1748-49.
149. König, F. K.; Schneider, B. *Aerztl. Praxis* **1986**, *38*, 344-345.
150. Harenberg, J.; Giese, C.; Zimmermann, R. *Atherosclerosis* **1988**, *74*, 247-249.
151. Kandziora, *J. Aerztl. Forsch.* **1988**, *35* (1), 1-8.
152. Kandziora, *J. Aerztl. Forsch.* **1988**, *35* (3), 3-8.
153. Vorberg, G.; Schneider, B. *Brit. J. Clin. Prac.* **1990**, *44* (suppl. 69), 7-11.
154. Auer, W.; Eiber, A.; Hertkorn, E.; Hoehfeld, E.; Koehrle, U.; Lorenz, A.; Mader, F.; Merx, W.; Otto, G.; Schmid-Otto, B.; Taubenheim, H. *Br. J. Clin. Prac.* **1990**, *44* (suppl. 69), 3-6.
155. Brosche, T.; Platt, D.; Dorner, H. *Br. J. Clin. Prac.* **1990**, *44* (suppl. 69), 12-19.
156. Zimmermann, W.; Zimmermann, B. *Br. J. Clin. Prac.* **1990**, *44* (suppl. 69), 20-23.
157. Mader, F.H. *Arzneim. Forsch./Drug Res.* **1990**, *40*, 1111-1116.
158. Lehmann, B.; Brewitt, B. *Cardiol. Prac.* **1991**, June suppl., 18.
159. Semm, H. *Pharm. Rundschau* **1987**, *3*, 28-30.
160. Sas, G. *Aerztl. Praxis* **1988**, *93*, 2894-2896.
161. Rotzsch, W.; Richter, V.; Rassoul, F.; Walper, A. *Arzneim. Forsch./Drug Res.* **1992,** *42* (II), 1223-1227.
162. Lutomski, J. *Z. Phytother.* **1984**, *5*, 938-942.
163. Luley, C.; Lehmann-Leo, W.; Moeller, B.; Martin, T.; Schwartzkopff, W. *Arzneim. Forsch./Drug Res.* **1986**, *36*, 766-768.
164. Sitptrija, S.; Plengvidhya, C.; Kangkaya, V.; Bhuvapanich, S.; Tunkayoon, M. *J. Med. Assoc. Thailand* **1987**, *70*, 223-227.
165. Plengvidhya, C.; Sitprija, S.; Chinayon, S.; Pasatrat, S.; Tankeyoon, M. *J. Med. Assoc. Thailand* **1988**, *71*, 248-252.
166. Bordia, A. *Apothek. Magazin* **1986**, No. *6*, 128-131.
167. Kümmel, B. *Med. Welt* **1991**, *42* (suppl. 7a), 20.
168. Arora, R. C.; Arora, S.; Gupta, R. K. *Atherosclerosis* **1981**, *40*, 175.
169. Arora, R. C.; Arora, S.; Nigam, P. *Materia Med. Polona* **1985**, *17*, 48-50.
170. Lau, B. H. S.; Lam, F.; Wang-Cheng, R. *Nutr. Res.* 1987, *7*, 139-149.
171. Sharma, K. K.; Sharma, A. L.; Dwivedi, K. K.; Sharma, P. K. *Ind. J. Nutr. Dietet.* **1976**, *13*, 7-10.
172. *The United States Pharmacopeia,* 22nd ed.; United States Pharmacopoeial Convention: Rockville, Maryland, 1990, pp. 1577-1578.
173. *The United States Pharmacopeia,* 22nd ed.; United States Pharmacopoeial Convention: Rockville, Maryland, 1990, pp. 1788-1789.
174. Koch, H. P.; Hahn, G. *Knoblauch*; Urban and Schwarzenberg: München,Baltimore, 1988, p. 112.
175. Bordia, A. *Cardiol. Prac.* **1989**, *7* (suppl. 10), 14.
176. Bordia, A. *Dtsch. Apoth. Ztg.* **1989**, *129* (suppl. 15), 16-17.

RECEIVED March 24, 1993

Chapter 22

Rohitukine and Forskolin

Second-Generation Immunomodulatory, Intraocular-Pressure-Lowering, and Cardiotonic Analogues

Noel J. de Souza

Hoechst India Ltd., Bombay 400 080, India

Rohitukine is the anti-inflammatory principle of *Dysoxylum binectariferum*. Forskolin is the unique adenylate cyclase-activating principle of *Coleus forskohlii*. The cyclic AMP-dependent effects it is known to display are now supplemented by newly discovered cyclic AMP-independent effects. Synthetic and structure-activity relationship studies leading to the development of antiinflammatory, immunomodulatory and anticancer rohitukine analogs, and intraocular pressure-lowering and cardiotonic forskolin analogs are described. A pathway to a forskolin-based antithrombotic phytotherapeutic product is also presented.

Our updated findings on the discovery and development of novel human medicinal agents from plants are described herein. These findings represent the current status of a venture that commenced in India over 20 years ago in the 1970s. The objectives, strategies, and results of this sustained mission-oriented venture have been adequately documented (*1-5*). The venture was sustained in an era when related natural products-oriented efforts were generally at their ebb in the pharmaceutical industry and there was regrettably seen the discontinuation or diminution of ventures worldwide, like that of the Roche Research Institute, Australia, the National Cancer Institute, U.S.A., and the Ciba Research Centre, India. Even in today's era of the 1990s, with the revival of such programs at the National Cancer Institute, U.S.A. and in industry, the realistic picture to be faced is that about the only plant-based novel chemical entities that have emerged as innovative medicines on the market after reserpine are the anticancer drugs, vincristine, vinblastine, etoposide and taxol. Our venture at Hoechst, aimed at converting plant constituents into novel drugs, owes its sustenance principally to our discovery of molecules like rohitukine and forskolin. The new chemical entities (NCEs) and a potential phytotherapeutic product which have been derived from them are shown in Table I.

Rohitukine-Derived HL 275 and HIL 8276

Compound HL 275 (designated in the earlier literature as L 86 8275) moved into development as a potential anticancer agent as a consequence of an evolution of events commencing with the isolation of the naturally-occurring chromone

0097–6156/93/0534–0331$06.00/0

Table I. Potential Human Medicinal Agents from Plants

Agent	*Therapeutic Category*
Rohitukine-Derived	
– HL 275	Anticancer
Forskolin-Derived	
– HIL 568	Antiglaucoma
– NKH 477[a]	Cardiotonic
– Phytotherapeutic product	Antithrombotic

[a]NKH 477 is a product of Nippon Kayaku Co. Ltd., Tokyo, Japan.

alkaloid, rohitukine, and the production of a synthetically-derived successor of rohitukine, compound HIL 8276. Rohitukine (Figure 1) was identified by us as the antiinflammatory and immunomodulatory principle of the Indian plant, *Dysoxylum binectariferum* (*6,7*). This plant is phylogenetically related to the Ayurvedic plant, *D. malabaricum*, used for rheumatoid arthritis. In view of the remarkable biological profile of rohitukine, a total synthesis of the molecule was accomplished (*6*). A number of rohitukine analogs was thus synthesised, a relationship between structure and antiinflammatory and immunomodulatory activity of the analogs was derived, and an optimal analog, HIL 8276, was selected as a candidate for development. HIL 8276 (Figure 1) is the optically active 2-phenyl analog of rohitukine.

Antiinflammatory and Immunomodulatory Profile of HIL 8276. The profile of HIL 8276 is summarised in Table II. The protective effect of HIL 8276 in different inflammatory models, such as carrageenin-induced rat paw oedema, cotton pellet granuloma in rats, and croton oil-induced oedema (topical), lies in the potency range of non-steroidal antiinflammatory drug substances (NSAIDS) like indomethacin. It also inhibits the reverse passive Arthus (RPA) reaction in rats as do the corticosteroids, but with a potency five-fold greater than that of hydrocortisone. Like NSAIDS and disease-modifying antirheumatic drug substances (DMARDS), it also inhibits neutrophil adhesion and, to a minor extent, neutrophil chemotaxis. Its immunomodulatory profile is characterised by an inhibition of delayed-type hypersensitivity (DTH), and of plaque-forming cells, but with no effect on adjuvant-induced polyarthritis, macrophage stimulation, or experimental allergic encephalomyelitis.

Table II. Antiinflammatory/Immunomodulatory Properties of HIL 8276

Bioassay Test Model	*Pharmacological Effect*[a]
1. Carrageenin paw oedema	ED_{50}: 2.3 mg/kg p.o.
2. Cotton pellet granuloma	3.0 mg/kg p.o. x 7, 25% inhibition
3. Croton oil oedema	0.5 mg/ear, 51% inhibition
4. RPA reaction	2.5 mg/kg p.o., 50.8% inhibition
5. Neutrophil adhesion	2.0 mg/kg p.o.x 5, 54.4% inhibition
6. Neutrophil chemotaxis	5.0 μM, 14.3% inhibition
7. Oxazolone-induced DTH	5.0 mg/kg x 5, 35% inhibition
8. Plaque-forming cell	5.0 mg/kg x 5, 27% inhibition

[a]Hoechst India Limited Product Development Data

In studies that were performed to determine the mode of action of HIL 8276, it was found that the compound does not inhibit cyclooxygenase, lipoxygenase, phospholipase, or collagenase. It is not a PAF-antagonist, it does not inhibit release of histamine from mast cells, and it displays only mild inhibition of polymorphonuclear-generated superoxide (IC_{50} = 36 μM) and of phagocytosis (IC_{50} = 70 μM). It shows, however, good inhibition of tyrosine kinase (IC_{50} = 6.0 μg/ml) (Czech, J.; Dickneite, G.; Sedlacek, H.H.. Unpublished data; Behringwerke, Marburg, Germany).

Protein Kinase Inhibition and Antiproliferative Activities of HL 275. Numerous oncogenes and human proto-oncogenes encode for protein products with protein kinase activity. Therefore, the inhibition of tyrosine kinase by HIL 8276 has opened up a new avenue of investigation for this compound series. A number of rohitukine analogs was subjected to a SAR analysis using the following three test systems: (a) inhibition of the oncogene product, $pp60^{v\text{-}src}$, which possesses tyrosine kinase activity, (b) epidermal growth factor(EGF)-receptor tyrosine kinase inhibition and (c) inhibition of the EGF-stimulated proliferation of the human tumor cell line, ChaGO (3-{4,5-dimethylthiazol-2-yl}-2,5-diphenyl tetrazolium bromide assay). The rohitukine analogs tested had the following structural variations in the flavone ring: (i) 5- and 7-hydroxy or -methoxy groups, and (ii) 2-alkyl or -aryl substituents. The optimal structural requirements for tyrosine kinase inhibition were found to be 5- and 7-OH groups and a 2-ortho-substituted phenyl substituent. An optimal compound was assessed to be the analog HL 275, (-)-cis-5,7-dihydroxy-2-(2-chlorophenyl)-8-[4'- (3'-hydroxy-1'-methyl)piperidinyl]-4*H*-1-benzopyran-4-one (Figure 1), which displayed EGF-receptor tyrosine phosphokinase inhibition (TPK, IC_{50}: 10 μg/ml), serine phosphokinase inhibition (PKA, IC_{50}: 58 μg/ml), *in vitro* antiproliferative activity against murine and human tumor cell lines, and *in vivo* activity on a broad spectrum of human tumors transplanted into nude mice *(8,9)*.

In depth biological evaluation of HL 275 has shown that it has potent growth-inhibitory activity for a series of breast and lung carcinoma cell lines (IC_{50}: 25-160 nM). It is not toxic to stationary cells, yet it potently, but reversibly, inhibits the growth of cells in exponential growth phase, in which, within 12 hours of drug addition, DNA synthesis is inhibited by 90%, protein synthesis by 80%, and RNA synthesis by 40-60%, without compromise of overall cellular ATP levels. It can selectively block cell cycle progression *in vitro*, in both the G1 and G2 phases of the cycle. These results have contributed to HL 275 being considered for further preclinical or early clinical studies in its development as an anticancer agent *(10-12)*.

Forskolin - Derived Products

Forskolin: Cyclic AMP-Dependent and Cyclic AMP-Independent Effects. Forskolin (Figure 2), the labdane diterpenoid isolated from the Indian herb, *Coleus forskohlii*, is a unique, potent adenylate cyclase activator *(1-5, 13-14)*. Its uniqueness lies in its action at the catalytic sub-unit of the enzyme, an ability not possessed at the time of its discovery by any known substrate of the enzyme *(13,14)*. Global interest in the compound is reflected in the rapid, exponential growth of publications on this subject since 1981, which now total about 5,000. In view of the cyclic AMP-dependent effects produced by forskolin, it was considered for development as an agent for the treatment of congestive cardiomyopathy, glaucoma and asthma *(15)*.

Recently, forskolin has been demonstrated to produce actions that are not mediated by increased levels of cyclic AMP. In this regard, pathways of transmembrane signalling and intracellular protein trafficking, other than adenylate cyclase activation, have been shown to be used by forskolin through inhibition of membrane transport proteins and channel proteins not involving the production of cyclic AMP. Thus, forskolin produces cyclic AMP-independent effects through modulation of nicotinic acetylcholine receptor channel desensitization, modulation of voltage-dependent potassium channels, reversal of multidrug resistance, and inhibition of the glucose transport protein *(16-18)*.

Derivation of structure-activity correlations based on adenylate cyclase activation of several semi-synthetic forskolin analogs and cyclic AMP-dependent physiological

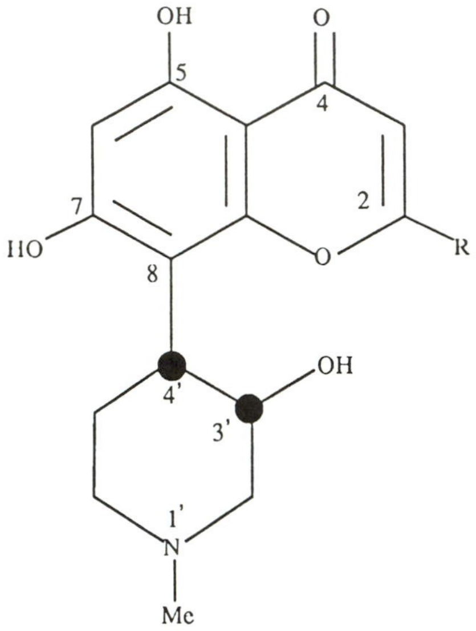

Rohitukine, R = Me

HIL 8276, R = C_6H_5

HL 275, R = C_6H_4 . o - Cl

Figure 1. Structures of rohitukine and its analogs, HIL 8276 and HL 275.

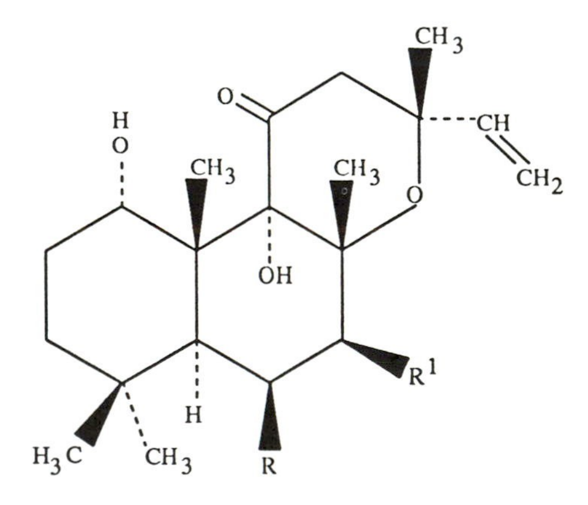

Forskolin, R = OH, R^1 = $OCOCH_3$

HL 706, R = $OCOCH_2N$ (piperidine ring) . HCl, R^1 = OH

NKH 477, R = $OCOCH_2CH_2N(CH_3)_2$. HCl, R^1 = $OCOCH_3$

Figure 2. Structures of forskolin and its analogs, HL 706 and NKH 477.

effects enabled the realization that the alpha interface, the region lying between the alpha face of forskolin and the active site of adenylate cyclase, appears to be where specific interactions occur. Two points of contact critical to the drug-receptor interaction are the 1-alpha- and the 9-alpha-OH groups of forskolin. It is mainly the 1,9-locus and the olefinic locus that are proposed in this interface to have a bearing on the order of potency and activity profile of forskolin and its analogs (*2, 3*).

In contrast, for the cyclic AMP-independent effect of inhibition of the glucose transport protein by forskolin and its analogs, the site of action for this activity is found to exhibit structural requirements that are different from those of the site responsible for the activation of adenylate cyclase. Similarities in the structures of glucose and forskolin have led to the postulation that forskolin binds to the glucose transporter through interactions determined by the beta-face of forskolin, involving the C-6 hydroxyl, C-7 acetoxy and C-8 ring oxygens (*19*).

Based on this understanding of the sites of forskolin action, forskolin analogs and photoaffinity labels have been identified and synthesized which exhibit specificity for adenylate cyclase, the glucose transporter, and the P-glycoprotein multidrug transporter (*20*). On the forefront of potential medicinal agents from forskolin, two potent adenylate cyclase activators, HIL 568, and the Nippon Kayaku product, NKH 477, continue to be under development, the progress of which is described in the following sections.

Forskolin-Derived Compound HIL 568: A Potential Anti-Glaucoma Agent. Glaucoma encompasses a group of disorders in which high intraocular pressure (IOP) causes optic nerve damage resulting in the loss of visual function. Current treatment of glaucoma is directed toward the reduction of IOP to a level at which progressive optic nerve damage is arrested. A unifying hypothesis concerning the regulation of IOP invokes a reduction in net aqueous humor inflow into the eye, mediated by increased cyclic AMP levels in the ciliary epithelium. Thus, exogenously applied substances and endogenous circulating hormones may regulate aqueous inflow (and IOP) via a final common pathway - the adenylate cyclase receptor complex in the ciliary epithelium (*21*).

As predicted, forskolin, on topical application in rabbits, monkeys, healthy human volunteers, and patients with open-angle glaucoma, effectively reduced IOP (*22-25*). It thus presented itself as a novel drug for the treatment of glaucoma, with advantages of a low potential to develop tolerance, a non-induction of miosis, an increase of intraocular blood flow, and an absence of systemic effects. Two major disadvantages of the compound were the hyperemic effects it produced, and its propensity in eye-drop formulations to be converted to the isomeric 7-deacetyl-6-acetyl analog. In programs to find a "second-generation" analog that would be devoid of these deficiencies, a compound that emerged as a potential development candidate was delta^5-6-deoxy-7-deacetyl-7-methylaminocarbonylforskolin, HIL 568 (Figure 3, *26*). Devoid of the 6-hydroxy group, the molecule does not permit a migration of the 7-acetyl moiety, and was then found to be stable in aqueous-based formulations.

In rat striatal membrane preparations, HIL 568 was found to be both a potent displacer of ^{3}H-forskolin from forskolin binding sites (Ki= 3.62 x 10^{-8} M) and an effective stimulator of adenylate cyclase relative to forskolin (108% of the activity of forskolin at a concentration of 300 μM).

HIL 568 (0.25%), applied topically in the normal New Zealand White rabbit, effectively reduced IOP as shown by a maximal reduction in outflow pressure of 36% at 2 hours with a significant reduction lasting for at least six hours post-administration (Figure 4). At this dose, no effects on pupillary diameter were observed, with only mild hyperemia in evidence.

In a comparative study of HIL 568 with the currently used anti-glaucoma agents, timolol and pilocarpine, the fall in IOP produced in White Belgian (WB) rabbits by HIL 568 was five-fold higher than that of timolol, and twenty-fold higher than that of pilocarpine (Figure 5).

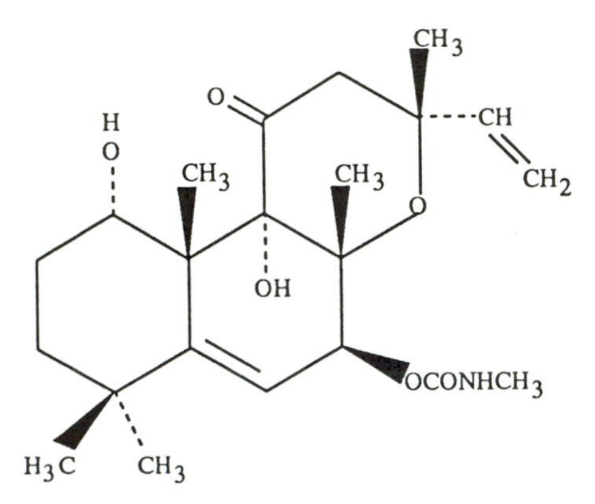

Figure 3. Structure of HIL 568

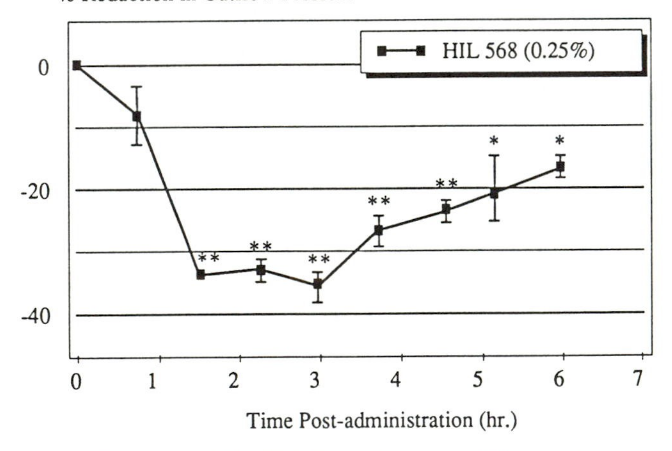

Figure 4. HIL 568: IOP effects in NZW rabbits.

To study the potential for tolerance, WB rabbits were treated with HIL 568 (0.1%) once daily for 21 days. Figure 6 summarizes the results of this study by comparing the percent change in IOP at one hour post-administration on the cited days over the initial value for the day. A maximum fall of 34.4% in IOP was observed on day 21 in the treated eyes in contrast to no significant change in the untreated eyes. Over the 21-day period, there was a gradual daily decrease of IOP, with a 19% reduction on day 21 over the value on day one (Hoechst India Limited Product Development data not shown).

In summary, HIL 568 is a second-generation forskolin analog which effectively reduces IOP in NZW and WB rabbits; it is stable in aqueous based formulations; it did not display tolerance following a 21-day administration in rabbits; and it has a low propensity to produce ocular or peripheral side effects. HIL 568 is a compound with a high potential for development as an anti-glaucoma agent.

Forskolin-Derived Compound NKH 477: A Potential Cardiotonic Agent. The clinical evaluation of forskolin was conducted in two groups of patients, with one having congestive cardiomyopathy, and the other having coronary artery disease, and convincingly substantiated the anticipated improvement of cardiac function and increase of myocardial contractility (*27*). In acting at a site distal to the beta-adrenergic receptor, it was considered to be an attractive new drug prototype for use in indications characterized by receptor desensitization. The poor solubility of forskolin in physiologically acceptable media for intravenous application was, however, a limiting factor in biological and clinical studies.

In programs designed to find acceptable water-soluble analogs of forskolin as potential cardiotonic agents, alterations of the forskolin molecule were made principally at the 6- and 7-positions, in consonance with the proposed model of forskolin action, which advocated that changes at these positions would be compatible with the retention of adenylate cyclase stimulating activity. Water-soluble aminoacylforskolin analogs, for instance, 6-(piperidinoacetyl)-7-deacetylforskolin hydrochloride (HL 706) (Figure 2; *28*) and 6-(3-dimethylaminopropionyl)forskolin hydrochloride (NKH 477) (Figure 2; *29*) which exhibited the desired biochemical and cardiovascular profile in *in vitro* and *in vivo* animal studies, were identified. The considerations which led the group at Nippon Kayaku Ltd., Japan to the selection of NKH 477 for clinical studies, currently in progress in Japan, to evaluate it for the treatment of heart failure have been adequately described (*29*).

Forskolin-Based Phytotherapeutic Product: A Potential Antithrombotic Agent. Our investigations with *C. forskohlii* have thus led to new chemical entities with a high potential for introduction as human medicinal agents. A question we have further addressed is whether forskolin-based herbal products can now be developed, in view of the knowledge that (a) *C. forskohlii* is identical with the Ayurvedic plant *Makandi*, and (b) there is a current resurgent consumer demand for phytotherapeutic products. An answer to the question necessitates that the requirements of different national/international drug authorities be met in respect of registration of new phytotherapeutic medicines. The main requirements to be met deal with the establishment of prior traditional use, standardization methodology, clinical efficacy and safety limits.

With respect to a forskolin-based phytotherapeutic product, such requirements have been met as follows:

(a) The Ayurvedic and ethnomedical basis for *Makandi* is documented in the Sanskrit literature (*30*). *Garmar* or *Mainmulli* (vernacular synonyms for *Makandi*) is a cultivated plant, readily available in season at different Indian locations for the preparation of condiments.

(b) Assay methods have been developed for estimation of forskolin and its naturally-occurring analogs through use of a triad of retention times in a combination of

IOP - Lowering Activity In Rabbits

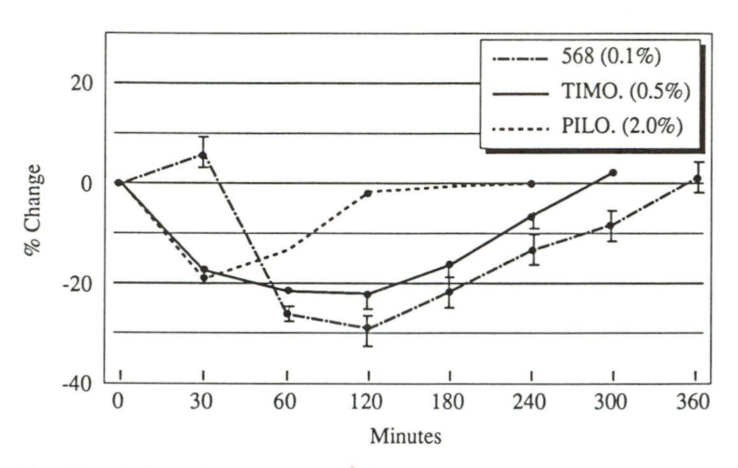

Figure 5. HIL 568 vs. timolol and pilocarpine. IOP-lowering activity in rabbits.

IOP at 60 Mins.

% Change

Vehicle HIL 568

10
0
-10
-20
-30
-40

3 6 9 12 15 18 21

Days

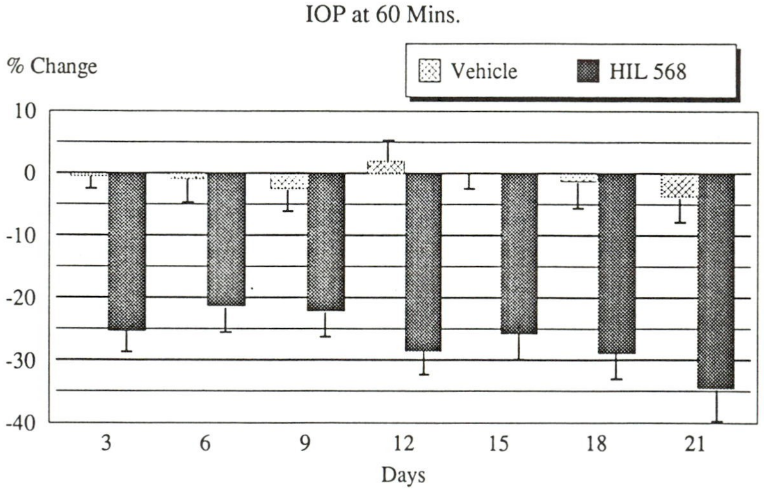

Figure 6. Chronic treatment of HIL 568. IOP at 60 mins.

GLC/HPLC methods (*31*). Through a more recent GC/MS method to make such quantitative determinations at ppm levels in the plant material (for forskolin, the quantification limit currently is 10 ppm), sensitive methodologies are now available for standardisation of plant extracts and herbal dosage forms, as well as for bioavailability studies (de Souza, N. J., *J. Ethnopharmacol.*, in press).

(c) A *C. forskohlii* extract, standardised with respect to its forskolin content, was studied for its antiplatelet aggregatory effects in a controlled study in rats. Oral administration of the extract (480 mg/kg, once a day) at a dose equivalent to 20 mg/kg of forskolin for a period of 15 days, produced a significant inhibition of platelet aggregation from day 5 to day 15, comparable to the effect produced by forskolin (20 mg/kg) or a standard drug, dipyridamole (10 mg/ kg). The data support a potential antithrombotic activity of the extract (de Souza, N. J., *J. Ethnopharmacol.*, in press). The consolidated data provide an adequate basis for a forskolin-based phytotherapeutic product to be considered for development as an antithrombotic agent. The generation of toxicity data and controlled pre-clinical and clinical data for a specified product would complete fulfillment of the requirements needed for registration.

Acknowledgments

I wish to thank the persons from within and outside of the Hoechst organisation, whose names are cited in the references and within the manuscript, for their excellent contributions to the development of the products; and especially to Dr. E. Baltin and Dr. W. Badziong of Hoechst India Limited for their kind support.

Literature Cited

1. de Souza, N. J. *Proceedings of the Third Asian Symposium on Medicinal Plants and Spices, Colombo, Sri Lanka*; UNESCO: Paris, France, 1977; pp 83-92.
2. de Souza, N. J. In *Forskolin: Its Chemical, Biological and Medical Potential*; Rupp, R. H.; de Souza, N. J.; Dohadwalla, A. N., Eds.; Hoechst India Limited: Bombay, 1986; pp 5-11 and 39-47.
3. de Souza, N. J. In *Innovative Approaches in Drug Research*; Harms, A. F., Ed.; Elsevier Science Publishers: Amsterdam, 1986; pp 191-207.
4. de Souza, N. J.; Shah, V. In *Economic and Medicinal Plant Research*; Wagner H.; Hikino, H.; Farnsworth, N. R., Eds.; Academic Press Limited: London, 1988; Vol. 2; pp 1-16.
5. de Souza, N. J. *Abstracts of the International Congress of Phytotherapy, Seoul, Korea*; The Pharmaceutical Society of Korea and Aloe Research Foundation, U.S.A.: Seoul, 1991; p 58.
6. Naik, R. G.; Kattige, S. L.; Bhat, S. V.; Alreja, B.; de Souza, N. J.; Rupp, R. H. *Tetrahedron* **1988**, *44*, 2081-2086.
7. Lakdawala, A. D.; Shirole, M. V.; Mandrekar, S. S.; Dohadwalla, A. N. *Asia Pacific J. of Pharmacol.* **1988**, *3*, 91-98.
8. Naik, R. G.; Lal, B.; Rupp, R. H.; Sedlacek, H. H.; Dickneite, G.; Czech, J. *Patentanmel dung Nr.* 89119710.5, 1989.
9. Sedlacek, H. H.; Hoffmann, D.; Czech, J.; Kolar, C.; Seemann, G.; Guessow, D.; Bosslet, K. *Chimia* **1991**, *45*, 311-316.
10. Kaur, G.; Stettler-Stevenson, M.; Worland, P. J.; Seber, S.; Sedlacek, H. H.; Czech, J.; Myers, C.; Sausville, E. A. *Proc. Am. Assoc. Cancer Res.* **1992**, *33*, 526.
11. Worland, P. J.; Kaur, G.; Stettler-Stevenson, M.; Seebers, S.; Sartor, O.; Sausville E. A. *Proc. Am. Assoc. Cancer Res.* **1992**, *33*, 512.
12. Kaur, G.; Stetler-Stevenson, M.; Sebers, S.; Worland, P.; Sedlacek, H. H.; Myers, C.; Czech, J.; Naik, R. G.; Sausville, E. *J. Natl. Cancer Inst.* **1992**, *84*, 1736-1740.

13. Metzger, H.; Lindner, E. *IRCS Med. Sci Biochem.* **1981**, *9*, 99.
14. Seamon, K. B.; Padgett, W.; Daly, W. J. *Proc. Natl. Acad. Sci. U.S.A.* **1981**, *78*, 3363-3367.
15. Rupp, R. H.; de Souza, N. J.; Dohadwalla, A. N., Eds. *Forskolin: Its Chemical, Biological and Medical Potential*; Hoechst India Limited: Bombay, India, 1986.
16. Laurenza, A; McHugh-Sutkowski, E.; Seamon, K.B. *Trends Pharmacol. Sci.* **1989**, *10*, 442-447.
17. Morris, D. I.; Robbins, J.; Ruoho, A. E.; McHugh-Sutkowski, E.; Seamon, K. B. *J. Biol. Chem.* **1991**, *266*, 13377-13384.
18. Lippincott-Schwartz, J.; Glickman, J.; Donaldson, J.G.; Robbins, J.; Kreis, T. E.; Seamon, K. B.; Sheetz, M. P.; Klausner, R. D. *J. Cell Biol.* **1991**, *112*, 567-577.
19. Joost, H. G.; Habberfield, A. D.; Simpson, I.A.; Laurenza, A.; Seamon, K. B. *Molec. Pharmacol.* **1988**, *33*, 449-453.
20. Morris, D. I.; Speicher, L. A.; Ruoho, A. E.; Tew, K. D.; Seamon, K. B. *Biochemistry* **1991**, *30*, 8371-8379.
21. Sears, M. L.; Mead, A. *Intl. Ophthalmol.* **1983**, *6*, 201-212.
22. Caprioli, J.; Sears, M. L. *Lancet i* **1983**, 958-960.
23. Caprioli, J. In *Forskolin: Its Chemical, Biological and Medical Potential*; Rupp, R. H.; de Souza, N. J.; Dohadwalla, A. N., Eds.; Hoechst India Limited: Bombay, 1986, pp 153-174.
24. Witte, P. U. In *Forskolin: Its Chemical, Biological and Medical Potential*; Rupp, R. H.; de Souza, N. J.; Dohadwalla, A. N., Eds.; Hoechst India Limited: Bombay, 1986; pp 175-182.
25. Pinto-Pereira, L. In *Forskolin: Its Chemical, Biological and Medical Potential*; Rupp, R. H.; de Souza, N. J.; Dohadwalla, A.N., Eds.; Hoechst India Limited; Bombay, 1986; pp 183-191.
26. Khandelwal, Y.; Moraes, G.; Lal, B.; Aroskar, V.A.; Dohadwalla, A.N.; Rupp, R.H. *Indian Patent No.* 168401, 1988; *German Patent Appl. No.* P 37 40 625.6, 1987.
27. Linderer, T.; Biamino, G. In *Forskolin: Its Chemical, Biological and Medical Potential*; Rupp, R. H.; de Souza, N. J.; Dohadwalla, A. N., Eds.; Hoechst India Limited: Bombay, 1986; pp 109-113.
28. Khandelwal, Y.; Rajeshwari, R.; Rajagopalan, R.; Swamy, L.; Dohadwalla, A. N.; de Souza, N. J.; Rupp, R. H. *J. Med. Chem.* **1988**, *31*, 1872-1879.
29. Hosono, M.; Takahira, T.; Fujita, A.; Fujihara, R.; Ishizuka, O.; Ohoi, I; Tatee, T.; Nakamura, K. *Eur. J. Pharmacol.* **1990**, *183*, 2110P.
30. Vaishya Shaligram, L. In *Shaligram Nighantu Bhushane*; 7-8 parts of Brahan Nighantu Ratnakar; Sri Krishnadas Khemraj: Bombay, 1896; p 1210.
31. Inamdar, P. K.; Khandelwal, Y.; Garkhedkar, M.; Rupp, R. H.; de Souza, N. J. *Planta Med.* **1989**, 386-387.

RECEIVED March 23, 1993

Author Index

Affiliation Index

Subject Index

C

D

H

I

M

N

T

Production: Meg Marshall
Indexing: Deborah H. Steiner
Acquisition: Barbara C. Tansill
Cover design: Alan Kahan

Printed and bound by Maple Press, York, PA